Adsorption Processes for Water Treatment

Adsorption Processes for Water Treatment

Samuel D. Faust

Osman M. Aly

Butterworths
Boston London Durban Singapore Sydney Toronto Wellington

Library of Congress Cataloging-in-Publication Data
Faust, Samuel Denton, 1929–
Adsorption processes for water treatment.

Bibliography: p.
Includes index.
1. Water—Purification—Adsorption. 2. Water chemistry. I. Aly, Osman M. II. Title.
TD449.5.F38 1986 628.1′64 85-26927
ISBN 0-409-90000-1

Butterworth Publishers
80 Montvale Avenue
Stoneham, MA 02180

10 9 8 7 6 5 4 3 2 1

Printed in the United States of America

Contents

Preface and Acknowledgments xi

1. Elements of Surface Chemistry 1
Principles of Adsorption 1
Types 1
Entropy Changes 2
Adsorption: Interaction Forces 3
Dispersion Forces 3
Potential Energy Curves 4
Surface Tension 5
Thermodynamics of Adsorption: Gibbs Adsorption Isotherm 7
Unimolecular Surface Films 10
Insoluble Surface Films 10
Expanded and Gaseous Films 11
Surface Films of Soluble Substances 12
Surface Layers and Gas Laws 13
Adsorption Equilibria 14
Generalizations 14
Langmuir Adsorption Isotherm 16
Freundlich Adsorption Isotherm 18
BET Adsorption Isotherm 20
Henry's Law: Linear Adsorption Isotherm 21
References 22

2. Adsorption Models 25
Theories 25
Background 25
Net Adsorption Energy Concept 26
Correlation of NAEC to Adsorption Isotherms 33
Polanyi Adsorption Potential Theory 34
Solvophobic Theory 45
Theory of Correspondence for Adsorption 53

Applications 58
References 63

3. Kinetics of Adsorption 65
Batch Systems 67
Single Compounds 67
Systemic Factors Affecting Rates of Adsorption: Single Compounds 74
Diffusion-Controlled Processes 85
Mixed Systems 94
Fixed Beds 100
Early Mass Transfer Models 100
Mass Transfer Parameters 106
Late Mass Transfer Models 107
References 121

4. Batch Systems and Fixed and Fluidized Beds 123
Batch Contact Systems 123
Generalizations 123
Determination of Carbon Dosage in Batch Systems 124
Dynamic Systems (Column Adsorbers) 129
Generalizations 129
The Breakthrough Curve and Mass Transfer Zone 130
Types of Adsorption Systems 134
Fixed Bed Contactors 134
Pulsed Bed Contactors 135
System Configurations 138
Mathematical Modeling for Fixed Bed Adsorbers 138
Generalizations 138
Fixed Bed Model 140
Process Design of Adsorbers 141
Generalizations 141
Analysis of Data 143
Major Design Approaches 143
Application of Design Methods 155
References 163

5. Physical and Chemical Properties of Carbon 167
Preparation and/or Activation 167
Source Materials 167
Activation Processes 168
General Physical Properties 170
Overall Characteristics 170
Surface Area 172
Porosity 178

Surface Functional Groups 180
Generalizations 180
Surface Acidity 182
Effects on Adsorption 185
Electrophoretic Characteristics 188
References 191

6. Adsorption of Organic Compounds 193
Nonspecific Organic Carbon 193
Analytical Considerations 193
Equilibrium Studies 207
Pilot Plant Studies 207
Pretreatment 207
Breakthrough 208
Organic Priority Pollutants 210
Carbon Tetrachloride 211
Trihalomethanes 211
Fulvic and Humic Acids 217
Equilibrium Studies 217
Column Studies 219
Phenolic Compounds 224
Equilibrium Studies 226
Column Studies 232
Petrochemicals 238
Equilibrium Studies 238
Column Studies 239
Organic Pesticides 240
Equilibrium Studies 241
Column Studies 246
Compounds Causing Tastes and Odors 247
Polycyclic Aromatic Hydrocarbons 251
Miscellaneous Organic Compounds 254
Equilibrium Studies 254
Column Studies 259
Organics in Mixed Systems 262
Equilibrium Studies 262
Column Studies 276
References 283

7. Removal of Inorganic Compounds 287
Arsenic 287
Aqueous Chemistry 287
Removal 288
Barium 289
Aqueous Chemistry 289
Removal 289

Cadmium 289
Aqueous Chemistry 289
Removal 290
Chromium 294
Aqueous Chemistry 294
Removal 294
Copper 298
Aqueous Chemistry 298
Removal 298
Fluoride 299
Health Concerns 299
Removal 300
Lead 300
Aqueous Chemistry 300
Removal 300
Mercury 302
Aqueous Chemistry 302
Removal 304
Radionuclides 310
Health Concerns 310
Removal 312
Selenium 312
Aqueous Chemistry 312
Removal 313
Zinc 313
Aqueous Chemistry 313
Removal 313
Chlorine Compounds 313
Chlorine 313
Chloramines 320
Packed Bed Reactors 322
Hydrogen Sulfide 325
Aqueous Chemistry 325
Removal 326
References 326

8. Operational, Pilot Plant, and Case Studies 329
Taste and Odor Removal 329
Nitro, West Virginia 329
Omaha, Nebraska 332
Sulfide Odors 336
Little Falls, New Jersey 337
Removal of Organic Compounds 338
Chloroform and Total Organic Carbon 338

Haloforms and Pesticides 342
Chlorinated Benzenes, Total Organic Carbon, and Total Organic Halogens 346
Organic Solvents 352
Trihalomethanes and Precursors 353
Treatment of Groundwaters 370
Large-Scale Systems 371
Small-Scale Systems 374
Pretreatment for Removal of Organic Compounds 379
Chemical Coagulation 379
Ozonation 385
Aeration 388
Chlorination 392
Design and Economic Considerations 393
Microbiology of Granular Activated Carbon 405
Regeneration Processes 412
Coagulation 425
Backwashing Granular Activated Carbon Filters 425
References 428

9. Biological Activated Carbon Treatment of Drinking Water 433
Pretreatment: Chemical Oxidation 434
Ozone 435
Chlorination 443
Performance 445
Operating Experiences and Case Studies 449
The Mülheim Process 449
Mery-sur-Oise and Choisy-le-Roi, France 452
The Düsseldorf Process 453
Rotterdam, The Netherlands 457
Design and Operational Considerations 460
Preliminary Design 460
Detailed Design 461
Economic Considerations 461
References 469

10. Adsorption by Macroreticular Resins 471
Physical and Chemical Properties 471
Types 471
Physical Properties 472
Removal of Individual Compounds 474
Humic Substances 474
Trihalomethanes 476

Phenols 479
Detergents 483
Comparative Studies: Resins and Granular Activated Carbon 484
Humic Substances 484
Trihalomethanes 488
Phenols 492
Miscellaneous Compounds 495
References 501

Index 503

Preface and Acknowledgments

Adsorption processes for water treatment have had a long and productive history. Charcoal, the forerunner of modern activated carbon, whose ability to purify water dates back to 2000 B.C., was used for medicinal purposes by ancient Egyptians. Carbon's adsorptive properties for gases were reported first by Scheele in 1733 and later by Fontana in 1777. In 1785, Lowitz observed the ability of charcoal to decolorize many liquids, including brandy. This discovery led to the widespread application of wood and bone chars in the refining of cane and beet sugars. Also, Lowitz established the first use of powdered charcoal for the removal of bad tastes and odors from water on an experimental basis in 1789–1790.

Early application of carbon in a water treatment plant to remove chlorophenolic tastes was reported by Baylis in the United States and Sierp in Germany in 1929. The Hackensack (New Jersey) Water Company employed powdered activated carbon (PAC) to remove odors in water caused by wastes from an alcohol denaturing plant in 1929. By 1932, 400 water treatment plants were using carbon to control odors in drinking water. The modern era of the application of granular activated carbon (GAC) began, perhaps, in the mid-1960s at Nitro, West Virginia, where the Kanawha River contained a variety of petrochemical compounds.

Today, both PAC and GAC are employed throughout the world in the water and wastewater treatment industry. The objectives range far beyond treating tastes and odors, to the current emphasis on removal of organic compounds that may be human carcinogens. Certainly, adsorption technology will play a major role in the treatment of drinking water and wastewaters. This book describes and details the current state of the art for carbon and resin adsorptive processes for potable water treatment. Some wastewater applications are mentioned also.

The authors acknowledge the patience and support of their families who were neglected during the long hours of preparation. We appreciate the typing skills of Ms. Carolyn Smith, who labored into the early hours of the morning. As always, we are grateful to our professional colleagues whose works are cited herein. Also, we appreciate the excellent and efficient cooperation of the editorial staff of Butterworths.

Samuel D. Faust
Changewater, N.J.

Osman M. Aly
West Deptford, N.J.

1

Elements of Surface Chemistry

Since the 1960s, an increased awareness of the occurrence of many synthetic and natural organic substances in natural waters has led to the emergence of adsorption by activated carbon and/or porous synthetic resins, as one of the most effective methods of removing these substances from drinking and wastewaters.

PRINCIPLES OF ADSORPTION

Types

Adsorption is a surface phenomenon that is defined as the increase in concentration of a particular component at the surface or interface between two phases. In any solid or liquid, atoms at the surface are subject to unbalanced forces of attraction normal to the surface plane. These forces are merely extensions of the forces acting within the body of the material and are ultimately responsible for the phenomenon of adsorption. In discussing the fundamentals of adsorption, it is useful to distinguish between *physical adsorption,* involving only relatively weak intermolecular forces, and *chemisorption,* which involves essentially the formation of a chemical bond between the sorbate molecule and the surface of the adsorbent. Although this distinction is conceptually useful, many cases are intermediate and it is not always possible to categorize a particular system unequivocally [1].

Physical adsorption can be distinguished from chemisorption according to one or more of the following criteria:

1. Physical adsorption does not involve the sharing or transfer of electrons and thus always maintains the individuality of interacting species. The interactions are fully reversible, enabling desorption to occur at the same temperature, although the process may be slow because of diffusion effects. Chemisorption involves chemical bonding and is irreversible.
2. Physical adsorption is not site specific; the adsorbed molecules are free to cover the entire surface. This enables surface area measurements of solid

adsorbents. In contrast, chemisorption is site specific; chemisorbed molecules are fixed at specific sites.

3. The heat of physical adsorption is low compared to that of chemisorption; however, heat of adsorption is not usually a definite criterion. The upper limit for physical adsorption may be higher than 20 kcal/mol for adsorption on adsorbents with very narrow pores. The heat of chemisorption ranges from over 100 kcal/mol to less than 20 kcal/mol. Therefore, only very high or very low heats of adsorption can be used as a criterion for this type of adsorption process [2].

Chemisorption is characterized mainly by large interaction potentials that lead to high heats of adsorption that approach the value of chemical bonds. This fact, coupled with other spectroscopic, electron spin resonance, and magnetic susceptibility measurements, confirms that chemisorption involves the transfer of electrons and the formation of true chemical bonding between the adsorbate and the solid surface [3]. Because chemisorption involves chemical bonding, it often occurs at high temperatures and is usually associated with activation energy. Also, the adsorbed molecules are localized on specific sites and, therefore, are not free to migrate about the surface.

Entropy Changes

The adsorption of solutes from solution by solid adsorbents depends on physical adsorption rather than chemisorption and is, therefore, the focus of this discussion. The heat of adsorption provides a direct measure of the strength of the bonding between sorbate and surface. Physical adsorption from the gas phase is invariably exothermic, as may be shown by a simple thermodynamic argument. Since the adsorbed gas molecules have a lower degree of disorder—that is, lower entropy than the surrounding molecules—the process of adsorption is accompanied by a decrease in entropy. The entropy change on adsorption,

$$\Delta S = S_{ads} - S_{gas}, \tag{1.1}$$

is necessarily negative. For significant adsorption to occur, the free energy change of adsorption, ΔG, must be negative, and since:

$$\Delta G = \Delta H - T\Delta S, \tag{1.2}$$

this requires ΔH to be negative or an exothermic change. This is usually true also for adsorption from the liquid phase, although some exceptions are possible [1].

ADSORPTION: INTERACTION FORCES

Dispersion Forces

Physical adsorption on nonpolar solids is attributed to forces of interactions between the solid surface and adsorbate molecules that are similar to the van der Waals forces (attraction-repulsion) between molecules. The attractive forces that involve the electrons and nuclei of the system are electrostatic in origin and are termed *dispersion forces.* These forces exist in all types of matter and always act as an attractive force between adjacent atoms and molecules no matter how dissimilar. They are always present regardless of the nature of other interactions and often account for the major part of the adsorbate-adsorbent potential [3,4]. The nature of the dispersion forces was first recognized in the 1930s by London [5]. Using quantum mechanical calculations, he postulated that the electron motion in an atom or molecule would lead to a rapidly oscillating dipole moment. At any instant, the lack of symmetry of the electron distribution about the nuclei imparts a transient dipole moment to an atom or molecule that would average zero over a longer time interval. When in close proximity to a solid surface, each instantaneous dipole of an approaching molecule induces an appropriately oriented dipole moment in a surface molecule. These moments interact to produce an instantaneous attraction. These forces are known as dispersion forces because of their relationship, noted by London [5], to optical dispersion. The dipole-dispersion interaction energy can be determined by:

$$E_D = -\frac{C}{r^6}, \tag{1.3}$$

where E_D = dispersion energy or potential, C = a constant, and r = distance of separation between the interacting molecules.

In addition to dipole-dipole interactions, other possible dispersion interactions contributing to physical adsorption include dipole-quadrapole and quadrapole-quadrapole interactions. If these two are included, the total dispersion energy becomes [3,5]:

$$E_D = -\frac{C}{r^6} - \frac{C'}{r^8} - \frac{C''}{r^{10}}, \tag{1.4}$$

where C' = a constant for dipole-quadrapole interactions and C'' = a constant for quadrapole-quadrapole interactions.

The contribution to E_D from the terms in Equation 1.4 clearly depends on the separation, r, between the molecules; therefore, the dipole-dipole interactions will be most significant. Quadrapole interactions involve symmetrical molecules with atoms of different electronegativities like CO_2. This molecule has no dipole moment but does have a quadrapole ($^{-}O - {}^{+}C^{+} - O^{-}$) that can lead to interactions with polar surfaces.

When an adsorbate molecule comes very close to a solid surface molecule to allow interpenetration of the electron clouds, a repulsive interaction will arise, which is represented semiempirically by the expression:

$$E_R = \frac{B}{r^{12}}, \tag{1.5}$$

where E_R = repulsion energy and B = a constant. The total potential energy of van der Waals interactions is the sum of the attractive energy and the repulsion energy:

$$E = -\frac{C}{r^6} + \frac{B}{r^{12}}. \tag{1.6}$$

The inverse sixth energy term falls rapidly with increasing r but not nearly as rapidly as the repulsion term. Thus, the dispersion energy is more important than the repulsion at longer distances.

Potential Energy Curves

The potential energy curve for van der Waals interactions between He atoms is illustrated in Figure 1-1. At separations of more than 3.5 Å, the first term in Equation 1.6 predominates. The atoms attract each other and the energy of the two atoms falls to a stabler level as the atoms move closer. If they come closer than 3 Å, however, a strong electron pair repulsion predominates over the London attraction and a large amount of energy is required to push the atoms together. Accordingly, the potential energy curve in Figure 1-1 rises. A balance between attraction and repulsion exists at a 3 Å separation, and the two atoms are 18.2 cal/mole stabler than two isolated atoms [6].

The dispersion interaction between a solid and an external atom or molecule can be determined in a manner analogous to that employed between a pair of atoms or molecules. The usual procedure is to use the property of additivity of the dispersion forces and assume that the atoms of the solid can be thought of as force centers and that the energy of interaction, E, of an external atom with the solid can be expressed as:

$$E = \Sigma E_j, \tag{1.7}$$

where E_j = the energy of interaction between the external atom and the jth atom of a solid. The expression of E_j is usually assumed to be of the form in Equation 1.3 or perhaps 1.6. This approach has been used by several investigators [1,3].

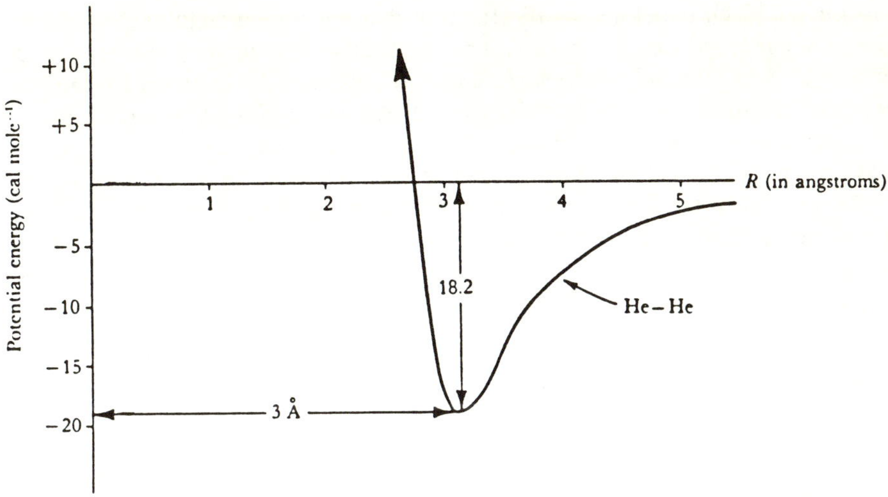

Figure 1–1 Potential energy curve for van der Waals attraction between He atoms. Reproduced from Dickerson et al. [6], courtesy of W.A. Benjamin, Inc.

It should be noted that while the additivity of the dispersion energies is strictly valid, it does not necessarily follow that the repulsive contributions are additive, although such an assumption is often made in theoretical calculations of this kind [1].

SURFACE TENSION

The surface of a liquid in contact with its vapor has different properties from those of the bulk phase. A molecule in the interior of a liquid is surrounded on all sides by neighboring molecules of the same substance and, therefore, is attracted equally in all directions. A molecule at the surface, however, is subject to a net attraction toward the bulk of the liquid, in a direction normal to the surface, because the number of molecules per unit volume is greater in the bulk of the liquid than in the vapor. Because of the unbalanced attraction, the surface of a liquid always tends to contract to the smallest possible area. To extend the area of the surface, work must be done to bring the molecules from the bulk of the liquid into the surface against the inward attractive force. The surface portion of a liquid, therefore, has a higher free energy than the bulk liquid.

The work required to increase the area by 1 cm^2 is called the *surface free energy.* As a result of the tendency to contract, a surface behaves as if it were in a state of tension, and it is possible to ascribe a definite value to this *surface tension,* γ, which is defined as the force in dynes acting at right angles to any line of 1 cm length in the surface. The work done in extending the area of a surface by 1 cm^2 is equal to the surface tension, which is the force per centimeter opposing the increase, multiplied by 1 cm, the distance through which the point of application of the force is moved. It follows, therefore, that the surface energy, in ergs per square centimeter, is numerically equal to the surface tension in dynes per centimeter. In more general terms, the work, W, done by the surface in extending its area, A, by an amount, dA, is:

$$dW = -\gamma dA = -dG, \tag{1.8}$$

whence

$$dG = \gamma dA, \tag{1.9}$$

where dG is the change in free energy.

Since the surface energy is a Gibbs free energy, the surface enthalpy, ΔH, can be evaluated from change of surface tension γ with temperature. From the Gibbs-Helmholtz equation:

$$\Delta G = \Delta H - T\left(\frac{\partial G}{\partial T}\right)_P. \tag{1.10}$$

The surface enthalpy is given by:

$$\Delta H = \gamma - T\left(\frac{\partial \gamma}{\partial T}\right)_P. \tag{1.11}$$

For water at 20°C, $\gamma = 72.75$ erg/cm^2, $(\partial\gamma/\partial T)_P = -0.148$ (the negative sign is due to the decrease of γ with increase in temperature), and so $\Delta H = 116.2$ erg/cm^2. This is the decrease in enthalpy associated with the destruction of 1 cm^2 of liquid surface.

The addition of a solute to a liquid may alter the surface tension considerably. In the case of aqueous solutions, solutes that can markedly lower the surface tension of water are organic compounds that contain both a polar hydrophilic group and a nonpolar hydrophobic group—for example, organic acids, alcohols, esters, ethers, ketones, and so on. The hydrophilic group makes the molecule reasonably soluble while the hydrocarbon residues have low affinity for water and little work is required to bring them from the interior to the surface. Solutes that lower the surface tension tend to accumulate preferentially at the surface, and hence, there will be a greater proportion of the solute at the interface than the bulk of the solution. This represents a case of adsorption of the solute at the surface of the solution, and the solute is said to be *positively adsorbed* at the interface.

Electrolytes, salts of organic acids, bases of low molecular weight, and certain nonvolatile electrolytes usually increase the surface tension of aqueous solutions above the value for pure water. These increases are much smaller than the decreases produced by organic acids and similar compounds. The observed increases are attributed to ion-dipole interactions of the dissolved ions that tend to pull the water molecules into the interior of the solution. Additional work must be done against the electrostatic forces to create a new surface. The surface layers in such solutions have lower concentrations of the solute than in the bulk solution. The solute is said to be *negatively adsorbed* at the interface [7].

THERMODYNAMICS OF ADSORPTION: GIBBS ADSORPTION ISOTHERM

The thermodynamics of the surface was originally formulated by Gibbs in 1878, and subsequent studies were directed mostly to gas/liquid, gas/solid, and liquid/solid interfaces [4,7,9]. Considerable progress has been made since the 1950s in understanding the thermodynamics of adsorption from solutions, largely because of the work of Hill [10] and Everette [11,12]. The following discussion is based on the method given by Gibbs since more detailed information can be obtained from his formulations [8].

The system considered for Gibbs thermodynamic analysis consists of two phases separated by plane interface. An imaginary surface, constructed within the interfacial region and parallel to the boundary planes, that locates the extent of the separate phases is called the *surface phase, σ*. The *bulk phase,* the remainder of the solution, is assumed to be homogenous up to the dividing surface phase. A surface phase defined in this way has an area, *A,* but no thickness. It is strictly two dimensional and occupies a finite region of space within which the concentration is uniform and different from that of the bulk phase. The thermodynamic properties of the adsorbent are considered to be independent of the temperature and the concentration of the adsorbed molecules; that is, the adsorbent is considered thermodynamically inert. Under these conditions, the adsorbed molecules may be regarded as a distinct phase, and the effect of the adsorbent is limited to the creation of force field, the detailed nature of which need not be specified [1,12]. The system, therefore, is considered to be divided into two parts, one consisting of all that portion that comes under the influence of the surface forces, which is the *surface phase,* and the other is the remainder of the solution, the *bulk phase,* which is free from the surface effects.

From the thermodynamics of bulk phases, the fundamental relationship of the free energy, *G,* of a two-component system is given by:

$$G = \mu_1 n_1 + \mu_2 n_2, \tag{1.12}$$

where μ_1 and μ_2 = chemical potentials of components 1 and 2, respectively, and n_1 and n_2 = number of moles of components 1 and 2, respectively. Also:

$$dG = -SdT + VdP + \mu_1 n_1 + \mu_2 n_2. \quad (1.13)$$

where S = entropy, T = temperature, V = volume, and P = pressure.

In addition to the ordinary state variables describing the bulk phases, new state variables must be introduced to define the state of the surface phase. To allow for the possibility of a change in free energy resulting from an increase or decrease of the exposed surface, it is necessary to include a term, γA, where γ is the interfacial tension—that is, the interfacial energy per square centimeter—and A is the surface area. Therefore, the surface free energy, G^σ, can be obtained from Equation 1.12 by adding the γA term:

$$G^\sigma = \gamma A + \mu_1 n_1 + \mu_2 n_2, \quad (1.14)$$

and on differentiation:

$$dG^\sigma = \gamma dA + Ad\gamma + \mu_1 dn_1 + n_1 d\mu_1 + \mu_2 dn_2 + n_2 d\mu_2. \quad (1.15)$$

Alternatively, adding the term γdA, for an increase of dA in the surface area, to Equation 1.13, the differential of the surface free energy becomes:

$$dG^\sigma = -SdT + VdP + \gamma dA + \mu_1 dn_1 + \mu_2 dn_2, \quad (1.16)$$

and by comparing Equations 1.15 and 1.16 yields:

$$SdT - VdP + Ad\gamma + n_1 d\mu_1 + n_2 d\mu_2 = 0 \quad (1.17)$$

at constant temperature and pressure:

$$Ad\gamma + n_1 d\mu_1 + n_2 d\mu_2 = 0. \quad (1.18)$$

If n_1^o and n_2^o are the number of moles of the two components in the bulk phase that correspond to n_1 and n_2 in the surface phase, then according to the Gibbs-Duhem equation:

$$n_1^o d\mu_1 + n_2^o d\mu_2 = 0. \quad (1.19)$$

Multiplying Equation 1.19 by n_1/n_1^o and subtracting from Equation 1.18 gives:

$$Ad\gamma + (n_2 - n_1 n_2^o/n_1^o)\, d\mu_2 = 0. \quad (1.20)$$

Then:

$$-\frac{d\gamma}{d\mu_2} = \frac{n_2 - n_1 n_2^o/n_1^o}{A}. \quad (1.21)$$

The quantity n_2 is the number of moles of one component, like the solute, associated with n_1 moles of solvent in the surface phase, and $n_1 n_2^o/n_1^o$ is the corresponding number of moles associated with n_1 moles of solvent in the bulk phase. The right-hand side of Equation 1.21, therefore, may be regarded as the excess concentration of the solute per unit of surface area. This excess concentration is given the symbol Γ_2 and is called the *surface concentration* of solute per unit area of the interface. It is not strictly a concentration term since it is a number of moles divided by an area, but it is nevertheless a definite quantity defined by the right-hand side of Equation 1.21. Substituting for μ_2 in Equation 1.21 yields:

$$\Gamma_2 = -\frac{d\gamma}{d\mu_2} . \qquad (1.22)$$

Since μ_2 may be represented by:

$$\mu_2 = \mu_2^o + RT \ln a_2 , \qquad (1.23)$$

where a_2 is the activity of the solute, then at constant temperature:

$$d\mu_2 = RT \ln a_2 . \qquad (1.24)$$

Substituting into Equation 1.22 gives:

$$\Gamma_2 = -\frac{1}{RT}\frac{d\gamma}{d \ln a_2} = -\frac{a_2}{RT}\frac{d\gamma}{da_2} . \qquad (1.25)$$

These two equations are various forms of the Gibbs adsorption isotherm. The isotherm holds equally for either component in a binary system, although in practice it is usually applied to the solute. The subscript can, therefore, be neglected, giving:

$$\Gamma = -\frac{a}{RT}\frac{d\gamma}{da} . \qquad (1.26)$$

For dilute solutions, the activity may be replaced by molar concentration, c, and:

$$\Gamma = -\frac{c}{RT}\frac{d\gamma}{dC} . \qquad (1.27)$$

It can be seen that if a solute causes a decrease in surface tension with an increase of concentration—that is $d\gamma/dc$ is negative—then Γ is positive and there will be adsorption on the interface.

In deriving the Gibbs adsorption equation, no assumption was made about the nature of the system or the surface. Its most obvious application, however, is to a liquid solution with the interface between the liquid and its vapor (liquid/vapor interface) or to an interface between a solution and a liquid with which it is immiscible (liquid/liquid interface). The quantity then refers to the interfacial tension. There has not been a direct application of the Gibbs equation for adsorption by solids from solution because of the difficulty of determining the solid/liquid tension and the lack of simple mechanical methods for measuring the surface tension of a solid surface. In addition, the simple Gibbs model of a sharp boundary between the surface and bulk phases and the autonomy of the surface phase presents many limitations in complex systems. More detailed discussions of the subject can be found elsewhere [12,13]. However, the discussion of the Gibbs isotherm here serves to illustrate the tendency of a given solute to be adsorbed at an interface.

UNIMOLECULAR SURFACE FILMS

Insoluble Surface Films

Certain sparingly soluble substances that possess one polar group can spread on the surface of water to form films one molecule in thickness. These are called *unimolecular films,* or *monolayers.* The area covered by a spreading compound such as oleic acid on the surface of water can be varied at will by confining the film between movable barriers placed across a shallow tray filled with water on which the compound is placed. Langmuir [15] devised a method for the direct measurement of the force, the *surface pressure* exerted by a film using an apparatus called the *surface balance* [7,14]. The dependence of the force exerted by the film, f, in dynes/cm, on the average area occupied by each molecule, in $Å^2$, is known as $(f - A)$ curves. Figure 1-2 shows an $(f - A)$ curve for stearic acid on distilled water [14]. At large areas, the surface force is relatively small, and it increases very slowly with decreasing area until a value of 20.5 $Å^2$ per molecule is reached where the pressure begins to increase extremely rapidly on further compression of the film. Similar curves are obtained for a series of long-chain fatty acids, long-chain amides, alcohols, methyl ketones, monobasic esters, and other substances with polar groups [7]. In each case the minimum area per molecule is observed at 20.5 $Å^2$.

It is believed that such films consist of monolayers, with the molecules arranged more or less vertically with the polar group attached to the surface of the water and the hydrocarbon chain pointing outward. As the film is compressed, the oriented molecules are so closely packed that any further decrease in area demands the exertion of an appreciable surface force, as in the three-dimensional case where compression of a solid or liquid is much more difficult than compression of a gas [14]. The limiting area occupied by each molecule is determined by the cross-sectional area of the hydrocarbon chain and, therefore, will be independent of its length or of the nature of the polar end group provided that the

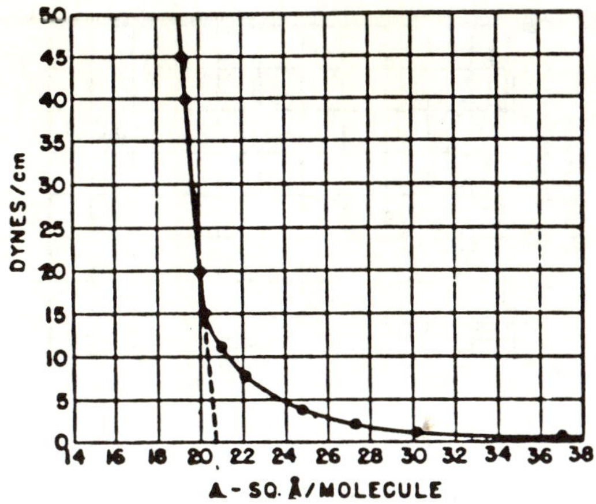

Figure 1–2 *f-A* isotherm at 20°C. Stearic acid on distilled water. Reproduced from Moore [14], courtesy of Prentice-Hall, Inc.

latter is sufficiently attracted by the water for the chains to be anchored to the surface [7,14].

Expanded and Gaseous Films

The films described in the last section, consisting of closely packed single layers of molecules, are called *condensed films* to distinguish them from *expanded films* that are formed as the temperature is raised. In these films the area occupied per molecule is greater than in the condensed films because of the mutual repulsion of the end groups [17].

Certain films, known as *liquid expanded,* have $(f - A)$ curves very similar to the pressure-volume $(P - V)$ curves obtained for a gas undergoing liquefaction in the neighborhood of the critical temperature. Figure 1–3 shows $(f - A)$ curves for a number of fatty acids $C_nH_{2n+1}COOH$ at 25°C [14]. There is a section at low pressures corresponding to the high compressibility of a gas. Then there is an intermediate region of areas over which the film exhibits a definite force equivalent to surface vapor pressure that is independent of the total area occupied. In this region a very small increase in pressure produces a large decrease in the area. Finally, there is a rapid rise in the $(f - A)$ curve that corresponds to the compression of a condensed phase. With lower molecular weights, the quasi-two-phase region (the flat portion of the curve) becomes less evident, and finally the curve appears to be gaseous throughout as in the dotted curve. At low pressures and, especially at elevated temperatures, when the molecules are relatively far

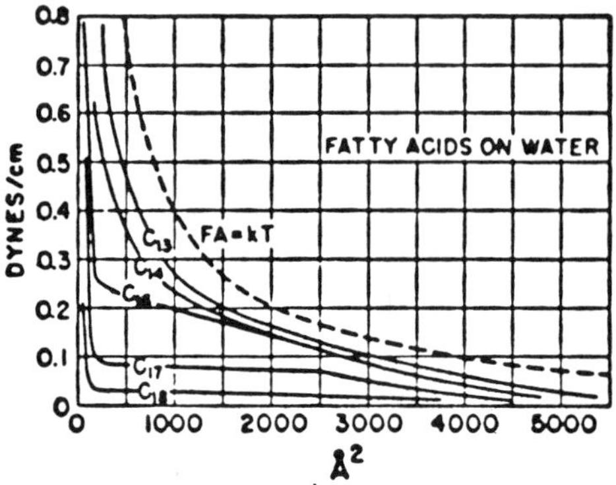

Figure 1–3 *f-A* isotherms at low pressures. Reproduced from Moore [14], courtesy of Prentice-Hall, Inc.

apart, *gaseous films* are formed, which behave like two-dimensional gases corresponding to three-dimensional gases. If the product of the force, f, and the average area, A, occupied by each molecule—that is, fA—is plotted against f, curves similar to the $PV - P$ curves for gases are obtained [7]. The molecules probably lie more or less flat on the surface. Therefore, substances with polar groups at each end, like esters of dibasic acids, form such films most readily.

Surface Films of Soluble Substances

Soluble substances that produce a marked decrease in the surface tension of water form monolayers at the surface. Therefore, condensed unimolecular films, with oriented molecules, are equivalent to an extreme case of adsorption of a soluble substance. The force, f, exerted by an insoluble film is equal to the difference between the surface tension of pure water and that of the water covered with the film. This definition can be extended to the surface layers of soluble substances so that f is equal to the difference between the surface tension of pure solvent, γ_o, and of the solution, γ, that is,

$$f = \gamma_o - \gamma. \tag{1.28}$$

Differentiating with respect to C:

$$\frac{df}{dC} = -\frac{d\gamma}{dC} \tag{1.29}$$

and

$$\frac{f}{C}\frac{d\ln f}{d\ln C} = -\frac{d\gamma}{dC} . \qquad (1.30)$$

Substituting $-\ d\gamma/dC$ in Gibbs Equation 1.27, then:

$$\Gamma = \frac{f}{RT}\frac{d\ln f}{d\ln C} . \qquad (1.31)$$

The area, A, occupied by a single molecule is $1/N\Gamma$, where N is the Avogadro number. Substitution in Equation 1.31 gives:

$$fA = \frac{RT}{N}\frac{d\ln C}{d\ln f} = k\mathrm{T}\frac{d\ln C}{d\ln f} , \qquad (1.32)$$

where k is the Boltzmann constant.

It is possible, therefore, from measurements of the surface tension of a solution at various concentrations, to plot fA against f, the f being equal to $\gamma_o - \gamma$ and the fA term evaluated from Equation 1.32. The curves obtained, like those of gaseous films of insoluble substances, are of exactly the same type as the $PV - P$ curves for gases. Surface films formed by spreading insoluble substances and those formed by the positive adsorption of solutes at the surface of a solution are basically alike.

Surface Layers and Gas Laws

The similarity between gaseous films of soluble and insoluble substances and ordinary three-dimensional gases is more than a qualitative one. If the molecules in the surface film are supposed to be free to move in any direction, within the two dimensions of the surface, and it is assumed further that the actual cross-sectional area of the molecules is small in comparison with the area they inhabit and that they do not attract each other, then it can be readily deduced from the kinetic theory of gases that [7]:

$$fA = kT. \qquad (1.33)$$

This equation is the quantitative equivalent of the gas law equation $PV = kT$, where V is the average volume occupied by a single molecule. As with gases, Equation 1.33 is obeyed at low values of the surface pressure, f, equivalent to low gas pressure. At higher surface pressures, an equation similar to that proposed for gases may be employed [7]. Thus:

$$f(A - b) = kTX, \qquad (1.34)$$

where b allows for the cross-sectional area of the molecules and X for their mutual attraction; as the attraction increases, X decreases. For a series of soluble acids, X decreases with increasing chain length, suggesting an increase of mutual attraction between the chains.

The fact that the two-dimensional gas equation holds for a solid unimolecular film suggests that even in such films the molecules have freedom of movement. In this connection, there is evidence that the molecules in certain adsorbed layers are able to move within the bounds of the surface.

ADSORPTION EQUILIBRIA

Generalizations

Adsorption from aqueous solutions involves concentration of the solute on the solid surface. As the adsorption process proceeds, the sorbed solute tends to desorb into the solution. Equal amounts of solute eventually are being adsorbed and desorbed simultaneously. Consequently, the rates of adsorption and desorption will attain an equilibrium state, called *adsorption equilibrium*. At equilibrium, no change can be observed in the concentration of the solute on the solid surface or in the bulk solution. The position of equilibrium is characteristic of the entire system, the solute, adsorbent, solvent, temperature, pH, and so on. Adsorbed quantities at equilibrium usually increase with an increase in the solute concentration. The presentation of the amount of solute adsorbed per unit of adsorbent as a function of the equilibrium concentration in bulk solution, at constant temperature, is termed the *adsorption isotherm*.

The shape of the adsorption isotherm gives qualitative information about the adsorption process and the extent of the surface coverage by the adsorbate. Brunauer classified adsorption isotherms into five basic shapes (Figure 1–4) [16]. Isotherms of Type I are associated with systems where adsorption does not proceed beyond the monomolecular layer. The other types of isotherms involve multilayer formation. The isotherms for adsorption from solution follow Type I, although under certain conditions multilayer adsorption may be encountered. Typical adsorption isotherms for adsorption from water systems are shown in Figure 1–5 [17]. Isotherms of this type are typical for adsorption by activated carbon from aqueous solutions when adsorption does not proceed beyond a monomolecular layer, whereas multilayer adsorption in these systems is not usually encountered.

The surface of activated carbon is heterogenous, not only in surface structure but also in the distribution of surface energy. During the course of adsorption, the heat of adsorption is not constant for each incremental increase in adsorption. Usually the initial portions of adsorbed solute have greater differential heats of adsorption than subsequent ones. Thus, a steep initial drop of the heat of adsorption with an increase of the amount adsorbed indicates that the first molecules to arrive at the bare surface are preferentially adsorbed on

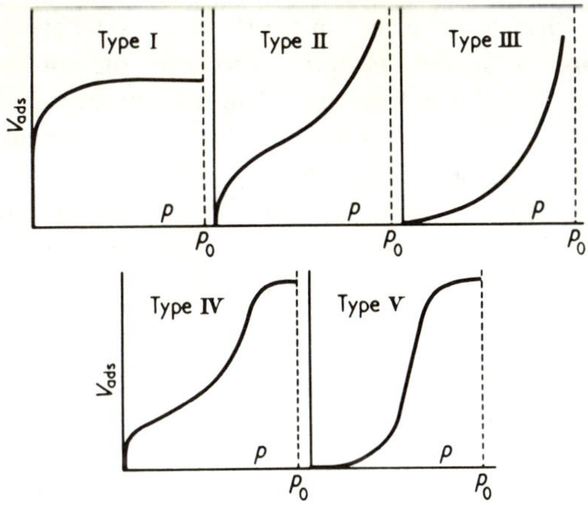

Figure 1–4 The five typical shapes of isotherms for physical adsorption. Reproduced from Brunauer et al. [16], courtesy of the American Chemical Society.

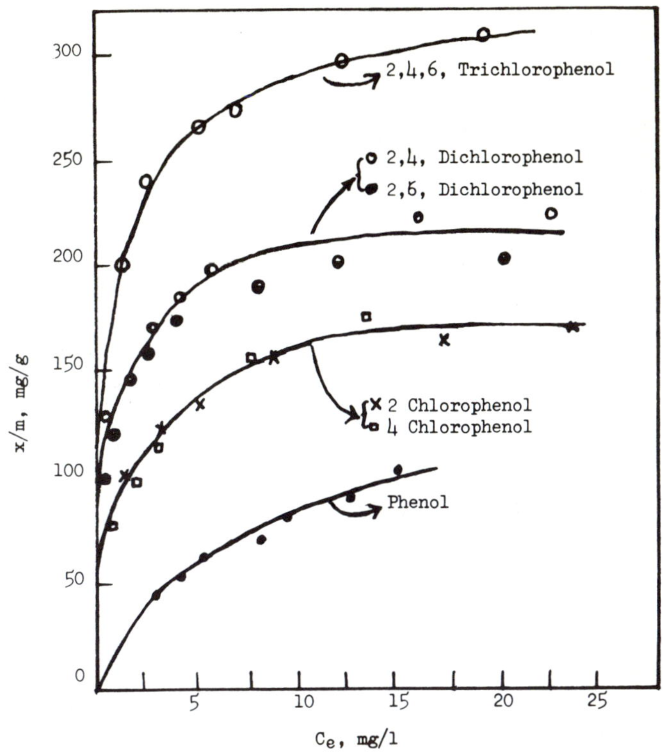

Figure 1–5 Adsorption isotherms of phenolic compounds on activated carbon. Reproduced from Faust and Aly [17], courtesy of Butterworth Publishers.

the most attractive sites or on positions on the surface where their potential energy will be a minimum [3]. As adsorption proceeds, the less active sites become occupied. Therefore, adsorption occurs on sites of progressively decreasing activity. Smooth adsorption isotherms are usually obtained because of the presence of a sufficiently large number of sites that may occur in patches of equal energy or randomly distributed sites of unequal energy [3]. Several models can be used for the description of the adsorption data, and Langmuir's and Freundlich's adsorption isotherms are the most commonly used.

Langmuir Adsorption Isotherm

The basic assumptions underlying Langmuir's model, which is also called the *ideal localized monolayer model,* are:

1. The molecules are adsorbed on definite sites on the surface of the adsorbent.
2. Each site can accommodate only one molecule (monolayer).
3. The area of each site is a fixed quantity determined solely by the geometry of the surface.
4. The adsorption energy is the same at all sites.

In addition, the adsorbed molecules cannot migrate across the surface or interact with neighboring molecules. The Langmuir equation was originally derived from kinetic considerations [18]. Later, it was derived on the basis of statistical mechanics, thermodynamics, the law of mass action, theory of absolute reaction rates, and the Maxwell-Boltzmann distribution law [3].

The kinetic derivation considered the adsorbed layer to be in dynamic equilibrium with the gas phase. A certain fraction of the molecules striking the bare sites will condense and be held by the surface forces for a finite time and are regarded as adsorbed; the remainder will be reflected. Those molecules striking sites that are already occupied will immediately re-evaporate as if they had been reflected. If the fraction of the site already filled is θ, then:

$$\text{Rate of adsorption} = k_a P(1 - \theta), \tag{1.35}$$

$$\text{Rate of desorption} = k_d \theta, \tag{1.36}$$

Where k_a = rate of adsorption, and k_d = rate of desorption from a fully covered surface.

At equilibrium the number of molecules in the adsorbed state at any instant is constant; therefore:

$$k_a P(1 - \theta) = k_d \theta \tag{1.37}$$

and

$$\frac{\theta}{1-\theta} = \frac{k_a}{k_d} P \tag{1.38}$$

Taking $k_a/k_d = b$, which is the adsorption equilibrium constant, Equation 1.38 becomes:

$$bP = \frac{\theta}{1-\theta} \tag{1.39}$$

or

$$\theta = \frac{bP}{1+bP}, \tag{1.40}$$

which is known as the Langmuir adsorption isotherm.

If V is the volume of gas adsorbed at pressure, P, and V_m is the volume adsorbed at infinite pressure—that is, when all the sites are occupied—then:

$$\theta = \frac{V}{V_m}, \tag{1.41}$$

and Equation 1.40 becomes:

$$V = \frac{V_m bP}{1+bP}. \tag{1.42}$$

This expression shows that V approaches V_m asymptotically as P approaches infinity. V_m is supposed to represent a fixed number of surface sites, and it should, therefore, be a temperature-independent constant while the temperature dependence of the equilibrium constant should follow a van't Hoff equation:

$$b = b_o e^{\frac{-\Delta H}{RT}}. \tag{1.43}$$

Since adsorption is exothermic (ΔH negative), b should decrease with increasing temperature.

For adsorption from solution by solid adsorbents, the Langmuir adsorption isotherm is expressed as:

$$X = \frac{X_m bC_e}{1+bC_e}, \tag{1.44}$$

where $X = x/m$, the amount of solute adsorbed, x, per unit weight of adsorbent, m; C_e = equilibrium concentration of the solute; X_m = amount of solute adsorbed per unit weight of adsorbent required for monolayer coverage of the surface, also called monolayer capacity; and b = a constant related to the heat of adsorption, $[b \propto (\exp(-\Delta H/RT)]$ [3].

Equation 1.44 indicates that X approaches X_m asymptotically as C_e approaches infinity. For linearization of the data, Equation 1.44 can be written in the form:

$$\frac{C_e}{X} = \frac{1}{bX_m} + \frac{C_e}{X_m} \; . \tag{1.45}$$

When C_e/X is plotted against C_e, a straight line, having a slope $1/X_m$ and an intercept $1/bX_m$, should result. Another linear form can be obtained by dividing Equation 1.45 by C_e:

$$\frac{1}{X} = \frac{1}{X_m} + \left(\frac{1}{C_e}\right)\left(\frac{1}{bX_m}\right) . \tag{1.46}$$

Plotting $1/X$ against $1/C_e$, a straight line, having a slope $1/bX_m$ and an intercept $1/X_m$, is obtained.

The monolayer capacity, X_m, determined from the Langmuir isotherm, defines the total capacity of the adsorbent for a specific adsorbate. Also, it may be used to determine the specific surface area of the adsorbent by utilizing a solute of known molecular area. Reliable X_m values can be obtained only for systems exhibiting Type I isotherms of the Brunauer's classification shown in Figure 1-4 [3].

It must be indicated that conformity to the algebraic form of Langmuir's equation does not constitute conformity to the ideal localized monolayer model, even if reasonable values of b and X_m are obtained. Therefore, a constant value of b may be a result of cancellation of variations in ΔH. In turn, constancy of ΔH may be a result of internal compensation of opposing effects such as attractive lateral interactions and surface nonuniformity [3]. In addition, the assumption that the area of the sites and, in turn, X_m are determined solely by the nature of the solid and are independent of the nature of the solute is contrary to what is encountered in adsorption systems. However, orientation of solute molecules on the surface of activated carbon has been shown to affect the monolayer capacity, X_m, as determined from the area occupied by the molecules [19,20]. Therefore, nonconformity with the physical model should not detract from the usefulness of the Langmuir isotherm for analytical description of adsorption systems that do not proceed beyond monomolecular layers and conform to Type I isotherms.

Freundlich Adsorption Isotherm

The Freundlich adsorption equation is perhaps the most widely used mathematical description of adsorption in aqueous systems. The Freundlich equation is expressed as [21]:

$$\frac{x}{m} = KC_e^{1/n}, \tag{1.47}$$

where x = the amount of solute adsorbed, m = the weight of adsorbent, C_e = the solute equilibrium concentration, and K and $1/n$ = constants characteristic of the system.

The Freundlich equation is an empirical expression that encompasses the heterogeneity of the surface and the exponential distribution of sites and their energies [3,22]. For linearization of the data, the Freundlich equation is written in logarithmic form:

$$\log \frac{x}{m} = \log K + \frac{1}{n} \log C_e. \tag{1.48}$$

Plotting $\log x/m$ versus $\log C_e$, a straight line is obtained with a slope of $1/n$, and $\log K$ is the intercept of $\log x/m$ at $\log C_e = 0$ (C_e=1). The linear form of the isotherm can be obtained conveniently by plotting the data on log-log paper (Figure 1-6). The value of $1/n$ obtained for adsorption of most organic compounds by activated carbon is <1. Steep slopes—that is, $1/n$ close to 1—indicate high adsorptive capacity at high equilibrium concentrations that rapidly dimin-

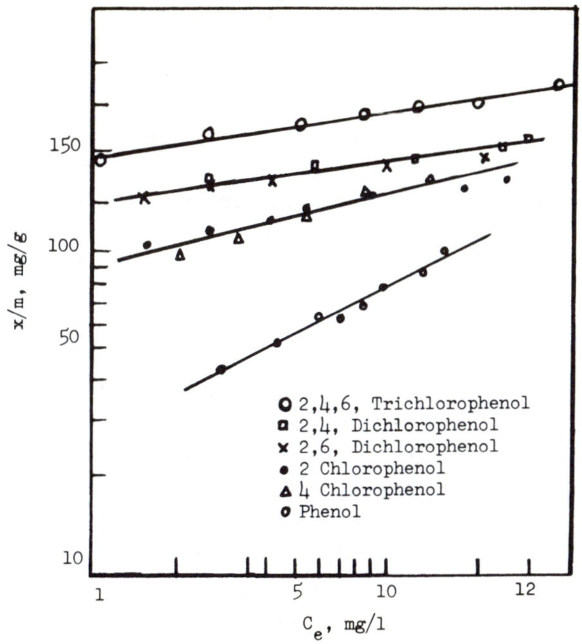

Figure 1-6 Logarithmic form of Freundlich adsorption isotherms for phenolic compounds on activated carbon [23].

ishes at lower equilibrium concentrations covered by the isotherm. Relatively flat slopes—that is, $1/n << 1$—indicate that the adsorptive capacity is only slightly reduced at the lower equilibrium concentrations. As the Freundlich equation indicates, the adsorptive capacity or loading factor on the carbon, x/m, is a function of the equilibrium concentration of the solute. Therefore, higher capacities are obtained at higher equilibrium concentrations.

The Freundlich equation can be used for calculating the amount of activated carbon required to reduce any initial concentration to a predetermined final concentration. By substituting $C_o - C_e$ in Equation 1.48 for x, where C_o = the initial concentration:

$$\log \left(\frac{C_o - C_e}{m} \right) = \log K + \frac{1}{n} \log C_e. \tag{1.49}$$

Equation 1.49 is useful for comparing different activated carbons in removal of different compounds or removal by the same carbon.

BET Adsorption Isotherm

The BET isotherm was developed by Brunauer, Emmett, and Teller (BET) for the generalization of the ideal localized monolayer treatment (Langmuir model) to account for multilayer adsorption [24]. The BET model is based on the simplifying assumptions that each molecule in the first adsorbed layer serves as a site for the adsorption of a molecule into the second, and so on. The concept of localization, therefore, prevails throughout the layers, and the forces of mutual interactions are neglected. The heat of adsorption, E, of the second and subsequent layers is assumed to be equal to the heat of liquefication of the bulk liquid and therefore different from the heat of adsorption of the first layer. The expression of the BET isotherm may be derived by an extension of the kinetic argument presented for the Langmuir isotherm or by a thermodynamic argument [1,3]. The resulting equation for the BET equilibrium isotherm is:

$$V = \frac{V_m BP}{(P_o - P)\,[1 + (B - 1)\,P/P_o]}, \tag{1.50}$$

where V and V_m have the same meaning as in the Langmuir isotherm, P_o = saturation vapor pressure of the saturated liquid sorbate, and B = a constant:

$$B = \frac{a_1 b_2}{a_2 b_1}\, e^{(E_1 - E_L)/RT}, \tag{1.51}$$

which can be simplified to:

$$B = e^{(E_1 - E_L)/RT}, \tag{1.52}$$

where a_1 and a_2 = rates of condensation on the first and second layers, b_1 and b_2 = rates of evaporation from the first and second layers, E_1 = first layer heat of adsorption, and E_L = heat of liquefication of the bulk phase.

The term $E_1 - E_L$ is known as the net heat of adsorption. Thus, the BET equation provides a measure of both the heat of adsorption and the surface area of the solid. Application of the BET equation to the adsorption from solution takes the form:

$$X = \frac{X_m BC}{(C_s - C_e)\,[1 + (B - 1)C_e/C_s]}\,, \tag{1.53}$$

where X, X_m, and C_e have the same meaning as in Langmuir's isotherm, and C_s = solubility of the solute in water at a specified temperature. Transforming Equation 1.53 to:

$$\frac{C_e}{X(C_s - C_e)} = \frac{1}{X_m B} + \frac{(B - 1)}{X_m B}\frac{C_e}{C_s} \tag{1.54}$$

shows that a plot of the left side against C_e/C_s should give a straight line having slope $(B-1)/X_mB$ and intercept $1/X_mB$.

Henry's Law: Linear Adsorption Isotherm

This represents the simplest isotherm in which the amount adsorbed varies directly with the equilibrium concentration of the solute. It is often called Henry's Law after the analogous isotherm for the solution of gases in liquids. The isotherm is described by:

$$X = K_h C_e \tag{1.55}$$

where $X = x/m$, the amount of solute adsorbed by unit mass of adsorbent; C_e = equilibrium concentration; and K_h = a constant.

This isotherm is obtained under conditions of low concentrations of solute. In such systems, the adsorbed layer is extremely dilute and the amount adsorbed is only a fraction of the monolayer capacity. Usually the linear relationship is observed at the lower concentration levels of a total adsorption isotherm. Therefore, the application of Henry's Law equation should be restricted to that region of the isotherm obtained from experiment. Almost all the adsorption isotherms are reduced to Henry's Law at low concentrations. For example, for the Langmuir adsorption isotherm, Equation 1.44, at very low concentrations, bC_e is small compared with unity so the equation reduces to:

$$X = X_m b C_e, \tag{1.56}$$

which is Henry's Law.

Most of the adsorption studies reported in the literature have been conducted in distilled water systems. However, inorganic salts have been shown to affect the adsorptive capacity of activated carbon for certain solutes. Snoeyink et al. [25] and Zogorski [26] reported on the enhancement of adsorptive capacity of activated carbon for some phenolic compounds at high pH values (anionic species) in the presence of inorganic salts. This effect was suggested to be due possibly to a reduction of the repulsive forces between adsorbed molecules and the carbon surface or between anions adsorbed on the surface. Although the concentrations of the inorganic salts used in these studies were too high to be encountered in drinking water supplies, the implication of the potential effects of inorganic ions on the carbon's adsorptive capacity should not be overlooked. Indeed, Weber et al. [27] showed that the presence of low concentrations of calcium and magnesium salts enhances the adsorption of humic acids on activated carbon possibly because of the formation of an ion-humate-carbon complex. Adsorption isotherms of humic acid in tap water systems showed higher carbon adsorptive capacities than those in distilled water systems. Therefore, adsorption studies for the application of activated carbon in water treatment plants should be conducted using the natural water or a synthetic medium of equivalent composition. In addition, the isotherms should be conducted within the concentration range corresponding to the levels likely to be encountered for the compound of interest since extrapolation of the isothermal data can lead to erroneous results.

REFERENCES

1. Ruthven, D.M. *Principles of Adsorption and Adsorption Processes.* John Wiley & Sons, New York (1984).
2. Lowell, S. *Introduction to Powder Surface Area.* John Wiley & Sons, New York (1979).
3. Young, D.M., and A.D. Crowell. *Physical Adsorption of Gases.* Butterworths, London (1962).
4. Osipow, L.I. "Physical Adsorption on Solids," Chapter 4 in *Principles and Applications of Water Chemistry: Proceedings of the 4th Rudolfs Conference.* S.D. Faust and J.V. Hunter, eds. John Wiley & Sons, New York (1976), 75.
5. London, F. *Trans. Faraday Soc. 33,* 8 (1937).
6. Dickerson, R.E., et al. *Chemical Principles.* W.A. Benjamin, New York (1970).
7. Glasstone, S. *Textbook of Physical Chemistry.* Van Nostrand Co., Princeton (1959).
8. Gibbs, J.W. *Collected Works.* Dover, New York (1961).
9. Kipling, J.J. *Adsorption from Solutions of Non-Electrolyte.* Academic Press, London (1965).
10. Hill, T.L. *J. Chem. Phys. 17,* 507, 520 (1949).
11. Everette, D.H. *Trans. Faraday Soc. 46,* 453 (1950).
12. Ottewill, R.H., et al. *Adsorption from Solution.* Academic Press, London (1983).
13. Goodrich, F.C. "Capillary Thermodynamics Without a Geometric Gibbs Convention," Chapter 1 in *Adsorption from Aqueous Solutions.* Advances in Chemistry Series 79. American Chemical Society, Washington, D.C. (1969), 1.
14. Moore, W.J. *Physical Chemistry,* 2nd ed. Prentice-Hall, Englewood Cliffs, N.J. (1960).

15. Langmuir, I. *J. Amer. Chem. Soc. 39,* 1848 (1917).
16. Brunauer, S., et al. *J. Amer. Chem. Soc. 62,* 1723 (1940).
17. Faust, S.D., and O.M. Aly. *Chemistry of Water Treatment.* Butterworths, Stoneham, Mass. (1983).
18. Langmuir, I. *J. Amer. Chem. Soc. 37,* 1139 (1915).
19. Puri, R.P. "Carbon Adsorption of Pure Compounds and Mixtures from Solution Phase," Chapter 17 in *Activated Carbon Adsorption of Organics from the Aqueous Phase,* vol. 1. I.H. Suffet and M.J. McGuire, eds. Ann Arbor Science Publishers, Ann Arbor, Mich. (1980), 353.
20. Morris, J.C., and W.J. Weber, Jr. "Adsorption of Biochemically Resistant Materials from Solution." *Environmental Health Series* AWTR-9. U.S. Department of Health, Education, and Welfare, Washington, D.C. (1964).
21. Freundlich, H. *Colloid and Capillary Chemistry.* Methuen and Co., London (1926).
22. Sips, R. *J. Chem. Phys. 16,* 490 (1948).
23. Aly, O.M., and S.D. Faust. "Sorption of Phenolic Compounds from Aqueous Solutions." Paper presented at the Kendall Award Symposium, American Chemical Society, 163rd National Meeting, Boston (1972).
24. Brunauer, S., et al. *J. Amer. Chem. Soc. 60,* 309 (1938).
25. Snoeyink, V.L., et al. *Environ. Sci. Technol. 3,* 918 (1969).
26. Zogorski, J.S. "The Adsorption of Phenols onto Granular Activated Carbon from Aqueous Solutions." Ph.D. dissertation, Rutgers University, Department of Environmental Science, New Brunswick, N.J. (1975).
27. Weber, W.J., Jr., et al. "Potential Mechanisms for Removal of Humic Acids from Water by Activated Carbon," Chapter 16 in *Activated Carbon Adsorption of Organics from the Aqueous Phase,* Vol. 1. I.H. Suffet and M.J. McGuire, eds. Ann Arbor Science Publishers, Ann Arbor, Mich. (1980), 317.

2

Adsorption Models

In drinking water treatment, adsorption of organic solutes has been a primary concern in recent years. This water quality problem has been discussed thoroughly by Faust and Aly [1]. The removal of these solutes from water, for the most part, is accomplished by adsorption onto carbon and/or resin surfaces; that is, adsorption is assumed to be the mechanism inasmuch as the surface, after saturation, is rarely examined for organic molecules. There is, of course, a concomitant decrease in the concentration of solute in the aqueous phase of the treatment process. See Chapter 1 and Osipow, Young and Crowell, London, and Lowell for discussions of the principles of adsorption [2,3,4,5].

THEORIES

Background

Adsorption of various substances onto carbon surfaces is an exceedingly complex process. Chapter 1 described equilibrium isotherms for the adsorption of single and, on occasion, mixed solutes on carbon. These isotherms as well as kinetic data are important for an understanding of the process. Also, it is necessary to have this information for design purposes because of the dynamics of adsorption processes.

In a description of adsorption dynamics, three consecutive steps are proposed frequently for porous adsorbents like carbon [6]: first, there is transport of the solute from bulk solution through a liquid film to the carbon's exterior surface. Second, there is solute diffusion into pores of the adsorbent except for a small quantity of adsorption on the external surface. Parallel to this step is the intraparticle transport mechanism of surface diffusion. Third, there is adsorption of the solute on the interior surfaces of the pores and capillary spaces of the adsorbent. This last step is considered to be an equilibrium reaction. This sequence leads to an adsorptive capacity for solutes (mainly organics in this chapter) on carbon.

In turn, many factors influence the adsorptive capacity for specific organic solutes of a homologous series [6]. First, there are the adsorbate properties of group functionality, branching or geometry, polarity, hydrophobicity, dipole moment, molecular weight and size, and aqueous solubility. Second, there are solution conditions such as pH, temperature, adsorbate concentration, ionic strength, and competitive solutes. Third, there is the nature of the adsorbent with its surface area, pore size and distribution, distribution of functional groups such as the number of O^-H sites per $Å_2$, and ash content.

Since the mid-1970s, numerous models have been offered in an attempt to explain adsorption of organics onto activated carbon. These range from mathematical and conceptual descriptions of breakthrough curves (mass transfer zone) to studies of the thermodynamic properties of solvents, adsorbents, and adsorbates [7]. Some of these models are described in the following section and in Chapter 3.

Net Adsorption Energy Concept

The net adsorption energy concept (NAEC) is based on the theory that predicts an increase in solvency as the solubility parameter, δ, of the solvent approaches that of the solute; that is, maximum solvency results when the values of solute and solvent are the same. Conception of this theory is credited to Hildebrand et al. [8] who proposed that, for relatively nonpolar solutes (j) and solvents (i), their interaction energy is mostly dispersive and is proportional to the number of electrons in j (nj) and to the polarizabilities of the electrons of j (α_j) and i (α_i). Assuming nj is approximately proportional to the molar volume of j (Vj) along with the definition that the total solubility parameters (δ_T^j and δ_T^i) are proportional to their polarizabilities (α_j and α_i), a molar interaction term, *Eij,* can be defined for breaking all the $i - j$ bonds. Conceptually, this bond breakage is similar to the vaporization of j from i:

$$i - j - i = i - i + j \tag{2.1}$$

where half as many bonds are formed as are broken. The energy change for this reaction is:

$$\Delta Ev = 2E_{ij} - E_{ii} \tag{2.2}$$

The major thermodynamic relationships for the solubility parameter are [9]:

Process	*Interaction Energy*	
Net electrostatic attraction of j to i	$E_{ij} = n_j \, \alpha_j \, \alpha_i$	(2.3)
Disruption of $j - i$ bonds per mole j	$E_{ij} = V_j \, \delta_T^j \, \delta_T^i$	(2.4)
Vaporization of j from i	$\Delta E_v = 2E_{ij} - E_{ii}$	(2.2)
Vaporization of pure j	$\Delta E_v = 2E_{ij} - E_{jj}$	
	$\Delta E_v = E_{jj} = V_j(\delta_T^j)^2$	(2.5)

Equation 2.5 contains the essence of the solubility parameter, which is expressed through the vaporization energy and molar volume.

This concept is one of many attempts to correlate physical-chemical parameters with adsorptive capacities for organic compounds on carbon [7,10]. The NAEC is a function of a compound's aqueous solubility, molecular weight, and density (ρ); the dispersion component of the solubility parameter for an organic compound; and the dispersion component of the solubility parameter for the activated carbon surface. Consequently, the net adsorption energy, E_T^A, is:

$$E_T^A = f\,(\text{aqueous solubility}, MW, \rho, \delta_d^j, \delta_d^s). \tag{2.6}$$

Important components of Equation 2.6 are the solubility parameters, δ, for the solute, j, and the surface, s (see Table 2–1 for the symbolism for the NAEC

Table 2–1 Symbols for the Net Adsorption Energy Concept

Symbol	*Definition*
a	Subscript for energy terms representing the *surface* in an interaction [12].
C_e	Equilibrium concentration of solute in solution, mol/l.
$\Delta \log C_e$	Difference in log equilibrium concentrations required to translate isotherms to new isotherm position, mol/l.
E_{ii}	Energy of interaction between solvent molecules ($i \rightarrow i$), kcal/mol.
E_{ji}	Energy of interaction between a solute (j) and a solvent (i), kcal/mol.
$E_{is}{}^A$	Energy of adsorption of a solvent (i) on a surface (s), kcal/mol.
$E_{js}{}^A$	Energy of adsorption of a solute (j) on a surface (s), kcal/mol.
$E_T{}^A$	Net adsorption energy, kcal/mol.
$E_{ji}{}'$	Net solute/solvent interaction energy, kcal/mol.
ΔE_{lsc}	Energy associated with liquid-solid chromatography, kcal/mol [12].
$(\Delta E^A)_{i/a}$	Energy of adsorption of a solute (i) on a surface (a), kcal/mol [12].
$(\Delta E^A)_{j/a}$	Energy of adsorption of a solvent (j) on a surface (a), kcal/mol [12].
$[(\Delta E^S)_{i/j} - (\Delta E^A)_{i/j}]$	Energy of affinity of a solute (i) with a solvent (j), kcal/mol [12].
i	Subscript for energy terms and superscript for component solubility parameter terms representing the *solvent* in an interaction (this study) [10]. Represents the *solute* in Karger et al. [12].

(cont.)

Table 2-1 (*continued*)

Symbol	*Definition*
j	Subscript for energy terms and superscript for component solubility parameter terms representing the *solute* in an interaction (this study) [10]. Represents the *solvent* in Karger et al. [12].
n	Ratio of the molar volumes of the solute and solvent [12].
q	Surface concentration of the solute on the carbon, mol/g.
Q_s	Experimentally determined heats of adsorption, kcal/mol.
s	Subscript for energy terms and superscript for component solubility parameter terms representing the *surface* in an interaction.
V_i and V_j	Molar volumes of the solvent and the solute, respectively, calculated by molecular weight/density, ml/mol (this study) [10].
X and Y	Number of molecules in the liquid phase and the surface phase, respectively, that interact with solute or solvent molecules.
δ_d, δ_o, δ_{in}, δ_a, and δ_b	Dispersion, permanent dipole orientation, induced dipole, acidic and basic components of the total solubility parameter, $(cal/ml)^{1/2}$.
δ_h and δ_P	Hydrogen bonding and polar components of the total solubility parameter, $(cal/ml)^{1/2}$.
δ_T	Total solubility parameter, $(cal/ml)^{1/2}$.
δ_d^j	Dispersion component of the total solubility parameter for a compound, $(cal/ml)^{1/2}$.
δ_d^s	Dispersion component of the total solubility parameter for a surface, $(cal/ml)^{1/2}$.
∞	Infinity—denotes that a solute is completely miscible in a solvent.

Source: Adapted from McGuire and Suffet [10], courtesy of Ann Arbor Science Publishers, Inc.

[10]). Many classification schemes have been proposed for organic compounds, but one scheme has emerged that has a thermodynamic basis and wide applicability [11,12,13]. The total solubility parameter (δ_T) is defined as the cohesive energy density of a compound in the pure state. The δ_T values for the least polar compounds, the alkanes, are between 6 and 7 $(cal/ml)^{1/2}$, and for the most polar compound, water, it is 23.5. Much of this concept has been applied to chromatography [11,12,13]. A comprehensive review of the solubility parameter was published by Barton [14].

Component parts of the total solubility parameter are:

$$\delta_T^2 = \delta_d^2 + \delta_o^2 + 2\delta_{in}\delta_d + 2\delta_a\delta_b, \tag{2.7}$$

where δ_d = the dispersion component, δ_o = the permanent dipole orientation component, δ_{in} = the induced dipole component, δ_a = the acidic component, and δ_b = the basic component. A hydrogen bonding component, δ_h, can be formed within Equation 2.7:

$$\delta_h^2 = 2\delta_a\delta_b. \tag{2.8}$$

Likewise, a polar component can be formulated:

$$\delta_p^2 = \delta_o^2 + 2\delta_{in}\delta_d. \tag{2.9}$$

Table 2–2 gives several organic solutes that were employed by McGuire and Suffet [10] in their development of the NAEC. These compounds are highly soluble in water and have medium to high polarity. The following discussion gives their adsorbability by carbon.

Several methods are available to calculate δ_T^2 values [10]. Good agreement is obtained in some cases, but there is variation in others (Table 2–3). Apparently, this variation is due to the widespread application of the solubility parameter—that is, in chromatography, reverse osmosis membrane rejection predictions, and polymer chemistry. One of the more difficult components to calculate is the polar term, δ_p, according to McGuire and Suffet [10]. Despite these limitations, the solubility parameter concept was applied to the transfer of organic solutes from a polar phase, water, to a nonpolar surface on activated carbon.

In Table 2–4, a schematic development of the component adsorption energies of a solute/solvent/surface system is given that was modeled after Keller et

Table 2–2 Organic Compound Properties

Compounds	*Molecular Weight*	*Boiling Point (°C)*	*Solubility (g/l)*	*Solubility Parameter (δ_T) (cal/ml)$^{1/2}$*
p-nitrophenol	139.11	279 (decomposes)	14[a]	12.9[b]
Nitromethane	61.04	100.8	110[c]	12.6[d]
Methyl ethyl ketone	72.12	79.6	210[c]	9.4[e]
n-butanol	74.12	117.2	66[c]	11.6[e]
1,4-dioxane	88.12	101	∞[c]	10.1[e]
Urea	60.06	decomposes	500[c]	22. [f]

Source: Reproduced from McGuire and Suffet [10], courtesy of Ann Arbor Science Publishers, Inc.
[a]Zogorski in [10].
[b]Calculated from heat of vaporization estimate.
[c]Approximated in this study at 25 ± 2°C in distilled water.
[d]Hildebrand and Scott in [10].
[e]Hoy in [10].
[f]Hansen. Saturated urea/water solution in [10].

Table 2–3 Comparison of Total and Component Solubility Parameters

Compounds	*Total,* δ_T *(cal/ml)*$^{1/2}$	*Nonpolar (dispersion),* δ_d*(cal/ml)*$^{1/2}$	*Polar,* δ_p *(cal/ml)*$^{1/2}$	*Hydrogen Bonding,* δ_h *(cal/ml)*$^{1/2}$
Nitromethane	12.6	7.1	10.4	0
	12.3	7.70	9.2	2.5
	—	—	—	—
	12.7	—	—	—
	12.9	7.3	10.6	0
Methyl ethyl ketone	9.4	7.1	6.2	0
	9.3	7.77	4.4	2.5
	9.4	6.88	4.52	4.634
	9.3	7.77	5.06	—
	9.5	7.1	6.3	0
n-butanol	11.6	7.4	8.8	1.4
	11.3	7.81	2.8	7.7
	11.6	7.82	4.05	7.547
	11.4	—	—	—
	—	—	—	—
1,4-dioxane	10.1	7.8	6.5	0
	10.0	9.3	0.9	3.6
	10.1	7.96	4.94	3.839
	9.9	8.55	4.65	—
	10.1	7.8	6.5	0

Source: Reproduced from McGuire and Suffet [10] courtesy of Ann Arbor Science Publishers, Inc.

al. [11]. It should be noted that the bracketed term in Equation 2.14 is uncertain in its formulation [10]. The significance of this term is that there are interactions between the solute and solvent that must be taken into consideration in the energy balance. This bracketed term is replaced by E_{ji} [10].

The transfer of a solute from the liquid to the solid phase is complicated by three sets of affinities [10]:

Solute for the surface ($j - s$)

Solvent for the surface ($i - s$)

Solute for the solvent ($j - i$)

Any affinity of the solute for the surface is counteracted by the other two. In any event, the NAEC of a solute/solvent/surface system is expressed:

$$E_T^A = E_{js}^A - E_{is}^A - E_{ji}' , \tag{2.15}$$

where the energy terms (E^A) for the respective affinities are given.

Table 2-4 Explanation of the Adsorption Mechanism

Adsorption system

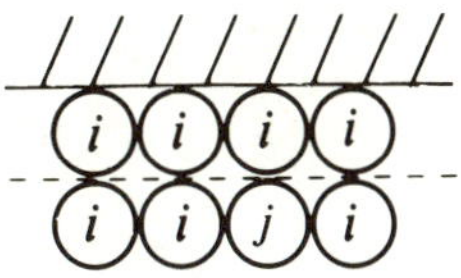

Solid phase

Surface phase

Liquid phase

Hole formation in the liquid phase

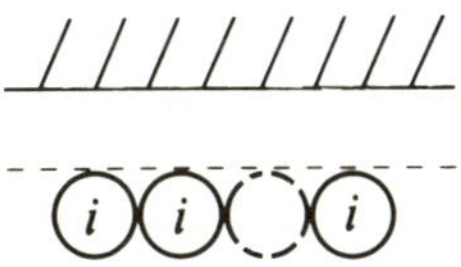

$$\text{Energy} = -XE_{ji} \tag{2.10}$$

Hole formation in the surface phase

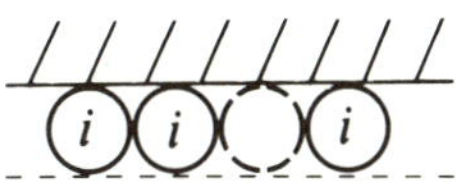

$$\text{Energy} = -YE_{ii} - E_{is}^{A} \tag{2.11}$$

Solute transfer into the surface phase

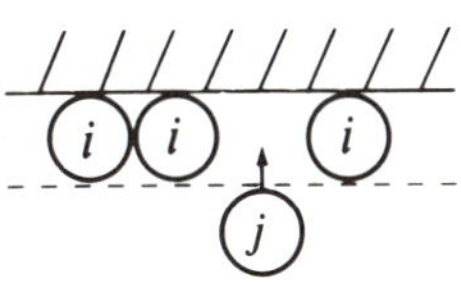

$$\text{Energy} = YE_{ji} + E_{js}^{A} \tag{2.12}$$

Solvent transfer into the liquid phase

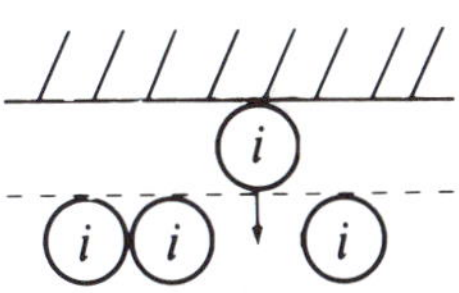

$$\text{Energy} = XE_{ii} \tag{2.13}$$

Net adsorption energy

$$E_T^{A} = E_{js}^{A} - E_{is}^{A} - [(X-Y)E_{ji} - (X-Y)E_{ii}] \tag{2.14}$$

Source: Reproduced from McGuire and Suffet [10] courtesy of Ann Arbor Science Publishers, Inc.

As a solute molecule approaches an adsorption site in the surface phase, it will encounter a surface occupied entirely by solvent molecules. Consequently, a solvent molecule must be desorbed from the surface first that requires energy. Affinities of the solute and solvent molecules for the surface phase have the same mathematic form:

$$E_{js}^{A} = V_j(\delta_d^j \delta_d^s + \delta_p^j \delta_p^s + \delta_a^j \delta_b^s + \delta_a^s \delta_b^j) \tag{2.16}$$

and

$$E_{is}^{A} = V_i(\delta_d^i \delta_d^s + \delta_p^i \delta_p^s + \delta_a^i \delta_b^i + \delta_a^s \delta_b^i). \tag{2.17}$$

An additional energy requirement is affinity of the solute for the solvent phase, E'_{ji}, which is in the form of an energy of adsorption since it is a surface phenomenon [10]. There is, also, the possibility of an adsorbed solute molecule reentering the solvent phase. In any event, solubility and/or affinity of an organic molecule for water is controlled mainly by formation of H-bonds between the two. Consequently, the δ_a and δ_b terms in Equation 2.16 are the only significant ones. Therefore:

$$E'_{ji} = V_j(\delta_a^j \delta_b^i + \delta_a^i \delta_b^j). \tag{2.18}$$

The equations for E_{js}^{A} and E_{is}^{A} may be simplified because interactions in the adsorption of nonpolar and polar compounds onto a nonpolar surface are controlled almost entirely by dispersion forces [13]. Consequently, Equations 2.16 and 2.17 are reduced to:

$$E_{js}^{A} = V_j(\delta_d^j \delta_d^s) \tag{2.19}$$

and

$$E_{is}^{A} = V_i(\delta_d^i \delta_d^s). \tag{2.20}$$

It is fairly obvious from Equations 2.19 and 2.20 that a reasonably accurate value of δ_d^s for the carbon's surface is required. This was accomplished by calculating the surface's solubility parameter from solutes with known component solubility parameters [11]. Another solution came from experimental determination of heats of adsorption, Q_s (kcal/mol), for many compounds on a nonpolar surface (not necessarily carbon) [15]. Substitution of Q_s for E_{js}^{A} and canceling zero terms in Equation 2.16, the δ_d^s can be calculated:

$$\delta_d^s = \frac{Q_s}{V_j \delta_d^j} . \tag{2.21}$$

A δ_d^s value of 12.2 ± 7.3% (cal/ml)$^{1/2}$ was calculated from Equation 2.21 for carbon [10].

Table 2-5 A Comparison of Calculated Net Adsorption Energies

Compounds	$E_{js}{}^{A}$ (kcal/mol)	$E_{is}{}^{A}$ (kcal/mol)	$E_{ji}{}'$ (kcal/mol)	$E_{T}{}^{A}$ (kcal/mol)
p-nitrophenol	11.9	1.4	1.1	9.4
Nitromethane	4.6	1.4	1.3	1.9
Methyl ethyl ketone	7.8	1.4	2.1	4.2
n-butanol	8.3	1.4	2.2	4.7
1,4-dioxane	8.1	1.4	4.0	2.7
Urea	4.8	1.4	2.1	1.3
Carbon tetrachloride	9.9	1.4	0	8.5
n-decane	18.1	1.4	0	16.7

Source: Reproduced from McGuire and Suffet [10] courtesy of Ann Arbor Science Publishers, Inc.

To calculate the E_{is}^{A} and E_{ji}' terms in Equation 2.15, it is critical to have accurate estimates of the solubility parameters for water. These values were obtained [8]: $\delta_T = 23.5$, $\delta_d = 6.3$, $\delta_p = 15.3$, $\delta_h = 16.7$, $\delta_a = 11.8$, and $\delta_b = 11.8$. Substitution of these data into Equation 2.20 yields an E_{is}^{A} value of 1.4 kcal/mole, which remains constant for all calculations in the carbon/water system. The E_{ji}' term in Equation 2.15 is calculated by using the given δ_a and δ_b values for water in Equation 2.18 along with the appropriate values for organic compounds [10]. Now, E_T^A values are calculated from Equation 2.15 for organic compounds of interest (Table 2–5). Two points emerge from this table: first, the E_{js}^{A} term (affinity of solute for the carbon surface) constitutes the major portion of the total energy, which is expected, and second, nonpolar compounds have the highest E_T^A values.

Correlation of NAEC to Adsorption Isotherms

Adsorption isotherms were developed for correlative purposes. These are seen in Figure 2–1 from McGuire and Suffet [10] and in Figure 2–2 from Zogorski and Faust [16] and Radke and Prausnitz [17]. A pattern of adsorption emerges from these two figures. Well-adsorbed compounds show very shallow slopes, and the plots lie in the upper surface concentration values (q). The more poorly adsorbed compounds have steeper slopes and lower q versus C_e relationships.

Table 2–6 gives the surface concentrations (q) when C_e is 10^{-3} M for most of the organics seen in Figures 2–1 and 2–2. As expected [10] there was a poor correlation between δ_T and q, but there was a good correlation between E_T^A and q, as seen in Figure 2–3 [18]. This graph shows a trend of greater adsorptive capacity with the higher E_T^A values. It is linear on a log-log plot for the most part and becomes curvilinear at lower E_T^A values.

The relationship in Figure 2–3 can be used, at least, to predict the surface capacity for organic compounds with the higher E_T^A values. An approximation

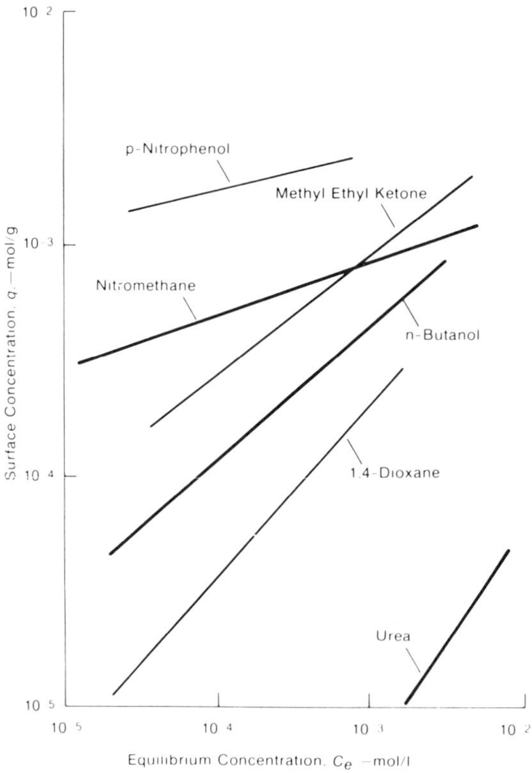

Figure 2–1 Adsorption isotherms on Filtrasorb 400 carbon at 25 ± 2°C and 0.2 mol/l phosphate buffer [18] (*p*-nitrophenol, pH = 4.0, no buffer; nitromethane, pH = 8.0; methyl ethyl ketone, pH = 8.0; *n*-butanol, pH = 4.2; 1,4-dioxane, pH = 8.0; urea, pH = 7.0). Reproduced from McGuire and Suffet [10], courtesy of Ann Arbor Science Publishers, Inc.

of an adsorption isotherm can be determined from a series of similar curves for various C_e values of a given compound. McGuire and Suffet [10] suggest that caution must be used when predicting adsorptive capacities of other neutral compounds. For example, the straight-chain and branched alkanes with their high molecular weights and low solubilities in water will not correlate well because a C_e of 10^{-3} M is well above their solubility limits. It appears that the NAEC is limited to neutral organic molecules of roughly the same molecular size, solubility, and molecular weight.

Polanyi Adsorption Potential Theory

According to Manes [19] the Polanyi adsorption potential theory [20,21] "has the advantages of unparalleled power and of simplicity in concept and applica-

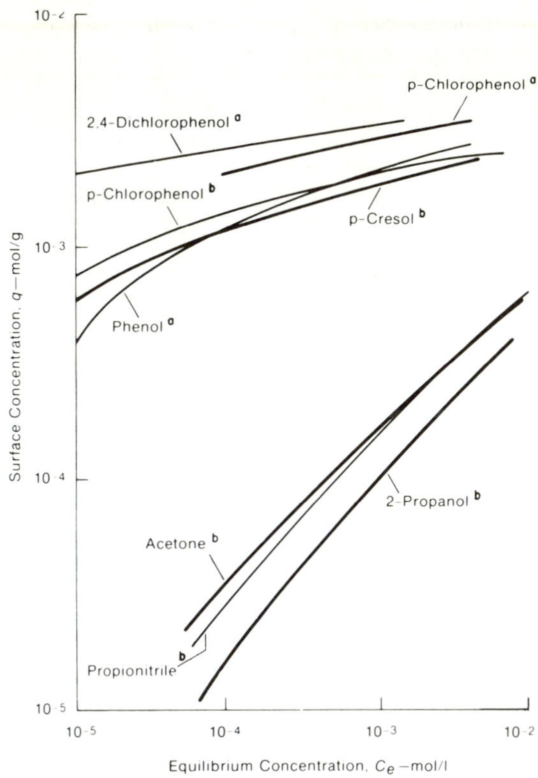

Figure 2–2 Adsorption isotherms from the literature [18]. (a) Zogorski and Faust [16] Columbia LCK carbon, neutral compounds, temperature 20°C, 1.00-mm size. (b) Radke and Prausnitz [17], Filtrasorb 300 carbon, neutral compounds, temperature 25°C, 10 × 32 mesh. Reproduced from McGuire and Suffet [10], courtesy of Ann Arbor Science Publishers, Inc.

tion.'' Manes has been a proponent of the Polanyi theory of adsorption since two important consequences of this theory were stated:

1. that for many individual adsorbates on a single carbon sample, including gases, vapors, and liquid and solid solutes in water, all of the individual adsorption isotherms, from trace concentrations to saturation, may be readily calculated from the experimental isotherm of any one of them; and
2. that the competitive adsorption isotherms of multiple solid solutes from water, again over the complete concentration range of interest, may be calculated from the individual single-component isotherms. [19]

An interpretation of the first statement is that performance of an activated carbon for a specific contaminant may be predicted from a wide choice of calibrating isotherms. This should lead to more precise specifications of activated

Table 2-6 Net Adsorption Energies, Total Solubility Parameters and Adsorption Capacity Data for Selected Organic Compounds

Compounds	*Total Solubility Parameter* $(cal/ml)^{1/2}$	*Net Adsorption Energy (kcal/mol)*	*Amount of Compound Adsorbed*[d] *(mol/g)*
p-nitrophenol[a]	12.9	9.4	2.4×10^{-3}
Nitromethane[a]	12.6	1.9	8.4×10^{-4}
Methyl ethyl ketone[a]	9.4	4.2	8.8×10^{-4}
n-butanol[a]	11.6	4.7	4.2×10^{-4}
1,4-dioxane[a]	10.1	2.7	2.0×10^{-4}
Urea[a]	22.0	1.3	5.7×10^{-6}
p-chlorophenol[b]	10.7	9.3	2.7×10^{-3}
Phenol[b]	11.3	6.7	2.1×10^{-3}
Propionitrile[c]	10.7	2.0	1.6×10^{-4}
2-propanol[c]	11.4	1.6	1.0×10^{-4}
Acetone[c]	9.6	1.2	1.6×10^{-4}
p-cresol[c]	11.1	9.4	1.9×10^{-3}
p-chlorophenol[c]	10.7	9.3	2.1×10^{-3}

Source: Reproduced from McGuire and Suffet [10] courtesy of Ann Arbor Science Publishers, Inc.
[a]Compounds used in this study.
[b]Zogorski and Faust [16].
[c]Radke and Prausnitz [17].
[d]At $C_e = 10^{-3}$ mol/l.

carbon for specific applications. The second statement reduces considerably the uncertainty of feasibility studies for activated carbon removal of a mixture of contaminants.

The Polanyi model postulates a fixed volume—that is, *adsorption space*—that is close to the adsorbent's surface where adsorption occurs. These spaces would be micropores of various sizes and of unspecified shapes for activated carbon. A schematic model of this concept is seen in Figure 2-4 [19]. Polanyi [20,21] defined the adsorption potential, ϵ, of any molecule within the attractive force field of the solid surface (the sum of attractive forces of all contributing atoms of the solid) as the work required to remove a molecule to infinity from its location in the adsorption space. ϵ is related to an excess of the adsorption energy over the condensation energy. It was postulated that ϵ would be independent of temperature, which is consistent with London-force attractions [20,21]. Now, it is assumed that the value of ϵ varies continuously in the adsorption space from a maximum value to zero (see Figure 2-4). This leads to equipotential surfaces of decreasing values of ϵ within each region of the adsorption space. A plot of ϵ versus the cumulative volume in which a value of ϵ is equalled or exceeded yields a so-called characteristic curve from which the energy abscissa is informative for predicting adsorption isotherms.

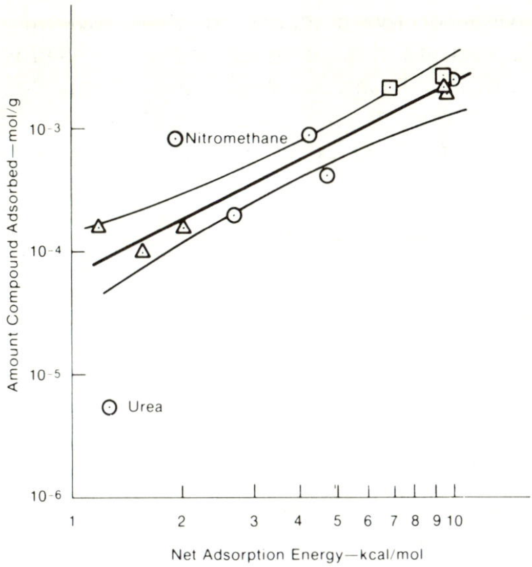

Figure 2–3 Relationship between the amount of compound adsorbed by activated carbon (at $C_e = 10^{-3}$ mol/l) and the net adsorption energy. ⊙: data from this study; ⊡: Zogorski and Faust [16]; △: Radke and Prausnitz [17]. See tabulation for compound identifications; curves indicate the 95% confidence interval. Reproduced from McGuire et al. [18], courtesy of the American Water Works Association.

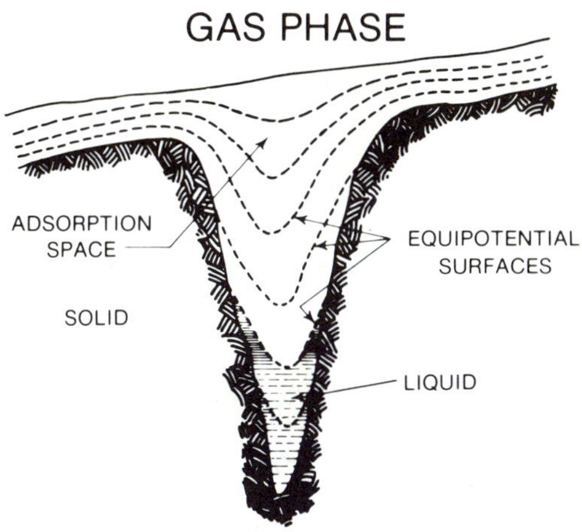

Figure 2–4 Schematic model of a region of a heterogeneous adsorbent, showing equipotential surfaces. In adsorption of a vapor, the vapor liquefies up to the equipotential surface at which the adsorption potential required for condensation is equalled or exceeded. In adsorption of a solute from water, the space is filled with water except where the solute condenses to an equipotential surface. Reproduced from Manes [19], courtesy of Ann Arbor Science Publishers, Inc.

Two alternative adsorptive processes are considered. In a gas-phase adsorption, the partial pressure of a gas or vapor above a solid is assumed to be set at any equilibrium value from zero to saturation. In the Polanyi model, the adsorption potential concentrates the gas or vapor above its bulk equilibrium pressure (or concentration) to an extent that is determined by the Boltzmann equation:

$$\epsilon = RT \ln (P^*/P) = RT \ln (C^*/C), \qquad (2.22)$$

where C^* and C are the respective concentrations at a point in the adsorption space and in bulk gas, and P^* and P are the corresponding pressures. The spatial variation of C^* or P^* is determined for a given set of equilibrium conditions by the concomitant variation of ϵ. In a similar manner, the concentration of a solute in water will vary through the adsorption space somewhat differently than described by Equation 2.22. For real gases, condensation to a liquid occurs whenever the adsorption potential suffices to concentrate it to a saturation concentration at the experimental temperature [19]. The Polanyi criterion for condensation of a gas is:

$$\epsilon \geq RT \ln (P_s/P), \qquad (2.23)$$

where P_s is the saturation pressure of the liquid at its equilibrium temperature (°K). The driving force for adsorption is:

$$\Delta G = -\epsilon + \ln(P_s/P). \qquad (2.24)$$

Condensative adsorption occurs whenever $\Delta G \leq 0$. When the vapor-phase adsorption isotherm data are replotted as liquid volume adsorbed per gram of carbon versus $RT \ln (P_s/P)$, a characteristic curve results [19]. Equation 2.22 is the link between the experimental adsorption isotherm and the concept of a characteristic curve (Figure 2–5) [19].

Characteristic curves have an additional property: they are identical for all adsorbates on a specific carbon, which is common to both liquid- and gas-phase systems, except for the factor ϵ. Once they are determined, these curves are expected to be the same on all carbons; that is, once the adsorption isotherms for a series of adsorbates have been determined on one carbon, then it is necessary to run the isotherm of any one on a second sample to calculate all the others.

The Polanyi model is applicable to liquid-solid systems as well, with condensation of the solid occurring in the adsorption space whenever the adsorption potential suffices to concentrate the solute to saturation. However, the driving force energy for adsorption is reduced by an equal volume of solvent. This is necessary because the condensing solute must displace an equal volume of adsorbed solvent from the adsorption space. The condition for condensation is:

$$\epsilon_{sl} = \epsilon_s - \epsilon_l \frac{V_s}{V_l} = RT \ln (C_s/C), \qquad (2.25)$$

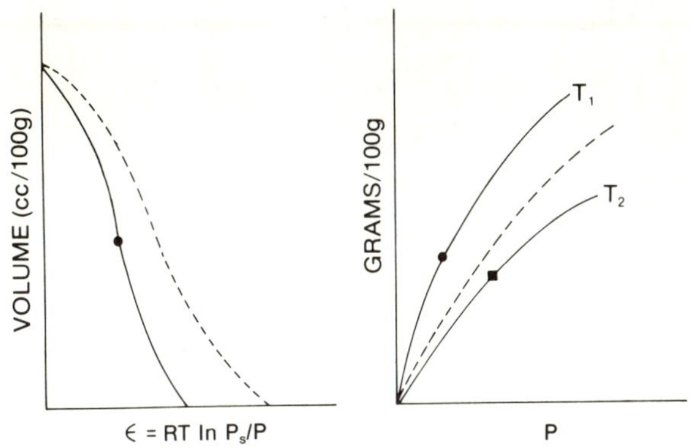

Figure 2–5 Characteristic curves and isotherms. The solid lines on the right represent two gas-phase isotherms of a single adsorbate on one carbon but at two different temperatures, T_1 and T_2. The dashed line represents the isotherm of a different adsorbate on the same carbon. On the left, the solid line is the characteristic curve for one adsorbate, and the dashed line is for the other adsorbate. To transform an isotherm point to a characteristic curve point, the pressure is converted to $RT \ln P_s/P$, using the vapor pressure appropriate to the equilibrium temperature. To transform the ordinate (mass/mass of carbon) to volume/mass carbon, one divides by the liquid density. (For liquid-phase adsorption the calculations are, in principle, similar, with $RT \ln C_s/C$ replacing $RT \ln P_s/P$.) In the diagram, two isotherm points converge to one point on the characteristic curve. Note that the characteristic curves for both adsorbates are the same except for the abscissa scale. Reproduced from Manes [19], courtesy of Ann Arbor Science Publishers, Inc.

where the net energetic driving force is ϵ_{sl}. ϵ_s and ϵ_l are the adsorption potentials of the solute and solvent as individual compounds. Equation 2.25 may be rewritten:

$$\frac{\epsilon_{sl}}{V_s} = \frac{\epsilon_s}{V_s} - \frac{\epsilon_l}{V_l} . \tag{2.26}$$

This form of the equation indicates that a plot of adsorbate volume against $(RT/V) \ln (C_s/C)$ gives a correlation curve of the same form as any single vapor phase adsorbate on the same carbon; that is, the abscissa is the solute minus the solvent. This gives a relationship between gas-phase and liquid-phase adsorption isotherms [19].

Polanyi adsorption indicates that individual correlation curves on a single carbon are related through a single characteristic abscissa scale factor that must be determined empirically for each adsorbate but that is the same for all carbons [19]. Figure 2–6 shows schematic plots of the correlation curves of individual adsorbates A and B on carbons 1 and 2. Curves 1A and 2A are assumed to be the calibrating correlation curves that are calculated from experimental isotherms for vapors or for solids from aqueous solution. Curves 1B and 2B represent solid solutes. The ordinate scale is typically in cc/100 g of carbon, which represents

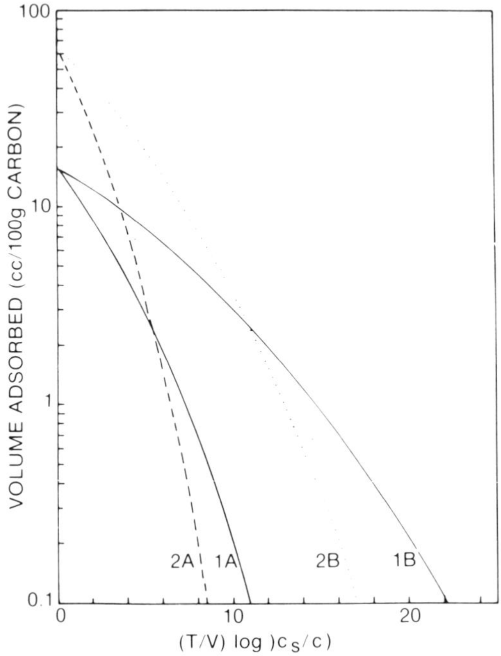

Figure 2–6 Schematic diagram of correlation curves of components A and B on carbons 1 and 2. For adsorption of solids in water, it is assumed that the adsorbate densities have been adjusted to make the limiting adsorbate volumes coincide on any single carbon, conveniently to the limit for liquid or gas adsorption. If curve A is the calibration curve for each carbon, and if curve 1B is an experimental curve for component B on carbon 1, then one can readily estimate curve 2B from 2A by applying the same abscissa scale factor that relates 1A to 1B. Reproduced from Manes [19], courtesy of Ann Arbor Science Publishers, Inc.

the adsorbate volumes. The abscissa scale is the (T/V^*) log (C_s/C) for solid solutes, where V^* is the molar volume calculated on the basis of adjusted densities. The solid lines represent curves that were calculated from experimental data on constituents H and B on carbon 1. The dashed line represents an experimental correlation curve for A on carbon 2, and the dotted line is an estimated correlation curve for B on carbon 2 from which an estimated isotherm may be calculated [19].

According to Manes [19], if every abscissa point on curve B on carbon 1 is equal to that of A but is multiplied by some constant factor, f_{BA}, that is the same on all carbons, then this factor is determined by a comparison of the correlation curves of A and B on carbon 1. Given f_{BA} and curve A on carbon 2, an estimated correlation curve can be constructed for B on carbon 2. Inasmuch as B and carbon 2 are representative of a wide variety of adsorbates and carbons, then this estimation method for adsorption isotherms should have wide application. It is not necessary to determine the adsorption isotherm of B over the entire capacity range; rather, its limiting capacity and the scale factor are sufficient. Likewise, if a carbon is intended for use over a very narrow capacity range, it is necessary only to calibrate it over that range of interest.

The Polanyi model has been adapted to competitive adsorption by Manes and co-workers [19]. The free energy driving force for adsorption of either of two solid solutes must be negative:

$$\Delta G_i = -(\epsilon_{sl})_i + RT \ \ln (C_s/C_i) \leq 0. \tag{2.27}$$

Neither constituent will adsorb unless the net energy of adsorption (corrected for solvent competition) is sufficient to overcome the negative entropy change associated with condensation to the saturation concentration. In this case, C_s is for the adsorbate in the presence of the equilibrium concentration of the other. This modifies the definition of C_s that incorporates any effects of interaction between the adsorbates in solution. On the one hand, if in the adsorption space, conditions for condensation are satisfied for only one solute, it will adsorb noncompetitively, and its correlation curve for that volume range will be identical, as in the absence of competition. On the other hand, if both solutes are individually adsorbable within the adsorption space, then the only one that adsorbs is that with the higher adsorptive driving force per unit volume, $-\Delta G/V^*$, where ΔG is defined by Equation 2.27 and V^* is the effective molar volume of the adsorbate (see earlier). The criterion for adsorptive displacement of component 1 by component 2 is [19]:

$$(RT/V_2^*) \ln (C_{s2}/C_2) \leq \epsilon_2/V_2^* - \epsilon_1/V_1^* + (RT/V_1^*) \ln (C_{s1}/C_1), \tag{2.28}$$

where the ϵs are the net adsorption potentials, ϵ_{s1}. Again, given the correlation curves for two individual adsorbates in water, the competitive correlation curve of the more strongly adsorbing solute thence gives the adsorption of the other. This form of competitive adsorption is somewhat different than that described by the Langmuir equation (see Chapter 1).

Advantages and Limitations of the Polanyi Adsorption Model

Perhaps the most distinguishing feature of the Polanyi model lies in the assumption of a heterogeneity of adsorption energies over the adsorption space or surface. This implies that adsorption occurs in different regions of the adsorbent's surface as the concentration (presumably the adsorbate) is increased. This model contrasts with the Langmuir model, which assumes a uniform energy surface. There is sufficient evidence to support a variable energy surface of the Polanyi model [19].

Another advantage is that the Polanyi model assumes that the adsorbate's properties do not differ from the corresponding bulk condensed phase except for its lower energy [19]. Some adsorption models assume ideal behavior of the adsorbate that may be valid for low concentrations but certainly not high concentrations.

Another significant feature of the Polanyi approach is the simplification that any interactions between solute and solvent and between other solutes in solution are reflected in their solubilities; that is, the solubilities that are used in calculations are those of each component in the presence of the other. This is evident in competitive adsorption [19].

Several limitations may be cited for the Polanyi model [19]. One comes from the assumption that all pores are accessible to the adsorbate. This is simply not the case. There are pores small enough to exhibit exclusion of large adsorbate molecules. Another limitation is that it is essentially limited to London-force adsorption [19]. The Polanyi model does not refer to adsorption by specific surface groups. Much evidence supports this point also. In itself, London-force adsorption is not necessarily weak, and strong adsorption must be due to chemisorption [19]. In dilute solutions (μg/l) adsorption is readily ascribed to London-force adsorption without reference to dipole moments.

A third limitation is that the Polanyi approach does not account for the effect of any ash content on adsorption; that is, adsorption of some constituents occurs with various basic oxides in carbon. Insoluble salts may be formed or chemisorption may occur at the oxygen sites.

Confirmations of the Polanyi Adsorption Model

Manes and co-workers [22,23] have studied and confirmed the Polanyi model rather extensively for many solute-solvent systems. Only their aqueous systems are reported here. Isotherms for adsorption of 1,2-dichloroethane, diethyl ether, ethyl acetate, methylene chloride, and propionitrile from water, 25°C, onto activated carbon were reported [22]. It is important to note that they used Pittsburgh Activated Carbon grade CAL (Calgon Corp.) that was pulverized to pass a 200 mesh screen. Consequently, Manes and co-workers were working with powdered, not granular, carbon, raising the question of applicability of the Polanyi model to the latter. Figure 2–7a shows adsorption isotherms for the five partially miscible solvents onto the CAL carbon. Correlation curves (Figure 2–7b) were calculated by taking points at equal intervals from the appropriate isotherms in Figure 2–7a.

In the liquid-phase adsorption of a solute from a relatively weakly adsorbed solvent, a graph of adsorbate volume against ϵ_{sl}/V should be the same as the hydrocarbon correlation line (vapor phase) except that the scale factor γ_{sl} is [22]:

$$\gamma_{sl} = \gamma_s - \gamma_l, \tag{2.29}$$

where the γs are scale factors for solute-liquid, solute, and liquid, respectively. It is necessary to estimate γ_l as the difference between γ_s for the solute and γ_{sl} for adsorption from water. According to theory, γ_l should be the same for each of the five systems. The data in Table 2–7 confirm this.

The gas-phase correlation curve (small dashes) for water, calculated from the average value of γ_l, is seen in Figure 2–7c [22]. Also given is the correlation curve (solid line) calculated from the vapor-phase adsorption isotherm of water

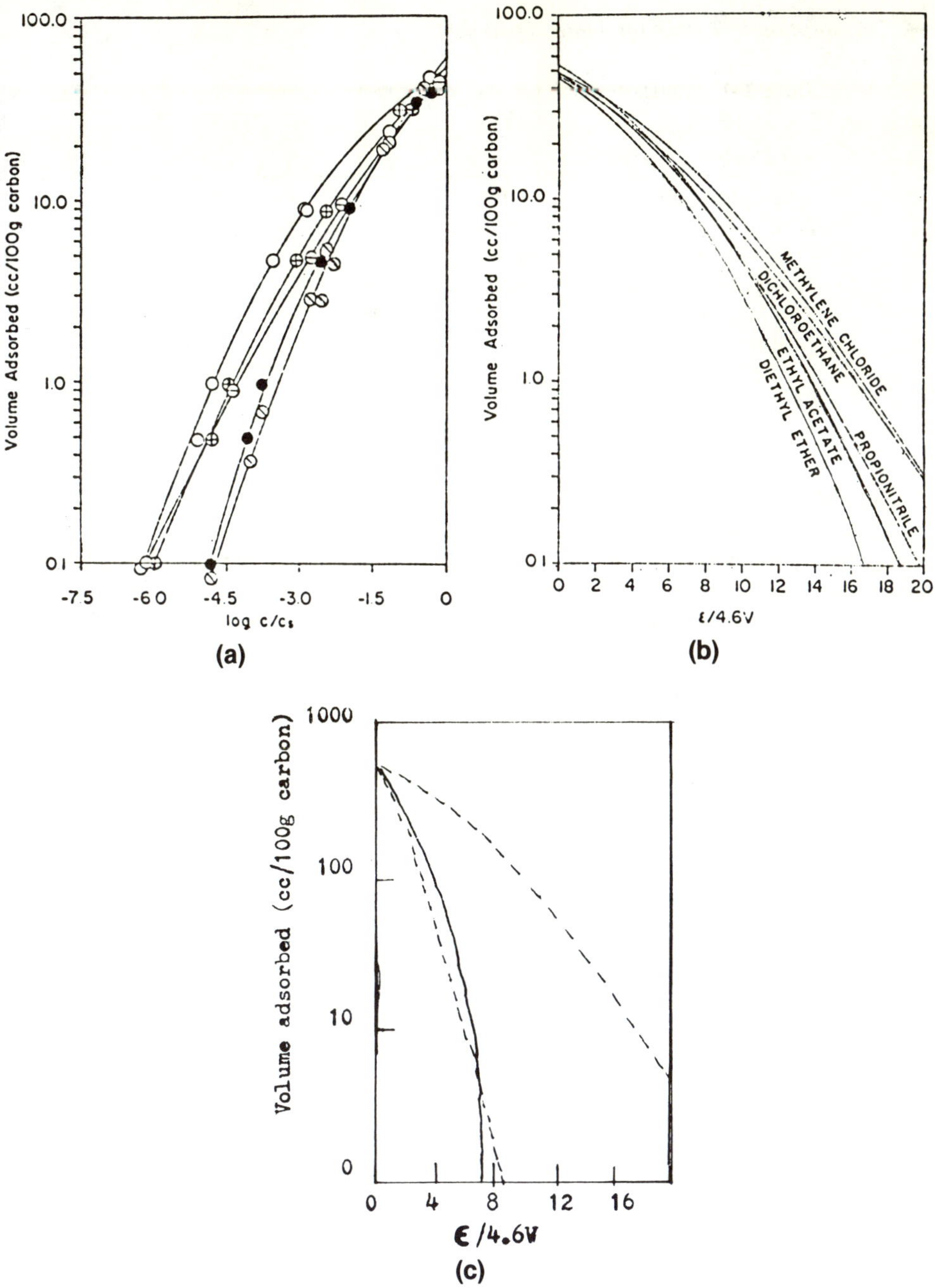

Figure 2–7 (a) Adsorption isotherms for 1,2-dichloroethane (⊖), diethyl ether (⊕), ethyl acetate (○), methylene chloride (●), and propionitrile (⦸); plotted as volume adsorbed versus log relative concentration. Temperature is 25°C. (b) Volume of 1,2-dichloroethane, diethyl ether, ethyl acetate, methylene chloride, and propionitrile plotted as a function of adsorption potential. (c) Correlation curves for water: ——, calculated from adsorption isotherm; ----, calculated from $\gamma_1 = .28$; ---, calculated from refractive index. Reproduced from Wohleher and Manes [22], courtesy of the American Chemical Society.

Table 2-7 Values for Solutes and Water

Compound	γ_s	γ_{sl}	γ_l
1,2-dichloroethane	1.12	0.80	0.32
Diethyl ether	0.91	0.62	0.29
Ethyl acetate	0.96	0.69	0.27
Methylene chloride	1.08	0.83	0.25
Propionitrile	0.95	0.70	0.25
Average			0.28

Source: Reproduced from Wohleher and Manes [22], courtesy of the American Chemical Society.

on the same carbon. Agreement between these two curves is a measure of the accuracy of the experimental results [22].

Additional confirmation of the Polanyi adsorption potential theory comes also from Manes and co-workers for mixed systems [24,25]. An example is seen in Figure 2-8 for *p*-nitrophenol (PNP) - 26% saturated benzamide - 33% sat-

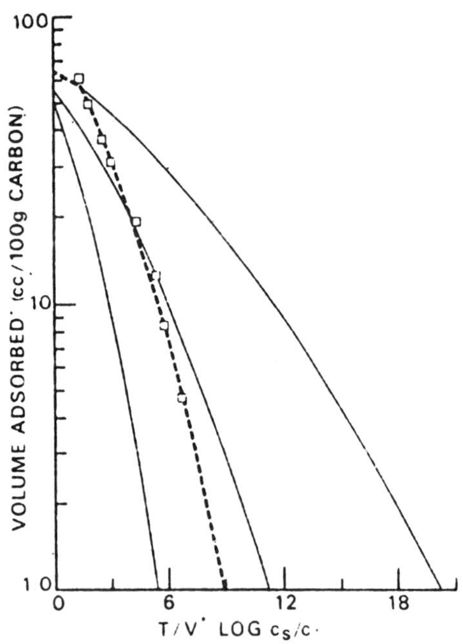

Figure 2-8 Comparison of experiment and theory for the system PNP-26%, saturated benzamide-33%, saturated glucose. The solid curves are the analogues of Figure 2-6 for the three individual components (reading from right to left). The dotted curve is constructed by subtracting the abscissa of the benzamide curve from the PNP curve. Its abscissa is $T/V \log C_s/C$, where the concentrations are the equilibrium and saturation concentrations in the multicomponent system. The points are experimental points for PNP. Reproduced from Rosene and Manes [24], courtesy of the American Chemical Society.

urated glucose on PAC. The theoretical curves are shown as solid and dotted lines, with agreement with the theory shown by proximity of the experimental points to the dotted lines. The agreement in Figure 2–8 is excellent. There were, however, several adjustments made in the experimental data to accomplish this level of agreement. Also, applicability of data generated by these experimental systems is questionable when the more pragmatic water treatment is considered. This point is considered again in this chapter.

Solvophobic Theory

The solvophobic theory is concerned with the thermodynamic aspects of adsorption of organic compounds by nonpolar adsorbents from very dilute aqueous solutions. The solvophobic theory is used for calculation of the solvent effect on the adsorptive process from available physicochemical parameters [26,27]. Also, the solvophobic theory describes the tendency of a solvent medium to influence aggregation or dissociation of molecules that have considerable microsurface areas exposed to the solvent [27]. Symbols and nomenclature for this theory are given in Table 2–8.

The reversible association reaction of the adsorbate molecules, S_i , with the activated carbon adsorbent, C, at the surface produces an adsorbed complex, S_iC. The effect of solvent on this reaction is obtained by subtraction of the standard free energy (ΔG) for the reaction in the gas phase from that in the solution phase under the conventional standard state. This process is seen in the schematic in Figure 2–9, which gives a net free energy, ΔG^{net}(solvent effect) [27,28]. Originally, this was conceived as a two-step dissolution process [29]. First, a hole or cavity is prepared in the solvent to accommodate the solute, carbon, or adsorbed complex molecule. Second, this molecule interacts with the solvent after it is placed into the cavity. Mathematically, this proces is:

$$\Delta G^{\text{net}}_{\text{solvent effect}} = \Delta G^{\text{assoc}}_{\text{solvent}} - \Delta G^{\text{assoc}}_{\text{gas}} = RT \ln [K^H_{S_iC}/K^H_{S_iC}/K^H_C] \qquad (2.30)$$

or

$$\Delta G^{\text{net}}_{\text{solvent effect}} = \Delta G^{\text{net}}_{j,S_iC} - \Delta G^{\text{net}}_{j,S_i} \text{ » } \Delta G^{\text{net}}_{j,C} \,, \qquad (2.31)$$

where $K^H_i =$ the appropriate Henry's Law constant for the ith species ($K^H_i = P_i/X_i$, where P_i = pressure and X_i = mole fraction), and j = each type of interaction. The overall standard free energy change for association with the solvent is:

$$\Delta G^{\text{assoc}}_{\text{solvent}} = \Delta G^{\text{assoc}}_{\text{gas}} + [\Delta G_{\text{cav}} + \Delta G_{\text{vdw}} + \Delta G_{\text{es}} + \Delta G_{\text{mix}} + \Delta G_{\text{red}}]^{\text{net}}_{S_iC - S_iC} - RT \ln (RT/P_oV), \qquad (2.32)$$

where each interaction term (in the bracket on the right-hand side of Equation 2.32) is specified: cavity, van der Waals, reduced electrostatic effects because of

Table 2-8 Symbols and Nomenclature for the Solvophobic Theory

Symbol	*Nomenclature*
a_i	Distance of closest ionic approach.
A_i	Surface area of species *i*.
b_i	Ionic radius.
b	Langmuir isotherm constant.
D	$[2(\epsilon - 1)/2\epsilon + 1]$.
e	Electronic charge.
g	Ratio of microsurface area change of reaction to solute hydrophobic surface area.
g'	Ratio of microsurface area change of reaction to solute total area.
HSA	Hydrophobic surface area of molecule.
I	Ionization potential.
k	Boltzmann's constant.
k_i^H	Henry's constant for the ith species.
K_i^e	Correction factor for energy of expanding highly curved solute cavity surface.
K^e	Same as above, for solvent.
K_a	Geometric ratio of surface area to volume$^{2/3}$.
$K_{(solvent)i}$	Equilibrium constant for the adsorption association reaction.
K_f	Freundlich isotherm constant.
K_{HA}	Acid dissociation constant.
$\log P$	Logarithm of octanol-water partition coefficient.
MR	Molar refraction.
MW	Molecular weight.
n	Freundlich isotherm constant.
n_D	Index of refraction.
N	Avogadro's number.
p	Partial pressure.
$\bar{p}_i$	Adsorption capacity.
P_0	Pressure.
$[P]$	Parachor.
q	Alternative correction factor for surface energy at curved surface.
$Q°$	Langmuir isotherm constant.
R	Universal gas constant.
TSA	Total surface area of solute cavity.
V	Molar volume.
W	Curvature correction factor.

Table 2-8 *(continued)*

Symbol	*Nomenclature*
X	Mole fraction.
z	Ionic valency.
Greek	
α	Atomic polarizability.
α°	Ionic degree of dissociation.
γ	Interfacial surface energy.
ΔA	Microsurface area change of reaction.
ΔG	Gibbs free energy change of reaction.
ϵ_i	Static dielectric constant of solute *i*.
ϵ_0	Permittivity of vacuum.
θ	Experimental adsorption capacity constant.
κ	Debye screening length.
λ	Ratio of volume of carbon-solute complex to that of solute.
μ_i	Static dipole moment.
ν_i	Molecular volume.
ρ	Density.
Subscripts	
cav	Cavity formation.
es	Electrostatic interactions.
gas	Gas phase process.
j	*j*th interaction.
mix	Mixing effects.
red	Dispersive reduction due to solvent.
solvent	Solvent phase.
solvent effect	Net effect of solvent on process.
vdw	van der Waals interactions.
Superscripts	
assoc	For adsorption association process.
net	Sum of all interactions.

Source: Reproduced from Altshuler and Belfort [27], courtesy of the American Chemical Society.

the presence of the solvent, and polymer mixing. The last term in Equation 2.32, ΔG_{red}, results from an entropy or free volume reduction. The term $\Delta G^{\text{assoc}}_{\text{solvent}}$ is related to an experimental equilibrium constant:

$$K_{\text{solvent},i} = \frac{X_{S_iC}}{X_{S_i}X_C}, \tag{2.33}$$

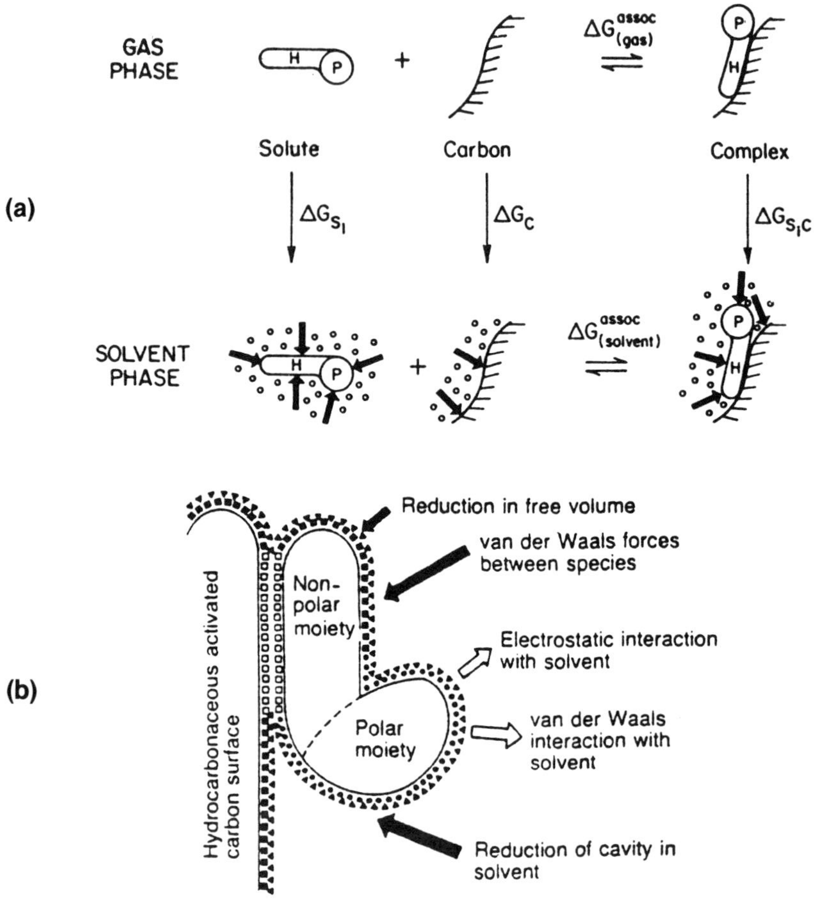

Figure 2–9 (a) A pictorial representation of the association adsorption reaction in the gas and liquid phase. Reproduced from Altshuler and Belfort [27], courtesy of the American Chemical Society. (b) Surface area reduction of amphiphilic solute adsorbed onto activated carbon surface. Reproduced from Horvath and Melander [28], courtesy of the American Chemical Society.

that, in turn, is related to the experimental adsorption capacity $\overline{P}_i$ for solute S_i. Each term in Equation 2.32 can be calculated from known physicochemical parameters in the literature or from explicit formulas [27]. The dominant interactive forces and the appropriate parameters are summarized in Table 2-9.

The electrostatic term in Equation 2.32 considers simple dipoles as well as ionic species [27]. Table 2–10 summarizes the electrostatic formulas where Equation 2.34 is a form of the Debye-Hückel equation that is apparently appropriate for organic compound removal from dilute aqueous solutions by activated carbon [27].

Table 2-9 Interactive Free Energy Changes

Interaction Process	*Free Energy Designation*	*Interaction Forces*	*Important Parameters*
Cavity formation	$\Delta G^{\text{net}}_{\text{cav}.i}$	Surface forces	Solute molecular surface area, TSA Corrected macroscopic surface tension, γ_i
Solvent-solute interaction, van der Waals	$\Delta G^{\text{net}}_{\text{vdw}.i}$	Dispersive (attractive) forces	Ionization potential and molecular volume, ν_i
Electrostatic	$\Delta G^{\text{net}}_{\text{es}.i}$	Electronic interaction	Inverse molecular volume, ν_i^{-1} (1) Simple dipole. Solute dipole moment, μ_i, and dielectric constant of solvent, ϵ, (2) Charged species. Ionic charge, ze, and ionic strength, I.
Mixing effects (polymers)	$\Delta G^{\text{net}}_{\text{mix}}$	Free energy of mixing	
Dispersive reduction due to solvent	$\Delta G^{\text{net}}_{\text{red}}$	Masking of the dispersive forces	
Free volume reduction (cratic)	$RT \ln (RT/P_0V)$	Entropic reduction	Solvent molar volume, V, and pressure, P

Source: Reproduced from Altshuler and Belfort [27] courtesy of the American Chemical Society.

Table 2-10 Unitary Free Energy Change for Electrostatic Interactions

Interactive Process	*Limitations*	$\Delta G_{es,i}$	*Equation Number*
Dipole-dipole[a]	Simple dipoles with no formal charges	$-\dfrac{N}{2}\dfrac{\mu_i}{\nu_i}DP$	(2.34)
Charged species[b]	Very low ionic strength, $I \leq 0.1$	$\dfrac{Z^2e^2N}{2\epsilon}\left[\dfrac{1}{b_i} - \dfrac{\kappa}{1+\kappa a_i}\right]$	(2.35)

Source: Reproduced from Altshuler and Belfort [27] courtesy of the American Chemical Society.

[a] μ_i = static dipole moment of i, $D = [2(\epsilon-1)/2\epsilon+1]$, ϵ = static dielectric constant of the solvent, $P = [4\pi\epsilon_0(1-D(\alpha_i/\nu_i)]^{-1} \approx [4\pi\epsilon_0(1-(\alpha_i/\nu_i)]D \approx 1$, α_i = polarizability of i, ϵ_0 = permittivity constant.

[b] Ze = electronic charge of the ion, b_i = ionic radius, a_i = distance of closest approach. κ = Debye screening parameter and is obtained from $\kappa^2 = 8\pi e^2 I/\epsilon\kappa T$, where the ionic strength $I = 1/2\, \Sigma n_i Z_i^2$, where n_i is the number of ions of species i per unit volume.

A comprehensive model can be written from Equation 2.32 after substitution for each relevant free energy term and after making a few assumptions about the nonpolar nature and large surface area of the sorbent [27]. The standard unitary free energy change for the overall process is:

$$\Delta G^{\text{assoc}}_{\text{solvent}} = \Delta G^{\text{assoc}}_{\text{gas}} - \Delta G_{\text{vdw},S_i} + \Delta G_{\text{es},S_i} + N\,(K^e_{S_iC} A_{S_iC}\,\gamma_{w,S_iC} - K^e_{S_i} A_{S_i}\,\gamma_{w,S_i} - K^e_C A_C \gamma_{w,C}) - RT \ln (RT/P_0 V). \tag{2.36}$$

Another equation contains a curvature correction (K^e_i) for the highly curved rather than flat surface of solute molecules. K^e_i values for species i can be estimated from:

$$K^e_i = 1 + (V/V_i)^{2/3}\,(K^e - 1), \tag{2.37}$$

where K^e = the microscopic cavity factor for pure solvent. When solute molecules are large relative to solvent molecules, then ($1 > V/V_i$), $K^e_i \rightarrow 1$, and $W \rightarrow 0$. Consequently, $K^e_{S_iC} = K^e_C = 1$ for a large adsorbent volume (S_iC and C) relative to the solvent's molecular volume. With these assumptions, the fourth term on the right-hand side of Equation 2.36 is replaced with $N\gamma(A_{S_iC} - K^e_{S_i} A_{S_i} - A_C)$, whereupon:

$$\Delta G^{\text{assoc}}_{\text{solvent}} = \Delta G^{\text{assoc}}_{\text{gas}} - \Delta G_{\text{vdw},S_i} + \Delta G_{\text{es},S_i} - N\gamma \Delta A, - N\gamma V^{2/3}\,(K^e - 1)\,\frac{A_{S_i}}{V_i^{2/3}} - RT \ln (RT/P_0 V). \tag{2.38}$$

The fifth term on the right-hand side of Equation 2.38 concerns the shape and corrections for molecular curvature.

Simplifications of Equation 2.38 is possible for specific situations [27]. For example, with adsorption of different solutes—like a homologous series—at constant temperature and flow rate and with the same solvent (water) and carbon, the unitary free energy change in the solvent is:

$$\Delta G^{\text{assoc}}_{\text{solvent}} = K'_1 + \Delta G_{\text{es},i} - K'_2\,\Delta A, \tag{2.39}$$

where the constants K'_j are given by:

$$K'_1 = \Delta G^{\text{assoc}}_{\text{gas}} - \Delta G_{\text{vdw},i} - N\gamma V^{2/3}\,(K^e - 1)\,\frac{A_i}{V_i^{2/3}} - RT \ln (RT/P_0 V) \tag{2.40}$$

and

$$K'_2 = N\gamma. \tag{2.41}$$

$\Delta G_{es,i}$ values are given by Equation 2.34 (Table 2–10) for dipole-dipole interactions and by Equation 2.35 (Table 2–10) for charged ionized species interactions.

Confirmation of the Solvophobic Theory

It is desirable, of course, to confirm these adsorption theories and models by experimental data. In the case of the solvophobic theory, this was accomplished more or less [27]. Equation 2.38 was used to calculate the contribution of each free energy term to the overall standard unitary free energy change for the adsorptive process (on pulverized carbon) as a function of carbon number for normal aliphatic alcohols and organic acids. Adsorption data were obtained from two independent sources [30,31]. The appropriate equations are:

$$\ln \bar{p}_i = \text{NET} - (\Delta G_{\text{gas}}^{\text{assoc}} + \theta)/RT, \tag{2.42}$$

where:

$$\text{NET} = \text{VDW} + \text{ES} + \text{CAV} + \text{CRAT} \tag{2.43}$$

$$\text{NET} = +\,\Delta G_{\text{vdw},i}/RT - \Delta G_{\text{es},i}/RT + [4.8364 N^{1/3} V^{2/3} (K^e - 1)\,\gamma + N\gamma\Delta A]/RT + \ln\,(RT/P_o V), \tag{2.44}$$

where:

$$A_{S_i} = 4.8364 v_{S_i}^{2/3} \text{ (for a sphere).} \tag{2.45}$$

Figure 2–10 shows the correlation and/or plot of an adsorption capacity term versus molecular weight, total surface area, and the solvophobic free energy changes (NET and VDW from Equation 2.43) (Q'_m = adsorption extent and Q^o,b = Langmuir constants [30,31]). Linearity of the plots indicates compliance to the solvophobic theory. The r^2 values indicate the statistical correlation between data sets.

Another measure of the confirmation of the solvophobic theory is its ability to predict the adsorbability of compounds whose experimental values are not available. For one example, an adsorbability of 3.1606 g-mol/g for 3-methyl-1-pentanol was estimated, whereas the experimental value was 2.8112 g-mol/g for an error of 12% [27].

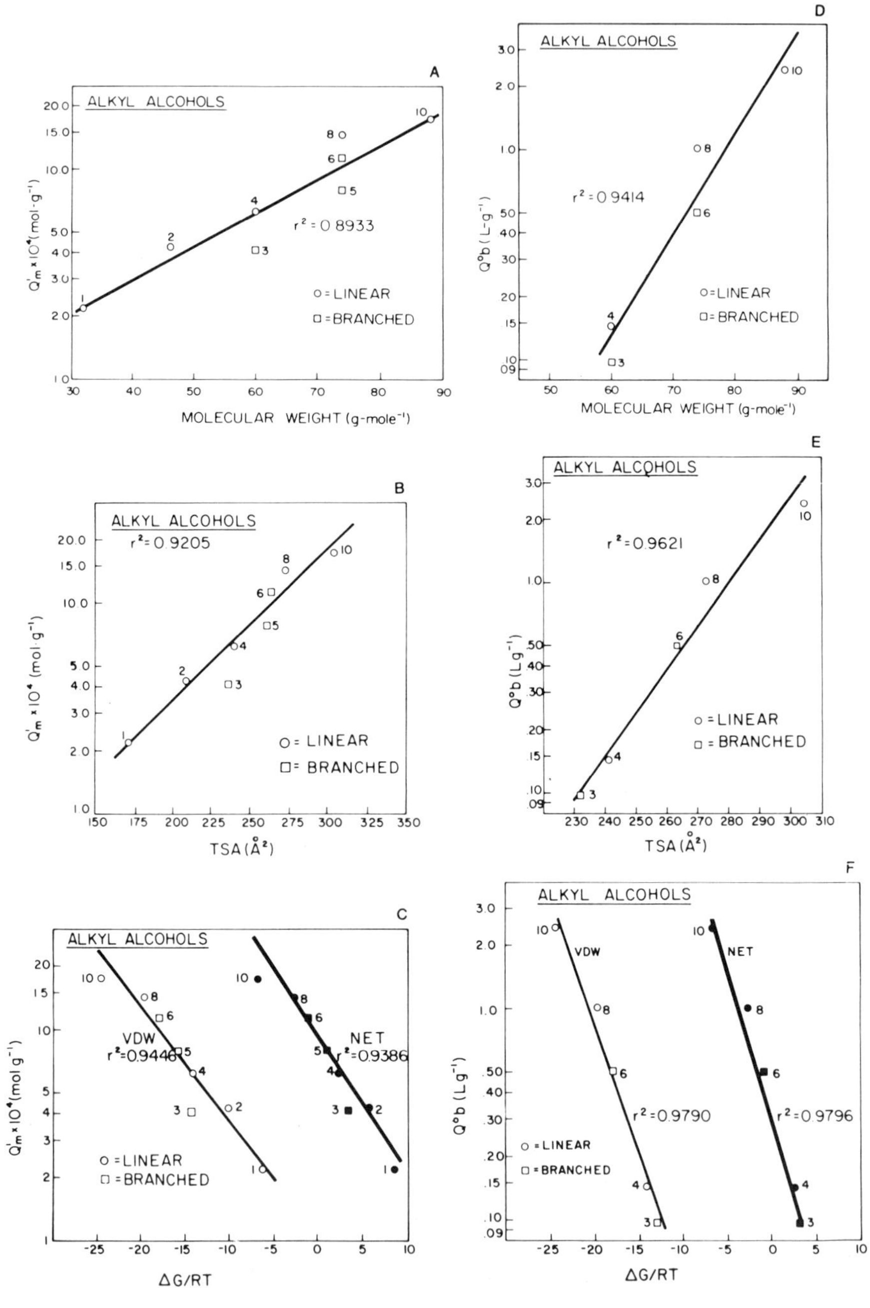

Figure 2–10 Plot of logarithm of the extent of adsorption for different sets of data versus molecular weight, total surface area (TSA), and solvophobic free energy changes (NET and VDW) for linear and branched aliphatic alcohols (Q_m^1 and Q^0b adsorption data from references 30 and 31, respectively). Reproduced from Altshuler and Belfort [27], courtesy of the American Chemical Society.

Theory of Correspondence for Adsorption

Much of the preceding discussion is concerned with generation of characteristic adsorption isotherms for vapor-solid interfaces with the intent to condense isotherms of different vapors into a single curve. Furthermore, this is applied also to solute-solid interfaces. If a characteristic curve is developed, then *affinity coefficients* are defined and correlated with certain physical constants of the adsorbate. Consequently, it would be feasible to predict adsorption isotherms from a singly measured isotherm of a reference compound.

Another extension of the Polanyi theory has been researched by Myers and Sircar [32] and is labeled the *theory of correspondence* for adsorption from dilute solutions on heterogeneous adsorbents. This theory was originally proposed for adsorption of pure vapors. Extension of this concept to liquid systems is complicated by competition of the solvent for available surface area. This problem was met by three assumptions:

1. The solute is only sparingly soluble in the solvent, e.g., no more than a few mole percent.
2. The solute is strongly adsorbed relative to the solvent so that S_1

The Gibbsian model of the liquid-solid interface was adopted for the theory of adsorption correspondence [33]. The surface excess for adsorption of solute 1 or solvent 2 is:

$$n_i^e = n\,(x_i - x_{i0}), \tag{2.46}$$

where n = total amount adsorbed per unit mass of adsorbent, x_i = the mole fraction of solute or solvent in the adsorbed phase, and x_{i0} = the corresponding mole fraction in the bulk liquid phase at equilibrium. It follows that:

$$n_1^e + n_2^e = 0, \tag{2.47}$$

whereupon a positive surface excess of solute (n_1^e) implies an equal in magnitude but negative surface excess of solvent (n_2^e). Excess Gibbs free energy for adsorption from solution is [34]:

$$\Delta G^e = n\,(g - g_0) - \Sigma n_i^e \mu_i^\circ, \tag{2.48}$$

where g = the molar Gibbs free energy of the adsorbed phase and g_0 is the value for the equilibrium solution. ΔG^e is negative for the feasible process of adsorption. Now, the Gibbs free energy of the adsorbed phase is [34]:

$$G = ng = \gamma A + n_1\mu_1 + n_2\mu_2. \tag{2.49}$$

Consequently, free energy of the adsorbed phase is the sum of contributions from the solid surface (γA), the solute ($n_1\mu_1$), and the solvent ($n_2\mu_2$). In Equation 2.49,

γ is the surface tension of the liquid-solid interface, and A is the surface area per unit mass of adsorbent. For a porous adsorbent like carbon, A is not meaningful, but γA is.

The molar Gibbs free energy of bulk liquid is [32]:

$$g_0 = x_{10}\mu_{10} + x_{20}\mu_{20}. \tag{2.50}$$

At equilibrium:

$$\mu_1 = \mu_{10}$$

and

$$\mu_2 = \mu_{20}. \tag{2.51}$$

Substituting Equations 2.49 to 2.51 into 2.48 gives:

$$\Delta G^e = \gamma A + n_1^e(\mu_1 - \mu_1^o) + n_2^e(\mu_2 - \mu_2^o). \tag{2.52}$$

For the adsorption of pure solvent, $n_1^e = n_2^e = 0$ and:

$$\Delta G = \gamma_0 A, \tag{2.53}$$

where γ_0 refers to the surface tension of the solvent-solid interface. It is necessary to subtract the solvent's properties from Equation 2.52. ΔG^e is redefined now relative to adsorption of the pure solvent:

$$\Delta G^e = (\gamma - \gamma_0)A + n_1^e(\mu_1 - \mu_1^o) + n_2^e(\mu_2 - \mu_2^o) \tag{2.54}$$

so that $\Delta G^e = 0$ for adsorption of the solvent alone. Inasmuch as the solute is present at low concentrations, the mole fraction of the solvent is close to unity, and $\mu_2 \simeq \mu_2^o$. Equation 2.54 reduces to:

$$\Delta G^e = (\gamma - \gamma_0)A + n_1^e(\mu_1 - \mu_1^o). \tag{2.55}$$

The solute's activity is given by Henry's Law since its content is assumed to be less than a few mole percent:

$$\mu_1 = \mu_1^o + RT \ln a_1^o, \tag{2.56}$$

where:

$$a_1 = \gamma_1^\infty x_{10} = \frac{x_{10}}{x_{10}^{sat}} = \frac{c_1}{c_{10}^{sat}} . \tag{2.57}$$

The γ_1^∞ = the activity coefficient at infinite dilution and is assumed to be constant up to the mole fraction at saturation, x_{10}^{sat}. Equation 2.55 becomes:

$$\Delta G^e = (\gamma - \gamma_0) A + n_1^e RT \ln a_1. \quad (2.58)$$

It can be shown that the surface excess of solute, n_1^e, is the actual amount of solute in the adsorbed phase, n_1, whereupon Equation 2.58 becomes:

$$\Delta G^e = (\gamma - \gamma_0) A + n_1 RT \ln a_1. \quad (2.59)$$

Surface tension of the interface is calculated from the Gibbs adsorption isotherm [34]:

$$-A d\gamma = n_1 d\mu_1 + n_2 d\mu_2. \quad (2.60)$$

Inasmuch as the selectivity, $s_{1,2}$, of the adsorbent for the solute relative to the solvent is large, the second term on the right-hand side of Equation 2.60 can be neglected in comparison to the first term. Equation 2.60 becomes:

$$-A d\gamma = n_1 d\mu_1 = n_1 RT d\ln a_1. \quad (2.61)$$

Equation 2.61 is integrated over the limits of immersion in pure solvent to a particular loading of solute. This integration is combined with Equation 2.59, which eventually leads to:

$$\Delta G^e/RT = \int_{n_1=0}^{n_1} \ln a_1 dn_1 \quad (2.62)$$

This is a well-defined thermodynamic quantity where, for example, Henry's Law for adsorption is obeyed at very low solute loadings:

$$n_1 = K a_1. \quad (2.63)$$

In this range, the integral is:

$$\Delta G^e/RT = K \int_{x=0}^{a_1} \ln x dx = K(a_1 \ln a_1 - a_1) = n_1(\ln a_1 - 1). \quad (2.64)$$

From Equation 2.59, it is observed that, at saturation of the solute ($a_1 = 1$):

$$\lim_{a_1 \to 1} \Delta G^e = (\gamma_{sat} - \gamma_0) A \equiv \Delta G, \quad (2.65)$$

where ΔG^e at saturation is the free energy of immersion of the adsorbent in pure liquid solute that, in turn, is relative to the free energy of immersion in pure solvent. Both free energies are negative. It follows that, at saturation of the solute, the remaining solvent is displaced from the adsorbent's surface due to the immiscibility of the solute-solvent pair and preference of the surface for the solute [32]. For porous carbon, a fractional filling of the micropores is defined:

$$\theta = n_1/m_1, \tag{2.66}$$

where m_1 = the saturation capacity of the adsorbent for the solute.

Surface heterogeneity is an important assumption in the theory of correspondence for adsorption [32]; that is, the surface has a distribution $f(E)$ of energies such that adsorption of a solute molecule on a portion of the surface with energy E is:

$$\theta = \theta(a_1, T, E). \tag{2.67}$$

In this case, E is a positive quantity corresponding to the decrease in energy for transfer of 1 mole of solute from the liquid phase to the adsorbed phase. Thus, the right-hand side of Equation 2.67 is a local adsorption isotherm as opposed to $\theta(a_1, T)$, the overall adsorption isotherm for the entire surface. In addition, E is the energy exchange process of replacing solvent molecules by solute molecules. The total adsorption isotherm comes from Equation 2.67 by integration [32]:

$$\theta(a_1, T) = \int_{E_{min}}^{E_{max}} \theta(a_1, T, E) f(E) dE, \tag{2.68}$$

where $f(E)$ is the energy density function. Eventually, the adsorption isotherm equation becomes:

$$\ln a_1 = J(T)\psi(\theta), \tag{2.69}$$

where $J(T)$ is the temperature function and $\psi(\theta)$ is the energy distribution function. It should be noted that Equation 2.69 applies to physical adsorption and does not apply to solutes that exhibit specific chemical interactions with the surface.

In Equation 2.69, it is desirable to calculate the function, ψ, the energy distribution on the surface. This is done by substituting Equations 2.66 and 2.69 into 2.62:

$$\Delta G^e = m_1 RTJ \int_0^\theta \psi(\theta) d\theta. \tag{2.70}$$

The free energy of immersion in pure solute, relative to the free energy of immersion in pure solvent, is:

$$\Delta G = m_1 RTJ \int_0^1 \psi(\theta) d\theta. \qquad (2.71)$$

The ratio of Equations 2.70 and 2.71 leads to the cancellation of m_1 and J and also leads to a proportionality to the ψ function. In turn, the following equation emerges:

$$K(\theta) = \frac{m_1 RT \ln(c_1/c_1^{sat})}{\Delta G}. \qquad (2.72)$$

Consequently, the dimensionless K function depends on θ only, resulting in the coincidence of adsorption isotherms of different solutes after transformation according to Equation 2.72. "This coalescence into a single universal curve is called the principle of correspondence, which should be obeyed for adsorption of dilute solutes on heterogeneous adsorbents. Different adsorbents have similarly shaped but different K functions according to their energy distributions" [32,p.71]. Equation 2.72 can be used to predict adsorption isotherms of a single solute using the K function of another solute as a reference. The required data for each solute are solubility (c_1^{sat}) of the solute in the solvent, saturation capacity (m_1), and free energy of immersion (ΔG). Adsorption isotherms at temperatures other than the reference temperature are calculated from the heat of immersion of the solute relative to the solvent (ΔH) by use of the Gibbs-Helmholtz relation [32].

A test of the principle of correspondence comes from the data of Wohleher and Manes [22], who measured adsorption of several organic solutes from aqueous solution at 25°C on Pittsburgh CAL activated carbon (see earlier discussion). The appropriate data are given in Table 2–11. The free energies of immersion were calculated with Equation 2.62. The saturation capacities, m_1, of the solutes in Table 2–11 were obtained by extrapolating their adsorption iso-

Table 2–11 Properties of Solutes Adsorbed from Aqueous Solution at 25°C on Pittsburgh CAL Activated Carbon

Property	*Diethyl Ether*	*Ethyl Acetate*	*Methylene Chloride*	*Propionitrile*
Molecular weight	74.12	88.11	84.93	55.08
Solubility, mole fraction	0.00604	0.01783	0.00338	0.03320
$-\Delta G, J/g$	35.4	42.6	53.7	48.8
m, mmol/g	4.03	4.50	7.38	7.54
$-\Delta G/mRT$	3.55	3.81	2.94	2.61

Source: Reproduced from Myers and Sircar [32] courtesy of the American Chemical Society.

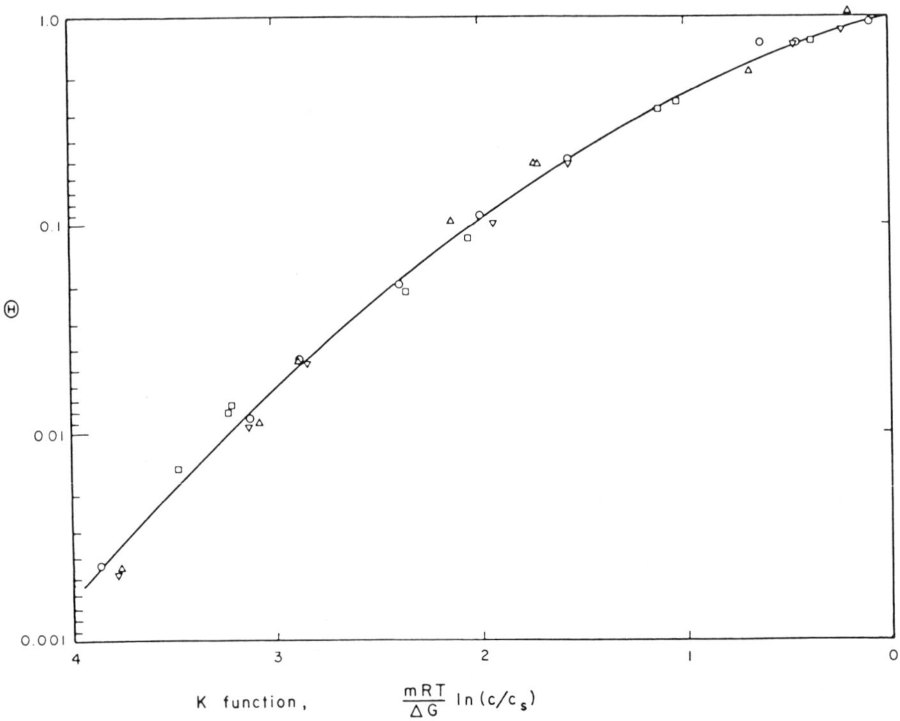

Figure 2–11 Test of principle of correspondence for adsorption of organic solutes from aqueous solution at 25°C on Pittsburgh CAL activated carbon. Key: ○, diethyl ether; △, ethyl acetate; ∇, methylene chloride; and □, propionitrile. Reproduced from Myers and Sircar [32], courtesy of the American Chemical Society.

therms to saturation. The theory of correspondence was tested for these data by comparing *K* functions calculated from Equation 2.72. As seen in Figure 2–11, there is good agreement between the experimental loading and the *K* function (solid line).

APPLICATIONS

The usefulness of adsorption models lies in their ability to predict isotherms for many compounds from a few model compounds. The current environmental concern is removal of organic contaminants from wastewaters and drinking waters. Arbuckle [35] has presented an intriguing analysis of the Polanyi, the NAEC, and the solvophobic theories as applied to Freundlich *K* constants for nine organic alcohols on pulverized F-400 carbon. In this study, Polanyi and net adsorption energy parameters were calculated and correlated with the Freundlich

K constants for nine alcohols. The resultant correlation line was used to predict *K* values for thirteen additional organic compounds.

Let us consider the good news first. Figure 2–12a is a Polanyi plot of the nine alcohols versus their Freundlich *K* values. In this case, the *K* value is a loading in terms of cm^3/100 g of adsorbent, which is correlated with $(T/V) \log (C_s/C)$, where T = the absolute temperature, V = the molar volume in cm^3/mole, C_s = the solute's solubility, and C = the solute's concentration. It is necessary to adjust the abscissa and ordinate first so all solutes fall on the same curve. The abscissa adjustment was obtained by dividing $(T/V) \log (C_s/C)$ by $\chi/0.236 - 0.28$, where χ is a refractive index function that is calculated from the refractive index (η_D) for each compound. Having done this, Figure 2–12a shows a regression line with a correlation coefficient of .994, which is somewhat misleading because both *x-y* terms have molar volume as a part of the parameter. Polanyi adsorption potentials were calculated for the remaining thirteen organics: three aldehydes, four acetates, five ketones, and one ether, and a *K* value was predicted from the alcohol regression line (Figure 2–12b). Differences between the actual and predicted values in mmol/g quantities were determined, squared, and summed ($\Sigma = 0.4665$) [35]. The data appear to fit well, but the logarithmic scale makes large absolute differences appear to be small.

Arbuckle [35] proceeded to calculate net adsorption energies for the nine alcohols with a subsequent correlation with the Freundlich *K* values (Figure 2–13a). Two distinct regions were found that were statistically significant. Net adsorption energies were calculated for thirteen additional compounds, and pre-

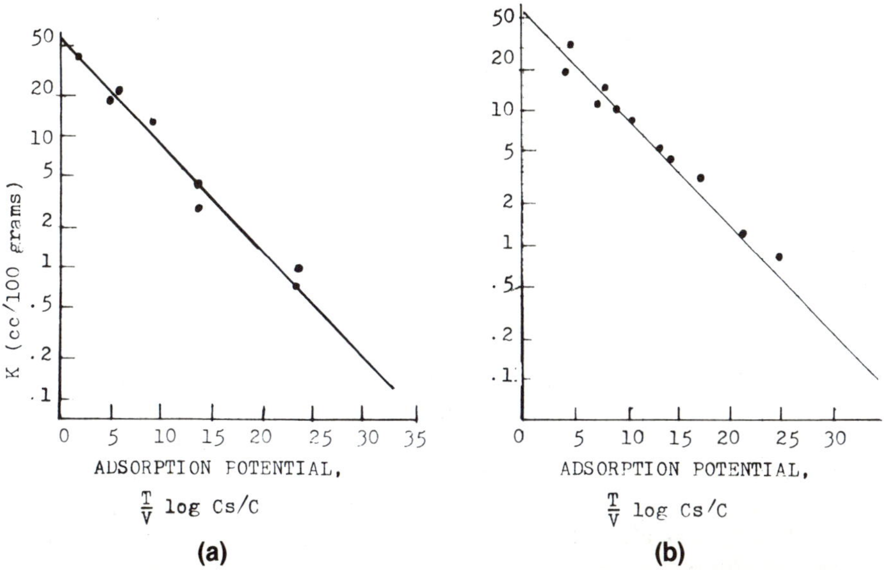

Figure 2–12 (a) Polanyi plot of alcohols. (b) Polanyi prediction line with actual data. Reproduced from Arbuckle [35], courtesy of the American Chemical Society.

(a)

(b)

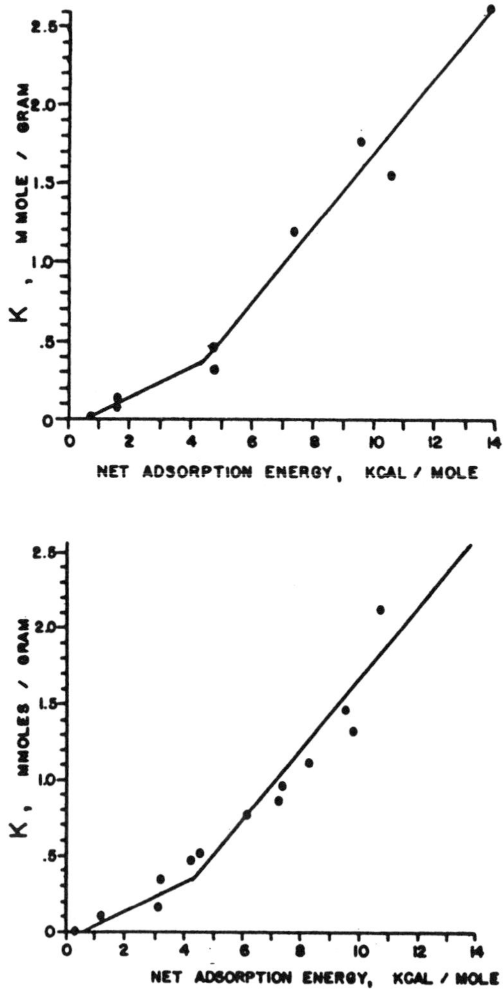

Figure 2–13 (a) Net adsorption energy plot of alcohols. (b) Adsorption energy line and actual data. Reproduced from Arbuckle [35], courtesy of the American Chemical Society.

dicted loadings were obtained from the alcohol curves and plotted (Figure 2–13b). The differences were squared and summed with a $\Sigma = 0.2624$, which is about half the sum for the Polanyi method. The plot in Figure 2–13b shows a good prediction, but as the adsorption energy increases, greater deviation results. As with the Polanyi theory, the net adsorption energy concept did not represent the branched and linear alcohol relationship. Poor fit came also from ethyl butyl ketone and isopropyl ether.

Prediction of isotherms is the major intent of adsorption models. Figures 2–14a and b show Freundlich plots using the Polanyi and net adsorption energy

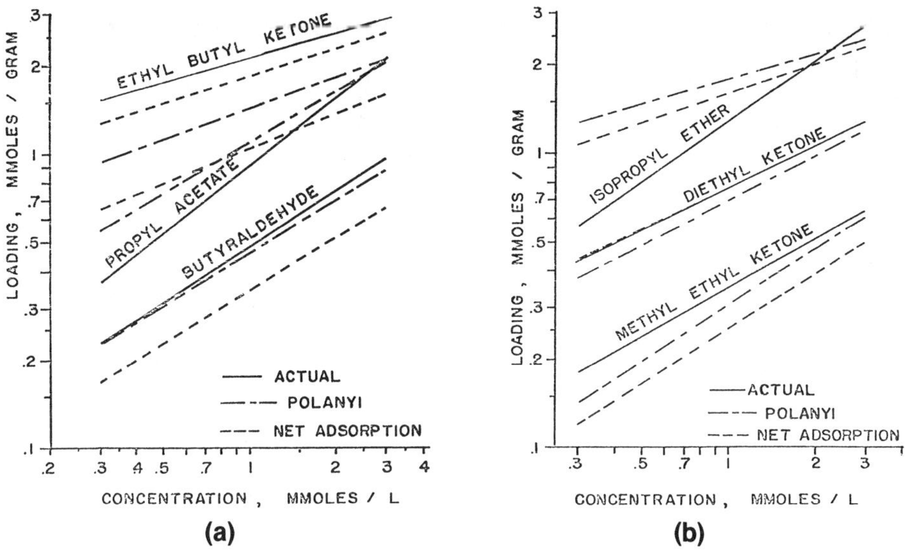

Figure 2–14 (a) Predicted and actual isotherms. (b) Predicted and actual isotherms. Reproduced from Arbuckle [35], courtesy of the American Chemical Society.

methods [35]. That these two models gave poor fits is seen by the dashed lines lying below the solid (or actual) line. No explanation was offered for the poor prediction. There are undoubtedly many sources of errors that range from inaccurate solubility parameters to experimental errors in Freundlich adsorption isotherms. Of the three adsorption models, the net adsorption energy concept was the most accurate. An attempt was made to apply the solvophobic model, but it was extremely inaccurate for the compounds and experimental conditions of Arbuckle [35].

As cited previously, Belfort [27] is a proponent of the solvophobic interaction theory to explain and to predict the adsorbability of organic compounds on activated carbon. In a publication that preceded Arbuckle's [35], Belfort [36] attempted to correlate various physicochemical parameters of some 25 organic compounds with their molar adsorptivity on pulverized GAC. These latter data were taken from Giusti et al. [30]. Figures 2–15a, b, and c show the plots of molar adsorptivity (Q_m') against the Hammet and Taft polarity parameter, σ^{**}. As the σ^{**} value or the polarity decreases, Q_m' increases and reaches a maximum Q_m' asymptotically for $\sigma^{**} = -0.134$ in Figure 2–15a for 11 organic alcohols. (Numbers in Figures 2–15a, b, and c refer to a table of data in Belfort [36]). Similar observations are seen in Figure 2–15b for ketones and aldehydes and in Figure 2–15c for organic acids. Figure 2–15a shows three correlation lines: one is for the normal alkyl alcohols [compounds 1 to 6 (10 and 11 excepted)], lines 7 to 8 are for isoalcohols, and line 9 is for a tertiary alcohol. Belfort [36] interprets the dashed lines (Figure 2–15a) of 4–8–9 as a cross correlation for the butanols that describes the effect of branching on adsorption. Arbuckle [35]

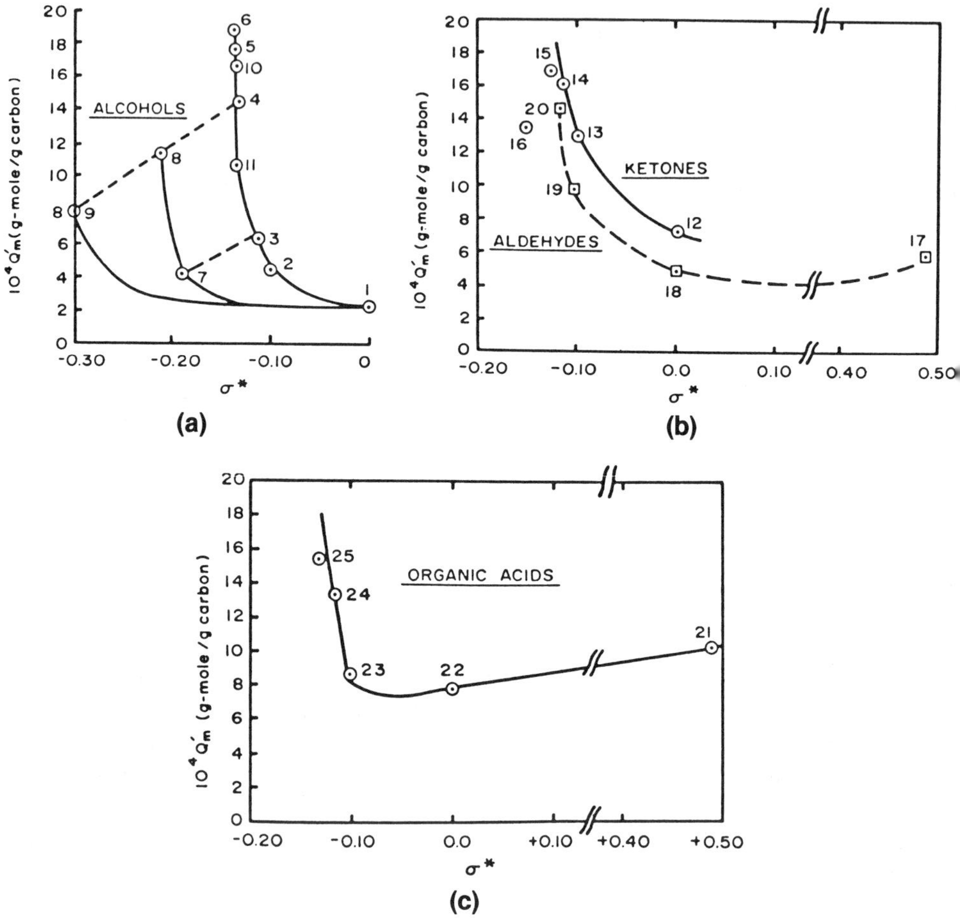

Figure 2–15 (a) Effect of polar parameter on molar adsorption on activated carbon. Adsorption data from Giusti et al. [30]. (b) Effect of polar parameter on molar adsorption on activated carbon. Adsorption data from Giusti et al. [30]. (c) Effect of polar parameter on molar adsorption on activated carbon. Adsorption data from Giusti et al. [30]. Reproduced from Belfort [36], courtesy of the American Chemical Society.

disagrees with this interpretation in rather strong terms. There is an obvious inconsistency here. Q'_m cannot be increasing and decreasing with polarity of these alcohols. Rather, the decrease of Q'_m with increased branching is consistent with other studies on granular carbons where micropores predominate the surface. Pulverized or powdered carbon was used in Belfort's study [36]. Similar Q'_m versus σ^{**} correlations were observed for ketones and aldehydes (Figure 2–15b) and for organic acids (Figure 2–15c). The asymptotic trend toward a $-.130\ \sigma^{**}$ value appears for the three groups of organics.

REFERENCES

1. Faust, S.D., and O.M. Aly. *Chemistry of Water Treatment.* Butterworths, Stoneham, Mass. (1983).
2. Osipow, L.I. "Physical Adsorption on Solids," in *Principles and Applications of Water Chemistry.* S.D. Faust and J.V. Hunter, eds. John Wiley & Sons, New York (1967), p. 75.
3. Young, D.M., and A.D. Crowell. *Physical Adsorption of Gases.* Butterworths, London (1962).
4. London, F. *Trans. Faraday Soc. 33,* 8 (1937).
5. Lowell, S. *Introduction to Powder Surface Area.* John Wiley & Sons, New York (1979).
6. Weber, W.J. Jr., and B.M. Van Vliet. "Fundamental Concepts for Application of Activated Carbon in Water and Wastewater Treatment," in *Activated Carbon Adsorption of Organics from the Aqueous Phase,* Vol. 1. I.H. Suffet and M.J. McGuire, eds. Ann Arbor Science Publishers, Ann Arbor, Mich. (1980), p. 15.
7. McGuire, M.J., and I.H. Suffet, *J. Amer. Water Works Assoc. 70,* 621 (1978).
8. Hildebrand, J.H., and R. L. Scott. *The Solubility of Nonelectrolytes,* 3rd ed. Dover, New York (1964).
9. Snyder, L. *Chemtech 9,* 750 (1979).
10. McGuire, M.J., and I.H. Suffet. "The Calculated Net Adsorption Energy Concept," in *Activated Carbon Adsorption of Organics from the Aqueous Phase,* Vol. 1. I.H. Suffet and M.J. McGuire, eds. Ann Arbor Science Publishers, Ann Arbor, Mich. (1980), p. 91.
11. Keller, R. A., et al. "Use of the Solubility Parameter in Predicting Chromatographic Retention and Elutropic Strength," in *Gas Chromatography 1970.* R. Stock, ed. Institute of Petroleum, London (1971).
12. Karger, B.L., et al. *J. Chromatog. 125,* 71 (1976).
13. Karger, B.L., et al. *Anal. Chem. 50,* 2126 (1978).
14. Barton, A.F. *Chem. Rev. 75,* 731 (1975).
15. Avgul, N.N., and A.V. Kiselev. "Physical Adsorption of Gases and Vapors on Graphitized Carbon Blacks," in *Chemistry and Physics of Carbon,* Vol. 6. P.L. Walker, Jr., ed. Marcel Dekker, New York (1970).
16. Zogorski, J.S., and S.D. Faust. *Proc. 2nd World Cong. Water Res.,* Vol. 2. New Delhi, India (1975), p. 89.
17. Radke, C.J., and J.M. Prausnitz. *Ind. Eng. Chem. Fund. 11,* 445 (1972).
18. McGuire, M.J., et al. *J. Amer. Water Works Assoc. 70,* 565 (1978).
19. Manes, M. "The Polanyi Adsorption Potential Theory and Its Applications to Adsorption from Water Solution onto Activated Carbon," in *Activated Carbon Adsorption of Organics from the Aqueous Phase,* Vol. 1. I.H. Suffet and M.J. McGuire, eds. Ann Arbor Science Publishers, Ann Arbor, Mich. (1980), p. 43.
20. Polanyi, M. *Verh. Inst. Physik Ges. 18,* 55 (1916).
21. Polanyi, M. *Z. Physik 2,* 111 (1920).
22. Wohleher, D.A., and M. Manes. *J. Phys. Chem. 75,* 61 (1971).
23. Wohleher, D.A., and M. Manes. *J. Phys. Chem. 75,* 3720 (1971).
24. Rosene, M.R., and M. Manes. *J. Phys. Chem. 81,* 1646 (1977).
25. Manes, M., and M. Greenbank. "Adsorption of Multicomponent Liquids from Water onto Activated Carbon: Convenient Estimation Methods," Chapter 2 in

Treatment of Water by Granular Activated Carbon. M.J. McGuire and I.H. Suffet, eds. Advances in Chemistry Series 202. American Chemical Society, Washington, D.C. (1983), p. 9.
26. Melander, W.M., and C. Horvath. "Thermodynamics of Hydrophobic Adsorption," in *Activated Carbon Adsorption of Organics from the Aqueous Phase,* Vol. 1. I.H. Suffet and M.J. McGuire, eds. Ann Arbor Science Publishers, Ann Arbor, Mich. (1980), p. 65.
27. Altshuler, G., and G. Belfort. "Selective Adsorption of Organic Homologues onto Activated Carbon from Dilute Aqueous Solutions: Solvophobic Interaction Approach—Branching and Predictions," in *Treatment of Water by Granular Activated Carbon.* M.J. McGuire and I.H. Suffet, eds., Advances in Chemistry Series 202. American Chemical Society, Washington, D.C. (1983), p. 29.
28. Horvath, C. and W. Melander. *Amer. Lab. 10,* 22 (1978).
29. Sinanoglu, O. In *Molecular Associations in Biology.* B. Pullman, ed. Academic Press, New York (1968), p. 427.
30. Giusti, D.M., et al. *J. Water Pollut. Cont. Fed. 46,* 947 (1974).
31. Arbuckle, W.B., and R. J. Romagnoli. *CEP Symp. Ser. (1979) 76,* 77 (1980).
32. Myers, A.L., and S. Sircar. "Theory of Correspondence for Adsorption from Dilute Solutions on Heterogeneous Adsorbents," in *Treatment of Water by Granular Activated Carbon.* M.J. McGuire and I.H. Suffet, eds. Advances in Chemistry Series 202. American Chemical Society, Washington, D.C. (1983), p. 63.
33. Sircar, S., et al. *Ind. Eng. Chem. Fund. 11,* 249 (1972).
34. Aveyard, R., and D.A. Haydon. *An Introduction to the Principles of Surface Chemistry.* Cambridge University Press, London (1973).
35. Arbuckle, W.B. *Environ. Sci. Technol. 15,* 812 (1981).
36. Belfort, G. *Environ. Sci. Technol. 13,* 939 (1979).

3

Kinetics of Adsorption

Adsorption of various compounds, in aqueous systems, onto or into activated carbon is a time-dependent process. This may be a lesser consideration when PAC rather than GAC is employed. It is imperative to have knowledge of the rate of uptake of whatever for an evaluation of design and performance of batch reactors, fixed beds, and fluidized beds. Adsorption kinetics is applied to the predictive modeling and design of fixed bed GAC adsorbers. Another consideration is the identification of the rate-determining step of the adsorptive process. For example, adsorption of organics from aqueous solution by porous carbon can be described by three consecutive steps [1,2]. As discussed later, there may be two additional steps.

The first step is transport of the adsorbate from bulk solution to the outer surface of the adsorbent granule by molecular diffusion. This is called *external, or film, diffusion.* There is no actual film surrounding the granule, but the term is used to describe the resistance to mass transfer at the surface of the particle. The mass transfer coefficient, K_f, in film theory, is related to the free liquid diffusivity of the solute, D_e, and thickness of the diffusional sublayer, δ, as follows:

$$K_f = D_e/\delta. \tag{3.1}$$

The concentration gradient in the liquid film around the granule is the driving force in film diffusion.

The second step, termed *internal diffusion,* involves transport of the adsorbate from the particle surface into interior sites by diffusion within the pore-filled liquid and migration along the solid surface of the pore (surface diffusion) [1,3,4]. Because these two transport processes act in parallel, the more rapid one will control the overall rate of transport. Keinath and Weber [4] found that for a typical water or wastewater sorption system, the rate parameter for surface diffusion was two to four orders of magnitude greater than the liquid phase pore diffusion.

The third step is adsorption of the solute on the active sites on the interior surfaces of the pores. Since the adsorption step is very rapid, it does not influence the overall kinetics [1,2]. The overall rate of the adsorption process, therefore, will be controlled by the slowest step, which would be either film diffusion or internal diffusion. However, control might also be distributed between intraparticle and external mechanisms in some systems [1,5].

The nature of the rate-limiting step in batch systems can be determined in a general way from properties of the solute and adsorbent. Rates of adsorption are usually measured by determining the change in concentration of the solute in contact with the carbon as a function of time. Linearization of the data is obtained by plotting the amount adsorbed per unit weight of adsorbent, X/M versus $t^{1/2}$ for the initial fraction of the reaction.

The adsorption rate, determined from the slope of the line with units of $mg/g/hr^{1/2}$ or $mmol/g/hr^{1/2}$ or equivalent, are not true reaction rates but relative rates useful for comparative purposes. For processes controlled by film diffusion, the adsorption rate is expected to be proportional to the first power of concentration. Zogorski and Faust showed that the adsorption of 2,4-dichlorophenol (DCP) by activated carbon is film diffusion controlled at concentrations less than 400 μmol/l and that a direct linear relationship exists between the initial concentration and adsorption rate [6]. At concentrations of 2,4-DCP greater than 500 μmol/l, the reaction rate was shown to be intraparticle diffusion controlled.

Variation of the rate with concentration was not linear, which is expected from diffusion theory [1]. Therefore, the concentration dependence of the adsorption rate can be used to identify the rate-limiting step. It is also recognized that the adsorption rate is film diffusion controlled during the initial stages of the adsorption process in batch reactors and initial breakthrough in fixed bed reactors. As the carbon becomes loaded with the adsorbate, the reaction rate becomes controlled by intraparticle diffusion [7,8].

For processes controlled by intraparticle diffusion, the size and configuration of the adsorbate molecule should affect the overall rate of adsorption. The larger the molecule or the more branched, the lower should be the rate at which it diffuses. Hence, the adsorption rate should be lower. Weber and Morris reported that a linear relationship existed between observed adsorption rates and molecular weights for a series of alkylbenzenesulfonates ranging from 2-hexylbenzenesulfonate to 2-tetradecylbenzenesulfonate [8]. An increase in the chain length by a -CH_2- group resulted in a reduction of the adsorption rate. An increase in the chain length apparently reduces mobility of the molecule and also increases its affinity to the carbon surfaces in accord with Traube's rule. Both effects result in a decrease of the surface diffusion inside the pores and, hence, an overall decrease in the observed adsorption rate. Adsorbates that exhibit high affinity to activated carbon usually cause intraparticle diffusion control of the adsorptive process. This is attributed to a decrease of surface diffusion of adsorbates that is tightly held and, thus, is less flexible to migrate on the internal surfaces of the carbon. Branching of the adsorbate molecule also results in a decrease of the observed adsorption rate due to a decrease in the surface diffusion rate. Weber and Morris showed that 6-dodecylbenzenesulfonate, which contains a bifurcated chain, and the extensively branched technical benzenesulfonate exhibited lower adsorption rates than the less branched 2-dodecylbenzenesulfonate [8].

Characterization of the rate-limiting step in adsorption by activated carbon can be obtained by determining the variation of the rate with the particle size of the adsorbent. For a film-diffusion-controlled process, the rate of adsorption is expected to vary as the reciprocal of the diameter of the particles for a given total weight of adsorbent. This is because the rate in this case is a first-order function of exterior surface area that, in turn, is inversely proportional to particle diameter [6,8]. Zogorski and Faust showed that the adsorption rate of 2,4-DCP, at concentrations less than 400 μmol/l, exhibited a linear relationship with the reciprocal of the activated carbon particle's diameter in the range of 0.16 to 1.41 mm [6]. This finding supports the conclusion that the adsorption rate was film diffusion controlled. Conversely, for internal-diffusion-controlled processes, the rate of adsorption is expected to vary as the reciprocal of some higher power of the particle's diameter. This relationship was demonstrated for the adsorption rates of benzenesulfonates on activated carbon, which are internal diffusion controlled. The adsorption rates were shown to vary as a function of the inverse of the diameter squared [6,8].

The nature of the rate-controlling step can also be obtained by determining the activation energy of the process. This can be achieved by studying the effect of temperature on the rates of adsorption that are expected to increase with an increase in temperature. This relationship was shown for adsorption of 2,4-DCP [6] and alkylbenzenesulfonates [8]. The activation energy can be computed from:

$$\ln \frac{k_1}{k_2} = \frac{Ea(T_1 - T_2)}{RT_1T_2}, \tag{3.2}$$

where k_1 and k_2 = observed adsorption rates at temperatures T_1 and T_2 in °K,
R = ideal gas law constant (1.987 cal/mol/°C).

The activation energy for internal-diffusion-controlled processes is expected to be in the range of 3 to 5 kcal/mol. The activation energy for adsorption of 3-dodecylbenzenesulfonate was reported to be 4.3 kcal/mol, which indicates a diffusion-controlled process [8].

BATCH SYSTEMS

Single Compounds

Much of the initial research on the kinetics of adsorption was conducted in the so-called detergent era of water pollution in the 1950s to 1960s [8]. These compounds, the alkylbenzenesulfonates, are large, with substantial molecular weights, and are persistent in ground and surface waters because of their resistance to biodegradation. Their major effect on water quality came from the foam created whenever water was agitated. Consequently, their removal from drinking water supplies was mandated through the establishment of a secondary drinking water standard [9]. One of the initial removal processes that was researched was GAC.

Research procedures for batch kinetic studies are similar to equilibrium systems wherein the carbon and adsorbate are shaken continuously for an appropriate period of time (see Chapter 6). Short-term studies usually take 24 to 48 hours, whereas long-term studies have taken as long as 4 weeks with, researchers hope, no attrition of the GACs particles.

Preliminary studies for adsorption of 3-dodecylbenzenesulfonate were conducted with 50/60 or 100/140 mesh (0.273 and 0.126 mm, respectively) Columbia LC carbon [8]. This is a hard carbon that resists attrition in long-term studies. Figure 3-1 shows a plot of typical data for adsorption of the detergent by 0.273 mm GAC. It is obvious that there is a relatively rapid initial rate of adsorption. This rate decreases markedly within 1 to 2 hours to give a gradual approach to an equilibrium condition "which is attained completely only after many days of contact between solution and carbon" [8, p.33]. Plots like that in Figure 3-1 are immediately followed by a linearization of the data by taking the square root of time and replotting (Figure 3-2). These linear plots suggest intraparticle diffusion to be the rate-limiting step in the overall kinetic process, which is discussed thoroughly in the following pages. Relative rate constants are taken from the slopes of straight lines, as seen in Figure 3-2. These k values do not have the usual dimensions for kinetic rate constants and cannot be considered as specific rate

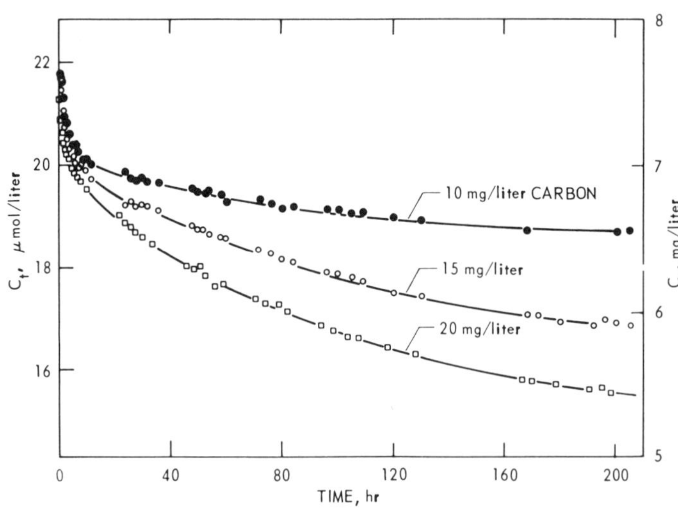

Figure 3-1 Rates of adsorption of 3-dodecylbenzenesulfonate on granular carbon. Reproduced from Weber and Morris [8], courtesy of the American Society of Civil Engineering.

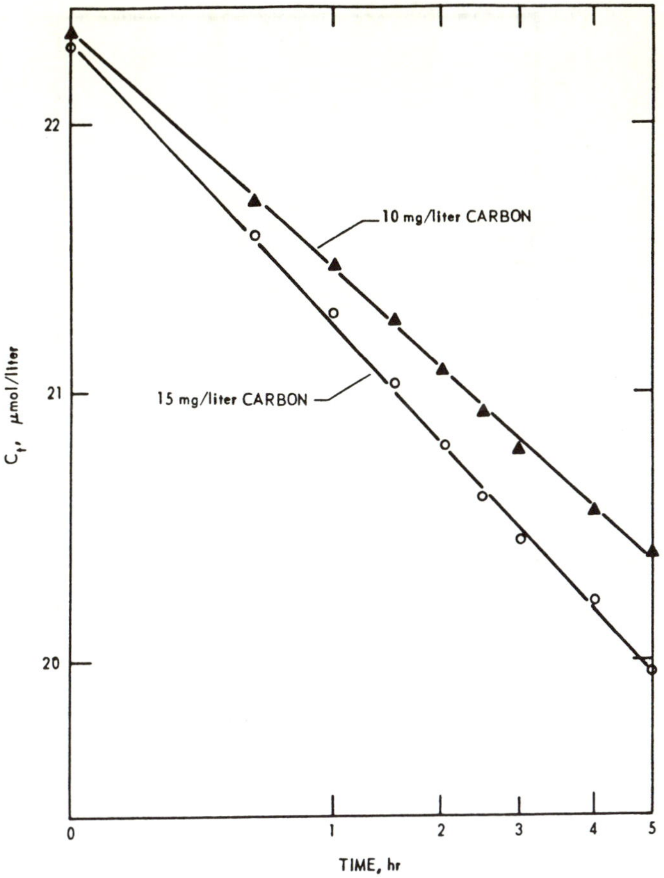

Figure 3–2 Rates of adsorption plotted on $t^{1/2}$ scale. Reproduced from Weber and Morris [8], courtesy of the American Society of Civil Engineering.

parameters. The k values are expressed in units of solute molarity per weight of carbon-$t^{1/2}$—that is, k = mmol/g/$t^{1/2}$. Rate constants for several alkylbenzenesulfonates are given in the following.

A study similar to the one discussed here was conducted on the kinetics of the adsorption of phenols onto GAC (Columbia LCK, 1.00 mm) [10]. The so-called phenol era of water pollution occurred during the 1960s and 1970s whereupon this class of organics was also an undesirable constituent of drinking water [9]. Experimental conditions of Zogorski [10] were similar to those of Weber and Morris [8]. A concentration versus $t^{1/2}$ plot for 2,4-DCP is seen in Figure 3-3 and is clearly not linear. This was interpreted at first to be contradictory to the data of Weber and Morris. However, it is not contradictory. Here, again, this point is discussed later in detail.

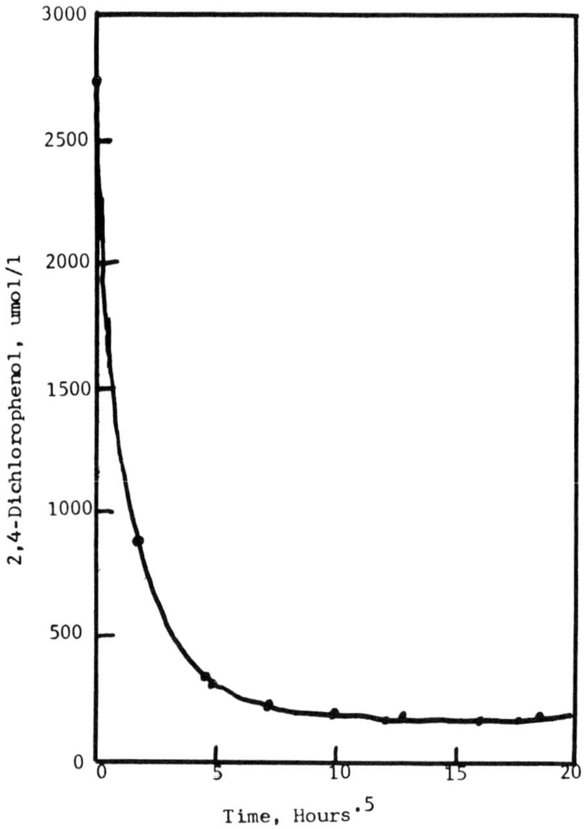

Figure 3–3 Rate of adsorption of 2,4-dichlorophenol on 1.00 mm Columbia LCK carbon (pH = 2.0, T = 20°C, and C_i = 2,730 μmol/l) [10].

Results similar to those of Zogorski [10] were obtained by Smith et al.[11] in a short-term (1.67 hours) study of 2,4-DCP on GAC. Two bituminous coal-based carbons (10 × 14 mesh) were employed.

Rate contants for the adsorption of PNP on several commerical GACs (12 × 40 mesh) and several carbons prepared from petroleum residues were reported [12]. The rate curves are seen in Figure 3–4 where Θ (fraction of equilibrium surface coverage) is plotted versus $t^{1/2}$. Subsequent rate constants are given in Table 3–1. It should be noted that the curves in Figure 3–4 show an initially slow rate of uptake that is opposite the data from alkylbenzenesulfonates and other phenols cited previously. No explanation is offered except the author described the adsorption as a "pseudo-first-order, diffusion controlled reaction" [12, p.313].

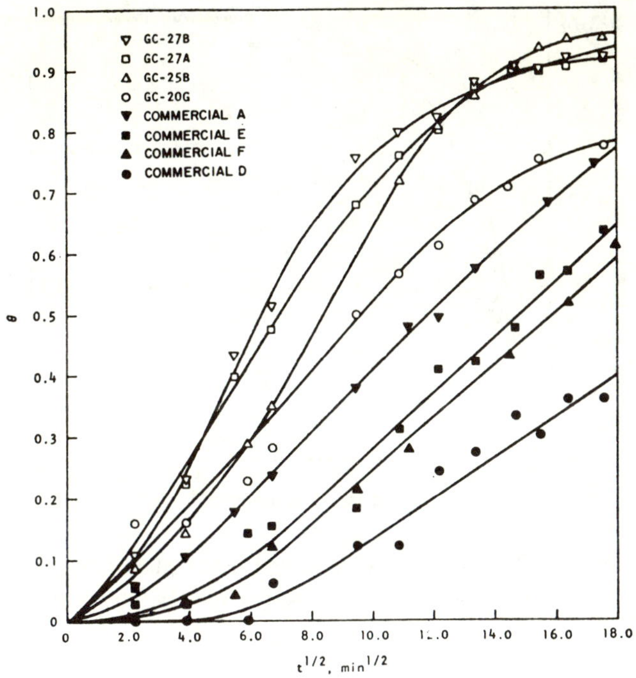

Figure 3–4 PNP adsorption kinetics, surface coverage versus $t^{1/2}$. Reproduced from Mattson [12], courtesy of the American Chemical Society.

Table 3–1 Rate Constants for PNP Adsorption

Carbon	$t^{1/2}\theta = 0.5$, $(min^{1/2})$	$[PNP]\theta = 0.5$ (μM)	k, $l/(mol\ min^{1/2})$
Commercial A	10.2, 11.9	166, 151	590, 557
Commercial E	14.9	130	517
Commercial D	13.1, 21.2	202,[a] 124	377, 380
GC-20 G	9.70, 9.60	159, 148	648, 704
GC-25A + 26	11.1	139	648
GC-25B	8.41	144	825
GC-27A	6.85	130	1,120
GC-27B	6.20	132	1,220
Commercial F	15.8	137	462

Source: Reproduced from Mattson [12] courtesy of the American Chemical Society.
Note: pH 2 $[PNP]_{t=0}$ = 0.2 m*M*, and 200 mg of carbon in 3.5 *M* solution.
[a]Starting [PNP] = 0.25 m*M* because of low rate constant [12].

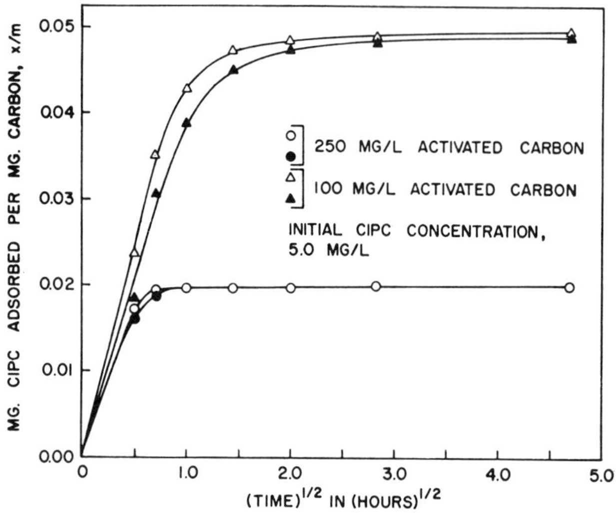

Figure 3-5 CIPC absorbed per unit weight of activated carbon 100 and 250 mg/l at pH 6.9 and 20°C. Reproduced from Schwartz [13], courtesy of the American Chemical Society.

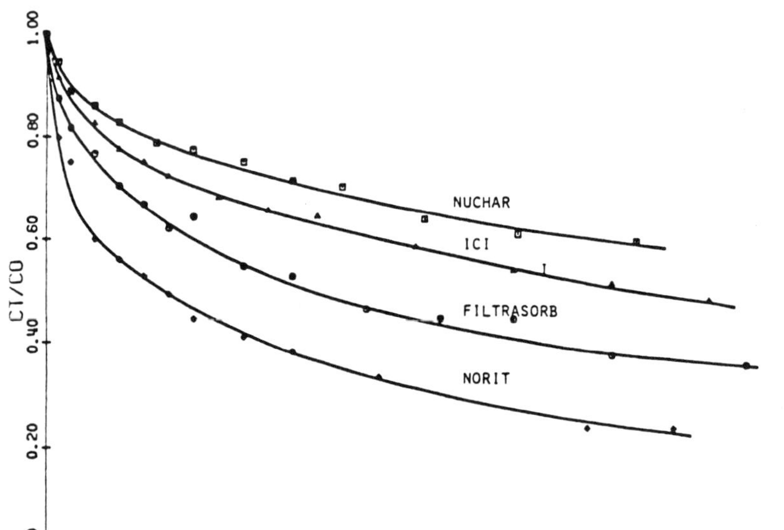

Figure 3-6 Rates of adsorption of humic acid on different commercial carbons in CMB reactors, C_o = 5 ppm, 16/20 U.S. sieve size carbon, pH = 7.0. Reproduced from Weber et al. [14], courtesy of Ann Arbor Science Publishers, Inc.

Another early study was concerned with the kinetics of adsorption of an organic pesticide, isopropyl N-(3-chlorophenyl) carbamate (CIPC), on PAC (Nuchar C-190) [13]. Here, the kinetics should be rapid because of the small particle size (median diameter, 1.65 mμ) of the PAC. This is seen in Figure 3–5 where the cumulative adsorption curves plateau within 1 to 4 hours. CIPC is, however, adsorbed poorly, with Freundlich constants of $1/n = .11$ and $K = 0.20$ mg/mgC at pH 6.9, 20°C, and 24 hours of contact.

Researchers in the mid-1970s and mid-1980s were quite concerned with the occurrence of trihalomethanes (THM) and their precursors in drinking water supplies [9]. Their rates of adsorption on GAC were consequently investigated in laboratory batch reactors. Figure 3–6 shows the rates of adsorption of humic acid on four brands of GACs [14]. These curves in batch reactors should be compared with the carbon column breakthrough curves in Chapter 6. The rates appear to be rapid and approach equilibrium conditions within a short period of time. Figure 3–7 exhibits the rates of adsorption of four THMs on 16/20 mesh F-400 [14]. These kinetics appear to be rapid also. Figure 3–8 gives the rates of adsorption of two phenols and two sulfonates on 16/20 mesh F-400 [14].

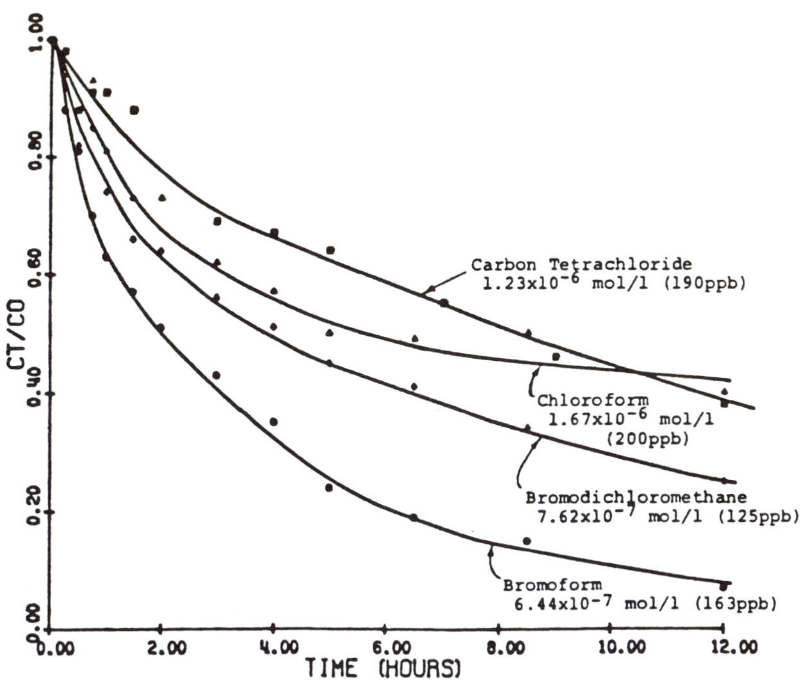

Figure 3–7 Rates of adsorption of typical volatile halogenated compounds in CMB reactors, 16/20 U.S. sieve size Filtrasorb 400 activated carbon; carbon = 0.19 g/l, pH = 7.0. Reproduced from Weber et al. [14], courtesy of Ann Arbor Science Publishers, Inc.

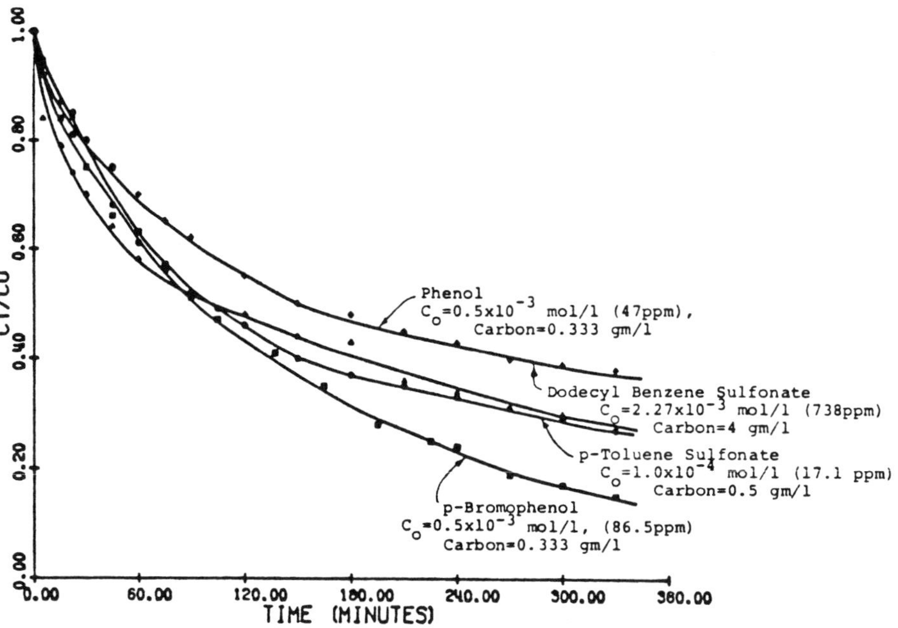

Figure 3-8 Rates of adsorption of typical wastewater compounds in CMB reactors, 16/20 U.S. sieve size Filtrasorb 400 activated carbon, pH = 7.0. Reproduced from Weber et al. [14], courtesy of Ann Arbor Science Publishers, Inc.

Systemic Factors Affecting Rates of Adsorption: Single Compounds

Hydrogen Ion Concentration

In general, the $[H^+]$ (or $[H_3^+O]$) affects adsorptive processes through dissociation of functional groups on the adsorbate and adsorbent (carbon, in this case). Figure 3-9 shows the effect of solution pH on the rate of uptake of 3-dodecylbenzenesulfonate on 0.126 mm Columbia LC carbon [8]. It appears that the uncharged molecular species of the sulfonate is adsorbed faster (pH 1.91 and 4.04) than the negatively charged species (pH 6.02, 7.79, and 8.73). For the latter, there probably are some repulsive effects between the negatively charged carbon's surface and the anionic sulfonate. For the acid pH values, very little competition is expected between H_3^+O ions and uncharged sulfonate for the carbon's surface.

There was a pronounced effect of $[H_3^+O]$ on the rate of adsorption of 4-chlorophenol onto Columbia LCK carbon [6]. This effect is seen in Figure 3-10 where there is a sharp decrease in the removal rate above a pH value of 7.5. The pKa value for this phenol is 9.4, whereupon the anionic form of the phenol is adsorbed at a rate considerably slower than the associated form. However, the

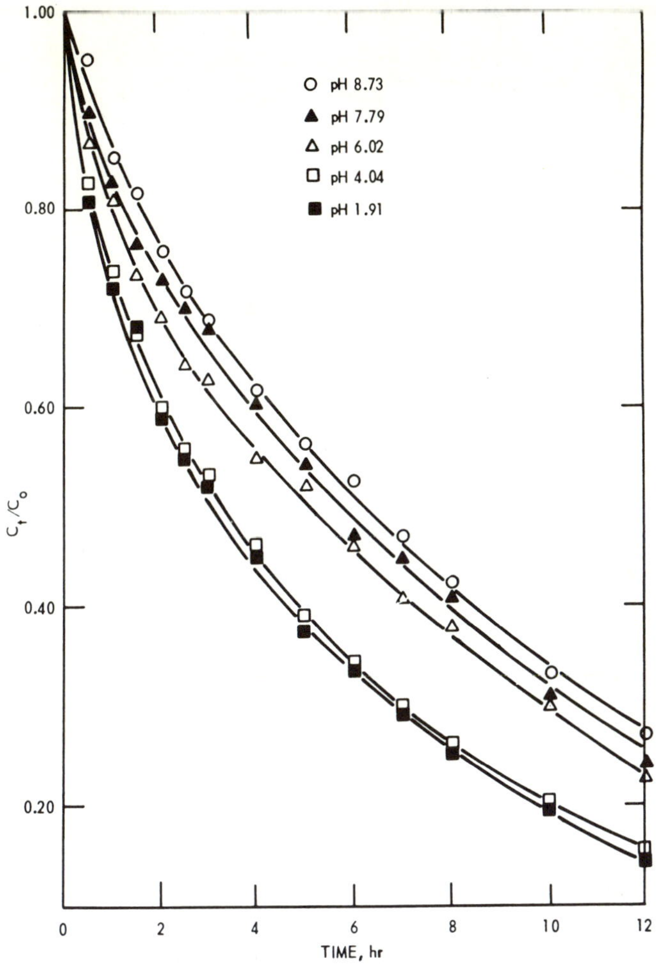

Figure 3–9 Effect of pH on rate of adsorption. Reproduced from Weber and Morris [8], courtesy of the American Society of Civil Engineering.

rate of adsorption at neutral pH values is not affected adversely where a treatment process is operated. Other phenolic compounds exhibited the same behavior [6,10].

Temperature

Inasmuch as carbon adsorption is a kinetic process, increasing or decreasing the system's temperature increases or decreases the rate of removal. The well-known Arrhenius equation describes this effect of temperature on reaction rate (see Equation 3.2) [15]. Some investigators offer this effect as evidence for a diffu-

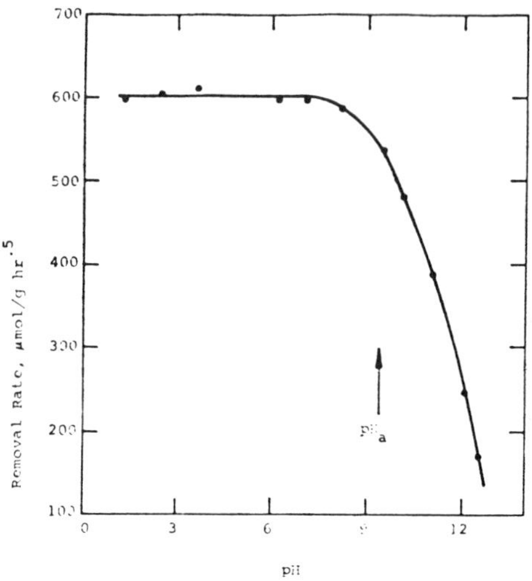

Figure 3–10 Variation in the rate of adsorption of 4-chlorophenol with hydronium ion concentration ($T = 20°C$, 500 mg/l of 0.71 mm size adsorbent, and initial adsorbate concentration = 285 μmol/l). Reproduced from Zogorski, et al. [6], courtesy of Academic Press, Inc.

sion-controlled process [8]. Figure 3–11 shows the semilog plot of relative rate constants versus $1/T$ for adsorption of 3-dodecylbenzenesulfonate on GAC [8]. The temperature range was 5 to 55°C. An *Ea* value of 4.27 kcal/mole was calculated from Equation 3.2 for this system. This value was compared to 3.5 kcal/mole "for large molecules similar in size to the sulfonated alkylbenzenes. It is apparent that the value 4.3 kcal/mole, lies within the expected range for diffusion-controlled processes" [8, p.34].

Similar temperature effects were observed on the kinetics of adsorption of phenol and 2,4-DCP on Columbia LCK carbon [6]. Figures 3–12a and b show the influence of temperature (5 to 43°C) on the removal rate constants for the two phenols, respectively. Raising the temperature from 10 to 30°C increased the removal rate of phenol by 21% and of 2,4-DCP by 28%. This temperature increase also caused a decrease in the t_{50} values by 36% for phenol and 39% for 2,4-DCP. Calculated *Ea* values for phenol and 2,4-DCP were 1.6 and 2.2 kcal/mole, respectively [6].

Concentration

According to one investigator, "prediction of the variation of rate of adsorption of alkylbenzenesulfonate on carbon as a function of solute concentration is difficult" [8, p.35]. If intraparticle diffusion is the mechanism, an equilibrium interfacial concentration should form rapidly that is followed by slow diffusion into the carbon particles. "Then, a simple dependence on solution concentration

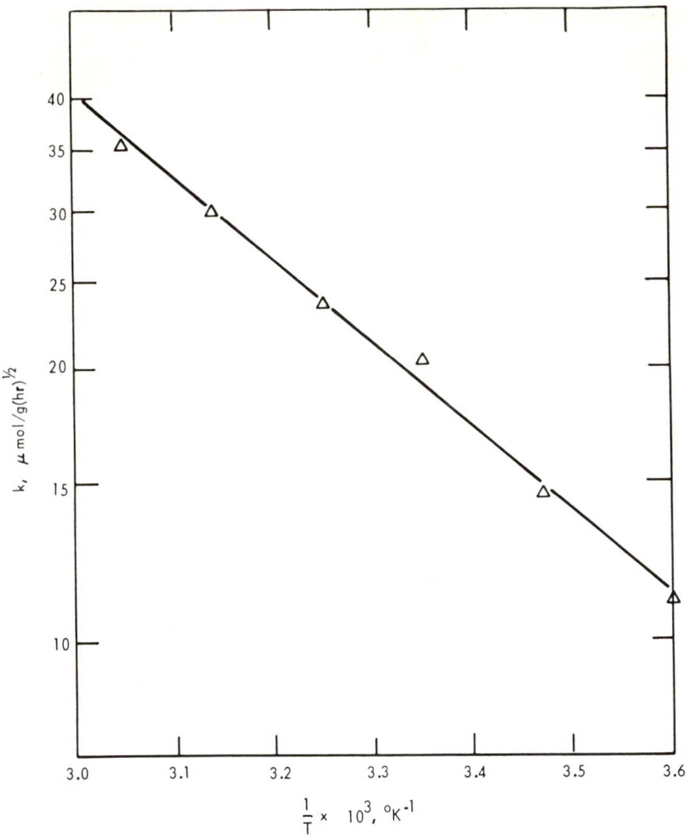

Figure 3–11 Determination of activated energy for adsorption of 3-dodecylbenzenesulfonate on granular carbon. Reproduced from Weber and Morris [8], courtesy of the American Society of Civil Engineering.

is not to be expected. . . . Moreover, the mathematical treatment of intraparticle diffusion does not lead to a simple algebraic relationship between external solute concentration and time of reaction even when a constant saturated external layer is maintained'' [8, p.35–36].

Rates of removal of alkylbenzenesulfonates were observed to increase nonlinearly with increasing concentration of these adsorbates that ranged from 8. to 30 μmol/l. This is seen in Table 3–2 where the k values were obtained from slopes of linear plots of C_o versus $t^{1/2}$. Constancy was obtained when the k values were squared and divided by C_o, which is supposed to show first-order time dependence. These data in Table 3–2 show also that adsorption, relative to C_o, is more rapid the more dilute is the solution [8].

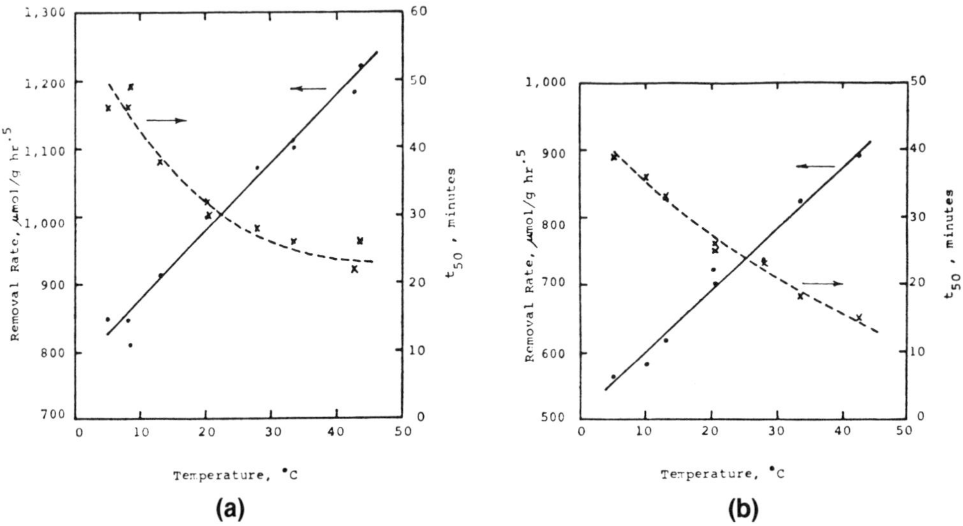

Figure 3–12 (a) Influence of temperature on the kinetics of adsorption of phenol (pH = 6.1, 500 mg/l of 0.71 mm size adsorbent, 0.05 M phosphate buffer, and initial adsorbate concentration = 595 μmol/l). (b) Influence of temperature on the kinetics of adsorption of 2,4-dichlorophenol (pH = 6.1, 500 mg/l of 0.71 mm size adsorbent, 0.05 M phosphate buffer, and initial adsorbate concentration = 340 μmol/l). Reproduced from Zogorski et al. [6], courtesy of Academic Press, Inc.

The influence of initial concentration, C_o, on the rate of adsorption of 2, 4-DCP and cyanonitrile (4-cyanophenol) was determined [6]. C_os ranged from 1.6 to 300. μmol/l for cyanonitrile and from 37. to 6,260. μmol/l for 2,4-DCP. The data are presented on log-log graphs for convenience and to condense a wide range of numbers. Slopes of the linear segments in Figures 3–13a and b are 0.965 and 0.936, respectively. A slope of 1.00 on a log-log plot would be indicative of a direct linear relationship. For all intents and purposes a direct linear relation exists between C_o and removal rate for C_o values less than 500. μmol/l.

Table 3–2 Influence of Initial Concentration of Solute on Rate of Adsorption

C_0 (μmol/l)	k (μmol/g hr$^{1/2}$)	$k/C_0^{1/2}$
8.1	62	21.8
16.1	88	21.9
23.1	99	20.6
30.0	112	20.4

Source: Reproduced from Weber and Morris [8] courtesy of the American Civil Engineering Society.
Note: 35 mg per l, 0.126 mm carbon, 30°C, 3-DBS.

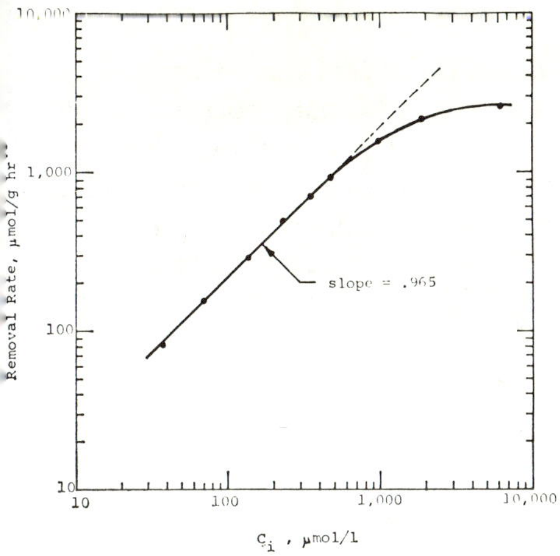

(a)

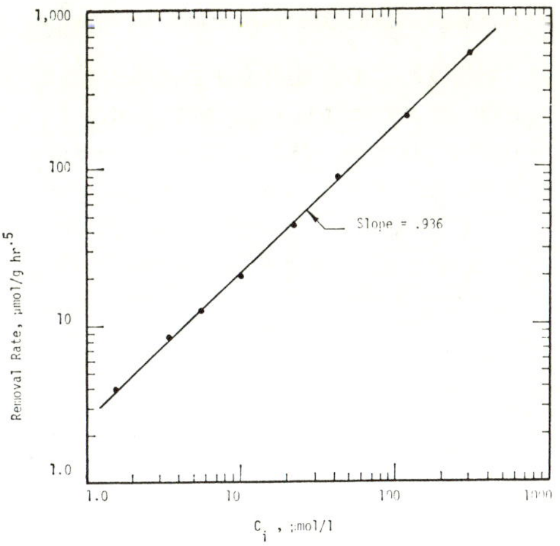

(b)

Figure 3–13 (a) Influence of the initial adsorbate concentration on the rate of adsorption of 2,4-dichlorophenol (pH=6.0, T = 20°C, 500 mg/l of 0.71 mm size adsorbent, and 0.05 M phosphate buffer). (b) Influence of the initial adsorbate concentration on the rate of adsorption of cyanonitrile (pH=6.05, T=20°C, 500 mg/l of 1.00 mm size adsorbent, and 0.05 M phosphate buffer). Reproduced from Zogorski, et al. [6], courtesy of Academic Press, Inc.

As seen in Figures 3–13a and b, the removal rate was increased as the C_o of the adsorbate was increased [6]. For example, raising the C_o of 2,4-DCP from 37.4 to 346. μmol/l caused an increase in the removal rate from 82. to 712. μmol/g/hr$^{1/2}$. However, the time (t_{50}) required for a 50% occurrence of ultimate adsorption was unaffected by C_o. Thus, raising C_o increased the quantity of adsorbate removed from solution per unit of time, but it did not affect the time required for removal of an equal fraction of the ultimate adsorption. This is in direct variation with other results [8].

Concentration dependence of the rate of adsorption is used to define the rate-limiting step in the reaction. When intraparticle transport limits the kinetics of an adsorption reaction, the variation in reaction rate is not expected to be linear, whereas the rates of strictly adsorptive reactions and simple diffusion-controlled processes are expected to be proportional to the first power of the adsorbate's concentration [16]. The results in Figures 3–13a and b strongly support the contention that the rate-limiting step in the adsorption of low molecular weight phenols (C_o < 500. μmol/l) is transport of the adsorbate from bulk solution to the external surface of the adsorbent [6]. This difference with previous research [8] is discussed in greater detail later. However, there is evidence in Figure 3–13a that, at C_os of 2,4-DCP > 500. μmol/l, a nonlinear relationship exists between removal rate and C_o of the adsorbate [6]. This relationship indicates a process controlled by intraparticle transport.

Molecular Size of the Adsorbate

The size of the diffusing molecule should affect the overall rate of adsorption; that is, other investigators contend that diffusion of adsorbate molecules within the micropore structure of GAC is the rate-determining step in the adsorptive process [8]. The larger the molecule, the slower should be the rate at which it diffuses and also the rate of adsorption. Evidence to support this statement is seen in Figure 3–14 where the length of the aliphatic hydrocarbon chain on the detergent molecule affects the rate of uptake (0.126 mm carbon) [8]; that is, the

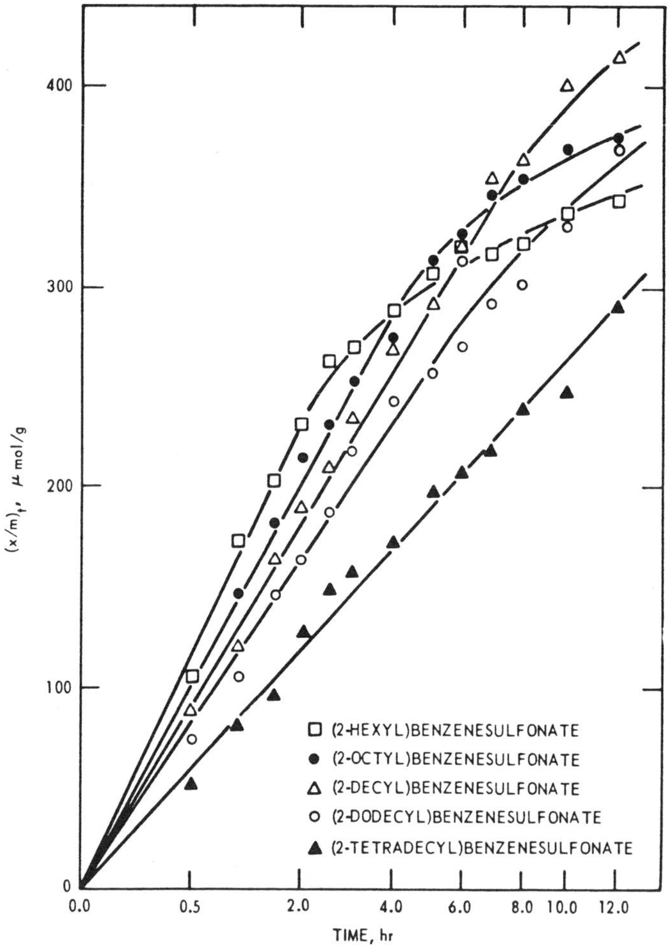

Figure 3–14 Relative rates of adsorption for an homologous series of alkylbenzenesulfonates. Reproduced from Weber and Morris [8], courtesy of the American Society of Civil Engineering.

longer the chain and also the higher the molecular weight, the slower the rate of adsorption. This is also seen in Figure 3-15 where a speedy phenol curve is included as a control [8].

A broad range of molecular sizes was investigated for their rates of adsorption whereupon a "remarkably linear dependence was found between the relative rate constant and the molecular weight" [8, p.36]. All these findings (Figures 3-14 and 3-16) are in accord with the so-called Traube's Rule [17], which states that the properties of molecules in a homologous series that have hydrophobic and hydrophilic components tend to change regularly with chain length.

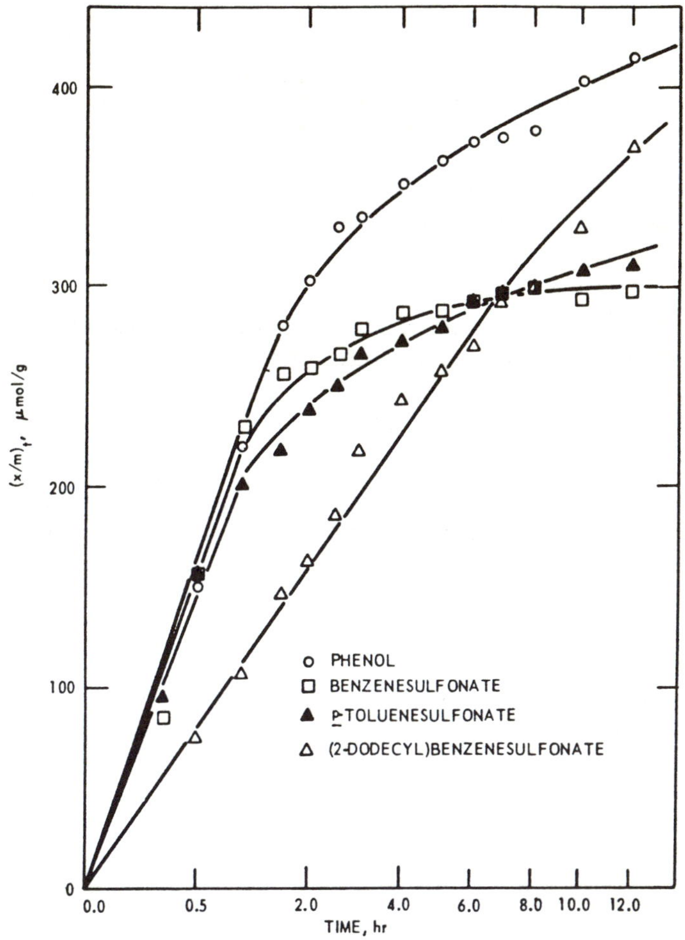

Figure 3-15 Relative rates of adsorption for phenol and several sulfonated alkylbenzenes. Reproduced from Weber and Morris [8], courtesy of the American Civil Engineering Society.

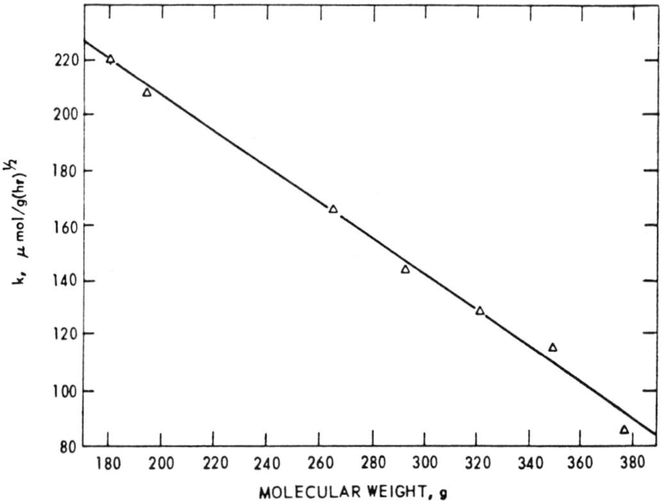

Figure 3–16 Effect of molecular size on rate of adsorption. Reproduced from Weber and Morris [8], courtesy of the American Society of Civil Engineering.

Therefore, the extent of adsorption from aqueous solution usually increases with the addition of CH_2 groups to the hydrocarbon chain.

Molecular Configuration of the Adsorbate

Here, again, another molecular property is that of configuration. The effects of this property on the removal rates of adsorption of alkylbenzenesulfonates were studied using compounds with the same molecular weight but with different structures [8]. These results are seen in Table 3–3 where, apparently, the structure of the alkyl chain attached to the benzene ring has a small but measurable effect

Table 3–3 Effect of Structural Differences in DBS on Rate of Adsorption

Solute	*k* ($\mu mol/g\ hr^{1/2}$)
2-DBS	118
3-DBS	106
6-DBS	96
Technical DBS	52

Source: Reproduced from Weber and Morris [8] courtesy of the American Civil Engineering Society.
Note: 23 to 25 μM initial solute concentrations, 30°C, 35 mg per l, 0.126 mm Columbia LC carbon.

on the adsorption rate. The position of the attachment of benzene on the side chain presumably affects the molecule's ability to diffuse through the carbon's micropores.

Similar studies were conducted with smaller-sized phenols [6]. These results are seen in Table 3-4 where the removal rates and t_{50} values are essentially the same for all compounds except 2,6-diisopropylphenol and 2,6-ditertbutylphenol. These phenols are small in comparison with giant detergent molecules, and configuration apparently has little or no effect on rate of adsorption.

Carbon Particle Size

Rates of adsorption on a solid surface are expected to vary with available surface area or, for a constant absorbent mass, with the particle size.

> This rate should then exhibit a monotonic increase with some function of the inverse of the diameter of the carbon particles. Mathematical considerations indicate that if the mechanism of uptake is simply one of adsorption on specific sites, the rate should vary reciprocally with the first power of the diameter, whereas for radial intraparticle diffusion the variation should be with the reciprocal of the square of the diameter (radial diameter of cylindrical particles). [8, p.37]

Figure 3-17 shows rates of adsorption of 3-dodecylbenzenesulfonate (25°C) by different sizes of granular Columbia LC carbon. This graph indicates "that the relative rates conform extremely well to the hypothesis that intraparticle diffusion is the rate-limiting step, exhibiting rate variation as a function of the inverse diameter squared" [8, p.37]. The explanation for this is that decreasing the

Table 3-4 Rate of Adsorption of Several Alkyl Substituted Phenols

Adsorbate	*Initial Concentration of Adsorbate (μmol/l)*	*Removal Rate*		
		(μmol/g hr$^{1/2}$)	*(g/g hr$^{1/2}$)*	*t_{50}[a] (min)*
Phenol	39.	74.8	7.03	31.
2-methylphenol	40.	73.4	7.94	32.
3-methylphenol	38.	72.8	7.88	33.
2,6-dimethylphenol	38.	71.8	8.78	34.
3,4-dimethylphenol	38.	74.8	9.14	30.
2,4,5-trimethylphenol	37.	72.0	9.81	30.
2,4,6-trimethylphenol	40.	77.4	10.5	31.
2,6-diisopropylphenol	38.	63.4	11.3	40.
2,6-ditertbutylphenol	39.	44.2	9.13	>75.

Source: Reproduced from Zogorski et al. [6] courtesy of Academic Press, Inc.
Note: Results are average of two duplicate experiments conducted at 20°C, pH = 6.3, 0.05 *M* phosphate buffer, and 500 mg/l of 0.71 mm Columbia LCK carbon.
[a]Assuming equilibrium concentration is negligible in comparison with initial concentration.

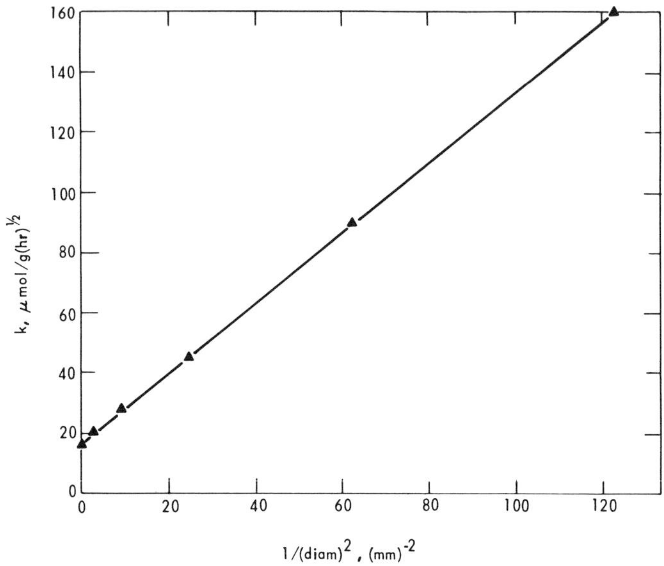

Figure 3–17 Effect of particle size of carbon on rate of adsorption. Reproduced from Weber and Morris [8], courtesy of the American Society of Civil Engineering.

particle size simply increases the specific surface area of a given weight of activated carbon (Figure 3–18).

Considerably different results were obtained when 2,4-DCP was adsorbed on various-sized particles of Columbia LCK carbon [6]. This study is summarized in Table 3–5. A linear relationship was realized when relative rate constants were

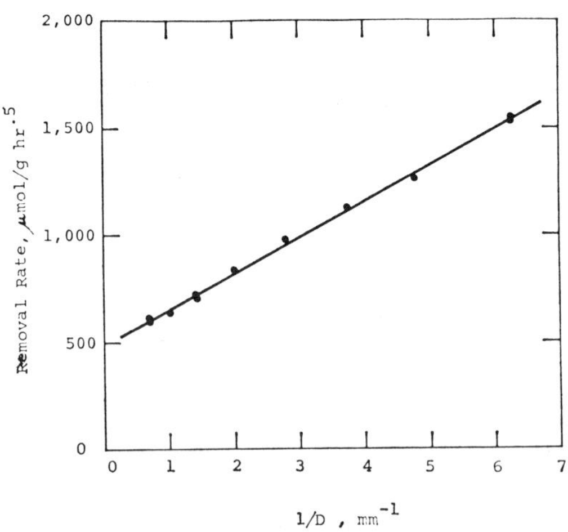

Figure 3–18 Relationship between removal rate and inverse of carbon particle size. Reproduced from Zogorski et al. [6], courtesy of Academic Press, Inc.

Table 3-5 Effect of Carbon Particle Size on the Rate of Adsorption of 2,4-DCP

Carbon Particle Size		*Removal Rate*		
(Mesh)	*(mm)*	*(μmol/g hr$^{1/2}$)*	*(g/g hr$^{1/2}$)*	t_{50} *(min)*
12 × 16	1.41	591.	0.0964	35.
12 × 16	1.41	615.	0.100	33.
16 × 20	1.00	630.	0.103	32.
20 × 30	0.71	722.	0.118	24.
20 × 30	0.71	701.	0.114	25.
30 × 40	0.50	839.	0.137	18.
40 × 50	0.36	983.	0.160	12.
50 × 60	0.27	1,120.	0.183	8.9
60 × 80	0.21	1,260.	0.205	6.5
80 × 100	0.16	1,530.	0.249	4.7
80 × 100	0.16	1,550.	0.252	4.6

Source: Reproduced from Zogorski et al. [6] courtesy of Academic Press, Inc.
Note: 20°C, pH = 6.15, 0.05 *M* phosphate buffer, initial adsorbate concentration = 345 μmol/l, and 500 mg/l of Columbia LCK carbon.

plotted against the reciprocal of the diameter raised to the first power. This finding supports the argument that external transport limits the rate of adsorption [16]. Here, again, it is important to remember that 2,4-DCP has a molecular size considerably smaller than the alkylbenzenesulfonates. This smaller size may overcome the size or configuration problem when diffusing into the carbon's pores.

Diffusion-Controlled Processes

Compelling evidence was cited earlier to support arguments for external or internal diffusion of the organic molecules onto or into GAC as the rate-limiting step in the overall kinetics of adsorption. Several characteristics of the adsorbent, adsorbate, and solution phase are important in the determination of the rate-limiting step. These include particle size of the adsorbent, concentration of the adsorbate, degree of mixing, affinity of the adsorbate for the adsorbent, molecular size and configuration of the adsorbate, and diffusion coefficients of the adsorbate in bulk solution and within the pores of the adsorbent. Usually, external transport is the rate-limiting step in systems that have poor mixing, dilute concentration of adsorbate, small particle sizes of adsorbent, and high affinity of adsorbate for adsorbent [16]. In contrast, the intraparticle step limits the overall transfer for those systems that have good mixing, large particle sizes of adsorbent, high concentration of adsorbate, and low affinity of adsorbate for adsorbent [16].

Certain properties of the adsorbate and adsorbent are useful indirectly and empirically in determining the nature of the rate-limiting step. These characteristics were derived from ion exchange theory and have been described in detail by Helfferich [16] and Weber and Morris [8]. Essentially, these indirect tests (described previously) consist of comparing observed dependencies between the rate of adsorption and particle size of adsorbent and between the rate of adsorption and initial concentration of the adsorbate with expected relationships derived from diffusion theory. Conclusions drawn from such indirect tests are frequently ambiguous [15].

Weber [4,8] based his early intraparticle diffusion explanation largely on the work of Edeskuty and Amundson [18]. These latter two authors developed mathematical equations to calculate intraparticle diffusion coefficients for phenol adsorption by "hugh" carbon particles. Only a brief description of these equations, boundary conditions, and assumptions is provided here. First, it was assumed "that the rate of adsorption is much faster than the rate of diffusion so that equilibrium obtains at each point inside the particle. In order to develop analytical formulas, the isotherm was taken to be a linear function of the concentration" [18, p.1699]. Second, only radial diffusion in the cylindrical carbon particles was considered. Axial diffusion was neglected mathematically inasmuch as the ends of the cylinders were made impervious to the diffusing material.

The experimental procedure, briefly, consisted of preparing uniform cylinders of carbon particles by means of a mechanical mold [18]. Grade SXA activated carbon was obtained from Carbide and Carbon Chemicals Corp. and was prepared in two sizes: 6/8 mesh (0.25 to 0.31 cm diameter) and 4/6 mesh (0.40 to 0.47 cm diameter). The ends of the cylinders were coated with a plastic material to prevent axial diffusion.

Typical adsorption isotherms and kinetic data are seen in Figure 3–19 for phenol. This figure shows clearly "that the whole isotherm is not available but that the initial section and the section at high concentrations are essentially linear" [18, p.1699]. Only the latter portion of the isothermal data was used. Curves A and B on Figure 3–19 were calculated from theory, whereas curve C is from experimentation.

Several parameters were developed from theoretical treatment that are related to the intraparticle diffusion coefficient [18]. There are:

$$\beta = \frac{W\alpha D\gamma}{\rho V}, \tag{3.3}$$

$$\gamma = \frac{k_1 + \alpha}{\alpha D}, \tag{3.4}$$

$$\omega_n = R\sqrt{-\gamma\rho_n}, \tag{3.5}$$

where W = total mass of particles, α = fractional void volume in particle, ratio of void volume to total volume of a single particle, ρ = particle density, V = volume of external solution, k_1 = constant in adsorption equilibrium isotherm, D = diffusion coefficient, ω_n = nth root of a transcendental equation, R = external diameter of cylinders, and ρ_n = transform parameter corresponding to

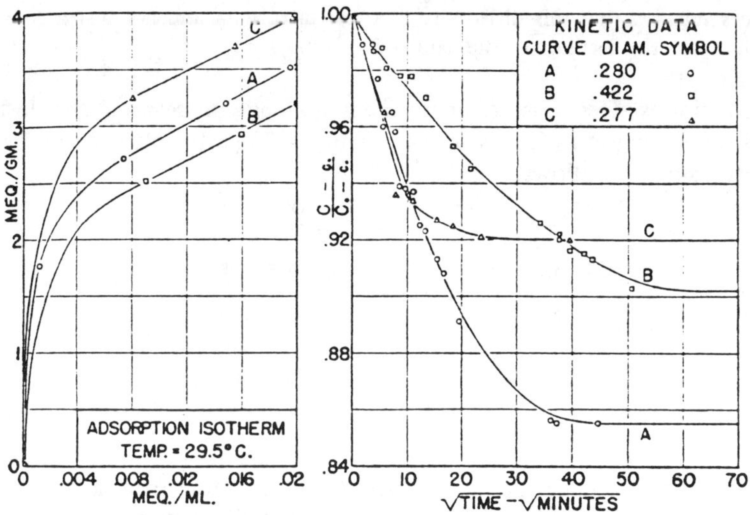

Figure 3–19 Plot of experimental data for runs 2 and 10. Reproduced from Edeskuty and Amundson [18], courtesy of the American Chemical Society.

t (time). The quantities W, ρ, V, α, and R were obtained by direct physical measurement, whereas k_1 is essentially the slope of the linear portion of the isotherm. The ω_ns were calculated by assuming D values.

Here it is appropriate to quote:

> It is clear that the only parameter which is unknown is the diffusivity. This diffusivity must be that within the particle and hence should be somewhat smaller than that in the free solution. That diffusivity which made the theoretical curve fit the data best was used and therefore for this paper the diffusivity is a parameter whose value is obtained from the experiment. For the particle sizes used in this work the same value of the diffusivity was satisfactory for all runs. Its value is 0.890×10^{-5} square centimeters per second, which is less than the literature value of 0.955×10^{-5} square centimeters per second. [18, p.1699]

Experimental data for two runs are plotted on the right-hand side of Figure 3–19. Curves A and B are the theoretical curves with the diffusion coefficient assumed. "It is seen that the agreement between theory and experiment is excellent" [18, p.1700]. Curve C is from experimental data for a run with completely uncoated particles. Additional evidence for the validity of the intraparticle diffusional coefficient for phenol came from:

> The fact, however, that data for all particle sizes can be approximated to quite accurately with the same diffusivity and that the activation energy computed from runs at a series of temperatures agrees in magnitude with that suggested by other

authors suggests that this diffusivity is a real diffusivity and not just a constant which makes the formula fit the data. [18, p.1700]

Diffusion is affected by temperature in a kinetic process and should follow the Arrhenius equation. This is seen in Figure 3-20 [18]. Values of D for other temperatures are as follows:

Temperature°*C*	D, $cm^2\ 10^{-5}/sec$
25.2	0.83
29.5	0.89
38	1.01
48	1.20

Additional evidence for intraparticle diffusion came from a kinetic study for adsorption of several straight-chain sulfonated alkylbenzenes [3]. A carbon from the National Carbon Co. was utilized at a mesh size of 100/140 with a mean diameter of 0.126 mm. From theoretical considerations of Langmuir's adsorption isotherm and Fick's First Law, an equation was derived that describes the conservation of diffusing mass in a porous substance:

$$\frac{\partial C}{\partial t} = \frac{1}{r}\frac{\partial}{\partial r}\left(rD\frac{\partial C}{\partial r}\right) - \frac{\partial S}{\partial t}, \tag{3.6}$$

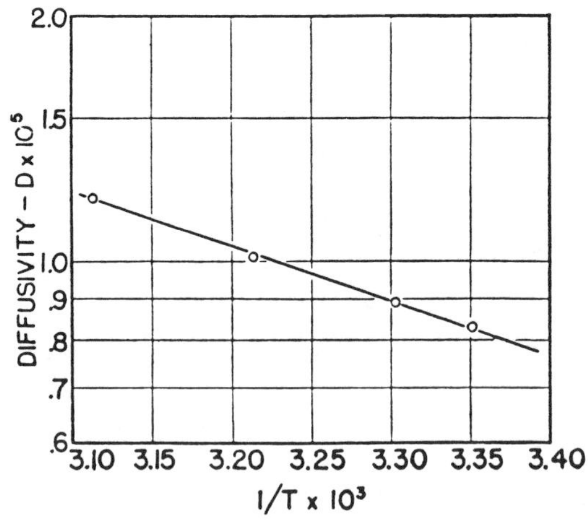

Figure 3-20 Plot of logarithm of diffusivity versus reciprocal absolute temperature for determination of activation energy. Reproduced from Edeskuty and Amundson [18], courtesy of the American Chemical Society.

where C = concentration of diffusing mass, t = time, r = any position in a cylindrical particle, D = diffusion coefficient, and S = concentration of adsorbed solute (from the Langmuir equation). Mathematical treatment of diffusion based on Equation 3.6 assumes that transport occurs radially and that the maximum length of path over which diffusion occurs is equal to the particle's radius. This statement is followed by a disclaimer in which the actual path may be longer due to the porous nature of carbon. Consequently, the D values calculated from Equation 3.6 may be lower than the actual diffusivities [3].

An additional assumption was made whereby the diffusivity, D, is independent of the particle's radial distance, r. This permits Equation 3.6 to be placed into a dimensionless form with introduction of several dimensionless quantities: $c = C/C_o$ (C_o = concentration of diffusing mass in solution outside the particle at $t = 0$), $s = S/C_o$, $\rho = r/a$ (a = unit length of cylinder), and $\tau = Dt^2/a$. This results in a dimensionless equation:

$$\frac{\partial c}{\partial \tau} = \frac{1}{\rho}\frac{\partial}{\partial \rho}\left(\rho\frac{\partial c}{\partial \rho}\right) - \frac{\partial s}{\partial \tau}. \tag{3.7}$$

The initial and boundary conditions for Equation 3.7 become

$$c = 0 \quad \tau = 0 \quad \text{for} \quad 0 \le \rho \le 1, \frac{A}{\pi a^2}(1 - c_1) = 2\int_0^1 (c + s)\rho \, d\rho,$$

where $c_1 \times C(1, \tau)$ [3]. It was also necessary to place the Langmuir adsorption isotherm into a dimensionless form:

$$s = \frac{\alpha \beta_c}{1 + \beta_c}, \tag{3.8}$$

where $\alpha = S_m/C_o$, $\beta = bC_o$, and $s = S/C_o$. These so-called adsorption parameters are known or calculated. A computer program was written to calculate C as a function of τ. Then values for c (C/C_o) were taken from experimental data for a range of values for t and substituted into $\tau = Dt/a^2$ to solve for values of D in each experimental run. These values are seen in Table 3–6 for ten detergents [3] and are the same order of magnitude for diffusivities for other compounds in aqueous solutions. There is a decrease in D values with an increase in the size of the alkylbenzenesulfonate molecule, which is expected for intraparticle transport [3]. Typical kinetic data for this study are seen in Figure 3–21.

The distinction between external and intraparticle diffusion-controlled processes can be made by performing an interruption test. This test has been used by Kressman and Kitchener [19] and Smith et al. [11]. In an interruption test, the adsorbent is periodically removed from the adsorbate solution for a brief period and then reimmersed. When concentration gradients are present, the pause

Table 3-6 Calculated Diffusion Coefficients and Related Parameters

Run Number	*Diffusing Solute*	*Initial Concentration of Solute,* C_0 *(μmol/l)*	*Area Served by Each Particle of Adsorbent,* A *(cm²)*	*Adsorption Parameter,* α	*Adsorption Parameter,* β	*Diffusion Coefficient,* D *(cm²/sec)* × 10^7
1	2-DBS	29.4	3.18	5,520	133.8	5.6
2	2-DBS	30.0	1.45	8,120	136.5	5.6
3	3-DBS	30.0	1.45	8,120	138.9	4.6
4	6-DBS	30.0	1.45	8,120	136.5	3.8
5	Benzenesulfonate	25.0	1.45	3,898	7.0	64
6	2-hexyl-	25.0	1.45	7,470	64.0	16
7	2-octyl-	25.0	1.45	8,526	72.5	12
8	2-decyl-	25.0	1.45	9,013	90.0	8.3
9	2-dodecyl-	25.0	1.45	9,744	113.8	5.8
10	2-tetradecyl-	25.0	1.45	10,718	169.0	3.2

Source: Reproduced from Weber and Rumer [3] courtesy of the American Water Resources Association.

gives time for the gradients to level off within the pores of the adsorbent. When intraparticle diffusion is the rate-limiting step, the rate of removal immediately after reimmersion is greater than the rate prior to interruption. The interruption period has no influence on the rate of adsorption after reimmersion when external transport is rate limiting. Thus, the interruption test is a simple test for de-

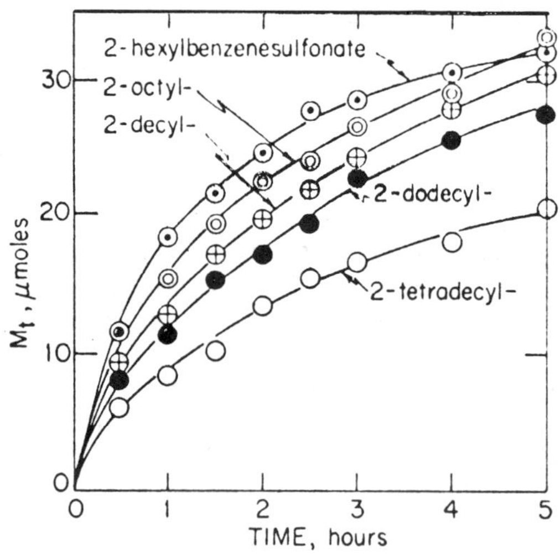

Figure 3-21 Uptake of sulfonated alkylbenzenes as a function of the square root of time. Reproduced from Weber and Rumer [3], courtesy of the American Water Resources Association.

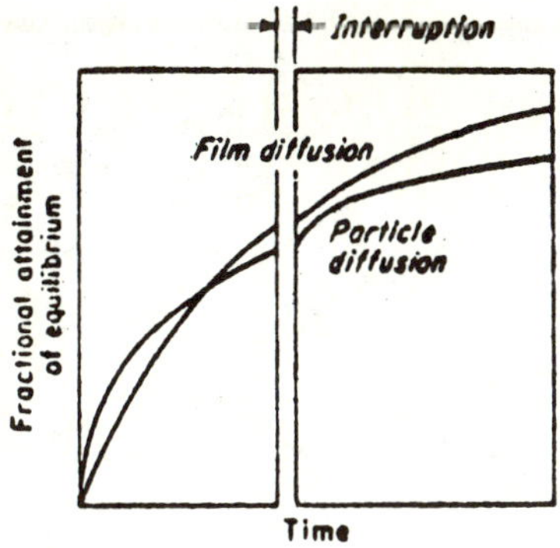

Figure 3–22 Interruption test with film diffusion control and particle diffusion control (schematic). With particle diffusion control the interruption causes a change in the momentary exchange rate. With film diffusion control there is no such change. Reproduced from Helfferich [16], courtesy of McGraw-Hill Book Co., Inc.

termining if large concentration gradients exist within the pores of the adsorbent (see Figure 3–22).

Two interruption tests were conducted with 2,4-DCP to define the nature of the rate-limiting step in the adsorption process of phenols [6]. The results of these interrupted tests are shown in Figure 3–23 in which the concentration of 2,4-DCP remaining in solution is plotted against accumulated immersion time. The upper section of the figure presents the results at an initial concentration of 2,4-DCP of 330 μmol/l (54 mg/l), while the lower section is for an initial concentration of 1,900 μmol/l (310 mg/l). Also shown in both the upper and lower portions of Figure 3–23 are results from continuous experiments conducted under the same conditions as the interruption test, except that these experiments did not contain an interruption period. The variation in cumulative removal of 2, 4-DCP between the interrupted and continuous experiments is due solely to the period of interruption. In all four experiments, the temperature was constant at 20°C.

The rate of adsorption of 2,4-DCP at an initial concentration of 330 μmol/l appears to be limited by external transport to the carbon surface. This observation is based on the lack of influence of the interruption period on the concentration profile immediately after reimmersion and the near coincidence of data points for the continuous and interrupted tests. From these two observations, it is evident that significant redistribution of the adsorbate did not occur within the pores of the adsorbent during the interrupted period. Subsequent results of experiments describing the effect of carbon particle size and initial adsorbate concentration on the rate of adsorption support this observation [10].

Results of the interruption tests conducted at an initial concentration of 2,4-DCP of 1,900 μmol/l are considerably different than those for the lower concentration. A marked increase in the cumulative removal of 2,4-DCP was

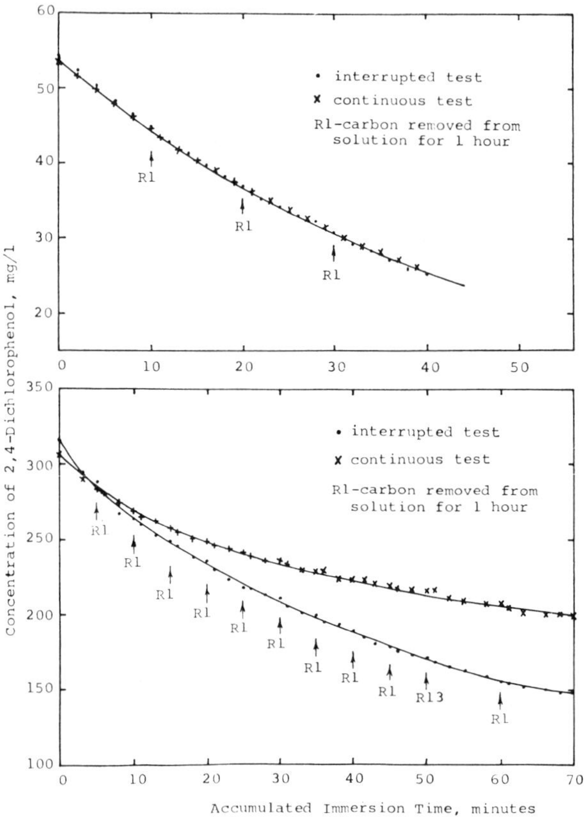

Figure 3-23 Interruption test for 2,4-dichlorophenol (pH = 6.0, T = 20°C, 500 mg/l of 1.00 mm size adsorbent, and 0.05 M phosphate buffer). Reproduced from Zogorski et al. [6], courtesy of Academic Press, Inc.

found in the interruption test conducted at the higher concentration (lower graph in Figure 3-23). After 60 minutes of accumulated immersion time, about 17% higher removal was observed in the interruption test in comparison to the continuous experiment. The same removal that was reached in 60 minutes in the interrupted test required 180 minutes of contact time in the continuous test. Thus, the interruption period allowed for the redistribution of the adsorbate within the pores of the adsorbent at the higher initial concentration. This redistribution resulted in an overall higher rate of adsorption in the interrupted test. These results indicate that external transport to the carbon surface is not the rate-limiting step in the adsorption process at the higher initial concentration of adsorbate but that the adsorption process is either a situation where either external export > internal or external transport ≈ internal transport. Further distribution between these two possibilities is difficult. An intraparticle-diffusion-limited process would exhibit a marked increase in the rate of removal immediately following reimmersion of the adsorbent. Close observation of the data presented in the lower portion of Figure 3–23 does indicate a slight increase in the rate of removal immediately after reimmersion. Whether or not this change is significant enough to state that the rate of intraparticle transport is considerably slower than external transport cannot be conclusively determined from the data shown in Figure 3–23. However, it is evident from the cumulative uptake data that intraparticle diffusion is of equal importance and possibly of greater importance than external diffusion in determining the overall rate of transfer of the adsorbate to the adsorption site.

It is interesting to return at this point to the detergent studies [8]. Interruption studies were conducted also in an effort to support the argument for intraparticle diffusion. These results are seen in Figure 3–24 and 3–25 where:

> [W]hen agitation was resumed, the rate of removal of solute was increased just barely, if at all, over the rate observed just prior to cessation of stirring. Even a slight increase, however, would indicate that intraparticle diffusion of the adsorbate continued during the period that the carbon remained at the bottom of the reaction vessel, while replenishment of the monolayer lagged behind. [8, p.35]

It is extremely difficult to observe the slight increase of removal after the interruption in Figure 3–24; it is more obvious in Figure 3–25.

There appears to be conflicting evidence that the rate-limiting step in the overall kinetic process is controlled by external or internal diffusion [6,8]. It may

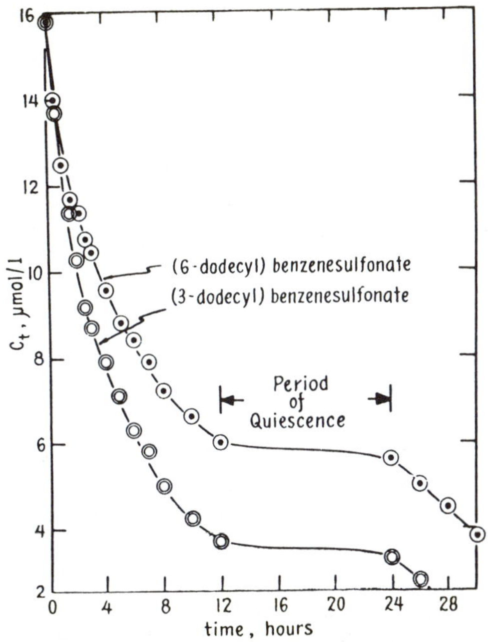

Figure 3–24 Effect of interrupted stirring on rate of adsorption. Reproduced from Weber and Morris [8], courtesy of the American Civil Engineering Society.

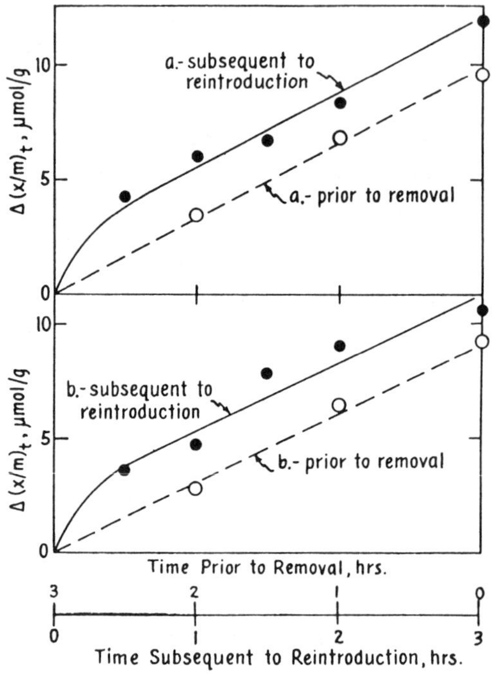

Figure 3–25 Apparent restoration of adsorptive characteristics. Reproduced from Weber and Morris [8], courtesy of the American Civil Engineering Society.

be that both arguments are valid because there are two molecules of different size and configuration. The alkylbenzenesulfonates range in size from 100 to 280 $Å^2$, whereas 2,4-DCP is in the 40 to 50 $Å^2$ range [20]. The smaller-sized phenol molecules should diffuse faster within porous carbon than the larger detergent molecules. Also, the initial concentration of the adsorbate plays a role. It appears that at high concentrations—hence, greater number of molecules—intraparticle diffusion predominates.

In a later study, an electrical analog provided a convenient means for visualizing the interaction of various mechanisms involved in moving adsorbates from the liquid into the solid phase [21]. The resistances to movement were categorized as diffusion of solute through solution, diffusion across the solid-liquid interface, diffusion through the internal pore structure, adsorption on internal surfaces, and migration of adsorbed molecules along the solid phase—that is, surface diffusion. The resistances act in parallel and in series. Consequently, the overall rate of adsorption may be controlled by a combination of both.

Mixed Systems

Rarely would there be a situation in which only a single compound would be adsorbed in a treatment process. Raw water supplies contain a variety of organic compounds that would be adsorbed at different rates and at different sites on carbon. This is usually considered to be a competitive situation where adsorption may be enhanced or reduced or may be adsorbed independently. Many of the studies cited earlier included mixed or competitive systems.

Early studies were concerned with the rates of contrasting sized molecules, phenol and dodecylbenzenesulfonate (DBS), on Columbia LC carbon [20,22]. Figure 3–26 shows the rates of adsorption of these two compounds at 30°C. We see that DBS had a greater effect on the rate of adsorption of phenol than the latter had on the rate of uptake of the surfactant. Here, again, intraparticle trans-

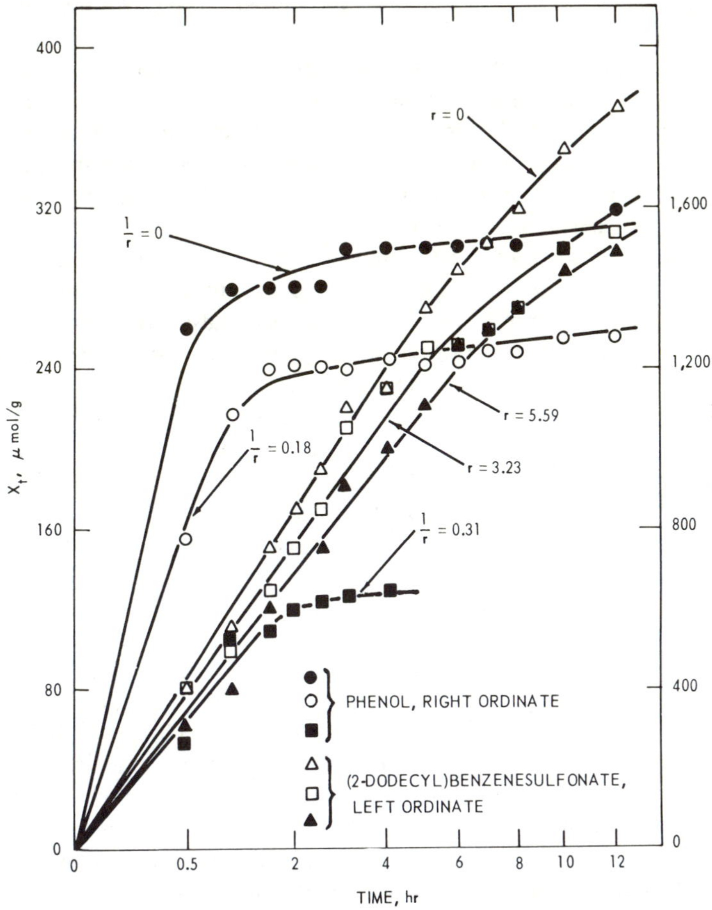

Figure 3–26 Competitive adsorption between phenol and (2-dodecyl)benzenesulfonate [20] [30°C, 35.0 mg/l, 0.126 mm Columbia Carbon,

$$r = \frac{(C_o)\ \text{phenol}}{(C_o)\ \text{ABS}}].$$

port of the adsorbate was invoked to explain the DBS-phenol interference; that is, it is plausible that the larger DBS molecule would block access to a pore for a phenol molecule.

Other competitive systems were bisolute mixtures of *o*-nitrochlorobenzene (NCB)-DBS and dodecylsulfate (DS_t)-DBS [22]. The effects of competitive rates of adsorption are seen in Figures 3–27a and b, respectively. NCB had little effect on the removal rate of DBS at $r_{n,a}$ values of less than 3 (Figure 3–27a). Conversely, there was a considerable depression on the rate of uptake of NCB by DBS. Again, molecular size of the respective molecules is an explanation for Figure 3–27a. To assess the effect that molecules of larger size might have on the removal rate of DBS, mixtures of DS_t and DBS were studied. The results are seen in Figure 3–27b where DS_t had a much greater effect on the rate of adsorption of DBS than either phenol or NCB.

There was, however, an interesting observation in the competitive rate studies; that is, the combined rate of removal for two compounds was greater than their individual rates [22]. Accordingly, the most likely explanation for this was that the overall diffusion pressure was increased by the higher concentration of molecules in the mixture. Whether or not a greater diffusion pressure was responsible was not determined, but there are certainly a greater number of molecules in a mixture that should increase the overall rate of removal.

Several experiments were conducted to delineate the interaction of mixed systems of phenolic compounds on their kinetics of adsorption [6]. Tables 3–7a and b show removal rates of the bisolutes, phenol and 4-methoxyphenol. The uptake of each compound was affected adversely by the presence of the other. 4-methoxyphenol exhibited more of an influence on the removal rate of phenol than vice versa. For example, increasing the concentration of phenol from 0 to 300 μmol/l decreased the removal rate of 4-methoxyphenol by only 11%, whereas the same increase of content for 4-methoxyphenol decreased the removal rate of phenol by 29%. Undoubtedly, the larger molecule of 4-methoxyphenol was responsible.

A mixture of six light halogenated hydrocarbons—chloroform, dichlorobromomethane, carbon tetrachloride, tetrachloroethylene, dibromochloromethane, and bromoform—was observed for their rate of removal on Nuchar WV-G (12 × 40 mesh) (24°C) [23]. The expected variation in uptake for individual compounds is seen in Figure 3–28. From the slopes of the lines in this figure, the relative rate constants were taken (Table 3–8). These compounds were adsorbed at relatively rapid rates: 53.5% of $CHCl_3$ was adsorbed within the first hour, whereas the rate was 83% for $CHBr_3$, 81% for $Cl_2C{=}CCl_2$, 80% for $CHClBr_2$, 69% for $CHCl_2Br$, and 70% for CCl_4. However, these rates are slow enough for rapid breakthrough from carbon columns (see Chapter 8).

Another kinetic study was conducted for the adsorption of carbon tetrachloride, tetrachloroethylene (PCE), and trichloroethylene (TCE) on Nuchar WV-G (12 × 40 mesh) (24°C) [24]. Removal rates are seen in Table 3–9 for single-compound solutions and TCE and PCE in a mixture. Individually, TCE and PCE were removed rapidly and much faster than CCl_4. However, their rates of uptake

(a)

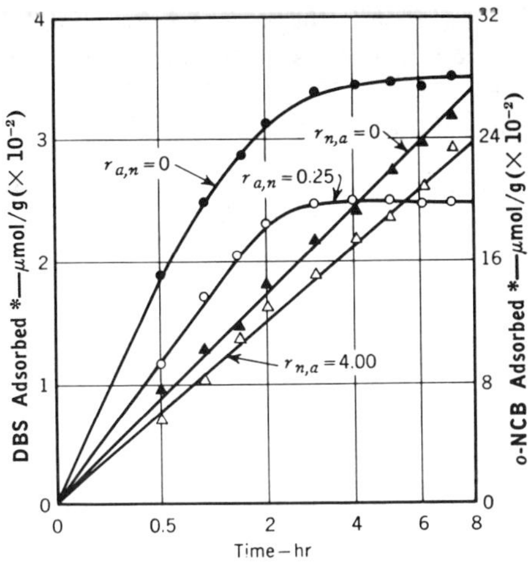

(b)

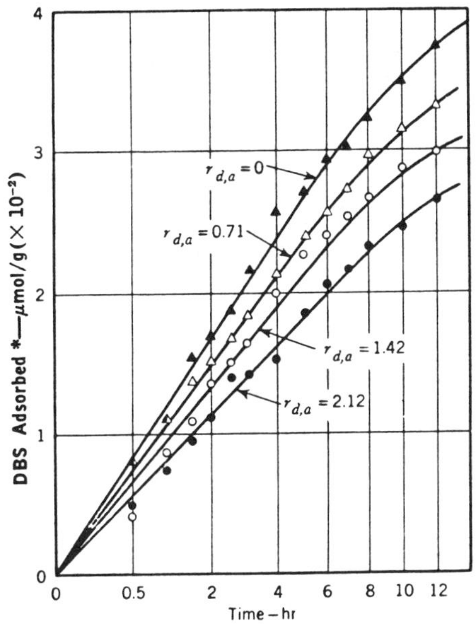

Figure 3–27 (a) Rates of Adsorption of *o*-NCB and DBS from mixed solutions on GAC. The data are for adsorption at 30°C on 35.0 mg/l of 0.126 mm carbon. Ratios of initial molar concentration of DBS to initial molar concentration of *o*-NCB are denoted by $r_{a,n}$; $r_{n,a}$ is the inverse of $r_{a,n}$. Reproduced from Weber and Morris [22], courtesy of the American Water Works Association. (b) Effect of DS_t on rate of adsorption of DBS. The amount of DBS adsorbed per gram of 0.126 mm carbon from mixed solutions with different ratios of initial molar concentration of DS_t to initial molar concentration of DBS, $r_{d,a}$, is plotted as a function of time from introduction of 35.0 mg/l activated carbon to each system. The experiments were carried out at 30°C. Reproduced from Weber and Morris [22], courtesy of the American Water Works Association.

Table 3–7a Variation in the Rate of Adsorption of 4-Methoxyphenol with Initial Concentration of Phenol

Initial Concentration of Adsorbates		*Removal Rate for*
4-Methoxyphenol ($\mu mol/l$)	*Phenol ($\mu mol/l$)*	*4-Methoxyphenol ($\mu mol/g\ hr^{1/2}$)*
302.	0	567.
306.	0	551.
300.	144.	532.
303.	293.	495.
288.	433.	458.
288.	572.	453.

Source: Reproduced from Zogorski et al. [6] courtesy of Academic Press, Inc.
Note: 20°C, pH = 6.0, 0.05 *M* buffer concentration, and 500 mg/l of 1.00 mm Columbia LCK Carbon.

Table 3–7b Variation in the Rate of Adsorption of Phenol with Initial Concentration of 4-Methoxyphenol

Initial Concentration of Adsorbates		*Removal Rate*
Phenol ($\mu mol/l$)	*4-Methoxyphenol ($\mu mol/l$)*	*for Phenol ($\mu mol/g\ hr^{1/2}$)*
302.	0	484.
304.	0	470.
299.	0	464.
298.	153.	400.
293.	303.	335.
288.	442.	302.

Source: Reproduced from Zogorski et al. [6] courtesy of Academic Press, Inc.
Note: 20°C, pH = 6.0, 0.05 *M* buffer concentration, and 500 mg/l of 1.00 mm Columbia LCK Carbon.

were reduced by 54% and 61%, respectively. The total rate of removal of TCE and PCE (7.0 μmol/g min$^{1/2}$) was not greater than their individual rates. This is at variance with other studies cited.

Similar results had been observed earlier with an alkylbenzenesulfonate, quinine, and 2-sec-butyl-4,6-dinitrophenol on 0.273 mm Columbia LC carbon [25]. Removal rates for these compounds in single and mixed solutions are given in Table 3–10. There was a considerable suppression of individual rates of adsorption in the mixture.

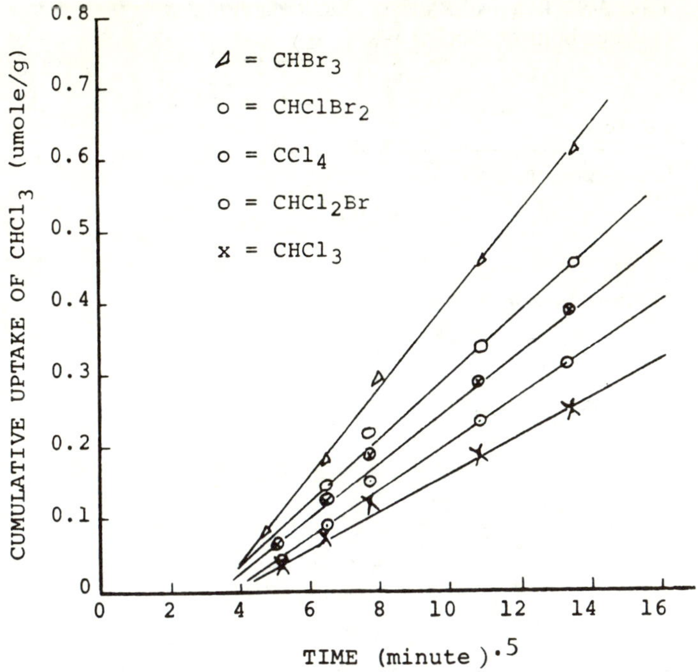

Figure 3–28 Cumulative removal versus $t^{1/2}$ for light halogenated hydrocarbons [23].

Table 3–8 Rate of Removal of Light Halogenated Hydrocarbons on GAC

Compound	*Rate of Removal (μmole/g min$^{1/2}$)*
$CHCl_3$	0.0250
$CHCl_2Br$	0.0325
CCl_4	0.0425
$Cl_2C{=}CCl_2$	0.0425
$CHClBr_2$	0.0450
$CHBr_3$	0.0625

Source: Youssefi [23].

Table 3-9 Removal Rates for Light Halogenated Hydrocarbons on Nuchar WV-G

Compound	*Rates of Removal ($\mu mole/g\ min^{1/2}$)*
Single-compound solution	
CCl_4	0.14
TCE	7.6
PCE	9
Mixed-compound solution	
TCE	3.5
PCE	3.5

Source: Benedict [24].

Table 3-10 Relative Rates of Adsorption from Single- and Multiple-Solute Systems

	Relative Rate Constant, k $[(\mu mol/g)^2/hr] \times 10^{-3}$	
Solute	*Single-Solute Solution*	*Multiple-Solute Solution*[a]
Sulfonated alkylbenzene	3.60	1.85
Quinine	7.23	1.60
2-*sec*-butyl-4,6-dinitrophenol	57.60	0.26

Source: Reproduced from Weber [25] courtesy of The Water Pollution Control Federation.
Note: 0.273-mm Columbia carbon, agitated nonflow system, 28°C.
[a]Mixture: 43.6 μmol/l sulfonated alkylbenzene
26.9 μmol/l quinine
31.8 μmol/l 2-*sec*-butyl-4,6-dinitrophenol

FIXED BEDS

Early Mass Transfer Models

The kinetics of adsorption in a fixed bed (downflow) of carbon becomes slightly more complex than in batch reactors. As seen in Chapters 6 and 8, contact times in packed beds of carbon are often very short, within an order of 7.5 minutes. Consequently, mass transfer of the adsorbate from the bulk liquid phase to the carbon's external and internal surfaces must be accomplished quickly. As such, the overall rate of adsorption may be limited by film diffusion, fluid-phase pore diffusion, internal surface diffusion, or reaction with the solid surfaces of the adsorbent [21]. The effective rate of adsorption will be controlled mostly by the step that exerts the greatest resistance to transfer—that is, the slowest step. Many researchers have reported several mathematical models, based on analogs from wherever, to calculate diffusion coefficients, mass transfer coefficients, and so forth. Many were developed to predict column breakthrough curves (see Chapter 4). Here, only the fundamental kinetics in a downflow, fixed bed of carbon is considered.

In an effort to describe the kinetics of adsorption, models commonly used to describe chromatographic separations in an adsorption column have been utilized [21]. The general mass balance for a given solution is written:

Rate in − Rate out + Rate of reaction = Rate of accumulation,

whereupon:

$$F/V_1 - F/V_1 + \Delta V_1 + 0 = \frac{d}{dt}\,[\epsilon\, C_F + C_S\,(1 - \epsilon)]\Delta V_1. \qquad (3.9)$$

Figure 3–29 shows the column segment over which the mass balance is made. In Equation 3.9, F = the total rate of movement of the adsorbate in the mobile phase, V_1 = the volume position in the bed, ΔV_1 = the incremental volume over which the mass balance is made, ϵ = the fractional void space, C_F = the fluid phase concentration of the adsorbate, C_S = the solid phase concentration, and t = time. Movement of the adsorbate in the mobile phase occurs by two mechanisms, convection and diffusion, as seen in:

$$F = V_o C_F - D_A A\,\frac{\partial C_F}{\partial z}, \qquad (3.10)$$

where V_o = volumetric flow rate, A = cross-sectional area of the column, z = distance from the column inlet, and D_A = apparent diffusivity of adsorbate in the system. This D_A is apparently a combination of molecular diffusivity and a

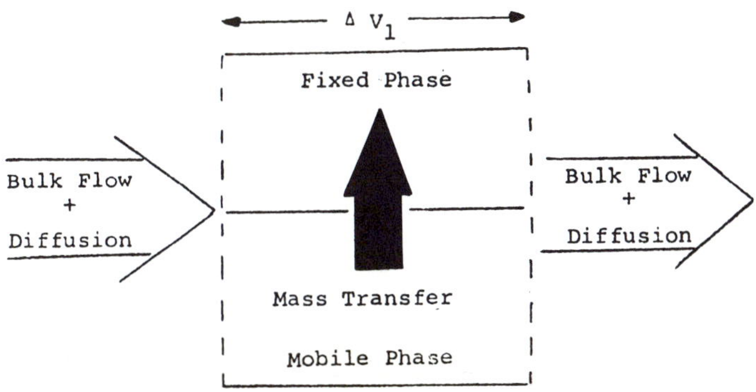

Figure 3–29 Activity within column segment ΔV_1 of fixed bed adsorber [21].

dispersion coefficient that considers back mixing and nonvelocity distribution [21].

Substituting Equation 3.10 into 3.9 and dividing through by ΔV_1, with the limit that ΔV_1 approaches zero, yields [22]:

$$-V_o \frac{\partial C_F}{\partial V_1} + D_A A \frac{\partial^2 C_F}{\partial V_1^2} = \frac{\partial}{\partial t} [\epsilon C_F + (1 - \epsilon) C_S]. \tag{3.11}$$

Now it is necessary to express the dependent variable, C_S, in terms of C_F.

First, an equilibrium adsorption model is considered. Neglecting the effects of longitudinal dispersion and the resistance to transport across the interface between the solid and liquid phases, the right-hand side of Equation 3.11 is written [21]:

$$\frac{\partial}{\partial t} [\epsilon\, C_F + (1 - \epsilon)\, C_S] = \frac{\epsilon \partial C_F}{\partial t} + (1 - \epsilon) \frac{\partial C_S}{\partial C_F} \cdot \frac{\partial C_F}{\partial t}. \tag{3.12}$$

Thus, the term $\partial C_S / \partial C_F$ is the slope of the equilibrium curve $f'(C_F)$. To solve for C_F as a function of time, t, and distance, z, the function is assumed to be [21]:

$$C_F = g(z - a' t), \tag{3.13}$$

where a' is a constant (g was not defined in the original publication, but it disappears eventually [22]). a' is defined by taking the partial derivatives of C_F with respect to t and z, neglecting longitudinal dispersion and permitting the independent variable to be z ($V_1 = A_z$), Equation 3.11 becomes [21]:

$$-V_o \frac{\partial C_F}{\partial z} = A[\epsilon + (1 - \epsilon) f'(C_F)] \frac{\partial C_F}{\partial t}. \tag{3.14}$$

Substitution of two equations for $\partial C_F / \partial z$ and $\partial C_F / \partial t$ into Equation 3.13 gives [21]:

$$a' = \frac{V_o}{A[\epsilon + (1 - \epsilon) f'(C_F)]}. \tag{3.15}$$

Taking the derivative of z with respect to t then gives the velocity of a particular concentration C_F moving down the column [21]:

$$\frac{dz}{dt} \bigg|\, C_F = \frac{V_o}{A[\epsilon + (1 - \epsilon) f'(C_F)]}. \tag{3.16}$$

At this point it is necessary to consider the Langmuir adsorption model [21]:

$$C_S = \frac{Q \rho b C_F}{1 + b C_F}, \tag{3.17}$$

where Q = the ultimate monolayer coverage of adsorbate on the solid phase, and b = the energy-related constant (ρ is presumed to be the density of the adsorbent). Taking the derivative of C_S with respect to C_F yields [21]:

$$\frac{dC_S}{dC_F} = \frac{Q \rho b}{(1 + b C_F)^2}. \tag{3.18}$$

With regard to the time-distance relation in Equation 3.16, velocity increases as well as the mobile phase concentration.

This increase causes a pile-up into a front as the bed's capacity becomes exhausted. Thus, the initial breakthrough of adsorbate exiting from the column will be the same as the inlet content. A mass balance of this front is [21]:

$$V_o C_{F_o} = [\epsilon C_{F_o} + (1 - \epsilon) C_{S_o}] \frac{dz}{dt} A, \tag{3.19}$$

whereupon:

$$\frac{dz}{dt}\bigg|_{\text{Front}} = \frac{V_o}{\left[\epsilon + (1-\epsilon)\dfrac{C_{S_o}}{C_{F_o}}\right] A}, \tag{3.20}$$

where C_{F_o} = inlet feed concentration and C_{S_o} = the solid phase concentration in equilibrium with C_{F_o}. Equation 3.20 is used to calculate the breakthrough time for a given flow rate, V_o, and bed volume provided the equilibrium is known. It should be noted that this model assumed Langmuirian adsorption whereupon the Freundlich relation is used in other models described later.

The effects of mass transfer resistances (reciprocal of mass transfer coefficients) are included in adsorption velocity models [26]. With regard to Equation 3.11, it is desirable to express the change in C_S in terms of C_F. Local mass transfer coefficients are defined for the steady-state condition [21]:

$$\frac{d\partial C_S}{d\partial t} = k_f S_o (C_F - C_{F_i}) \tag{3.21}$$

$$= k_s S_o (C_{S_i} - C_S) \tag{3.22}$$

$$= K_F S_o (C_F - C_F^*), \tag{3.23}$$

and

$$K_F = \left[\frac{1}{k_f} + \frac{1}{mk_s} \right]^{-1}, \tag{3.24}$$

where k_f = liquid-phase mass transfer coefficient, k_s = stationary-phase mass transfer coefficient, K_F = overall mass transfer coefficient, C_{F_i} = interfacial concentration in the moving phase, C_{S_i} = interfacial concentration in the stationary phase, C_F^* = imaginary moving-phase concentration in equilibrium with C_S, S_o = surface area to volume ratio for the particle, and m = average slope of dC_S/dC_F of the equilibrium distribution. For a linear isotherm, m becomes the actual slope dC_S/dC_F, but for the general situation m is an average slope [21]:

$$m = \frac{(C_S - C_{S_i})}{(C_F^* - C_{F_i})}. \tag{3.25}$$

The moving phase or external film resistance, k_f, is evaluated by several empirical correlation techniques [21]. In contrast, k_s must be evaluated by consideration of the movement of adsorbate within the solid phase. This involves calculating an effective intraparticle diffusion coefficient, D (see earlier discussion). Once D is evaluated, k_s for spherical particles is expressed empirically [21]:

$$k_S = 10 \frac{D}{D_p}, \tag{3.26}$$

where D_p = particle diameter [27].

Chemical reaction models are available also for inclusion of mass transfer resistances in Equation 3.11 [27,28]. To apply this type of model to fixed bed adsorbers, it is necessary to relate mass transfer models to a generalized second-order kinetics expression derived from the Langmuir equation [21]:

$$\frac{\partial C_S}{\partial t} = k \left[C_F(Q \rho - C_S) - \frac{C_S}{b} \right], \tag{3.27}$$

where k = forward rate constant, and b = ratio of forward to backward rate constants. The Langmuir equilibrium adsorption model was derived by setting dC_S/dt equal to zero. The term, Q, in Equation 3.27 is eliminated by introducing the constant, C_{S_o}, that represents the solid-phase concentration in equilibrium with the inlet fluid-phase concentration, C_{F_o}. Accordingly, the Langmuir equilibrium relationship gives [21]:

$$\frac{d\omega}{dt} = k \left[\frac{1}{b} + C_{F_o} \right] \left[\phi - \frac{b\, C_{F_o}}{(1 + b\, C_{F_o})} \omega \phi - \frac{\omega}{(1 + b\, C_{F_o})} \right]. \tag{3.28}$$

The dimensionless variables ω and ϕ are given by:

$$\omega = \frac{C_S}{C_{S_o}} \tag{3.29}$$

and

$$\phi = \frac{C_F}{C_{F_o}}. \tag{3.30}$$

In addition, substitution of the so-called equilibrium modulus, λ, given by:

$$\lambda = \frac{1}{1 + b\,C_{F_o}} = 1 - \frac{C_{S_o}}{Q\rho}, \tag{3.31}$$

reduces Equation 3.27 to a dimensionless form:

$$\frac{d\omega}{d\tau} = \phi(1 - \omega) - \lambda\omega(1 - \phi). \tag{3.32}$$

The dimensionless time parameter, τ, is:

$$\tau = k\left(\frac{1}{b} + C_{F_o}\right)t. \tag{3.33}$$

Thus, Equation 3.32 provides the necessary input for dC_S/dt in the mass balance given by Equation 3.11.

To relate Equation 3.32 to a chosen mass transfer controlling step, a generalized rate coefficient, κ, and a dimensionless coefficient, D_G, are substituted for $k(1/b + C_{F_o})$ in Equation 3.33 [21]:

$$\tau = \frac{\kappa}{D_G}t. \tag{3.34}$$

The coefficient, D_G, represents the ratio of the limiting distribution of solute between the solid and liquid phases [21]:

$$D_G = \frac{C_{S_o}}{C_{F_o}} \cdot \frac{(1 - \epsilon)}{\epsilon}. \tag{3.35}$$

For any mechanism postulated to describe the controlling mass transfer resistance, it is necessary to derive a second-order kinetic expression. This was accomplished by involving the equilibrium concentration of adsorbate in the driving force equation and using the Langmuir equilibrium model.

For example, a film diffusion model was written:

$$\frac{\partial C_S}{\partial t} = k_f S_o (C_F - C_F^*), \tag{3.36}$$

where C_F^* = the fluid-phase concentration on the inside of the liquid film in equilibrium with the average solid-phase concentration. Thus, C_F^* is given by the Langmuir model, Equation 3.17. Conducting the same substitution procedure as described earlier, for Equation 3.32, a general second-order kinetic expression, is obtained [21]:

$$\tau = \frac{k_f S_o C_{F_o}}{C_{S_o}\left[1 + (\lambda - 1)\dfrac{C_S}{C_{S_o}}\right]}\, t = \frac{\kappa}{D_G}\, t. \tag{3.37}$$

Consequently, the general rate coefficient, κ, is expressed in terms of k_f.

Mass Transfer Parameters

The ultimate objective in mathematical exercises like those described in the preceding section is to design an effective carbon bed. Confirmation of these models usually takes the form of a predicted breakthrough curve superimposed on an experimental breakthrough curve (see Chapter 4). The intent here, however, is to describe fixed bed adsorption in kinetic terms and the many variables in the rate-limiting step.

It is desirable to have direct experimental evidence for mass transfer and diffusion coefficients whether or not the kinetics are limited by film or pore diffusion or a combination thereof. Several mathematical models were tried for the latter case. A simplified film and pore diffusion model inadequately described the mass transfer process in adsorption of phenol and PNP on Columbia LC carbon (30/40 mesh) [21]. Evidence was obtained wherein film diffusion was important. Ignoring pore diffusion, a simple film transport model was employed to describe the initial rate data:

$$V\frac{dC_F}{dt} = -k_f A_o (C_F - C_{F_i}), \tag{3.38}$$

where V = volume served per particle ($V = \rho/W\ 4/3\ \Pi\ a^3$; W = carbon dosage, g/l; and a = diffusional path), C_F = fluid-phase concentration, C_{F_i} = interfacial fluid concentration, and A_o = particle surface area. C_{F_i} is assumed to be the local equilibrium concentration inside the film. It may be calculated from the adsorption equilibrium relationship provided that the quantity adsorbed at any time is uniformly distributed inside the particle. Dividing each side in Equation 3.38 by C_{F_o} and integrating yields [21]:

$$\int_1^{\phi} \frac{d\phi}{\phi - \frac{C_{F_i}}{C_{F_o}}} = -\frac{k_f A_o}{V} t \Big|_o^t , \tag{3.39}$$

where ϕ = dimensionless fluid-phase concentration (C_F/C_{F_o}). Neglecting C_{F_i} for initial adsorption, Equation 3.39 becomes [21]:

$$ln\ \phi = -\frac{k_f A_o}{V} t. \tag{3.40}$$

This assumption is valid when the distribution ratio of solid-to-liquid-phase concentrations increases inasmuch as C_{F_i} correspondingly decreases for any given amount.

Kinetic data were obtained (not given here) for the neutral species of phenol, PNP, and dinitrophenol (DNP) for identical C_{F_o} and ϕ_e values whereupon ln ϕ was plotted versus time in Figure 3.30. Mass transfer coefficients, k_f, of .82, .88, and .76 $\times$ 10^{-2} cm/sec were calculated for phenol, PNP, and DNP, respectively. These numbers were calculated by division of the initial slope by external carbon surface area per cubic centimeter of solution. These k_f values are offered here to illustrate an order of magnitude for mass transfer in a batch reactor. The authors of Figure 3–30 discuss in some detail the inadequacy of the simple film diffusion model to describe an entire approach to equilibrium because the increase in C_{F_i} is ignored [21]. It was argued that a series mechanism of film and pore diffusion would be more plausible. Interruption tests (see earlier section) were conducted with PNP to distinguish between film and intraparticle diffusion. No difference in rate of adsorption was observed when the interruption was as long as 11 hours. "Although the simple film model could only be used for about the first 150 minutes of the experiment, it appeared that external film resistance was still very important for at least an additional 120 minutes" [21, p.69].

Additional experiments were conducted for k_f values for phenol [21]. Results are seen in Table 3–11 where the order of magnitude for k_f values for phenol is the same as for DNP. However, the t^* (time over which the simple film model could only be applied) was 60 minutes for phenol and 120 minutes for DNP. This was explained by differences in their isotherms from which adsorption was more favorable for DNP.

Late Mass Transfer Models

Another mathematical model was developed to design fixed bed carbon adsorbers (see Chapter 4) [29–32]. Again, several parameters must be estimated for model

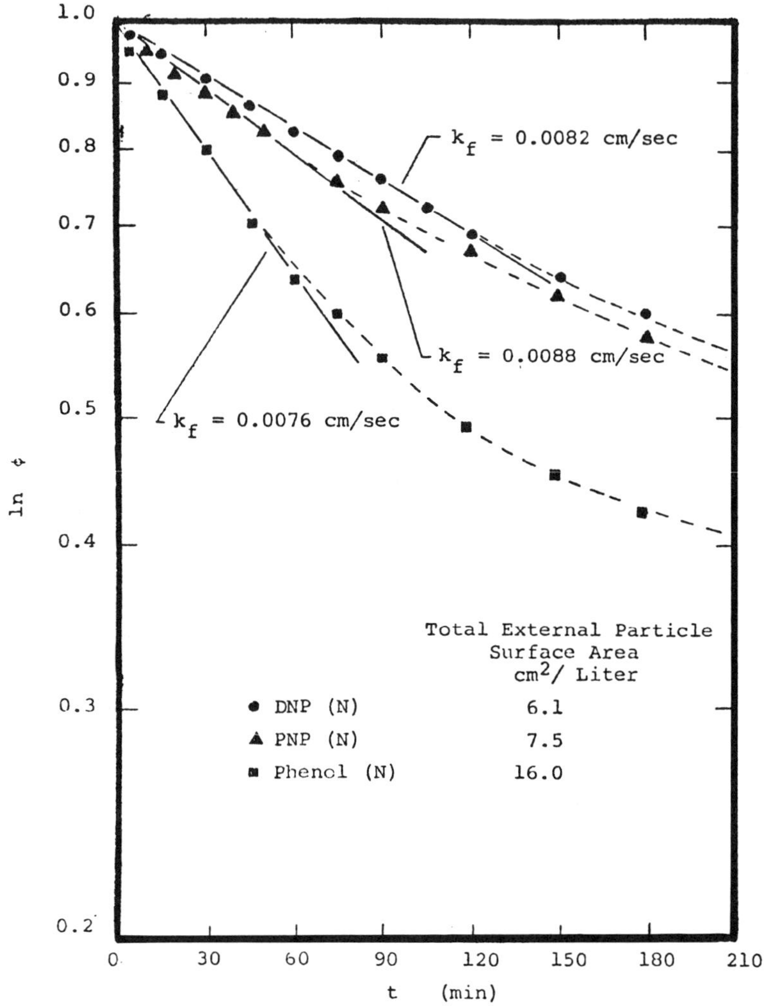

Figure 3–30 Simple film diffusion model [21]. Data for DNP (N), PNP (N) and phenol (N) (C_{F_o} = 1 × $10^{-4}M$, ϕ_e = 0.2, a = 0.025 cm, 1,250 rpm).

verification: film transfer coefficients, surface diffusion coefficients, and adsorption isotherm parameters. The model's framework incorporated mathematic descriptions of fluid diffusion resistance to mass transfer (i.e., film transfer), local adsorption equilibrium adjacent to the carbon surface, and surface diffusion as the predominant intraparticle mass transfer process. The model is given in Chapter 4.

The usual isotherm experiments were conducted with *p*-bromophenol (PBP), phenol, sodium *p*-toluenesulfonate (PTS), and technical-grade sodium DBS on

Table 3-11 Application of Simple Film Diffusion Model to Phenol and DNP Rates of Adsorption

Compound	C_{F_o} ($\times 10^4$ *M*)	C_{F_e} ($\times 10^4$ *M*)	t^* (*min*)	k_f (*cm/sec* $\times 10^2$)
Phenol	1.0	0.2	60	0.76
	1.0	0.5	60	0.77
DNP	1.0	0.2	120	0.82
	1.0	0.5	120	0.81

Source: DiGiano and Weber, Jr. [21].

F-400 activated carbon (16 × 20 mesh). Sealed bottles containing the adsorbates as single solutes and carbon were agitated in a tumbler for several weeks [30]. A three-parameter isotherm was employed that can be reduced to linear Freundlich or Langmuir isotherms [33]:

$$q = \frac{Ac}{1 + Bc^{\beta}}, \tag{3.41}$$

where the parameters, *A*, *B*, and β, "were determined by the best statistical representation of the experimental data" [30, p.186]. The other terms in Equation 3.41 are q = solid-phase concentration and c = fluid concentration. Kinetic studies were conducted in 3 batch reactors where mixing conditions were such that there was no cavitation or carbon particle attrition (?) and the effect of film transfer was kept at a minimum. The rate study duration was about 8 hours.

It shall suffice, at this point, to tabulate the surface diffusion coefficients, D_s, and the mass transfer coefficients, k_f, for the four adsorbates cited [30,31]. These two coefficients were determined in batch reactors, which is independent of a column study. Consequently, any column model is considered to be predictive [30]. D_s and k_f values were calculated statistically for the batch reactor data by "the principle axis method described by Brent" [34]. Table 3-12 shows the mass transfer parameters that were applied to single-component fixed bed predictions [31]. The Sherwood number, *Sh*, compares the liquid-phase to solid-phase mass transfer resistances. An equation is [30]:

$$Sh_b = \frac{k_f a_p}{D_s D_{g,b}}, \tag{3.42}$$

where a_p = particle radius and $D_{g,b}$ = solute distribution parameter. In Table 3-12, *Sh* values "ranged from a level indicative of very important surface diffusion resistance (17.0) to a level indicative of very important liquid diffusion resistance" (0.86) [31, p.435]. The D_s values are constant for a particular compound, whereas k_f values vary with the Sherwood number conditions.

Table 3–12 Mass Transfer Parameters Used for Single-Component Fixed Bed Predictions and Standard Deviations Determined from Model Data Comparisons

Fixed Bed Run Number	*Solute*	D_s *(cm²/sec)*	k_f *(cm/sec)*	*Sh*	*Standard Deviation (percent)*
1	Phenol	3.53×10^{-3}	3.71×10^{-3}	4.30	5.2
2	Phenol	3.53×10^{-8}	3.71×10^{-3}	4.31	3.8
3	Phenol	3.53×10^{-8}	4.61×10^{-3}	5.34	6.8
4	Phenol	3.53×10^{-8}	4.61×10^{-3}	5.31	8.0
5	*p*-bromophenol	8.56×10^{-8}	2.58×10^{-3}	0.857	5.7
6	*p*-bromophenol	8.56×10^{-8}	3.53×10^{-3}	1.17	4.5
7	*p*-bromophenol	8.56×10^{-8}	3.90×10^{-3}	1.30	6.2
8	*p*-bromophenol	8.56×10^{-8}	4.17×10^{-3}	1.33	6.0
9	*p*-toluenesulfonate	1.91×10^{-8}	2.20×10^{-3}	4.25	4.5
10	*p*-toluenesulfonate	1.91×10^{-8}	3.25×10^{-3}	6.28	6.2
11	*p*-toluenesulfonate	1.91×10^{-8}	3.61×10^{-3}	6.77	4.9
12	*p*-toluenesulfonate	1.91×10^{-8}	4.04×10^{-3}	7.67	5.6
13	DBS	4.68×10^{-9}	1.79×10^{-3}	10.7	8.2
14	DBS	4.68×10^{-9}	2.35×10^{-3}	14.1	4.5
15	DBS	4.68×10^{-9}	2.53×10^{-3}	15.3	6.2
16	DBS	4.68×10^{-9}	2.89×10^{-3}	17	2.8

Source: Reproduced from Crittenden and Weber [31] courtesy of the American Civil Engineering Society.

Mass transfer parameters were determined subsequently for phenol on two carbons and four macroreticular resins [35]. D_s and k_{fb} values are seen in Table 3–13 for a finite bath, completely mixed batch (CMB). The CMB finite-bath film transfer coefficient, k_{fb}, for the various adsorbents differs markedly, ranging from 7.04×10^{-3} cm/sec for BACM to a low of 1.01×10^{-3} cm/sec for Ambersorb XE-340. These different k_{fb} values were ascribed to different exterior surface characteristics of the adsorbents. External surface roughness and shape of the particles apparently affect the character of the hydrodynamic boundary layer immediately adjacent to the surface. In the context of film theory, an underlying definition of k_f is [35]:

$$k_f = \frac{D_l}{\delta}, \tag{3.43}$$

where D_l = molecular diffusivity in aqueous solution, and δ = conceptual thickness of the hydrodynamic boundary layer. The decrease in k_{fb} values for the ones measured for the activated carbons to lower values for the resins may be qualitatively related to the topographical features of the adsorbents. Activated carbons have η values $\sim$ 1 (see Table 3–13), which is due to their rough surfaces. The Ambersorb XE-340, the carbonized polymer, has a shiny acid-polished ap-

Table 3-13 Finite Bath Rate Coefficients for Phenol

Adsorbent	$D_s \times 10^9$ *(cm²/sec)*	$k_{fb} \times 10^3$ *(cm/sec)*	*Shape and Roughness Factor, η, Dimensionless*
BACM carbon	2.18[a]	7.04[b]	1.00[c]
Filtrasorb 400 carbon[d]	1.24	6.88	0.98
Ambersorb XE-348	0.696	4.88	0.69
Amberlite XAD-4	20.6	4.35	0.62
Ambersorb XE-347	0.572	1.32	0.19
Ambersorb XE-340	0.484	1.01	0.14

Source: Reproduced from Van Vliet et al. [35] courtesy of Pergamon Press.
Note: Stirrer speed = 1,200 rpm; adsorbent particle size fraction: 0.30 to 0.42 mm.
[a]D_s = effective intraparticle surface diffusion coefficient.
[b]k_{fb} = film transfer coefficient derived from CMB rate experiments.
[c]η: defined as the ratio of the k_{fb} for a particular adsorbent to k_{fb} for the BACM carbon.
[d]Based on the 100 mg l^{-1} initial concentration isotherm.

pearance that presumably produces the lowest η value. The other three resins apparently have rougher surfaces because their η values range from .19 to .69. The shape and roughness of a particle affect the k_{fb} values in Table 3-13 inasmuch as the local hydrodynamic regimes adjacent to the adsorbent particles must respond to the topography. "Nevertheless, the sequence of values does reflect, on a relative scale, the character of a particle surface as it affects external mass transport dynamics" [35, p.1719].

Evaluation of film transfer coefficients and effective intraparticle diffusivities discussed earlier has involved correlation procedures and separate batch reactor experiments for adsorption rates. Accordingly, these methods introduce and compound significant errors in the subsequent predictive modeling of fixed bed adsorbers for process design. A more accurate and reliable approach was developed for the determination of mass transport parameters [7,36]. A method has been proposed whereby the characteristics of breakthrough curves from specially designed microcolumns yield "more accurate and reliable simultaneous determination of both film transfer coefficients and effective surface diffusivities" [7, p.1542]. This approach eliminates the previously used correlation procedures.

Mathematical models were derived from mass balances for the solid and fluid phases in fixed bed carbon adsorbers (and/or ion exchange columns) [7]. Here again, spherical particles and plug flow are assumed, and effective surface diffusivity describes the rate-controlling component of intraparticle diffusion. The solid-phase mass balance is [7]:

$$\frac{\partial q}{\partial t} = D_s \left(\frac{\partial^2 q}{\partial r^2} + \frac{2\partial q}{r\partial r} \right) . \tag{3.44}$$

The liquid-phase mass balance is:

$$-D_p \frac{\partial^2 C}{\partial z^2} + v_i \frac{\partial C}{\partial z} + \left(\rho \frac{1-\epsilon}{\epsilon}\right) \frac{\partial q}{\partial t} + \frac{\partial C}{\partial t} = 0 \tag{3.45}$$

The Freundlich adsorption isotherm is:

$$q_s = K_F C_s^{\,n}. \tag{3.46}$$

The initial and boundary conditions are as follows:

$$C = C_o + \left(\frac{D_p}{v_i}\right) \frac{\partial C}{\partial z} \quad \text{at } z = 0 \tag{3.47}$$

$$\frac{\partial C}{\partial z} = 0 \quad \text{at } z = L \tag{3.48}$$

$$\frac{\partial q}{\partial r} = 0 \quad \text{at } r = 0 \tag{3.49}$$

$$\frac{\partial q}{\partial r} = \frac{k_f}{D_s \rho}(C - C_s) \quad \text{at } r = R \tag{3.50}$$

$$q = 0 \quad \text{at } t = 0 \tag{3.51}$$

$$C = 0 \quad \text{at } t = 0 \tag{3.52}$$

where: q = solid-phase concentration (mM/g carbon),
D_s = effective surface diffusivity (cm^2/sec),
r = radial distance (cm),
z = longitudinal distance in column (cm),
t = time (sec),
C = fluid-phase concentration (mM/l),
C_s = fluid-phase concentration near the particle surface (mM/l),
q_s = solid-phase concentration at the particle surface (mM/g carbon),
K_F = constant for the Freundlich isotherm,
n = exponent for the Freundlich isotherm,
k_f = film transfer coefficient (cm/sec),
D_p = dispersion coefficient (cm^2/sec),
v_i = interstitial fluid velocity (cm/sec),
R = radius of particles (cm),
ϵ = bed void fraction (dimensionless),
ρ = particle density (g/cm^3).

The following dimensionless transforms may be performed to simplify these equations:

$$\bar{r} = r/R \tag{3.53}$$

$$\bar{q} = q/q_o \tag{3.54}$$

$$\bar{c} = C/C_o \tag{3.55}$$

$$\bar{t} = t/(\tau D_g) \tag{3.56}$$

$$D_g = \frac{q_o \rho\,(1 - \epsilon)}{C_o \epsilon}$$

$$= \text{solute distribution parameter} \tag{3.57}$$

$$N_d = \frac{D_s D_g \tau}{R^2} = \text{surface diffusion modulus} \tag{3.58}$$

$$Pe = \frac{v_i L}{D_p} = \text{Peclet number} \tag{3.59}$$

$$Sh = \frac{k_f R}{D_g D_s} \left(\frac{1 - \epsilon}{\epsilon} \right) \tag{3.60}$$

$$= \text{modified Sherwood number,}$$

where: C_o = initial fluid-phase concentration (mM/l),
q_o = solid phase concentration in equilibrium to C (mM/g carbon),
τ = retention time for column (sec),
L = column length.

Equation 3.44 then becomes:

$$\frac{\partial \bar{q}}{\partial \bar{t}} = N_d \frac{1}{\bar{r}^2} \frac{\partial}{\partial \bar{r}} \left(\bar{r}^2 \frac{\partial \bar{q}}{\partial \bar{r}} \right); \tag{3.61}$$

Equation 3.45 becomes:

$$- \frac{1}{Pe} \frac{\partial^2 \bar{c}}{\partial \bar{z}^2} + \frac{\partial \bar{c}}{\partial \bar{z}} + \frac{\partial \bar{q}}{\partial \bar{t}} + \frac{1}{D_g} \frac{\partial \bar{c}}{\partial \bar{t}} = 0\,; \tag{3.62}$$

and Equation 3.46 becomes:

$$\bar{q}_s = \bar{c}_s^n. \tag{3.63}$$

The transformed initial and boundary conditions are:

$$\bar{c} = 1 + \frac{1}{Pe}\frac{\partial \bar{c}}{\partial \bar{z}} \qquad \text{at } \bar{z} = 0 \tag{3.64}$$

$$\frac{\partial \bar{c}}{\partial \bar{z}} = 0 \qquad \text{at } \bar{z} = 1 \tag{3.65}$$

$$\frac{\partial \bar{q}}{\partial \bar{r}} = 0 \qquad \text{at } \bar{r} = 0 \tag{3.66}$$

$$\frac{\partial \bar{q}}{\partial \bar{r}} = Sh(\bar{c} - \bar{c}_s) \qquad \text{at } \bar{r} = 1 \tag{3.67}$$

$$\bar{q}\,(\bar{r},\bar{z}) = 0 \qquad \text{at } \bar{t} = 0 \tag{3.68}$$

$$\bar{c}(\bar{z}) = 0 \qquad \text{at } \bar{t} = 0 \tag{3.69}$$

Equations 3.61 to 3.69, the Freundlich exponent, n, and the dimensionless groups, D_g, Pe, Sh, and N_d, represent the dynamics of breakthrough curves for fixed bed adsorbers. The parameters, n and D_g, are determined experimentally, and Pe is negligible for adsorbers of reasonable length [30,31,32].

Subsequently, D_s and k_f values were determined from microcolumn breakthrough curves [7]. Inasmuch as film transfer and surface diffusion occur simultaneously along the entire s-shaped profile, it was reasoned that, initially, film transfer would dominate when the column length was made short enough [37]. Consequently, the film transfer coefficient was determined independent of the surface diffusivity from the initial portion of a breakthrough curve from a microcolumn. After k_f is determined, D_s values were obtained using regression analysis for the entire breakthrough curve [7].

Table 3–14 gives physical parameters for the microcolumn kinetic study using PBP, PTS, and DBS [7]. Previous studies had shown that PBP adsorption on activated carbon was controlled primarily by film transport, whereas DBS adsorption was controlled primarily by surface diffusion [30,31,32]. Adsorption of PTS was somewhere between that of PBP and DBS [32]. F-400 activated carbon, 16 × 20 mesh size, was used in the microcolumns.

Table 3–15 shows mass transfer coefficients for the three compounds obtained from microcolumns [7]. Because k_f is dependent on flow rate, studies were conducted for flow rates ranging from 2.5 to 10.0 gal/min/ft^2. Confirmation and/or validity of the D_s and k_f values were obtained by experimental data and fitted microcolumn breakthrough curves based on calculated values of k_f and D_s (not shown here). There was close agreement between the experimental data and the calculated curves [7]. Long-column experiments were conducted also.

In the continuing effort to perfect diffusion-controlled models for adsorption of organic compounds on or in GAC, a dual particle diffusion (DPD) model

Table 3–14 Physical Parameters for the Microcolumn Kinetic Studies

Run	*Solute*	C_o *(mm/l)*	D_g	ϵ	v_i *(cm/sec)*	L *(cm)*
0619A	PBP	0.480	3 901	0.492	1.40	6.13
0619B	PBP	0.480	3 786	0.500	0.76	4.05
0619C	PBP	0.480	4 785	0.442	0.39	4.50
0703A	PTS	0.091	3 097	0.471	1.46	6.16
0703B	PTS	0.091	3 091	0.470	0.74	3.89
0703C	PTS	0.091	3 217	0.461	0.37	3.51
0830A	DBS	0.265	6 198	0.522	1.31	5.90
0803B	DBS	0.265	8 183	0.453	0.74	4.00
0803C	DBS	0.265	6 685	0.503	0.38	4.00

Source: Reproduced from Liu and Weber [7] courtesy of the Water Pollution Control Federation.

was developed [37]. Previous models (presented earlier) predicted a greater throughput volume for removal of a greater quantity of material than was actually removed. This was ascribed to slow, or hindered, diffusion phenomena as the adsorbent approached saturation. More sophisticated models are able to predict adsorber performance only to a breakthrough level of 0.7 to 0.8. These slow diffusion phenomena are explained in terms of steric effects within the nonuniform pore structure of the carbon. The pore walls apparently reduce the freedom of movement of larger molecules, which results in restricted diffusion. In any event, another model was developed from mass balance considerations [37] (see Chapter 4 for equations).

It is relevant, however, to consider a conceptual representation of the carbon particle in the DPD model. This is seen in Figure 3–31 where the essential

Table 3–15 Mass Transfer Coefficients Obtained from the Microcolumn Kinetic Studies

Run	*Solute*	D_s *(cm³/sec × 10⁹)*	k_f *(cm/sec × 10³)*
0619A	PBP	20.0	5.50
0619B	PBP	17.0	4.50
0619C	PBP	13.0	3.38
0703A	PTS	6.4	4.60
0703B	PTS	6.0	3.60
0703C	PTS	6.2	3.20
0830A	DBS	0.85	3.20
0830B	DBS	0.95	2.50
0830C	DBS	0.85	2.50

Source: Reproduced from Liu and Weber [7] courtesy of the Water Pollution Control Federation.

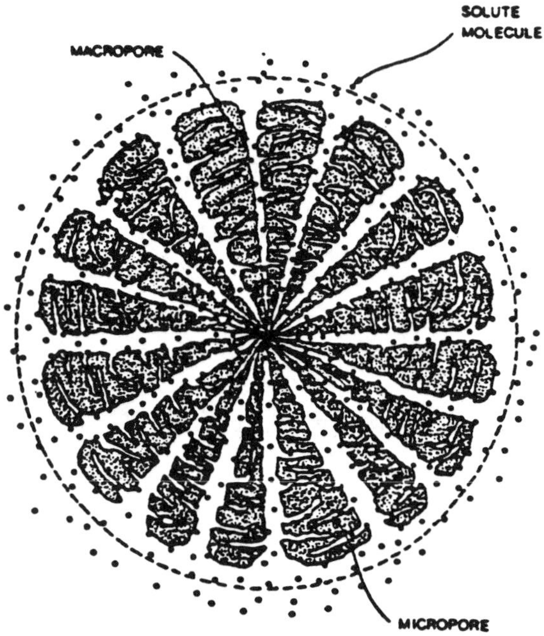

Figure 3–31 Conceptual representation of the macropore and micropore structure of a porous adsorbent. Reproduced from Weber and Liang [37], courtesy of Environmental Progress.

features of the model are the macro- and micropores that were defined arbitrarily as follows:

> Macropores—those which are large relative to the molecular size of the solute and which are readily accessible to the solute. Solute diffusion rates are unhindered in these pores and mass transport proceeds comparatively rapidly.
>
> Micropores—those pores which are similar in size to that of the solute molecule. Solute diffusion in these pores may be severely retarded due to steric interactions. [37, p.168]

The mass transport and transformation processes in the DPD model involve:

> *boundary layer transport,* diffusion of adsorbate through a liquid film surrounding the adsorbent particle;
>
> *phase conversion,* transfer of adsorbate from dissolved to adsorbed state;
>
> *macropore surface transport,* diffusion of adsorbate in an adsorbate state along walls of larger "macropores" and in parallel;
>
> *micropore surface transport,* diffusion of adsorbate in an adsorbed state along walls of smaller "micropores." [37, p.169]

Table 3–16 Freundlich Isotherm Constants

	F-400		BACM	
Solute	*a*	*b*	*a*	*b*
Phenol	2.03	0.310	1.86	0.289
PCP	2.68	0.176	2.62	0.150
PNP	2.69	0.136	2.82	0.123
PBP	2.95	0.117	2.90	0.117

Source: Reproduced from Weber and Liang [37] courtesy of Environmental Progress.
Note: $q_s = aC_s^n$, where C_s and q_s are expressed in mmol/liter and mmol/g carbon, respectively.

Experimentally, four phenolic compounds—phenol, PCP, PNP, and PBP—were employed with F-400 and BACM carbons of 0.30 to 0.42 mm particle sizes. The former carbon is irregularly shaped, whereas the latter is very nearly spherical. Equilibrium studies (8 day) were utilized for Freundlich isotherms, whereas the microcolumn technique described earlier was employed to determine the mass transfer parameters.

Here, it is sufficient to present only the mass transfer parameters for the DPD model [37]. Table 3–16 shows the Freundlich constants for the phenolic compounds on the two carbons that were employed in the model. In Table 3–17, the $D_{s,a}$ values are the macropore diffusion coefficients, $D_{s,i}/R_i^2$ is a pseudo-effective micropore surface diffusion coefficient, f is the fraction of total adsorptive capacity available in the macropores, and $\beta = ta/ti$ is the ratio of time scales for diffusion processes in the macro- and microspheres, respectively—that is, the times required for penetration of these spheres by diffusion. Small values of β indicate that macropore diffusion is relatively rapid and that micropore diffusion probably dominates as the rate-controlling transport step as exhaustion is approached. Consequently, the smaller the value of β, the steeper is the breakthrough curve from fixed beds predicted by the DPD model [37].

Table 3–17 Parameters Obtained from the Microcolumn Model Calibration Studies

Adsorbent	*Solute*	k_f (cm/sec × 10^3)	$D_{s,a}$ (cm²/sec × 10^9)	$D_{s,i}/R_i^2$ (sec^{-1} × 10^6)	f	β
F-400	Phenol	5.80	68.9	0.31	0.71	0.00143
F-400	PCP	7.55	6.29	4.59	0.68	0.236
F-400	PNP	6.22	3.78	9.55	0.68	0.818
F-400	PBP	6.50	2.79	902.0	0.67	104.75
BACM	Phenol	6.89	85.9	0.98	0.71	0.00368
BACM	PCP	7.15	9.12	3.55	0.69	0.126
BACM	PNP	6.70	4.22	9.88	0.69	0.759
BACM	PBP	6.55	5.85	1,080.0	0.68	59.82

Source: Reproduced from Weber and Liang [37] courtesy of Environmental Progress.

A branched pore model was proposed to describe the intraparticle kinetics of adsorption of phenol and chlorophenol on F-400 (16 × 30 mesh) [38,39]. An idealized carbon particle is assumed to consist of two types of pores:

1. Macropores are defined as those pores in which diffusion rates are unhindered by the pore walls, and rapid initial uptake of adsorbate is assumed.
2. Micropores are those with radii of similar size to the diffusing compound. Here, transport rates are considerably lower than in the larger pores because of strong multidirectional adsorptive forces in these pores and by the proximity of the pore walls. There is a slow approach to equilibrium in the micropore region.

These definitions were a prelude to the development of a truly predictive intraparticle kinetic model whereas those previously published were not truly predictive [29–32]. Batch kinetic experiments were conducted prior to column modeling [38]. Equilibrium parameters for phenol and o-chlorophenol were described well by the Freundlich equation (20°C):

$$\text{phenol: } q_e = 78.1 C_e^{.212} \tag{3.70}$$

$$\text{o-chlorophenol: } q_e = 240. C_e^{.098} \tag{3.71}$$

Figure 3–32 is a conceptual diagram of the adsorption model for the dual rate kinetics [38]. For the first and external film transfer step, an equation is:

$$N_A = k_f(c - c_s), \tag{3.72}$$

where c = concentration in bulk liquid, c_s = liquid concentration at carbon particle surface, N_A = mass flux across liquid film, and k_f = external film mass transfer coefficient. In the second step, transfer of adsorbate from the liquid to the solid phases can be described by equating the fluxes at the outer surface of the particle [38]:

$$N_A = k_f (c - c_s) = D_S f \rho_c \left. \frac{\partial q_m}{\partial r} \right|_R , \tag{3.73}$$

in which D_s = surface diffusion coefficient, f = fraction of total adsorptive capacity available in macropore region, ρ_c = apparent density of carbon particle, q_m = solid-phase concentration in the macropores, and r = carbon particle radial variable. In the third step, subsequent diffusion through the macropore region, a surface diffusion mechanism describes the process that includes a term for transfer from macropores to micropores (R_b):

$$f \frac{\partial q_m}{\partial t} = \frac{f D_s}{r^2} \frac{\partial}{\partial r} \left(r^2 \frac{\partial q_m}{\partial r} \right) - R_b, \tag{3.74}$$

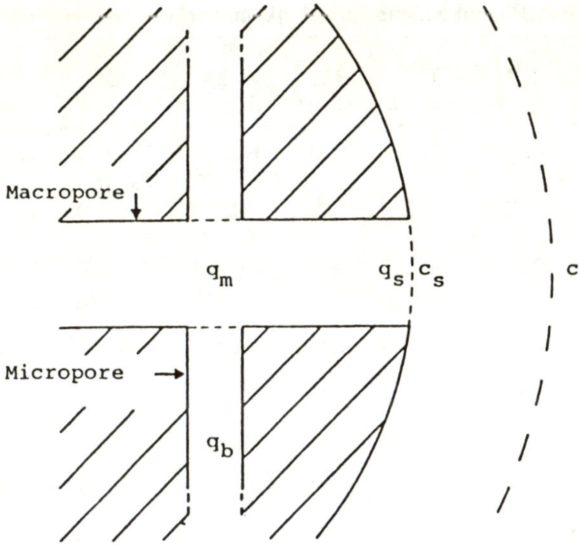

Figure 3-32 Conceptual diagram of adsorption model. Reproduced from Peel and Benedek [38], courtesy of the American Society of Civil Engineering.

in which t = time (sec). In the fourth and last step, a linear driving force is used to describe the transfer from macropores to micropores:

$$(1 - f) \frac{\partial q_b}{\partial t} = k_b (q_m - q_b) = R_b , \qquad (3.75)$$

in which q_b = solid-phase concentration in the micropores, and k_b = rate coefficient for transfer between the macropores and micropores. The intraparticle kinetic parameters were measured by fitting the proposed model to data from batch kinetic experiments [38]. Table 3-18 shows these parameters for phenol

Table 3-18 Batch Kinetic Model Parameters

Solute	*Surface Diffusion Coefficient,* D_s *(cm²/sec × 10⁸)*	*Micropore Transfer Coefficient,* k_b *(sec⁻¹ × 10⁶)*	*Fraction of Capacity (macropores)*	
			f_{50}	f_{100}
Phenol	5.32 [±6%]	1.59 [±8%]	0.708 [±1%]	0.758
o-chlorophenol	3.81 [±7%]	1.14 [±9%]	0.678 [±1%]	0.704

Source: Reproduced from Peel and Benedek [38] courtesy of the American Society of Civil Engineering.
Note: Numbers in brackets are approximate 95% confidence limits.

Table 3-19 Data of Phenol Kinetic Runs I, II, and III

Run Number	N *(r/min)*	W *(g)*	C_o *(mg/l)*	C_e *(mg/l)*	k_f *(cm/sec)*	D_s *(cm²/sec)*	k_b *(sec⁻¹)*	f
I	250	0.9898	96.3	27.0	1.33E-02 (±5%)	7.78E-08 (±10%)	1.44E-06 (±27%)	0.63 (±2%)
II	250	0.7216	96.3	41.0	1.82E-02 (±5%)	9.01E-08 (±9%)	1.87E-06 (±18%)	0.66 (±2%)
III	250	0.5393	94.7	55.0	2.01E-02 (±5%)	7.75E-08 (±7%)	1.80E-06 (±16%)	0.68 (±2%)

Source: Reproduced from Peel et al. [39] courtesy of the American Society of Civil Engineering.
Note: ±95% confidence limits on linear hypothesis.
Reactor volume equal to 2.3l in each case.

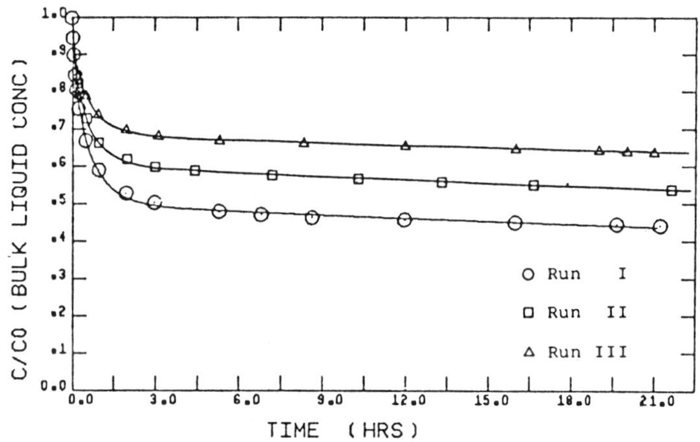

Figure 3-33 Batch kinetic experiments: phenol. Reproduced from Peel et al. [39], courtesy of the American Society of Civil Engineering.

and o-chlorophenol. The D_s and k_b values were applied directly to a carbon column adsorption model (see Chapter 4). Table 3-19 also shows the same values for phenol adsorption in a similar study. "The ability of the model to adequately fit the experimental data is clearly excellent" [39] (Figure 3-33).

REFERENCES

1. Weber, W.J. Jr. *Physicochemical Processes for Water Quality Control.* Wiley-Interscience, New York (1972).
2. Smith, J.M. "Kinetics of Adsorption," Chapter 2 in *Adsorption from Aqueous Solution.* Advances in Chemistry Series 79. American Chemistry Society, Washington, D.C. (1968).

3. Weber, W.J. Jr., and R.R. Rumer, Jr. *Water Resources Res. 1,* 361 (1965).
4. Keinath, T.M., and W.J. Weber, Jr. *J. Water Pollut. Cont. Fed. 40,* 743 (1968).
5. Fritz, W., et al. "Competitive Adsorption of Dissolved Organics on Activated Carbon," Chapter 9 in *Activated Carbon Adsorption of Organics from the Aqueous Phase,* Vol. 1. I.H. Suffet and M.J. McGuire, eds. Ann Arbor Science Publishers, Ann Arbor, Mich. (1980).
6. Zogorski, J.S., et al. *J. Colloid Interf. Sci. 55,* 329 (1976).
7. Liu, K.T., and W.J. Weber, Jr. *J. Water Pollut. Cont. Fed. 53,* 1541 (1981).
8. Weber, W.J. Jr., and J.C. Morris. *J. Sanit. Eng. Div. ASCE SA2,* 31 (1963).
9. Faust, S.D., and O.M. Aly. *Chemistry of Water Treatment.* Butterworths, Stoneham, Mass. (1983).
10. Zogorski, J.S. "The Adsorption of Phenols onto Granular Activated Carbon from Aqueous Solution." Ph.D. dissertation, Department of Environmental Science, Rutgers, The State University, New Brunswick, N.J. (October 1975).
11. Smith, S.B., et al. *Chem. Eng. Progr. Symp. Ser. 55,* 25 (1959).
12. Mattson, J.S. *Ind. Eng. Chem. Prod. Res. Devel. 12,* 312 (1973).
13. Schwartz, H.C. Jr. *Environ. Sci. Technol. 1,* 332 (1967).
14. Weber, W.J. Jr., et al. Chapter 8 in *Viruses and Trace Contaminants in Water and Wastewater.* J.A. Borchardt et al., eds. Ann Arbor Science Publishers, Ann Arbor, Mich. (1977).
15. Arrhenius, S. *Z. Physik. Chem. 4,* 226 (1889).
16. Helfferich, F. *Ion Exchange.* McGraw-Hill, New York (1962).
17. Adamson, A.W. *Physical Chemistry of Surfaces.* Wiley-Interscience, New York (1960).
18. Edeskuty, F.J., and N.R. Amundson. *Ind. Eng. Chem. 44,* 1698 (1952).
19. Kressman, T.R.E., and J.A. Kitchener. *Discuss. Faraday Soc. 7,* 90 (1949).
20. Morris, J.C., and W.J. Weber, Jr. "Adsorption of Biochemically Resistant Materials from Solution." Public Health Service Pub. 999-WP-11, Washington, D.C. (1964).
21. DiGiano, F.A., and W.J. Weber, Jr. "Mathematical Modeling of Sorption Kinetics in Finite and Infinite-Bath Systems." Grant WP-00706 (1969). Federal Water Pollution Control Administration, U.S. Department of the Interior, Washington, D.C. (1964).
22. Weber, W.J. Jr., and J.C. Morris. *J. Amer. Water Works Assoc. 56,* 447 (1964).
23. Youssefi, M. "The Determination, Formation, and Removal of Halogenated Pollutants." Ph.D. dissertation, Department of Environmental Sciences, Rutgers, The State University, New Brunswick, N.J. (1978).
24. Benedict, A.L. "The Adsorption and Desorption of Trichloroethylene, Tetrachloroethylene, and Carbon Tetrachloride." M.S. thesis, Department of Environmental Sciences, Rutgers, The State University, New Brunswick, N.J. (1982).
25. Weber, W.J. Jr. Vol. 1, *Proc. 3rd Internat. Conf. Water Pollut. Res. and Cont. Fed.,* Water Pollut. Cont. Fed. (1967).
26. Vermeulen, T. *Adv. Chem. Eng. 2,* 147 (1958).
27. Lightfoot, E.N., et al. In *New Chemical Engineering Separation Techniques.* H.M. Schoen, ed., Wiley-Interscience, New York (1962).
28. Thomas, H.C. *Ann. N.Y. Acad. Sci. 49,* 161 (1948).
29. Weber, W.J. Jr., and J.C. Crittenden. *J. Water Pollut. Cont. Fed. 47,* 924 (1975).
30. Crittenden, J.C., and W.J. Weber, Jr. *J. Environ. Eng. Div. ASCE EE2,* 185 (1978).
31. Crittenden, J.C., and W.J. Weber, Jr. *J. Environ. Eng. Div. ASCE, EE3,* 433 (1978).
32. Crittenden, J.C., and W.J. Weber, Jr. *J. Environ. Eng. Div. ASCE, EE6,* 1175 (1978).

33. Radke, C.J., and J.M. Prausnitz. *Ind. Eng. Chem. Fund. 11,* 445 (1972).
34. Brent, R.P. ''Algorithm for Finding Zeros and Extrema of a Function without Calculating Derivatives.'' Stanford University Technical Report C5-198, Stanford, Calif. (1971).
35. Van Vliet, B.M., et al. *Water Res. 14,* 1719 (1980).
36. Weber, W.J. Jr., and K.T. Liu. *Chem. Eng. Commun. 6,* 49 (1980).
37. Weber, W.J. Jr., and S. Liang. *Environ. Progress 2,* 167 (1983).
38. Peel, R.G., and A. Benedek. *J. Environ. Eng. Div. ASCE, EE4,* 797 (1980).
39. Peel, R.G., et al. *Amer. Inst. Chem. Eng. J. 27,* 26 (1981).

4

Batch Systems and Fixed and Fluidized Beds

The practical application of the adsorption process for the removal of contaminants from drinking waters or wastewaters is accomplished by contacting the adsorbent by water either in a batch-type contact unit operation or a continuous flow packed bed system.

BATCH CONTACT SYSTEMS

Generalizations

In batch contact unit systems, a predetermined amount of adsorbent is mixed with the water for a given contact period and subsequently is separated by sedimentation or filtration. PAC is the adsorbent most suitable for the batch-type contact process. The flexibility of this process makes it adaptable for many applications, particularly in drinking water treatment plants. Existing equipment such as mixing tanks, clarifiers, and filtration systems can be utilized. Only a minimum capital investment is usually required for slurry tanks and metering pumps.

Use of PAC in water treatment is more adaptable to applications where relatively low contaminant levels require small carbon dosages or low carbon usage rates of less than 400 lb/day. Also, when applications are periodic or seasonal in nature, such as removal of tastes and odors in some surface water supplies, the carbon addition can be adjusted to changes in water quality. The use of PAC in most modern water treatment plants processing surface waters is almost a routine practice. Almost 90% of water treatment plants worldwide that use activated carbon use it in the powdered form [1]. The particle size distribution is such that at least 80% of the carbon should pass number 325 mesh or measure less than 44 μm. The small particle size allows high rates of adsorption and, therefore, short contact periods to reach adsorption equilibrium.

Powdered carbon is usually applied to raw water in the form of a slurry at any treatment stage before filtration (in the batch-type system). The optimum point of application should allow adequate dispersion of the carbon and sufficient contact time to ensure maximum adsorption. The carbon slurry may be added to the mixing tank or a flocculation tank ahead of the settling tanks. This application point allows adequate contact time, and the carbon can be removed easily with the settled sludge. However, the carbon particles may become embedded in the coagulant flocs, leading to a reduction of adsorption efficiency. In addition, many organic compounds are adsorbed that would be removed by coagulation alone, therefore requiring more activated carbon. The cost of increased carbon use is offset in many instances by savings in the coagulant dose due to improvement in the flocculation process that is a result of a seeding effect of the carbon particles. Another point of application of PAC would be to the filter's influent. A small dose of a polymer usually is added as well to improve the filter's ability to retain carbon particles. Addition of carbon to the sand filter's influent should not normally exceed 4 mg/l; however, multimedia beds possess a high capacity for storage of solids. Therefore, carbon doses up to 10 mg/l can be tolerated for short durations.

Simultaneous addition of chlorine and activated carbon should be avoided because of carbon's dechlorination capability. In cases where chlorine addition is required, the activated carbon should be applied at a point upstream or ahead of the chlorine. This allows the use of lower carbon doses, reduces the potential of formation of THMs and other chlorination products, and avoids the potential loss of the carbon's adsorptive capacity due to formation of oxidized surface groups. The effects of other treatment processes should be considered also in the application of PAC. For example, the addition of softening chemicals produces high pH values that are not favorable for adsorption of most organic compounds, whereupon high carbon doses would be required [2].

In situations where large quantities of activated carbon are required to reduce the concentration of organic compounds, a split feed system, or *countercurrent application,* will reduce the carbon dosage significantly [3]. In this system, also called *divided feed system* or *multiple-stage batch system,* the water is contacted first with a dose of fresh carbon. After separation of the carbon, the partially treated water is further contacted with a second fresh dose of carbon. In the *countercurrent application,* a two-stage system consists of contacting the water first with a quantity of once-used carbon that is subsequently separated and discarded. The partially treated water is contacted with a dose of fresh carbon that, after separation, becomes the once-used carbon for treatment of the next batch of untreated water. This allows the carbon to reach equilibrium at a higher contaminant concentration that achieves a higher carbon loading, x/M, and lower carbon usage.

Determination of Carbon Dosage in Batch Systems

The first step in the evaluation of carbon dosages for batch systems is determining the contact period. This is determined by adding a known quantity of carbon

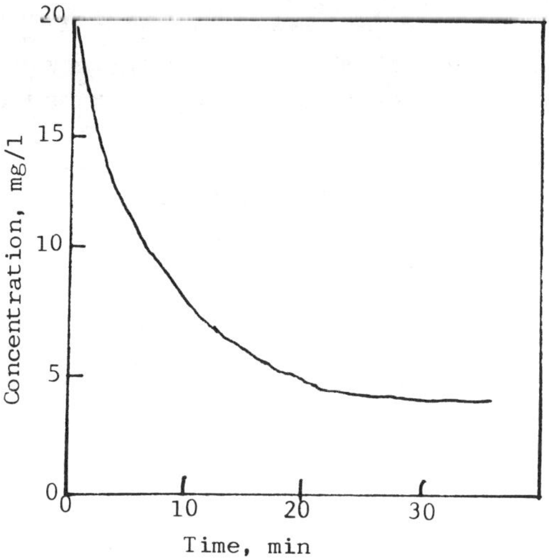

Figure 4–1 Rate of adsorption of phenol by activated carbon.

to a given volume of the water to be treated and placing the mixture in a stirring apparatus. A sample is removed at various time intervals, and the carbon is immediately removed either by filtration or centrifugation. The residual concentration of the contaminant subsequently is determined in each sample. The data are plotted to show the change in concentration with time. Adsorption equilibrium is achieved at the contact period where no significant change in concentration is observed with increased time. Figure 4–1 shows that a contact of 25 minutes is required to achieve equilibrium.

The carbon dosage or carbon usage rate can be determined by constructing an adsorption isotherm. This is run by adding known quantities of carbon to equal volumes of the water to be treated. The mixture is stirred for a period of time equal to the contact time determined in the previous experiment. The carbon is removed then by filtration, and the equilibrium concentration of the contaminant is determined. The data are plotted according to the Freundlich adsorption isotherm (see Chapter 1). According to the equation:

$$\frac{x}{M} = \frac{C_o - C_f}{M} = KC_f^{1/n} \quad . \tag{4.1}$$

The value of x/M at C_f can be read from the plot. The carbon dosage, M, required to reduce the initial concentration, C_o, to the desired final concentration, C_f, is calculated by rearranging Equation 4.1:

$$M = \frac{C_o - C_f}{x/M} . \tag{4.2}$$

The carbon dosage for countercurrent application, M_{cc}, which will achieve the same final concentration, C_f, as the larger single-stage dosage, M, is determined by solving simultaneous equations derived from the Freundlich equation [3]:

$$\frac{M_{cc}}{M} = \frac{C_f}{C_i} \tag{4.3}$$

and

$$\frac{M_{cc}}{M} = \frac{C_i - C_f}{C_o - C_f}\,, \tag{4.4}$$

where C_i is the intermediate concentration after the first-stage treatment and M is the single-stage dosage. These equations must be solved graphically by trial and error since they do not have an algebraic solution. Figure 4–2 gives a general solution as a family of curves, each of which shows the variation of the ratio M_{cc}/M, with the Freundlich adsorption isotherm slope, $1/n$, for a specific C_f value expressed as a fraction of C_o, for a two-stage countercurrent treatment [3].

Dosages for split treatment can be determined from the Freundlich equation [3]:

$$M_1 = \frac{C_o - C_1}{x/M} \tag{4.5}$$

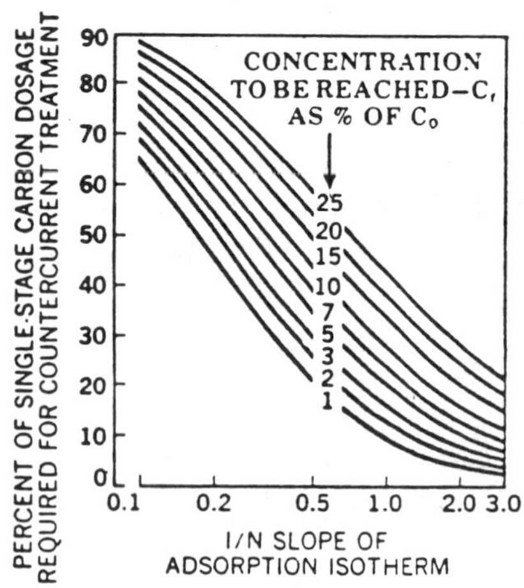

Figure 4–2 Two-stage countercurrent carbon dosage chart. Reproduced from Hutchins [3], courtesy of CRC Press, Inc.

and

$$M_2 = \frac{C_1 - C_f}{x/M}, \tag{4.6}$$

where C_1 is the equilibrium concentration after the first carbon dosage x/M is obtained from the adsorption isotherm at the corresponding C_i or C_f values. The optimum split dosage is computed by trial and error. The first dose should reduce C_o by 50% or more.

Example:

A treatment plant that processes 100,000 gal/day is required to reduce the concentration of a phenolic compound from an initial concentration of 20 mg/l to 2 mg/l using PAC.

First, contact time is determined from a rate study experiment shown in Figure 4-1, minimum = 25 minutes. Second, an adsorption isotherm is conducted and plotted according to the Freundlich isotherm (Figure 4-3). Calculate

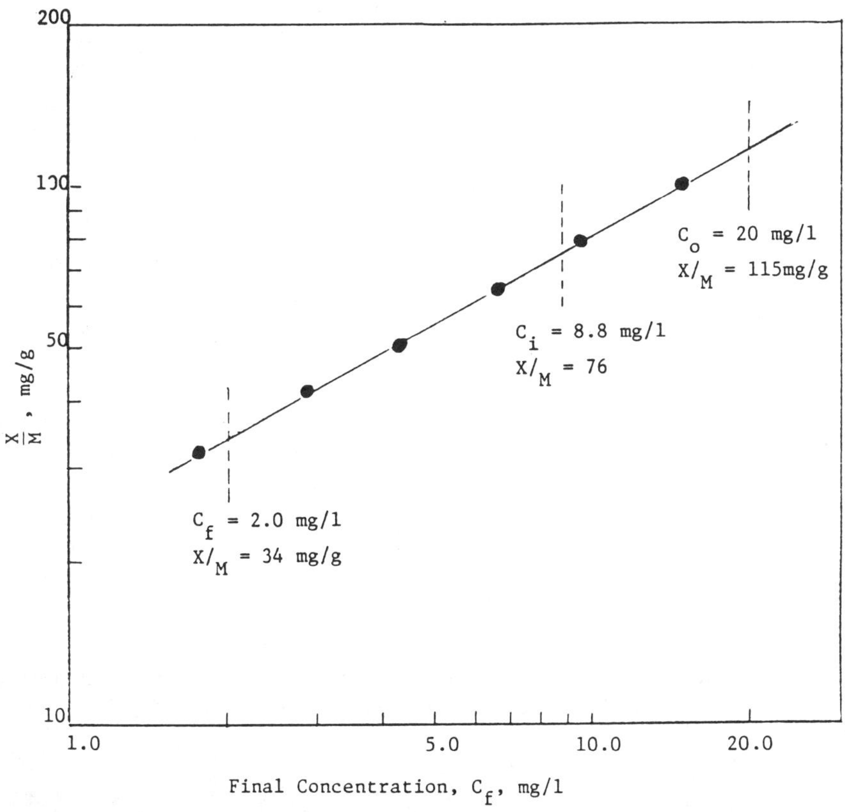

Figure 4-3 Freundlich isotherm for adsorption of phenol by activated carbon.

the carbon dosage and carbon usage rate for single-stage treatment, countercurrent treatment, and split treatment. The solution is as follows:

1. Single-stage treatment:

   The value of x/M at a final concentration, C_f, of 2.0 mg/l is obtained from the isotherm plot (Figure 4–3) = 34 mg/g. From Equation 4.2:

$$\begin{aligned}\text{Carbon dosage} &= \frac{20 - 2\ (\text{mg/l})}{34\ (\text{mg/g})}\\ &= 0.529\ \text{g/l}\\ &= 529\ \text{mg/l.}\end{aligned}$$

$$\begin{aligned}\text{Carbon usage rate} &= 0.529\ \text{g/l} \times \frac{3.785\ \text{l/gal}}{454\ \text{g/lb}} \times 100{,}000\ \text{gal/day}\\ &= 441\ \text{lb/day.}\end{aligned}$$

2. Countercurrent treatment:

   a. Slope of the adsorption isotherm, $1/n = 0.48$.
   b. Percent $C_f/C_o = 100 \times 2/20 = 10\%$.
   c. From Figure 4–2, the percent of single-stage carbon dosage required for countercurrent treatment is obtained by erecting a vertical line from $1/n = 0.48$ to the intersection with $C_f/C_o = 10\%$ and reading the corresponding value on the ordinate = 40%. From Equation 4.3:

$$\frac{M_{cc}}{M} = 0.4.$$

   Since the single stage dosage, $M = 529$ mg/l, the countercurrent dosage, $M_{cc} = 0.4 \times 529 = 211$ mg/l, and carbon usage rate $= 0.211 \times 3.785/454 \times 100{,}000 = 176$ lb/day.

3. Split treatment:

   The first carbon dosage is selected to be the value that would reduce the initial concentration by 50%, or $20 \times 0.5 = 10$ mg/1. From the isotherm plot x/M at $C_f = 10$ is 80. From Equation 4.5:

$$\begin{aligned}M_1 &= \frac{20 - 10}{80}\\ &= 0.125\ \text{g/l}\\ &= 125\ \text{mg/l.}\end{aligned}$$

   Similarly, the second carbon dosage should reduce the partially treated water to 2.0 mg/l. From the isotherm plot at $C_f = 2$, the value of $x/M = 34$:

$$\begin{aligned}M_2 &= \frac{10 - 2}{34}\\ &= 0.235\ \text{g/l}\\ &= 235\ \text{mg/l.}\end{aligned}$$

Total dosage = $M_1 + M_2 = 125 + 235 = 360$ mg/l. By repeating these computations for C_i/C_o = 80%, 40%, and 20%, the total carbon dosages are 451 mg/l, 345 mg/l, and 392 mg/l, respectively. Therefore, the lowest total carbon dosage is obtained at C_i/C_o = 40%; that is:

$$C_i = 0.4 \times 20 = 8 \text{ mg/l},$$

$$M_1 = \frac{20 - 8}{71} = 0.169 \text{ g/l} = 169 \text{ mg/l},$$

$$M_2 = \frac{8 - 2}{34} = 0.176 \text{ g/l} = 176 \text{ mg/l},$$

$$\text{Carbon dosage} = 169 + 176 = 345 \text{ mg/l},$$

$$\text{Carbon usage rate} = 0.345 \times 3.785/454 \times 100{,}000 = 288 \text{ lb/day}.$$

It is obvious from these examples that the countercurrent treatment results in 60% savings in carbon usage and that the split treatment results in 35% savings in comparison with single-stage treatment. However, the difference in carbon costs between the split and countercurrent treatments frequently does not justify the capital investment in additional sedimentation tanks and pumps required for the latter treatment. Therefore, the final selection of either method should be based on the economic feasibility for the individual plant.

DYNAMIC SYSTEMS (COLUMN ADSORBERS)

Generalizations

In most large-scale applications of the adsorption process, fluid flows continuously through a column of adsorbent or a packed bed where dynamic adsorption of the solute occurs. GAC is the most widely used adsorbent in packed beds for removal of organic contaminants from waters. Porous polymeric resins have been used in certain new applications (see Chapter 10). Particle size and hydraulic loading determine the pressure drop or head loss in the bed and, therefore, are critical factors in the economics of the process. The relationship for clean water passing through a bed of carbon is expressed [4]:

$$\Delta P = \frac{K \upsilon V L_c}{D_p^2 \, D_c}, \tag{4.7}$$

where ΔP = pressure drop in inches, K = constant, υ = viscosity in centipoise, V = flow rate in gallons per minute, L_c = bed depth in feet, D_p = mean particle diameter in millimeters, and D_c = column diameter in inches. Therefore, hydraulic head loss is related directly to flow rate and inversely to particle size. Particle size also has a pronounced effect on the kinetics of the adsorption process where rates of adsorption increase with the decrease in particle size. Commercially available particle sizes of GAC expressed in limiting U.S. Standard Sieve sizes include 8 × 16, 8 × 30, 10 × 30, 12 × 40, 14 × 40, and 20 × 40, with effective sizes ranging from 0.55 mm to 1.35 mm [2]. This size range provides a reasonable compromise between good hydraulic characteristics and higher adsorption rates.

The columnar operation allows more efficient utilization of the adsorptive capacity than the batch process. The adsorbent at the inlet end is contacted continuously by the solution at the initial solute concentration. The change in concentration at each subsequent layer is small. Therefore, the adsorbent in the column acts like a series of layers, with each layer in contact with a fresh solution of constant solute concentration. This procedure results in maximum loading of the carbon (high x/M on the adsorption isotherm) at constant solute concentration and is in contrast to the continuously declining solute concentration in batch processes.

The Breakthrough Curve and Mass Transfer Zone

When contaminated water is passed through a bed of adsorbent, a wave front or a mass transfer zone is formed by continuous adsorption of the solute in the bed. Figure 4–4 illustrates the change of concentration of adsorbed species on the surface of the adsorbent with bed depth [5]. The solute is rapidly adsorbed on the top layers of the bed until the amount adsorbed is in equilibrium with influent contaminant concentration. At this time, the adsorbent is loaded to ca-

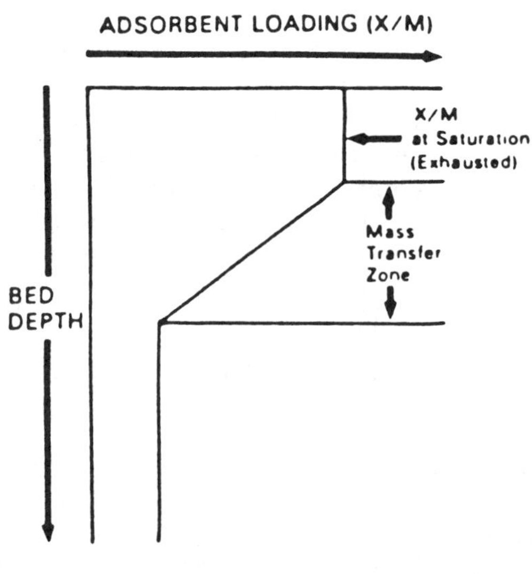

Figure 4–4 Mass transfer zone in a GAC adsorber [5].

pacity, and that portion of the bed is exhausted. Below this zone is a second zone where dynamic adsorption is occurring; that is, the contaminant is being transferred from the liquid solute to the adsorbed phase. This zone is called the *mass transfer zone,* and its depth is controlled by many factors, depending on the contaminant being adsorbed, characteristics of the adsorbent, and hydraulic factors. The depth of the mass transfer zone is a measure of physical/chemical resistance to mass transfer. Once formed, the mass transfer zone moves down through the adsorbent bed until it reaches the bottom, whereupon the effluent concentration of the contaminant in the aqueous phase begins to rise (Figure 4–5). Figure 4–5 shows the concentration gradient of adsorbed material (x/M) in an adsorber as the mass transfer zone moves down the column with time. As the mass transfer zone reaches the bottom of the column, breakthrough of the contaminant occurs, as noted by a detectable increase in effluent concentration (Fig-

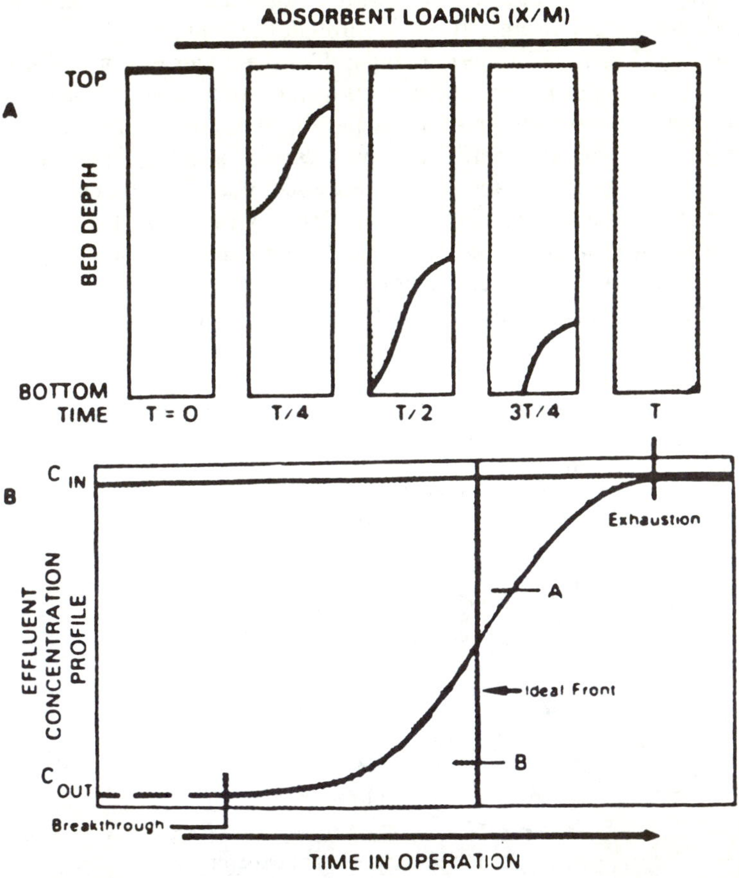

Figure 4–5 Typical adsorption column performance [5].

ure 4–5). When the adsorber is operated to exhaustion (at equilibrium, $C_{in} = C_{out}$), the breakthrough curve (plot of effluent concentration with time) takes on the classic S shape that is controlled by the shape and length of the mass transfer zone. Steeper slopes for the breakthrough curves are obtained for systems that exhibit high film transfer coefficients, high internal diffusion coefficients, or flat Freundlich adsorption isotherms—that is, smaller $1/n$ values [6].

The nature of the mass transfer zone is determined entirely by the equilibrium isotherm, although the shape of the concentration profile (breakthrough curve) may be modified significantly by kinetic effects [7]. Three general cases are distinguished that depend on whether or not the equilibrium relationship is linear, favorable, or unfavorable. The meaning of these terms can be understood by reference to Figure 4–6, which shows possible general forms for the equilibrium isotherm. For systems in which the equilibrium isotherm is linear or unfavorable, the mass transfer zone broadens continuously as the front propagates through the column. Such behavior is commonly referred to as a dispersive, or proportionate, pattern since the width of the mass transfer zone increases in direct proportion to the distance traveled through the column [7].

However, when the isotherm is favorable, the pattern of behavior is quite different. In the initial region, the mass transfer front spreads as it progresses, but some distance from the inlet it reaches an asymptotic form and attains a constant pattern form, and thereafter it progresses as a stable mass transfer zone with no further change in shape. The distance required to approach the constant pattern limit depends on the degree of nonlinearity of the isotherm and on the kinetics of adsorption. For most water and wastewater treatment applications, this distance is very small since many contaminants are invariably strongly adsorbed with a favorable isotherm and the concentration profile rapidly ap-

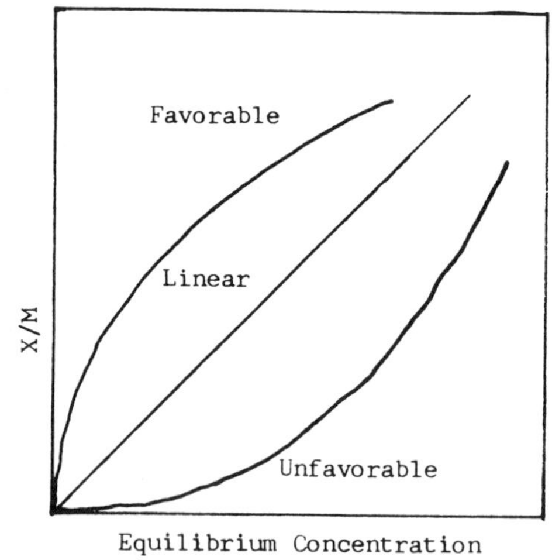

Figure 4–6 Equilibrium isotherm types.

proaches a constant pattern [7]. The constant pattern approximation is a very useful design tool since the calculation of the form of the mass transfer front under constant pattern conditions is very simple. The length of the mass transfer zone, L_m, is defined by [7]:

$$L_m = L\left(1 - \frac{t_b}{t_s}\right), \tag{4.8}$$

where L = total length of the bed, t_b = breakthrough time, and t_s = time to saturation. Both t_b and t_s may be determined from a single experimental breakthrough curve (Figure 4–7) and may also be expressed as:

$$t_b = \int_0^{t_b} \left(1 - \frac{C}{C_o}\right) dt \tag{4.9}$$

and

$$t_s = \int_0^{\infty} \left(1 - \frac{C}{C_o}\right) dt. \tag{4.10}$$

The length of the mass transfer zone is also called the *critical bed depth*, which is the theoretical depth of the adsorbent that is sufficient to prevent the effluent solute concentration from exceeding the desired breakthrough concentration at zero time.

The time or volume of water treated to the breakthrough point is decreased by increased particle size; increased solute concentration in the influent; increased pH value of the water in the case of adsorption of weak electrolytes; increased

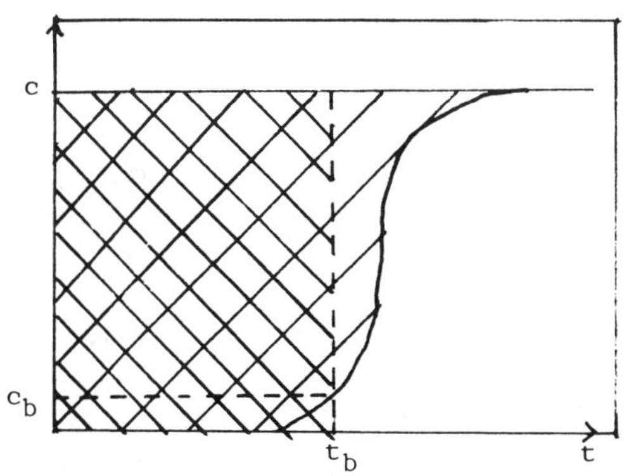

Figure 4–7 Breakthrough curves showing method of calculation of t_b and t_s. Reproduced from Ruthven [7], courtesy of John Wiley & Sons, Inc.

flow rate, usually expressed as gal/min/ft^2; and decreased bed depth [8–11]. The typical smooth S-shaped breakthrough curve described earlier is obtained for single-solute systems where the influent concentration to the carbon column is assumed to be constant. However, in natural water systems, there is a broad spectrum of organic contaminants, and a wide variation in the concentration of any single compound is not unusual. Competitive adsorption, displacement of weakly adsorbed compounds by strongly adsorbed ones (chromatographic effect), and concentration variations complicate the prediction of the shape and position of the breakthrough curves. Therefore, pilot plant tests that employ the carbon to be used and the water to be treated must be conducted at the treatment plant site so the appropriate design parameters for the carbon beds can be selected.

TYPES OF ADSORPTION SYSTEMS

Fixed Bed Contactors

The most common type of contactor is that used where the adsorbent bed remains fixed or stationary during operation [3,4]. The influent flows through the bed to the breakthrough point or until the adsorptive capacity of the bed is exhausted. The entire bed is then removed from the contactor and replaced by virgin carbon. Fixed beds can be operated in downflow or upflow modes, but downflow is more common.

Downflow fixed beds may perform two functions: adsorption and filtration. However, excessive head losses necessitate frequent backwashing, and the economic gain of combining the two processes may be offset by decreased functional life of the bed and higher operating costs. Downflow beds can be either of the gravity or pressure type, with the latter being the more prevalent.

Downflow gravity beds are designed similarly to concrete sand beds with provisions for the addition and withdrawal of carbon and enough sidewall depth to allow 50% expansion of the bed during backwash (Figure 4–8). The minimum carbon depth is 3 ft and the hydraulic loading ranges between 2 and 6 gal/min/ft^2 of the cross-sectional area of the bed.

Pressurized downflow contactors include a pressure steel vessel, underdrain and media support system, influent and effluent distributors, surface wash, and backwash conveyance system (Figure 4–9). The range of hydraulic loading is 2 to 10 gal/min/ft^2 of cross-sectional area with bed depths ranging from a minimum of 3 ft to a maximum of 40 ft. There are limitations to the size of pressure systems; the maximum vessel size for safe shipment is 12 to 13 ft diameter with an overall length not to exceed 60 ft. The major advantage of the pressurized contactors is their ability to operate under high flow rates. Therefore, space requirements will be less than for the gravity type.

In upflow fixed bed contactors, the influent enters through the bottom, and the treated effluent is drawn off at the top. The flow rate is kept low enough to allow the bed to remain stationary. The influent to this type of system should be

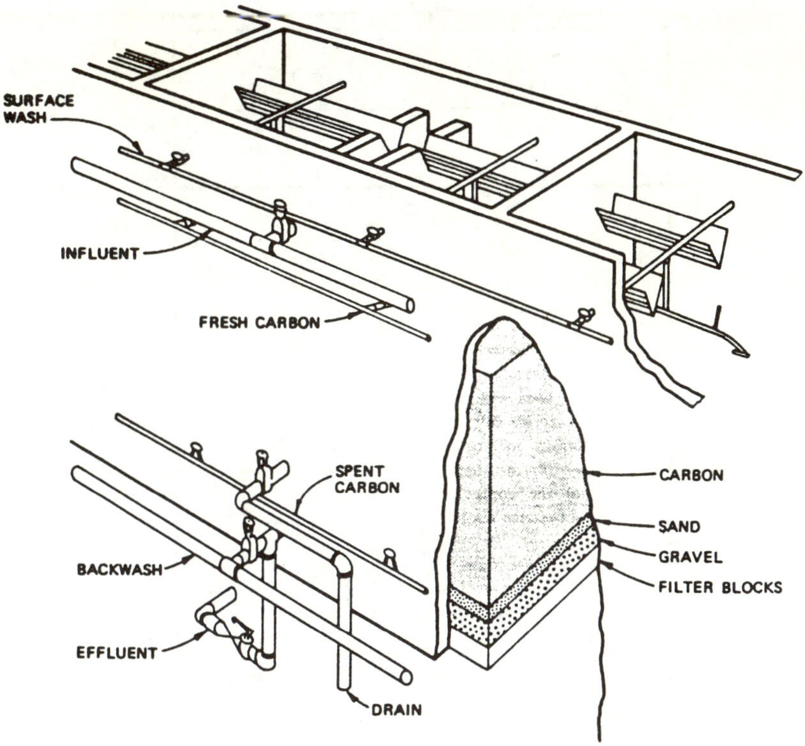

Figure 4–8 Typical downflow gravity contactor [4].

free of suspended solids, with turbidity less than 2.5 Jackson Turbidity Units. Usually high adsorption rates are encountered because of the smaller particle sizes (8 × 30). Also, higher adsorption efficiency is achieved since the mass transfer zone is not disturbed by backwashing because it is not normally allowed.

Another type of upflow contactor is the upflow expanded bed type. Increasing the flow rate in the upflow contactor will allow the bed to expand or become suspended (fluidized) by the influent fluid stream. Fluidization of the adsorbent bed separates the individual granules sufficiently to allow suspended solids to pass through. At about 10% expansion of an upflow bed, suspended solids will pass freely through the bed. The flow rate is usually about 6 gal/min/ft^2, and the adsorbent mesh size is 12 × 40 mesh. Pressure drop is usually low because the bed expands with the flow and backwashing is required occasionally.

Pulsed Bed Contactors

Pulsed bed contactors are also called moving bed contactors [12,13,14]. The influent enters the bottom of the contactor and flows upward through the carbon

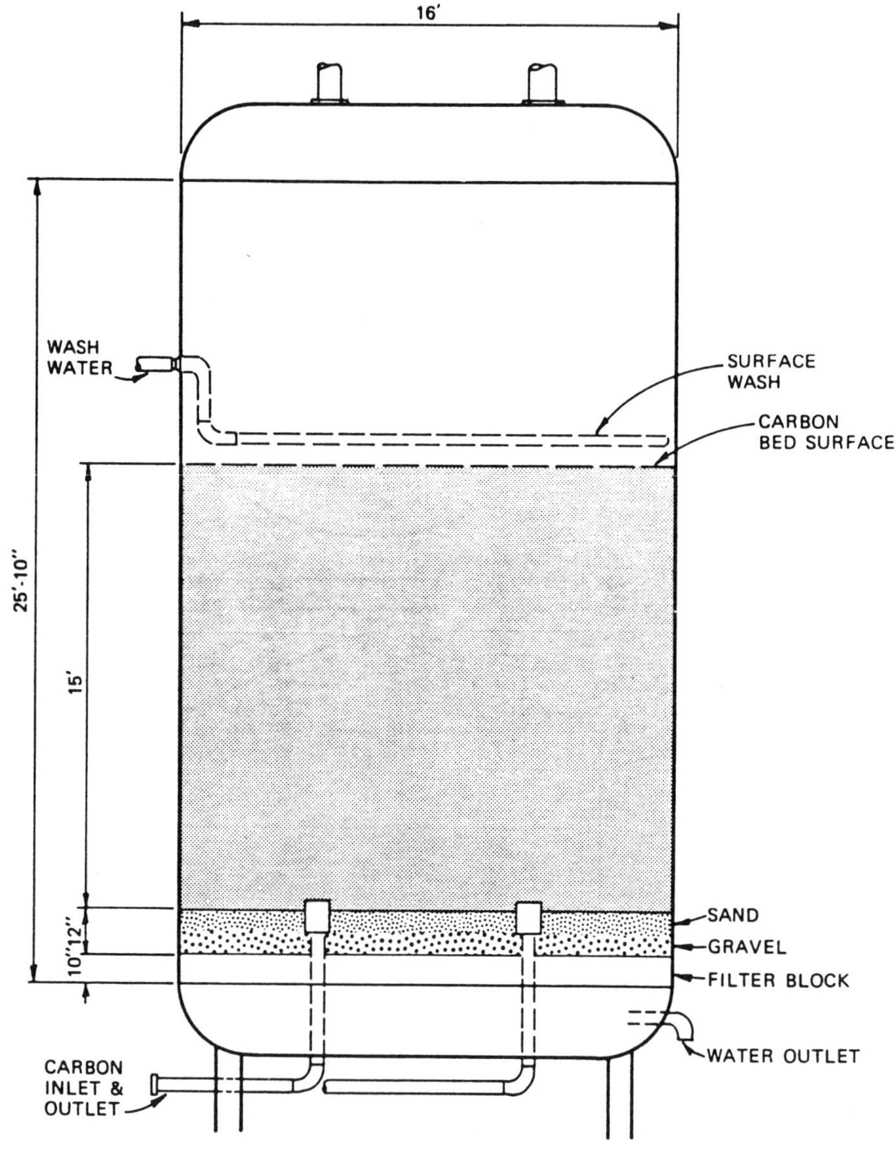

Figure 4-9 Pressurized downflow contactor [4].

bed while the spent carbon is removed from the bottom and an equal volume of fresh carbon is added to the top (Figure 4-10). The adsorption wave front is contained in the column since the most exhausted carbon is removed from the bottom while virgin carbon is simultaneously being added at the top. The pulsed bed system, therefore, performs like a series of stacked fixed bed columns operating in series. Removing a given volume of spent carbon from the bottom is

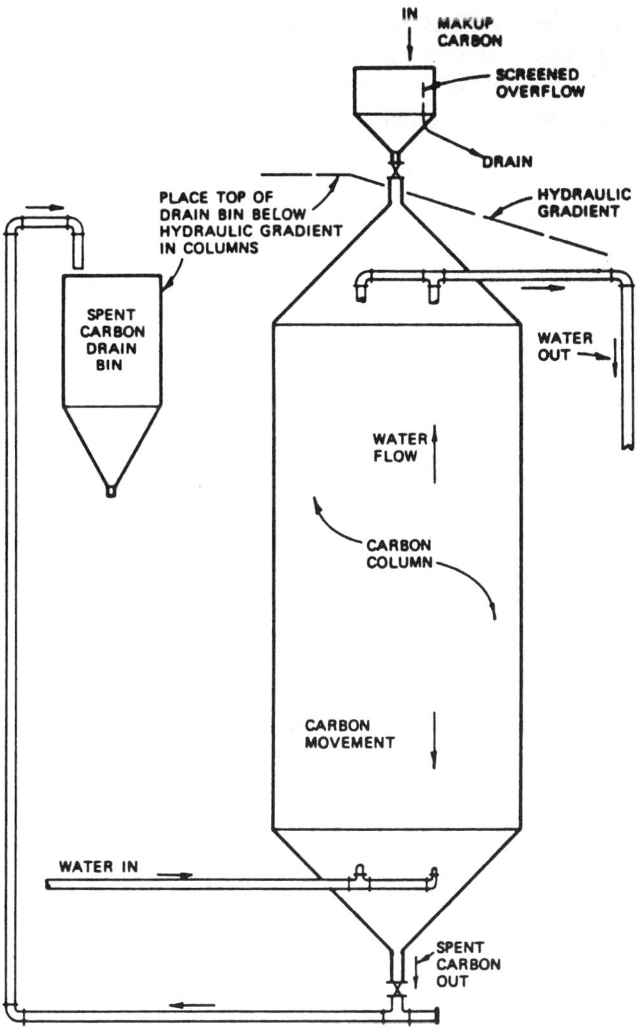

Figure 4–10 Pulsed bed contactor [4].

similar to removing the lead column in a series system [3]. To maintain a sharp mass transfer zone, plug flow of the carbon should be maintained as much as possible. These beds are normally operated with the column completely filled with carbon to prevent bed expansion during operation [3]. However, some pulsed bed systems allow bed expansion during operation, but efficiency of the system suffers because of mixing of the adsorption zone. The pulsed bed contactors can be operated on a continuous or semicontinuous basis, the latter being more common. In the continuous operation, fresh carbon is continuously added at the top while spent carbon is withdrawn from the bottom. In the semicontinuous operation, a predetermined volume of carbon is withdrawn (pulsed) at certain intervals, and an equivalent amount of fresh carbon is added at the top. Because of the efficient utilization of the carbon adsorptive capacity, these systems are used when the carbon usage rate is high.

System Configurations

The fixed and pulsed beds can be operated as single beds, multiple beds in parallel or series, or combined parallel and series [3,13]. Single fixed beds are selected if the mass transfer zone is short—that is, the breakthrough curve, from pilot experiments, is steep; the carbon usage rate is low and, therefore, the carbon bed will not be replaced for a long period of time; and the capital cost of an additional contactor is not justified because of the low operating costs.

Fixed beds in parallel are used if continuous operation of the adsorber is required, and the required size of the bed is too large to be economically feasible to fit in the available space or to result in a high pressure drop in the bed. The carbon in the single or parallel fixed bed is removed after breakthrough of the contaminant occurs. Therefore, the mass transfer zone would still be in the bed, and the carbon is not exhausted when it is replaced.

Fixed beds in series are preferred if the mass transfer zone is long—that is, the breakthrough curve is gradual and saturation of the bed occurs a long time after breakthrough; or the carbon usage rate is high, and therefore, it would be more economical to exhaust the carbon. When the carbon in the lead contactor in a series operation is exhausted, it is replaced by fresh carbon and the contactor is put on stream at the end of the series. The second contactor in the series becomes the lead contactor. This allows more efficient utilization of the adsorptive capacity of the carbon than single or parallel beds.

MATHEMATICAL MODELING FOR FIXED BED ADSORBERS

Generalizations

Mathematical models for predicting the dynamic behavior of column adsorbers have been developed by several investigators [8,15–25]. Also, the subject has been exhaustively reviewed by Ruthven (see also Chapter 3) [7]. These models allow prediction of the breakthrough curve from basic kinetic and equilibrium data. Therefore, they provide, in principle, a method for the design and projection of the performance of fixed bed adsorbers without the need to conduct extensive experimental work. The development of mathematical equations is not presented here, but a brief overview and a specific model is discussed.

The overall rate of adsorption by porous adsorbents is always controlled by mass transfer resistance rather than by the actual adsorption process on the interior surfaces. Two distinct diffusional resistances to mass transfer are encountered. The first is associated with transfer of the solute through the laminar fluid boundary layers surrounding the particle (film diffusion). The second is resistance to diffusion of the solute within the pores of the particle and/or along the pore wall surfaces (intraparticle or pore diffusion). The rate-limiting step depends on the system and conditions and can be either film diffusion or pore diffusion or both [7,8].

The starting point for the development of a mathematical model to describe the dynamic behavior of the system is the differential mass balance equations for an element of the adsorption column and for an adsorbent particle within such an element [7]. For an element of the bed through which a fluid stream containing concentration, $c(z,t)$, of adsorbable species is in a plug flow pattern, the differential fluid phase mass balance is:

$$-D_s\frac{\partial^2 c}{\partial z^2} + \frac{\partial}{\partial z}(vc) + \frac{\partial c}{\partial t} + \left(\frac{1-\epsilon}{\epsilon}\right)\frac{\partial \bar{q}}{\partial t} = 0. \tag{4.11}$$

The mass balance of an adsorbent particle yields the adsorption rate expression:

$$\frac{\partial \bar{q}}{\partial t} = f(q,c). \tag{4.12}$$

Although a single equation is presented here, the mass transfer rate expression is commonly a set of equations comprising one or more diffusion equations with associated boundary conditions, incorporating the equilibrium constraints to which the mass transfer rate expression must reduce after a sufficiently long time [7].

The dynamic response of a column is given by the solution $[c(z,t), \bar{q}(z,t)]$ to Equations 4.11 and 4.12 subject to the initial and boundary conditions imposed on the column. The response to a change in the feed composition involves a mass transfer zone or concentration front that propagates through the column with a characteristic velocity determined by the equilibrium isotherm. Location of the front at any time may be found from an overall mass balance, but to determine the form of the concentration front, Equations 4.11 and 4.12 must be solved simultaneously. The theoretical breakthrough curve may be calculated by numerical solution of the model equations using standard finite difference or collocation methods [7].

Use of these models requires a determination of the isotherm and mass transport parameters. The isotherm parameters are determined from a batch equilibrium study. The Freundlich adsorption isotherm is used most frequently.

The film transfer coefficient, k_f, is determined from correlations available in the literature [26,27], and the effective intraparticle diffusivity, D_s, is obtained from rate studies of adsorption in batch reactors. Recently, Weber and Liu developed a microcolumn method that can simultaneously and accurately determine both characteristic mass transport parameters, k_f and D_s, required for adsorber modeling and design. The value of k_f can be determined from the initial portion of the microcolumn breakthrough profile. The internal surface diffusion coefficient, D_s, can be determined from the entire breakthrough profile using regression analysis.

Fixed Bed Model

The model developed by Liu and Weber [28] is presented here. The equations are derived from mass balances for the solid and fluid phases involved in fixed bed adsorber or ion exchange columns. In addition to the traditional assumptions made for fixed bed models, such as spherical adsorbent particles and plug flow, effective surface diffusivity is used to describe the rate-controlling component of intraparticle diffusion. See Equations 3.44 to 3.69.

The solid-phase mass balance is:

$$\frac{\partial q}{\partial t} = D_s \left(\frac{\partial^2 q}{\partial r^2} + \frac{2\partial q}{r\partial r} \right). \tag{4.13}$$

The liquid-phase mass balance is:

$$-D_p \frac{\partial^2 C}{\partial z^2} + v \frac{\partial C}{\partial z} + \left(\rho \frac{1 - \epsilon}{\epsilon} \right) \frac{\partial q}{\partial t} + \frac{\partial C}{\partial t} = 0. \tag{4.14}$$

The Freundlich adsorption isotherm is:

$$q_s = K_F \, C_s^n . \tag{4.15}$$

Equation 4.13 then becomes:

$$\frac{\partial \bar{q}}{\partial \bar{t}} = N_d \frac{1}{\bar{r}^2} \frac{\partial}{\partial \bar{r}} \left(\bar{r}^2 \frac{\partial \bar{q}}{\partial \bar{r}} \right). \tag{4.16}$$

Equation 4.14 becomes:

$$-\frac{1}{Pe} \frac{\partial^2 \bar{c}}{\partial \bar{z}^2} + \frac{\partial \bar{c}}{\partial \bar{z}} + \frac{\partial \bar{q}}{\partial \bar{t}} + \frac{1}{D_g} \frac{\partial \bar{c}}{\partial \bar{t}} = 0, \tag{4.17}$$

and Equation 4.15 becomes:

$$\bar{q}_s = \bar{c}_s^n. \tag{4.18}$$

The transformed initial and boundary conditions are:

$$\bar{c} = 1 + \frac{1}{Pe}\frac{\partial \bar{c}}{\partial \bar{z}} \qquad \text{at } \bar{z} = 0, \tag{4.19}$$

$$\frac{\partial \bar{c}}{\partial \bar{z}} = 0 \qquad \text{at } \bar{z} = 1, \tag{4.20}$$

$$\frac{\partial \bar{q}}{\partial \bar{r}} = 0 \qquad \text{at } \bar{r} = 0, \tag{4.21}$$

$$\frac{\partial \bar{q}}{\partial \bar{r}} = Sh(\bar{c} - \bar{c}_s) \qquad \text{at } \bar{r} = 1, \tag{4.22}$$

$$\bar{q}(\bar{r}, \bar{z}) = 0 \qquad \text{at } \bar{t} = 0, \tag{4.23}$$

and

$$\bar{c}(\bar{z}) = 0 \qquad \text{at } \bar{t} = 0. \tag{4.24}$$

Equations 4.21 to 4.24, along with the Freundlich exponent, n, and the dimensionless groups, D_g, Pe, Sh, and N_d, represent the dynamics of breakthrough curves for fixed bed adsorbers. The parameters, n and D_g, can be determined experimentally from isotherm studies. Pe is negligible for adsorbers of reasonable length.

The mass transport parameters, k_f and D_s, can be determined by the microcolumn method described by Liu and Weber [28]. Values of k_f, the external film transfer coefficient, can be determined from the initial portion of the microcolumn breakthrough profile. The internal surface diffusion coefficient, D_s, can be determined from the entire breakthrough profile using regression analysis. These equations are solved numerically with a computer.

PROCESS DESIGN OF ADSORBERS

Generalizations

Theoretical mathematical models, although useful in predicting the performance of an adsorber, should be verified before their application to practical situations in water and wastewater treatment. It is always advisable to base the final design on detailed studies simulating the application. The most commonly used procedures for the design of GAC adsorbers are briefly discussed here.

First, a laboratory-scale batch equilibrium study should be conducted using the selected activated carbon and the water to be treated. The data are then plotted according to the Freundlich adsorption isotherm. The isotherm results provide information on the feasibility of using activated carbon to meet the treatment objectives and to determine the theoretical capacity and usage rate of the carbon. The theoretical capacity is determined by reading x/M at C_o from the isotherm plot. Figure 4–3 indicates that where $C_o = 20$ mg/l phenol, the corresponding value for $x/M = 115$ mg/g:

$$\left(\frac{x}{M}\right)_{C_o} = 115\,\frac{\text{mg}}{\text{g}} = 0.115\,\frac{\text{g}}{\text{g}} = 0.115\,\frac{\text{lb contaminant}}{\text{lb carbon}}$$

Usage rate is determined from:

$$\frac{C_o}{(x/M)_{C_o}} \times 8.337 = \text{lb of carbon/1,000 gal} \qquad (4.25)$$

$$\frac{20}{115} \times 8.337 = 1.45\ \text{lb/1,000 gal.}$$

Once the feasibility of using activated carbon is established, a pilot-scale continuous flow column study is conducted to simulate the full-scale operation and to determine breakthrough curves under different operating conditions. Usually three or more columns are used in series (Figure 4–11) to represent different bed depths and contact times. The inside diameter of the columns may range from 1 to 6 in. The columns are loaded with quantities of activated carbon predetermined from the usage rate. Flow rate may range between 2 and 10 gal/min/ft^2. The test should be continued until the breakthrough curves are fully devel-

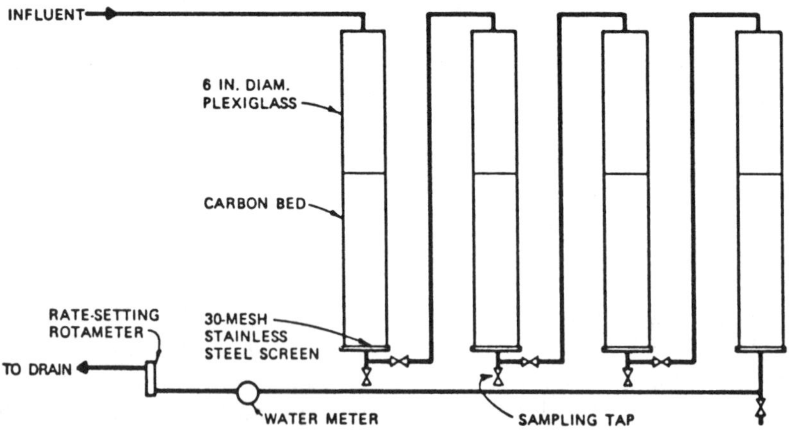

Figure 4–11 Downflow pilot carbon columns [4].

oped; that is, the final column effluent concentration approaches the influent concentration. Breakthrough curves are plotted from the test results for each column in the series.

Analysis of Data

The capacity of the carbon is determined from breakthrough curves by taking the area between the column influent and effluent to the point of breakthrough divided by the weight of carbon in the column. The capacity at complete column exhaustion is determined by taking the total area to the point where the effluent plot joins the effluent and dividing this value by the weight of carbon in the column (see Figure 4–7). The carbon's capacity can then be compared to batch isotherm values. The results of the column study are scaled up to a full-scale operation.

Major Design Approaches

Two major design approaches are used for the scale-up of the adsorber from the pilot-scale column study—namely, empty bed contact time and bed depth-service time.

Empty Bed Contact Time

The major design parameters to be evaluated are the contact time, or the empty bed contact time (EBCT), and the carbon usage rate [3,29,30]. Exhaustion curves are developed from the column data by plotting the effluent concentration versus the volume processed or processing time for each column. A horizontal line that intersects the breakthrough curves at the desired concentration is drawn to determine the breakthrough volume for each column. The EBCT is determined from:

$$\text{EBCT} = \frac{\text{bed volume (ft}^3) \times 7.48}{\text{hydraulic surface loading (gal/min/ft}^2) \times \text{bed surface area (ft}^2)} \tag{4.26}$$

where

$$\text{Bed volume} = \frac{\text{weight of carbon (lb)}}{\text{carbon bulk density (lb/ft}^3)}.$$

$$\text{EBCT} = \frac{\text{bed volume (ft}^3) \times 7.48}{\text{flow rate (gal/min)}}, \tag{4.27}$$

$$\text{EBCT} = \frac{\text{bed depth (ft)}}{\text{linear velocity (ft/min)}}, \tag{4.28}$$

$$\text{EBCT} = \frac{\text{bed depth (ft)} \times 7.48}{\text{hydraulic surface loading (gal/min/ft}^2\text{)}}. \tag{4.29}$$

The carbon dosage is determined from:

$$\text{Usage rate (lb/1,000 gal)} = \frac{\text{weight of carbon in column (lb)}}{\text{volume at breakthrough (gal)}} \times 1{,}000. \tag{4.30}$$

From the data, a plot of the carbon usage rate/1,000 gal treated versus EBCT is constructed. An example of the exhaustion curve is shown in Figure 4–12 from a pilot study for treating industrial wastewater. The data needed to construct the EBCT/usage curve are shown in Table 4–1, and the plot is shown in Figure 4–13. The EBCT used for the design of the contactor is taken from the plot at the EBCT at the lowest carbon usage rate. It can be noticed in Figure 4–13 that, after an initial rapid decrease in usage rate with increasing EBCT, the curve flattens and no significant reduction in usage rate is gained with increasing contact time.

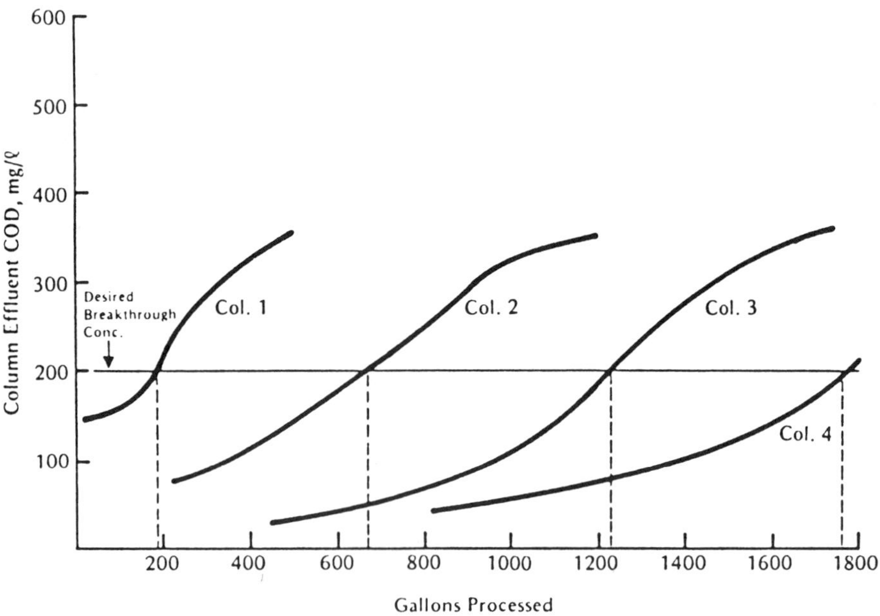

Figure 4–12 Column exhaustion curves. Reproduced from Perrich [30], courtesy of CRC Press, Inc.

Table 4–1 Carbon Usage Rates for COD Removal

Column	EBCT	Cumulative lb Carbon	Gallons Processed at Breakthrough	Carbon (lb/1,000 gal)	Usage per Day	Carbon Loading at Breakthrough[a]
1	14	8	192	41.7	834	0.20
2	28	16	672	23.8	476	0.35
3	42	24	1,232	19.5	390	0.43
4	56	32	1,767	18.1	382	0.46

Source: Reproduced from Perrich [30] courtesy of the CRC Press, Inc.
[a]lb COD removed per lb carbon.

Therefore, optimization of the process design requires economic evaluation of the benefits of increasing the EBCT with the slight gain in carbon usage. Once an EBCT is selected at the design flow rate, the size of the full-scale contactor can be determined. In a system designed for a series operation, the selected EBCT should allow the lead column to be completely exhausted while maintaining suitable effluent quality.

Bed Depth-Service Time

Bohart-Adams Model. The bed depth-service time (BDST) method is based on a model proposed by Bohart and Adams [31] that was later modified by Klotz [32] and Dole and Klotz [33]. The model assumes that the adsorption rate is propor-

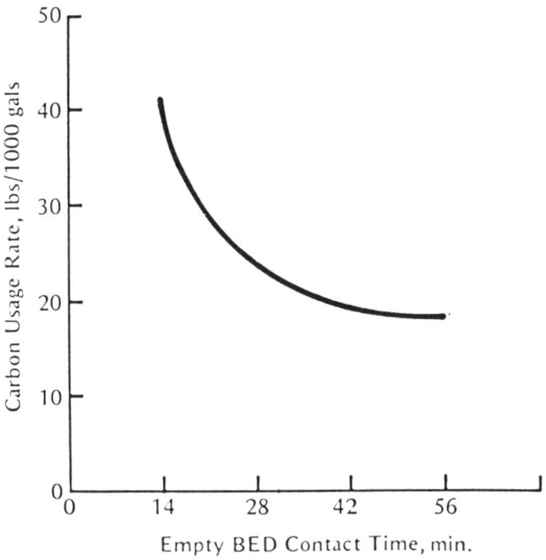

Figure 4–13 Carbon usage rate versus system residence time. Reproduced from Perrich [30], courtesy of CRC Press, Inc.

tional to both the residual carbon capacity and the remaining adsorbate concentration. According to Bohart and Adams, the BDST curves are described by [31,34,35]:

$$\ln\left(\frac{C_o}{C} - 1\right) = \ln[\exp(KN_oD/V) - 1] - KC_ot. \quad (4.31)$$

Rearranging:

$$t = \frac{N_o}{C_oV}\left[D - \frac{V}{KN_o} \ln\left(\frac{C_o}{C_B} - 1\right)\right], \quad (4.32)$$

where t = service time (hr), V = linear flow rate (ft/hr), D = depth of carbon bed (ft), D_o = critical depth of carbon bed (ft), K = rate constant (ft^3/lb/hr), N_o = adsorptive capacity (lb/ft^3), C_o = influent solute concentration (mg/l), and C_B = solute concentration at breakthrough (mg/l).

Equation 4.32 predicts linear BDST plots. The theoretical depth of carbon that is sufficient to prevent the effluent solute concentration to exceed C_B at zero time is called the *critical bed depth* (length of mass transfer zone) and may be calculated from Equation 4.32 when t = 0 [34]:

$$D_o = \frac{V}{KN_o}\left[\ln\left(\frac{C_o}{C_B} - 1\right)\right]. \quad (4.33)$$

The adsorptive capacity of the system, N_o, and the rate constant, K, can be evaluated from the slope and intercept of the plot of t versus D, which yields a straight line. The numerical value of the slope is N_o/C_oV whereupon N_o can be calculated. The rate constant is in turn calculated from the intercept by using:

$$b = \frac{1}{C_oK} \ln\left(\frac{C_o}{C_B} - 1\right). \quad (4.34)$$

Figure 4–14 shows a BDST plot, taken from Eckenfelder and Ford [34], that illustrates these relationships.

Data for the BDST method are obtained by conducting a pilot-scale study using at least three columns in series. The water to be treated should be used and applied at a predetermined hydraulic loading. The total carbon depth and flow rate should approximate those expected in the full-scale unit. Upflow or downflow mode of operation can be used. Breakthrough curves are developed by plotting effluent concentration or the percentage of solute remaining vesus service time for each column, and the breakthrough points then are determined. The breakthrough points, expressed as service time, are plotted against bed depth to give the BDST curve.

The service time at breakthrough should be a linear function of bed depth if the hydraulic flow rate and composition of the influent were reasonably con-

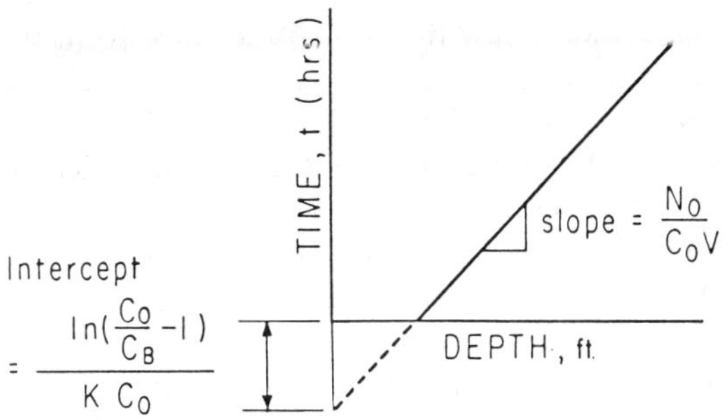

Figure 4–14 BDST plot from Bohart-Adams model. Reproduced from Eckenfelder and Ford [34], courtesy of Pemberton Press, Inc.

stant throughout the test. The coefficients in the Bohart-Adams equations are evaluated from the plot. The service times at different bed depths are calculated from Equation 4.32. An acceptable service time and the corresponding bed depth are selected based on judgment and/or experience. However, to optimize the full-scale design, an economic analysis should be conducted estimating the total annual cost (capital and operating and maintenance) for different bed depths and subsequently selecting the bed depth that incurs the lowest annual cost. Once the bed depth is determined, at the preselected flow rate used in the pilot study, the size of the full-scale contactor can be calculated.

Eckenfelder and Ford [34] employed the Bohart-Adams relationship to design carbon columns. A pilot study was conducted at different bed depths and flow rates, and the scale-up was within the experimental parameters without the need for extrapolation.

Hutchins' BDST Design Equations. The Bohart-Adams model has been modified by Hutchins [36] and extended for design at different flow rates and feed concentrations and for series and pulsed beds. Since the Bohart-Adams model gives a linear relationship between the bed depth and service time, the curve can be described by the equation of a straight line:

$$t = ax + b, \tag{4.35}$$

where t = service time at breakthrough (hr), x = bed depth (ft), a = slope = 1,990 N_o/C_oV (hr/ft), b = ordinate intercept = $16.018/KC_o \ln (C_o/C_B - 1)$ (hr), N_o = carbon efficiency (lb impurity/ft^3 of carbon), C_o = feed impurity concentration (ppm), V = linear flow rate (gpm/ft^2), K = adsorption rate constant (ft^3 of liquid treated per lb of impurity fed to the system per hr, required for the adsorption wave front to move through a bed depth equivalent to the critical bed depth), and C_B = impurity concentration in effluent at breakthrough (ppm).

The preceding is the BDST equation. The numerical constants vary depending on the units used to develop the data.

The BDST curve shows how the adsorption wave front moves through the carbon bed. The slope, *a,* of the line is the amount of time needed for the wave front to move through 1 ft of carbon.

The reciprocal of the slope, then, is the rate at which the carbon bed is spent. Multiplying this value by the adsorbent's apparent bulk density gives an estimate of the carbon use rate (or rate at which carbon must be regenerated) to continuously produce an acceptable product from, in this example, wastewater.

The intercept of the abscissa (x axis) in Figure 4–15 is the critical bed depth, X_o, defined as the minimum depth for obtaining a satisfactory effluent at time zero under the test operating conditions. The intercept, b, of the ordinate is a measure of the adsorption rate and is defined as the time required for the adsorption wave front to pass through the critical bed depth.

Changes in Flow Rate

The BDST equation developed from column tests at one linear flow rate can be modified to calculate equations for other rates by multiplying the original slope, a, by the ratio of the original and new rates:

$$a' = \frac{V \cdot a}{V'} \tag{4.36}$$

where a and a' are the old and new slopes, respectively, and V and V' are the old and new linear flow rates, respectively. This equation for the new condition is valid because changing the linear flow rate has little or no effect on the ordinate intercept, b. With this procedure, laboratory tests can be reliably scaled up to

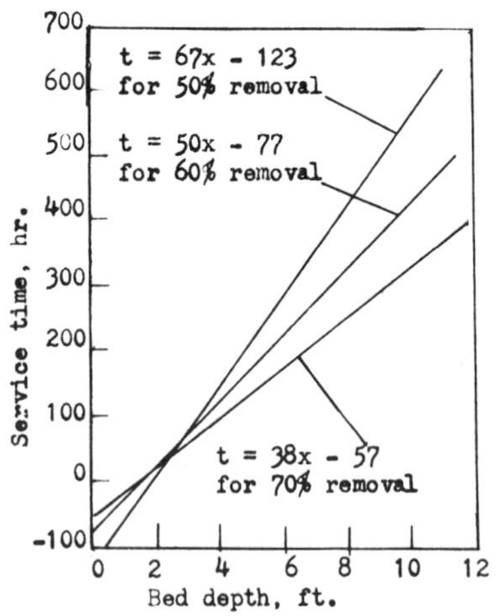

Figure 4–15 BDST curves show linear relationship between bed depth and service time. Reproduced from Hutchins [36], courtesy of the McGraw-Hill Book Co., Inc.

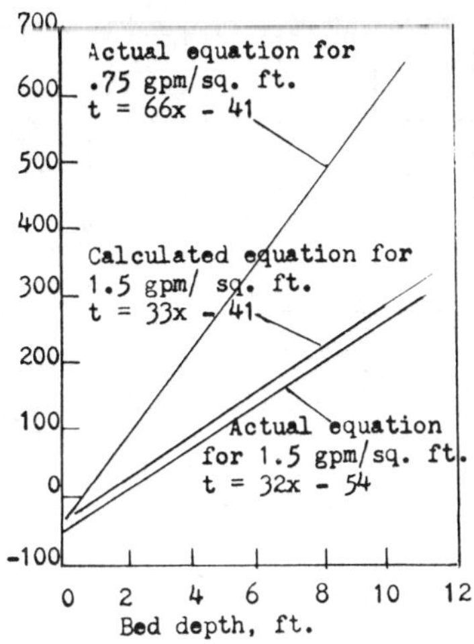

Figure 4–16 Actual versus calculated BDST curves for new flow rate show good agreement. Reproduced from Hutchins [36], courtesy of the McGraw-Hill Book Co., Inc.

other flow rates without running pilot tests in larger columns. For example, Figure 4–16 shows BDST equations obtained from testing phenol removal from water at linear flow rates of 0.75 and 1.5 gpm/ft^2. In both cases, phenol concentration was reduced from 100 ppm to 1 ppm. The slope for the higher rate calculated from data at 0.75 gpm/ft^2 is 33 hr/ft. This agrees very well with the slope of 32 hr/ft obtained in actual tests.

Changes in Concentration

The BDST equation can be modified to account for changes in the feed concentration according to the following relationships:

$$a' = \frac{C_o \cdot a}{C_1} \tag{4.37}$$

and

$$b' = b \left(\frac{C_o}{C_1}\right) \frac{\ln[(C_1/C_F) - 1]}{\ln[(C_o/C_B) - 1]}, \tag{4.38}$$

where C_o and C_1 are the old and new feed concentrations, and C_B and C_F are the old and new effluent concentrations.

Figure 4–17 shows BDST curves and equations for removing phenol from water at 100 and 200 ppm feed concentrations. The figure also contains an estimated curve for the higher concentration calculated from the BDST equation.

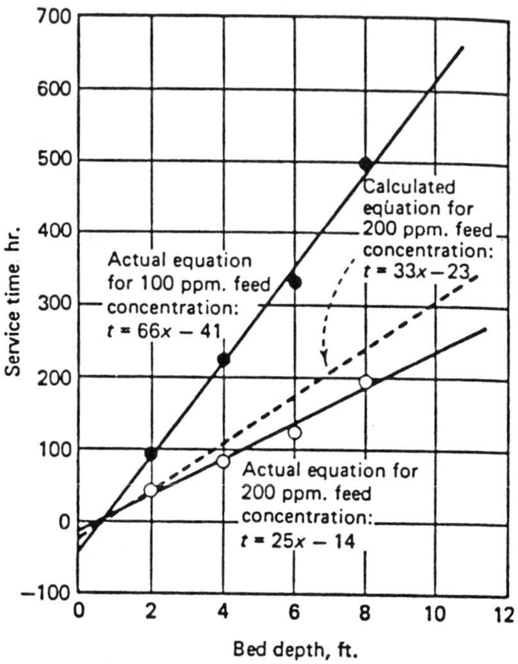

Figure 4–17 Effect of new feed concentration may also be estimated with BDST equation. Reproduced from Hutchins [36], courtesy of the McGraw-Hill Book Co., Inc.

The slope of the curve for the higher concentration was estimated by multiplying the slope at 100 ppm by the ratio of the two feed concentrations:

$$a' = 66 \cdot \frac{100}{200} = 33.$$

The intercept, b, for 100 ppm feed is shown on the graph as -41. An estimated value for the intercept at 200 ppm was calculated from the relationship:

$$b' = -41 \left(\frac{100}{200}\right) \frac{\ln[(200/1) - 1]}{\ln[(100/1) - 1]}$$

$$b' = -23.$$

In this case, the difference between actual and calculated equations is within experimental error. This method for estimating the effect of changing feed concentration appears to work well when removing one-component impurities from water. Additional work must be done to determine if it can be applied to more complex mixtures in process streams and wastewaters.

Modified BDST Equation for Series or Pulsed Beds

The BDST equation, calculated at instantaneous breakthrough points, describes how an adsorption wave front moves through a single fixed bed column. However, the equation may be varied to cover series and moving bed systems as well. With series and moving bed operations, a unit is removed or fresh carbon is added when the adsorption wave front begins to leave the system. Therefore, the rate at which the wave front moves will be the rate that determines when the next column should be taken off line or when a pulse of carbon must be removed.

Conducting an actual test is the most reliable way of determining the wave front rate. However, a good estimate can be made by applying the BDST equation, with an assumed average concentration of adsorbable impurities in the feed and the appropriate factors from Table 4–2.

The values in Table 4–2 were developed by assuming that the moving bed is pulsed or a column removed just as the wave front begins to exit from the system. This is the point at which the carbon is exhausted—that is, in equilibrium with the feed concentration. The data also assume that the impurity concentration in the adsorption zone is an exponential function of the bed depth in the column.

As the wave front moves through fresh carbon, the concentration of impurities entering the freshly added carbon increases. The average concentration fed to the freshly added carbon is the average of that fed from the time the adsorption wave front enters the freshly added carbon until it begins to exit. The factors in Table 4–2 are the fraction of the original impurity concentration that is assumed to be the average fed to freshly added granular carbon.

Consider a moving bed system operating at 4 gpm/ft^2, with a wastewater impurity of 129 ppm. Acceptable effluent is 38.7 ppm (70% removal). The BDST equation developed from a column test was:

$$t = 19x - 57.$$

Table 4–2 Factors for Modifying the BDST Equation

	Required Removal of Adsorbable Impurities, %					
Factor	*50*	*60*	*70*	*80*	*90*	*95*
Moving bed factors, pulsing 5% of bed	0.51	0.41	0.31	0.21	0.107	0.054
Series factors for number of columns						
2	0.70	0.63	0.54	0.44	0.31	0.22
3	0.59	0.50	0.40	0.30	0.18	0.11
4	0.56	0.46	0.37	0.26	0.144	0.080
5	0.55	0.44	0.35	0.24	0.133	0.072

Source: Reproduced from Hutchins [36] courtesy of the McGraw-Hill Book Co.

If we assume that the composition of adsorbable impurities at the wave front compares to a diluted feed with average concentration of 0.31 C_o and that the y axis intercept (−57) is not significant in a series or moving bed system, then we can estimate the slope of a BDST equation describing the movement of a wave front (38.7 ppm) through fresh or regenerated carbon:

$$a' = \frac{C_o \cdot a}{f \cdot C_o}$$

$$= \frac{129 \cdot 19}{0.31 \cdot 129} = 61.3 \text{ hr/ft.}$$

This example is shown in the sample calculations. The reciprocal of a' can usually be used to estimate the carbon use rate for series and moving bed operations, although this gives liberal rather than conservative values of carbon consumption.

Cost Estimates and Sizing

After defining the basic design equation, cost estimates can be made to optimize the size of the adsorption and regeneration units. The size of the adsorption system depends on carbon bed depth and linear flow rate. These two factors determine residence time, which is the major design parameter. The size of the regeneration system is based on the carbon use rate, a function of the BDST equation. Estimates for these separate operations—adsorption and regeneration—determine total investment and operating costs.

Examples of using the BDST equation for sizing are presented in the sample calculations. The initial estimate should be based on reasonable operating parameters, usually selected from prior experience. Then other situations are examined to find the best balance between investment and operating costs.

The relationships between residence time, investment, and operating costs (direct and indirect) are shown in Figures 4–18 and 4–19. These curves are based on a fixed bed system and the operating conditions listed in the sample calculations. Usually, residence time at minimum total operating cost fixes the optimum system size. In the example, this is 25 minutes, equal to a bed depth of 13 feet.

The curve for total operating costs usually has a minimum point at a certain residence time when working with single-column or parallel systems. This may not be true for series or moving bed installations because the carbon use rate is more a function of the reciprocal of the slope of the BDST equation than of the entire equation. In these cases, system selection must be based on the relative importance of capital and operating costs in the particular company or municipality.

Limitations

As with all data evaluation techniques, the BDST equation will have limited value if the column feed is not representative of the normal plant system. Also, if there

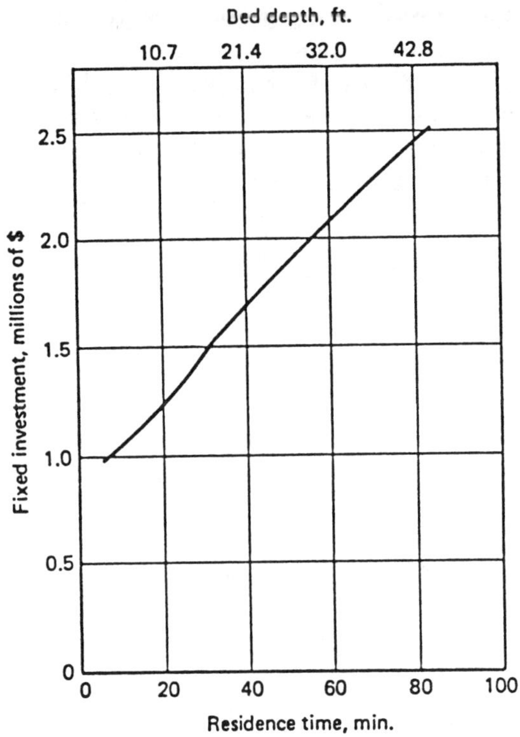

Figure 4–18 Investment in a fixed bed carbon system varies directly with residence time. Reproduced from Hutchins [36], courtesy of the McGraw-Hill Book Co., Inc.

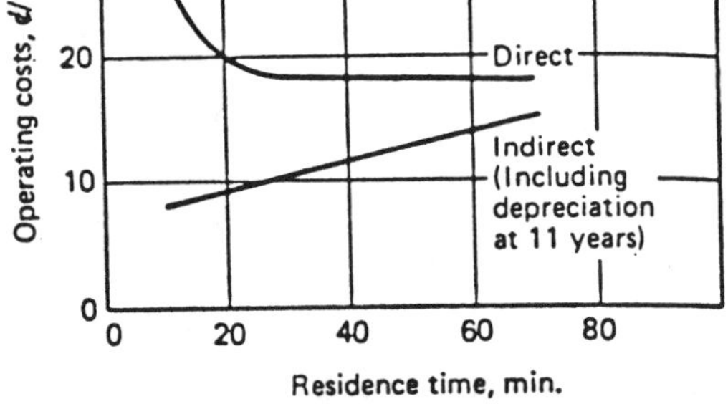

Figure 4–19 Residence time at minimum operating cost usually determines best system size. Reproduced from Hutchins [36], courtesy of the McGraw-Hill Book Co., Inc.

is chemical or biological activity in the test system, results may not be equivalent to those found in the plant-scale unit.

For the most accurate results, the adsorption wave front must move through the column at a constant rate. Conditions for this requirement—reasonably constant impurity concentration and composition and uniform linear flow rate—are not always met during testing. Care should be taken to run tests at the flow rate and bed depth most likely to be used in the final plant. Then, continued testing until unacceptable effluent is consistently obtained in the last column will give an estimate of the carbon use rate even under poor test conditions.

Hutchins' Method: Sample Calculations

Assume:

Slope, a: 19 hr/ft

Bed depth, x: 13 ft

Ordinate intercept, b: $-$ 57 hr

Feed impurity concentration, C_o: 129 ppm

Product impurity concentration, C_1: 38.7 ppm

Linear flow rate, v: 4 gpm/ft^2

Total flow: 5 million gal/day

Moving bed factor: 0.31

Carbon density: 23 lb/ft^3

A. Single-Column Fixed Bed: Adsorber System Size

1. Calculate cross-sectional area:

$$\frac{5{,}000{,}000}{4 \text{ gpm/ft}^2 \times 1{,}440 \text{ min/day}} = 868 \text{ ft}^2$$

2. Calculate volume of carbon required:

$$868 \text{ ft}^2 \times 13 \text{ ft} = 11{,}284 \text{ ft}^3$$

3. Calculate adsorber volume (allow 50% expansion):

$$11{,}284 \text{ ft}^3 \times 1.5 = 16{,}926 \text{ ft}^3$$

4. Make a preliminary cost estimate for an adsorption system with this volume.

B. Single-Column Fixed Bed: Regeneration Furnace System Size

1. Convert carbon volume to weight:

$$11{,}284 \text{ ft}^3 \times 23 \text{ lb/ft}^3 = 260{,}000 \text{ lb}$$

2. Use the BDST equation to calculate service time:

$$(19 \text{ hr/ft} \times 13 \text{ ft}) + (-57 \text{ hr}) = 190 \text{ hr}$$

3. Calculate carbon use rate:

$$260{,}000 \text{ lb}/190 \text{ hr} = 1{,}368 \text{ lb/hr}$$

4. Make a preliminary cost estimate for a regeneration furnace system of this capacity or the next largest available size.

C. Moving Bed: Adsorber System Size

1. Cross-sectional area same as fixed bed.
2. Volume of carbon same as fixed bed.
3. No allowance needed for bed expansion. Adsorber volume thus = 11,284 ft^3.
4. Make a preliminary cost estimate for an adsorption system with this volume.

D. Moving Bed: Regeneration Furnace System Size

1. Use the slope a to calculate a new slope, a', for the adsorption wave front passing through fresh carbon:

$$a' = (C_o \times a)/(f \times C_o)$$

$$a' = (129 \times 19)/(0.31 \times 129) = 61.3 \text{ hr/ft}$$

2. Calculate the rate at which the adsorption wave front passes through the fresh carbon:

$$1/61.3 \text{ hr/ft} = 0.0163 \text{ ft/hr}$$

3. Calculate the carbon use rate:

$$0.0163 \text{ ft/hr} \times 23 \text{ lb/ft}^3 \times 868 \text{ ft}^2 = 325 \text{ lb/hr}$$

4. The carbon-use rate is a minimum value. Make a preliminary cost estimate for the furnace system of the next largest available size.

Application of Design Methods

Design a GAC system to treat water containing 10 mg/l alkylbenzenesulfonates to a residual concentration of 0.5 mg/l at a flow of 20,000 gal/day.

Given:

1. The Freundlich adsorption isotherm developed from a batch equilibrium study was favorable.
2. A pilot-scale study using three columns having an inside diameter of 1 inch in series gave the results that follow [34].

Solution Using EBCT Method

1. A plot of carbon usage rate (lb/1,000 gal) versus EBCT is constructed as shown in Figure 4–20. Note that the carbon usage rate decreases rapidly with the increase in EBCT and that the curves for the through flow rates converge at high contact time.
2. Low carbon usage rate is obtained at a 15 minute contact time. Increasing

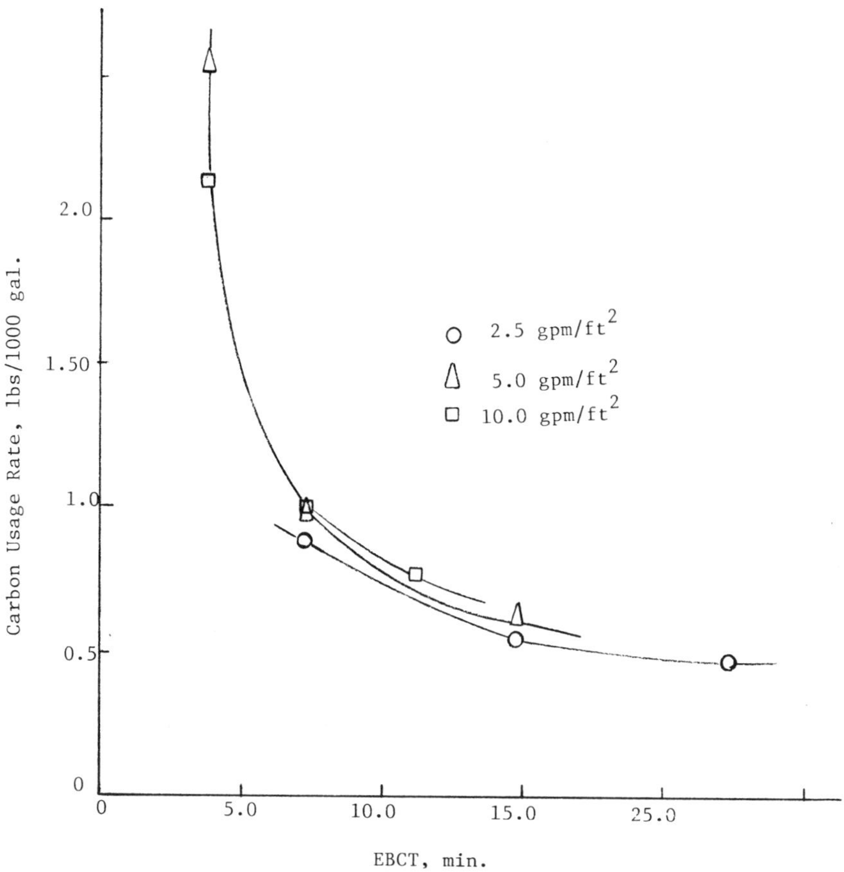

Figure 4–20 EBCT plot.

the EBCT beyond this value results in minor reduction in usage rate. Therefore, design EBCT = 15 minutes.

3. Volume of carbon (ft^3) = EBCT (min) × flow (ft^3/min).

$$\text{Flow (ft}^3\text{/min)} = \frac{\text{Daily flow (gal/day)}}{\text{min/day} \times 7.48 \text{ gal/ft}^3}$$

$$= \frac{20{,}000}{1{,}440 \times 7.48}$$

$$= 1.86 \text{ ft}^3\text{/min}$$

$$\text{Volume of carbon (ft}^3) = 15 \times 1.86 = 27.9 \text{ ft}^3$$

4. Weight of carbon (assume the bulk density of the carbon = 25 lb/ft^3):

$$\text{Weight of carbon} = 27.9 \times 25 = 697.5 \text{ lbs}$$

5. Adsorber sizing (assume linear flow rate = 2.5 gal/min/ft^2):

$$\text{Adsorber area (ft}^2) = \frac{\text{flow (gal/min)}}{\text{linear flow rate (gal/min/ft}^2)}$$

$$= \frac{20{,}000}{1{,}440 \times 2.5}$$

$$= 5.6 \text{ ft}^2$$

Assume a circular adsorber is used.

$$\text{Diameter of adsorber} = \sqrt{\frac{4 \times 5.6}{\pi}}$$

$$= 2.7 \text{ ft}$$

$$\text{Carbon bed depth} = \frac{\text{volume of carbon (ft}^3)}{\text{area of adsorber (ft}^2)}$$

$$= \frac{27.9}{5.6}$$

$$= 5 \text{ ft}$$

Sidewall height (ft) (allow 50% freeboard for bed expansion):

$$\text{Sidewall height} = 1.5 \times 5$$
$$= 7.5 \text{ ft}$$

Select the nearest commercially available sizes, and adjust the linear flow rate accordingly.

6. Carbon bed cycle time:

$$\text{Cycle time of bed (days)} = \frac{\text{lb carbon in adsorber}}{\text{lb carbon usage/day}}$$

Carbon usage rate determined from Figure 4.20 = 0.55 lb/1,000 gal

$$\text{Cycle time of bed} = \frac{697.5 \text{ lb}}{0.55 \text{ lb/1,000 gal} \cdot 20{,}000 \text{ gal/day}}$$
$$= 63.4 \text{ days}$$
$$= 1{,}522 \text{ hours}$$

To optimize the design, a higher EBCT can be used for sizing the adsorber, and an economic analysis is conducted. The EBCT that gives the lowest total annual cost should be selected.

Solution Using BDST Method: Bohart-Adams Design Model

1. Using the data of the pilot study (Table 4–3), plot the service time, t, against the bed depth for the different flow rates as shown in Figure 4–21.
2. The coefficients in the Bohart-Adams equation for carbon may be evaluated from this plot. The adsorptive capacity, N_o, can be calculated from the slope of the line:

$$N_o = (\text{slope})\, C_o V, \tag{4.39}$$

where:

$$C_o = \left(\frac{10 \text{ mg/l}}{10^6}\right)\left(\frac{62.4 \text{ lb}}{\text{ft}^3}\right)$$

$$V = \text{linear flow rate } \frac{\text{gal}}{\text{min} \times \text{ft}^2} \times \frac{\text{ft}^3}{7.4 \text{ gal}} \times \frac{60 \text{ min}}{\text{hr}} = \text{ft/hr.}$$

The rate constant, K, can be calculated from the intercept (Equation 4.34):

$$K = \frac{\ln(C_o/C_B - 1)}{C_o b} \tag{4.40}$$

From the data:

$$\ln\left(\frac{10}{0.5} - 1\right) = 2.94$$

The values of these coefficients are summarized in Table 4–4 and plotted in Figure 4–22 [34].

Table 4-3 Pilot-Scale Data for Adsorption of Alkylbenzenesulfonates from Water to 0.5 mg/l Breakthrough Point

Flow Rate (gal/min/ft²)	*EBCT (min)*	*Bed Depth (ft)*	*Cumulative Carbon Weight (g)*	*Breakthrough Volume (gal)*	*Breakthrough Time (hr)*	*Carbon Usage Rate (lb/1,000 gal)*
2.5	7.5	2.5	150	363	440	0.91
	15.0	5.0	300	1,216	1,480	0.54
	22.5	7.5	450	2,148	2,620	0.46
5.0	3.75	2.5	150	141	87	2.34
	7.5	5.0	300	730	440	0.91
	15.0	10.0	600	2,190	1,340	0.60
10.0	5.0	5.0	300	332	102	2.14
	10.0	10.0	600	1,380	420	0.96
	15.0	15.0	900	2,760	835	0.72

Source: Adapted from Eckenfelder and Ford [34].

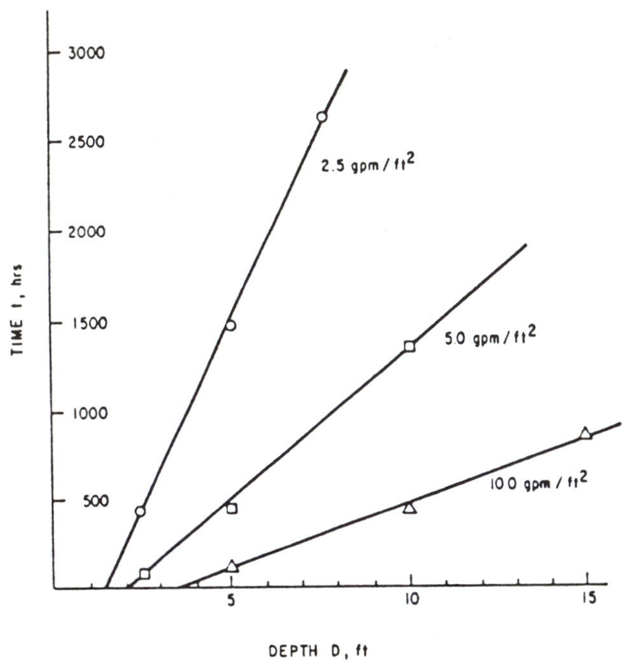

Figure 4-21 BDST plot. Reproduced from Eckenfelder and Ford [34], courtesy of the Pemberton Press, Inc.

Table 4-4 Equation Parameters

Flow Rate (gal/min/ft²)	*Slope*	*N_o (lb/ft³)*	*Intercept b*	*K (ft³/lb-hr)*	*D_o (ft)*
2.5	435	5.4	−630	7.5	1.46
5.0	170	4.2	−370	12.7	2.18
10.0	73	3.6	−250	18.8	3.48

Source: Reproduced from Eckenfelder and Ford [34], courtesy of Pemberton Press.

3. Select a design linear flow rate. For this example a linear flow rate of 2.5 gpm/ft² is selected.
4. The service time, t, can be calculated from Equation 4.32. The coefficients in the equation are obtained from Figure 4.22.

   At 2.5 gpm/ft², the following values are obtained:

$$K = 7.5 \text{ ft}^3/\text{lb/hr}$$

$$N_o = 5.4 \text{ lb/ft}^3$$

$$V = \frac{2.5 \text{ gpm}}{\text{ft}^2} \times \frac{\text{ft}^3}{7.48 \text{ gal}} \times \frac{60 \text{ min}}{\text{hr}}$$

$$= 20 \text{ ft/hr}$$

$$t = \frac{5.4}{0.000624(20)} \left[D - \frac{20}{7.5(5.4)} \times 2.95 \right] \tag{4.41}$$

Equation 4.41 contains two unknowns—namely, t and D. Assume a carbon bed depth of 5 ft is used. Substituting in Equation 4.41 gives:

$$t = 1{,}533 \text{ hours.}$$

5. Sizing of the adsorber:

$$\text{Adsorber area} = \frac{20{,}000 \text{ gal/day}}{1{,}440 \text{ min/day} \times 2.5 \text{ gal/min/ft}^2}$$

$$= 5.6 \text{ ft}^2.$$

$$\text{Diameter} = \sqrt{\frac{4 \times 5.6}{\pi}}$$

$$= 2.7 \text{ ft.}$$

Sidewall height (allow 50% bed expansion):

$$\text{Sidewall height} = 5 \times 1.5 = 7.5 \text{ ft.}$$

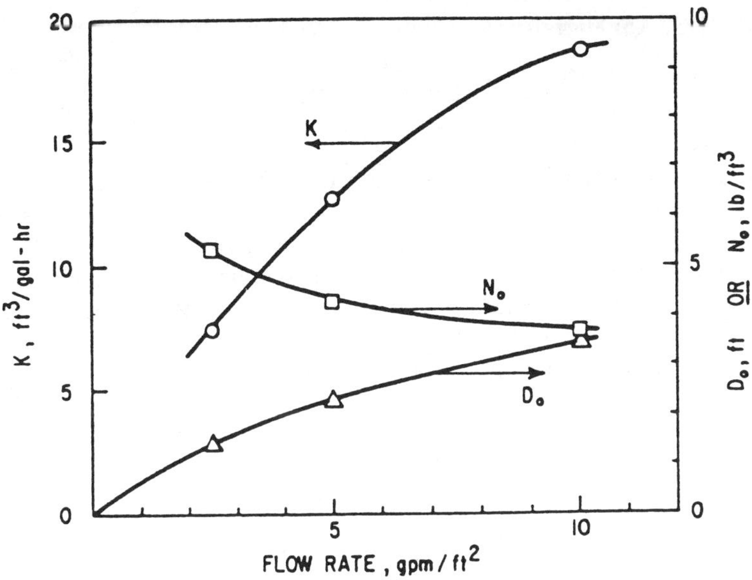

Figure 4–22 The effect of flow on the value of *K*, N_o, and D_o. Reproduced from Ecknefelder and Ford [34], courtesy of the Pemberton Press, Inc.

Summary of design:

Diameter of adsorber	= 2.7 ft
Depth of carbon bed	= 5.0 ft
Flow rate	= 2.5 gpm/ft²
Service time	= 1,533 hr

Note that this design yields the same results as the EBCT method. However, under the Bohart-Adams method, a judgment regarding acceptable bed depth had to be made. The design can be optimized by computing service times for different depths and conducting an economic analysis to select the depth and corresponding service time that give the lowest total annual cost.

Solution Using BDST Method: Hutchins' Design Methods for a Single Fixed Bed

1. The data from the pilot-scale study for a single flow rate of 2.5 gpm/ft² are plotted in a BDST relationship as shown in Figure 4–21.
2. The slope and intercept are determined from the plot to develop the BDST equation. From Table 4–4, the slope = 435 ft/hr and the intercept = −630 hr.

3. The BDST equation is:

$$t = 435x - 630 \tag{4.42}$$

where x = bed depth.

4. The most important design parameter that determines the bed depth is the EBCT. This can be determined from an economic analysis for different bed depths and their corresponding residence times and selecting the bed depth that gives the minimum total annual costs (capital and operating cost). For this example the optimum EBCT was found to be 15 minutes (compare the results under the EBCT method). The corresponding bed depth for this residence time can be obtained from the relationship:

$$\text{EBCT} = \frac{\text{bed depth (ft)} \times 7.48}{\text{linear flow rate (gpm/ft}^2)}$$

$$15 = \frac{\text{bed depth} \times 7.48}{2.5}$$

$$\text{Bed depth} = 5 \text{ ft.}$$

5. Adsorber sizing: Follow the same procedure as under the EBCT method. The results can be summarized:

Flow rate	= 2.5 gpm/ft^2
Bed diameter	= 2.7 ft
Bed depth	= 5.0 ft
Sidewall height	= 7.5 ft
Volume of carbon	= 27.9 ft^3
Weight of carbon	= 697.5 lb

6. Use the BDST equation (developed in step 2) to calculate service time:

$$\text{Service time} = 435(5) - 630$$
$$= 1{,}545 \text{ hr}$$

7. Carbon use rate:

$$\frac{697.5 \text{ lb}}{1{,}545} = 0.45 \text{ lb/hr}$$

Solution Using BDST Method: Hutchins' Design Method for a Pulsed Bed

1. Sizing of the adsorber is exactly the same as the fixed bed except no allowance is made for bed expansion.

2. Carbon use rate:
   a. Calculate a new slope, a', for the adsorption wave front passing through the fresh carbon using the slope, a, determined in the BDST equation for the fixed bed:

$$a' = (C_o a)(fC_o)$$

$$a' = (10 \times 435)/(0.054 \times 10)$$

$$= 8{,}055 \text{ hr/ft}$$

   (f is obtained from Table 4.2 for 10 − 0.5/10 = 0.95, or 95% removal).

   b. Calculate the rate at which the adsorption wave front passes through the fresh carbon:

$$1/8{,}055 \text{ hr/ft} = 0.000124 \text{ ft/hr}$$

   c. Calculate the carbon use rate:

$$0.000124 \text{ ft/hr} \times 25 \text{ lb/ft}^3 \times 5.6 \text{ ft}^2 = 0.017 \text{ lb/hr}$$

   (Note the considerable carbon saving when comparing the pulsed bed with the fixed bed.)

General Comment on Design Methods

The EBCT method is the simplest and most straightforward for sizing fixed bed adsorbers. The Bohart-Adams equation and Hutchins equation are the same except that Hutchins extended the model to apply to different flow rates, influent concentrations, and sizing of pulsed beds. However, both equations require the assignment of an appropriate bed depth for the final sizing of the adsorber. The bed depth must be obtained by conducting an economic analysis that ultimately yields the optimum EBCT. This illustrates that the single most important design parameter for carbon adsorbers is the EBCT. Irrespective of the design method used, the final sizing of the bed should be used on a pilot-scale study simulating the water characteristics, flow rate, and bed depth used in the scaled-up system.

REFERENCES

1. Sontheimer, H. *Translation of Reports on Special Problems of Water Technology, 9, Adsorption*. EPA-600/9-76-030, U.S. EPA, Washington, D.C. (1976), p. 67.
2. Faust, S.D., and O.M. Aly. *Chemistry of Water Treatment*. Butterworths, Stoneham, Mass. (1983).
3. Hutchins, R.A. "Development of Design Parameters," chapter 4A in *Activated Carbon Adsorption for Wastewater Treatment*. J.R. Perrich, ed. CRC Press, Boca Raton, Fla. (1981), p. 61.

4. *Process Design Manual for Carbon Adsorption.* Technology Transfer. U.S. E.P.A., Washington, D.C. (1973).
5. Symons, J.M., et al. *Treatment Techniques for Controlling Trihalomethanes in Drinking Water.* EPA-600/2-81-156. U.S. E.P.A., Cincinnati, Ohio (1978).
6. Butler, J.A.V., and C. Okrent. *J. Phys. Chem. 34,* 2841 (1930).
7. Ruthven, D.M. *Principles of Adsorption Processes.* John Wiley & Sons, New York (1984).
8. Weber, W.J. Jr. *Physicochemical Processes for Water Quality Control.* Wiley-Interscience, New York (1972).
9. Zogorski, J.S. "The Adsorption of Phenols onto Granular Activated Carbon from Aqueous Solution." Ph.D dissertation, Department of Environmental Science Rutgers University, New Brunswick, N.J. (1975).
10. Morris, J.C., and W.J. Weber, Jr. "Adsorption of Biochemically Resistant Materials from Solution, 1." Environmental Health Series, AWTR-9. U.S. Department of Health, Education, and Welfare, Washington, D.C. (1964).
11. Aly, O.M., and S.D. Faust. "The Adsorption of Chlorinated Phenols by Activated Carbon." Research report, Department of Environmental Sciences, Rutgers University, New Brunswick, N.J. (1970).
12. Petura, J.C. "Development of Design Parameters," Chapter 4B in *Activated Carbon Adsorption for Wastewater Treatment.* J.R. Perrich, ed. CRC Press, Boca Raton, Fla. (1981), p. 95.
13. Storm, D.W. "Contacting Systems," Chapter 5 in *Activated Carbon Adsorption for Wastewater Treatment.* J.R. Perrich, ed. CRC Press, Boca Raton, Fla. (1981), p. 127.
14. Schuliger, W.G. In *Carbon Adsorption Handbook.* P.N. Cheremisinoff and F. Ellerbusch, eds. Ann Arbor Science Publishers, Ann Arbor, Mich. (1980), p. 55.
15. Keinath, T.M., and W.J. Weber, Jr. *J. Water Poll. Control Fed. 40,* 741 (1968).
16. Erskine, D.B., and W.G. Schuliger. *Amer. Inst. Chem. Eng. Symp. Ser. 68,* 185 (1972).
17. DiGiano, F.A., and W.J. Weber, Jr. *J. Water Pollut. Cont. Fed., 45,* 713 (1973).
18. Brauch, V., and E.U. Schlunder. *Chem. Eng. Sci. 30,* 539 (1975).
19. Keinath, T.M. In *Mathematical Modeling for Water Pollution Processes.* Ann Arbor Science Publishers, Ann Arbor, Mich. (1975), p. 1.
20. Crittenden, J.C., and W.J. Weber, Jr. *J. Envir. Eng. Div., Amer. Soc. Civil Engineers 104,* 185 (1978).
21. Crittenden, J.C., and W.J. Weber, Jr. *J. Envir. Eng. Div., Amer. Soc. Civil Engineers 104,* 433 (1978).
22. Liu, K.T., and W.J. Weber, Jr. *Chem. Eng. Comm. 6,* 49 (1980).
23. Famularo, J.A., et al. *J. Water Pollut. Cont. Fed. 52,* 2019 (1980).
24. Roth, J.A., and A.S. Kodvassal. In *Application of Adsorption to Wastewater Treatment.* W.W. Eckenfelder, Jr., ed. Enviro Press, Nashville, Tenn. (1981), p. 69.
25. Peel, R.G., and A. Benedek. *J. Envir. Eng. Div. Amer. Soc. Civil Engineers 106,* 797 (1981).
26. Harriot, P. *Amer. Inst. Chem. Eng. J. 8,* 93 (1962).
27. Brian, P.L.T. et al. *Amer. Inst. Chem. Eng. J. 15,* 727 (1969).
28. Liu, K.T., and W.J. Weber, *J. Water Pollut. Cont. Fed. 53,* 1541 (1981).
29. Wagner, N.J., and R.J. Jula. "Activated Carbon Adsorption," Chapter 3 in *Activated Carbon Adsorption for Wastewater Treatment.* J.R. Perrich, ed. CRC Press, Boca Raton, Fla. (1981), p. 41.

30. Bilello, L.J., and P.B. DeJohn. "Total Process Design and Economics," Chapter 7 in *Activated Carbon Adsorption for Wastewater Treatment.* J.R. Perrich, ed. CRC Press, Boca Raton, Fla. (1981), p. 177.
31. Bohart, G.S., and E.Q. Adams. *J. Amer. Chem. Soc. 42,* 523 (1920).
32. Klotz, I.M. *Chem. Revs. 39,* 241 (1946).
33. Dole, M., and I.M. Klotz. *Ind. Eng. Chem. 38,* 1289 (1946).
34. Eckenfelder, W.W., and D.L. Ford. *Water Pollution Control—Experimental Procedures for Process Design.* Jenkins Book Publishing, Austin (1970).
35. Ford, D.L., and F.S. Manning. "Treatment of Petroleum Refinery Wastewater," Chapter 18 in *Carbon Adsorption Handbook.* P.L. Cheremisinoff and F. Ellerbusch, eds. Ann Arbor Science Publishers, Ann Arbor, Mich. (1980), p. 687.
36. Hutchins, R.A. *Chem. Eng.,* 133 (August 20, 1973).

5

Physical and Chemical Properties of Carbon

Inasmuch as the sources of material for preparing activated carbon are many and diverse, the physical and chemical properties are expected to vary considerably. Activated carbon's remarkable adsorptive properties are due to its porous nature. Surface areas of pore walls range from 400 to 1,800 m^2/g in various commercial brands. When PAC and GAC were utilized first in drinking water treatment for color, taste, and odor removal, little attention was given to material source and surface characteristics of carbon. One of the early mainstays of the industry was prepared from waste pulp from the manufacture of paper. The trend today is to vary the carbon's activation process whereupon surface properties can be altered for specific purposes. However, no single carbon can remove all undesirable organic and inorganic compounds from drinking water. The inability to remove a compound is sometimes surface related, sometimes structural, and sometimes related to the configuration of the adsorbate.

PREPARATION AND/OR ACTIVATION

Source Materials

That the source materials are diverse is seen in Table 5-1 [1]. For water treatment, fewer substrates are suitable for use. These would include coconut shells, peat, sawdust, wood char, lignin, petroleum coke, bone char, anthracite coal, and molasses. Some of these materials yield products of a unique porous nature that leads to specialized applications. For example, high-density coconut shells produce a carbon with a considerably finer pore size distribution and a greater apparent density in the final product than is produced from paper mill waste [1]. Thus, this carbon is preferred for applications like gas adsorption and is less suitable for wastewater treatment because of the fine pores.

Another factor is the degree to which the burnoff of carbon has been allowed. The extent of burnoff is a variable that determines the development of high surface area and adsorptive capacity [1]. Source materials do vary in their elemental and molecular composition. There is an assortment of types of acti-

Table 5-1 Source Materials that Have Been Studied for the Production of Activated Carbon

Bagasse	Kelp and seaweed
Beet-sugar sludges	Lampblack
Blood	Leather waste
Bones	Lignin
Carbohydrates	Lignite
Cereals	Molasses
Coal	Nut shells
Coconut shells	Oil shale
Coffee beans	Peat
Corncobs and corn stalks	Petroleum acid sludge
Cottonseed hulls	Petroleum coke
Distillery waste	Potassium ferrocyanide residue
Fish	Pulp-mill waste
Flue dust	Rice hulls
Fruit pits	Rubber waste
Graphite	Sawdust
	Wood

Source: Reproduced from Hassler [1], courtesy of the Chemical Publishing Co.

vated products with respect to organic functional groups or their heteroatomic content and to their ultimate overall chemical composition; that is, an activated carbon's surface is not all elemental carbon. Rather, the surface is rich in electrons from various elements, one of which is oxygen.

Activation Processes

Carbonization

The process, pyrolysis, is conducted usually in the absence of air. Temperatures below 600°C are preferable for producing chars suitable for steam activation [1]. However, this is apparently not a universal temperature. Sugar carbonized at 900°C produces an activated char. In a study of activation of anthracite, carbonization at 850°C in an atmosphere of steam produced a more porous structure that was readily activated [1]. In any event, this heating at a high temperature causes an extensive burnoff, or degradation of noncarbon impurities. If this heating step is conducted in the absence of oxygen, then oxidation of the material must follow the heating process to yield an active product. This oxidation occurs by exposing the product to air or other oxidants during the cooling operation. This is perhaps the most important step in the overall activation process.

Oxidation

Many activation processes involve a reaction with oxidizing gases of steam, air, and CO_2 at elevated temperatures [1]. The reactions are as follows:

$$H_2O + C_x \rightarrow H + CO + C_{x-1} \quad \text{(at 800 to 900°C)} \tag{5.1}$$

$$CO_2 + C_x \rightarrow 2CO + C_{x-1} \quad \text{(at 800 to 900°C)} \tag{5.2}$$

$$O_2 + C_x \rightarrow 2CO + C_{x-2} \quad \text{(at 800 to 900°C)} \tag{5.3}$$

$$O_2 + C_x \rightarrow CO_2 + C_{x-1} \quad \text{(below 600°C)} \tag{5.4}$$

Adsorptive powers and/or capacity of the product are influenced by the chemical nature and concentration of the oxidizing gas, the reaction temperature, the extent to which the activation is conducted, and the amounts and kinds of mineral ingredients in the char [1]. In the activation of many types of chars, steam is preferable to CO_2 and is better than air, as seen in Table 5-2 where adsorptive capacities are compared. These capacities are presumably related to surface area and porosity. The oxidation step selectively erodes the surface, increasing the surface area and porosity and leaving the remaining atoms arranged in configurations that have specific affinities [1]. Apparently not all adsorptive capacities develop simultaneously or at a uniform rate. A portion develops during the initial period of activation whereupon the capacity increases with time over a 60 minute period. This may have been observed with batch activation techniques. A continuous process is employed, with the carbon moving downward and countercurrent to an upward-moving gas stream.

There have been many kinetic studies of the C–O_2 reaction at elevated temperatures [2,3,4]:

$$2C_{(s)} + O_{2(g)} = 2CO_{(g)} \tag{5.5}$$

One study was conducted in the 600 to 800°C range [3], whereas another study was run in the 800 to 1,300°C range [4]. For the first study, the rate of reaction

Table 5-2 Adsorption by Activated Pine Wood Char

Activation Conditions		*Substance Adsorbed (g/g carbon)*			
Gas	*Temperature (°C)*	*Chrysoidine R*	*Ponceau R*	*Aniline Blue*	*Iodine*
Air	600	0.34	0.10	0.05	0.36
Air	740	0.16	0.08	0.05	0.40
Air	790	0.15	0.08	0.06	0.42
Air	860	0.14	0.08	0.06	0.42
Air	910	0.13	0.10	0.06	0.40
Steam	770	0.37	0.19	0.16	0.60
Steam	825	0.37	0.17	0.17	0.60
Steam	880	0.36	0.16	0.21	0.62
CO_2	880	0.32	0.12		

Source: Reproduced from Hassler [1], courtesy of the Chemical Publishing Co.
Note: All processes of activation conducted to 30% yield based on weight of original char. Adsorption evaluated at an equilibrium solution concentration of 0.10 g/l.

(Equation 5.5) was found to be one-half order with respect to the oxygen's partial pressure. From an Arrhenius plot of the rate data, an activation energy of 42 kcal/mol (of O_2) was calculated [3].

Following the activation process, carbons are washed with water for applications that require a low soluble ash. When a pH neutral carbon is needed, a suitable mineral acid is added to remove the various basic oxides. After neutralization, the carbon is washed thoroughly with water, and Na_2CO_3 is added to remove any traces of the acid.

GENERAL PHYSICAL PROPERTIES

Overall Characteristics

Some general characteristics of activated carbon are listed in Table 5-3, of which several are important in drinking water treatment. Some detailed information is given here for surface areas and porosity, for example. Occasionally, the ash content of carbon is a factor affecting treatment, especially the iron content. Table 5-4 shows the inorganic contents of six commercially available carbons in the 1950s. Perhaps these are not utilized today, but their ash contents would be typical. Table 5-5 gives the inorganic analysis of four carbons employed as gas adsorbents [6]. Table 5-6 also shows inorganic analyses for carbons employed in water treatment [7].

Table 5-3 Characteristics of Activated Carbon that May Require Consideration for Some Products and Processes

Density	Content of:
Particle size distribution	Sulfur
Porosity	Sulfides
Surface area	Sulfates
Filtration rate	Phosphates
Wettability	Chlorides
Dustiness	Iron
Ignition temperature	Copper
Electrical conductivity	Zinc
Oil retention	Calcium
Moisture	Magnesium
pH	Silica
Total ash	
Resistance to attrition	
Hardness	
Water-extractable inorganics	
Acid-soluble inorganics	

Source: Reproduced from Hassler [1], courtesy of the Chemical Publishing Co.

Table 5-4 Inorganic Impurities of Carbon Adsorbents

Number	*Starting Material*	*Sulfate Ash (%)*	*Sulfur (%)*	*Phosphorus (%)*	*Iron (%)*
Graphon	Spheron 6 (heat treated)	—	—	—	—
1	Wood	4.27	0.02	0.01	0.11
2	Black ash from paper mill liquor	3.06	1.60	0.02	0.06
3	Soft coal	22.70	1.40	0.09	1.32
4	Wood	1.80	0.40	0.65	0.02
4Δ	#1 (heat treated)	—	—	—	—
5	Lignite	15.70	0.61	0.03	0.14
6	Wood	2.47	0.40	0.007	0.03

Source: Reproduced from Graham [5], courtesy of the American Chemical Society.

Table 5-5 Analyses of Gas-Adsorbent Carbons

	Weight (%)			
Element	*Sargent A*	*Sargent C*	*Burrell*	*Pittsburgh*
Carbon	95.2	94.7	94.4	77.8
Hydrogen	0.91	1.07	1.70	1.62
Oxygen	0.94	2.12	1.46	2.25
Sulfur	0.09	—	—	—
Ash, total, as oxides	3.9	2.4	3.8	18.41
Aluminum	0.13	—	—	—
Calcium	0.77	—	—	—
Iron	0.07	—	—	—
Magnesium	0.27	—	—	—
Manganese	0.01	—	—	—
Potassium	0.29	—	—	—
Silicon	1.2	—	—	—
Sodium	0.26	—	—	—

Source: Reproduced from Wolff [6], courtesy of the American Chemical Society.

Table 5-6 Characteristics of Adsorbent Carbons

Carbon	*Carbon (%)*	*Ash (%)*	*Soluble HCl (%)*	*pH Water Extract*
Norit	93–96	4.5–6.5	2–3	7.8–8.3
Darco	65–70	25–30	0.5	4.5–6
Suchar	97–99	2–3.5	1.2–2.4	6.3–7
Nuchar W	95–98	2–3.5	1.3–2.5	7.5
Nuchar 2	90–95	5–10	3–7	7–8
Bone char	9–10	90		
Cliffchar	89–90	3–4		9.5–10.5

Source: Reproduced from Mantell [7], courtesy of the Robert E. Krieger Publishing Co.

Additional physical properties of currently available carbons are seen in Table 5-7 [8]. There are differences in surface area for these six carbons. Here, the ash contents are similar except for the Witco carbon, which may have been acid washed.

Tables 5-8 and 5-9 give several properties of some commercially available carbons that were used in adsorption of phenols from water [9,27]. Surface area measurements were made on a Micrometrics Automatic Surface Area Analyzer, Model 2200, using N_2 gas. A relative hardness scale is given also that indicates resistance to abrasion. Additional surface areas are seen in Table 5-10 [7].

Surface Area

That the surface area of GAC varies with the source of raw material has been seen. Some recent studies have investigated the effect of deashing treatment on surface areas of eight commercial carbons that contained about 2 to 5% ash [10]. This deashing treatment was hot HF-HCl that lowered the ash contents to 1.0 to 1.4%. Table 5-11 shows the effect of this treatment on the $N_{2(g)}$ equals BET surface areas that all increased significantly. In another procedure, some carbons were given different burnoffs in oxygen under low pressures of 10 to 20 Torr at 600°C, after which they were outgassed at 1,000°C to eliminate chemisorbed oxygen. The results from burnoffs are given in Table 5-12 where the surface areas were increased also. Adsorption isotherms of phenol from aqueous solutions were determined on two of the treated carbons. These are seen in Figure 5-1 where greater quantities of phenol were adsorbed on the deashed and burned off carbons.

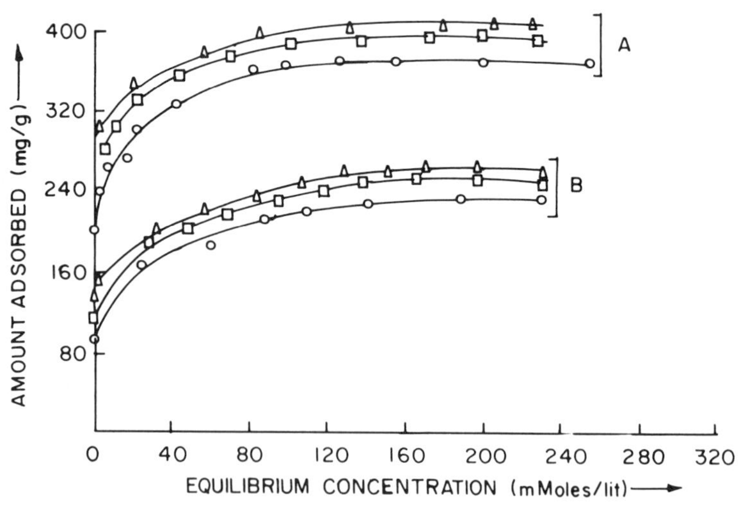

Figure 5-1 Adsorption isotherms of phenol on activated carbons A and B. Key: ○, before any treatment; □, after deashing; △, after burnoff (10.4% in A and 9.6% in B). Reproduced from Puri [10], courtesy of the American Chemical Society.

Table 5-7 Specifications of Several Commercially Available Carbons

Physical Properties and Specifications	*Atlas Darco*	*Calgon Filtrasorb 300*	*Calgon Filtrasorb 400*	*Westvaco*		*Witco 517*
				8 × 10	*12 × 40*	
Surface area, m^2/g (BET)	600–650	950–1050	1,000–1,200	850	850	1,050
Apparent density, g/cc	0.38	0.48	0.44	—	—	0.48
Density, backwashed and drained, lb/ft^3	24	26	25	30	30	30
Real density, g/cc	—	2.1	2.1	—	—	—
Particle density, g/cc	0.67	1.3–1.4	1.3–1.4	1.4	1.4	0.92
Effective size	—	0.8–0.9	0.55–0.65	0.90	0.65	—
Uniformity coefficient	—	1.9 or less	1.9 or less	1.8	1.6	—
Pore volume, cc/g	0.98	0.85	0.94	—	—	0.60
Mean particle diameter, mm	1.05	1.5–1.7	0.9–1.1	—	—	—
Sieve size (U.S. Std. Series)						
Larger than no. 8-max. %	—	8	—	2	—	—
Larger than no. 12-max. %	5	—	5	—	2	[a]
Smaller than no. 30-max. %	—	5	—	1	—	[a]
Smaller than no. 40-max. %	5	—	5	—	1	—
Iodine number	650	900	1,000	850	850	—
Abrasion no., minimum	—	70	75	70	70	85
Ash, %	—	8	8.5	7	7	0.5
Moisture as packed, max. %	—	2	2	2	2	—

Source: Reproduced from *Process Design Manual for Carbon Adsorption* [8].
[a] 12 × 30 (no further details given).

Table 5–8 Properties of the Nine Activated Carbons Evaluated for Adsorptive and Leachate Characteristics

Type of Carbon	*Purchased Mesh Size*	*Source Material*	*Apparent Density*[a]	*BET Surface Area (m^2/g)*		*pH*[d]	*Relative Hardness*[e]
				Reported[b]	*Measured*[c]		
Darco	12 × 40	Lignite	24.5	700	612	5.7	3
Darco G-60	20 × 40	N.S.	N.S.	N.S.	568	—	2
Norit	8 × 20	N.S.	16–19	600–800	742	7.0	3
Nuchar WL-V	8 × 30	Bituminous coal	30	1,000	1,115	6.4	2
Nuchar WV-W	8 × 30	Bituminous coal	36	850	802	6.2	2
Filtrasorb 300	6 × 40	Bituminous coal	26	950–1,050	990	6.5	1
Columbia LCK	12 × 28	Coconut shell	29	N.S.	1,030	6.7	1
Barney-Cheney	12 × 30	N.S.	30	N.S.	1,100	—	2
Cal-Char	N.S.[f]	N.S.	N.S.	N.S.	1,160	—	3

Source: Reproduced from Zogorski [9].
[a]lb/ft^3.
[b]Reported by manufacturer.
[c]Measured as part of present investigation (16 × 20 mesh).
[d]pH of 1 g/l solution of carbon after 1 hour of agitation.
[e]Lower values represent relative greater hardness or resistance to abrasion.
[f]Not specified.

Table 5-9 Properties of Ten Commercially Available Activated Carbons

Properties	*Barney Cheney NL 1266*	*Barney Cheney SC 1548*	*Barney Cheney SK 1301*	*Calgon CAL*	*Calgon RB*	*ICI Darco S-51*	*Westvaco Nuchar WV-L*	*Westvaco Nuchar S-A*	*Westvaco WV-DC*	*Witco Witcarb 950*
Surface area (m^2g^{-1})	750–800	900	925	1,100–1,300	1,100–1,300	650	1,000	1,400–1,800	1,500–1,700	1,000–1,100
Apparent density ($g\ cm^{-3}$)	0.18–0.25	0.48–0.53	0.48–0.53	0.44	0.41	0.24–0.29	0.43–0.47	0.38–0.41	0.24–0.29	0.46–0.53
Effective size (mm)	0.3–0.6	0.3–0.6	0.3–0.6	0.55–0.7			0.85–1.05	0.85–1.05	0.9–1.0	0.5–0.7
% passing No. 325					70	70				
Iodine	800	900	900	1,000	1,070	950	950	800	850	1,050
Maximum moisture %	2	2	2	2	2	8	2	10	10	1
Source	Hardwood acid washed	Nut shell	Nut shell	Coal	Coal	Lignite	Bituminous coal	Bituminous coal	Wood	Petroleum
Size	30 × 50	30 × 50	30 × 50	12 × 40	Powder	Powder	8 × 30	8 × 30	10 × 25	18 × 40
Mean pore radius (Å)	100–150	20	20	50	18	30	19	23	23	20

Source: Reproduced from Netzer and Hughes [27], courtesy of Pergamon Press.

Table 5-10 Surface Areas of Activated Carbon

Carbon	*Source*	*Surface Area (m^2/g)*
PCC SGL	Bituminous coal	1,000–1,200
PCC BPL	Bituminous coal	1,000–1,200
PCC RB	Bituminous coal	1,200–1,400
PCC GW	Bituminous coal	800–1,000
Columbia CXA/SXA	Coconut shell	1,100–1,300
Columbia AC	Coconut shell	1,200–1,400
Columbia G	Coconut shell	1,100–1,150
Darco S 51	Lignite	500–550
Darco G 60	Lignite	750–800
Darco KB	Wood	950–1,000
Hydro Darco	Lignite	500–600
Nuchar Aqua	Pulp mill residue	550–650
Nuchar C	Pulp mill residue	1,050–1,100
Nuchar (various)	Pulp mill residue	300–1,400
Norit (various)	Wood	700–1,400

Source: Reproduced from Mantell [7], courtesy of the Robert E. Kreiger Publishing Co.

Table 5-11 Effect of Deashing Treatment on Surface Areas of Carbons

	Ash (%)		*Surface Area (m^2/g)*	
Carbon	*Before* [a]	*After* [b]	*Before* [a]	*After* [b]
A	4.01	1.31	1,122	1,242
B	3.20	1.38	1,046	1,208
C	2.78	1.09	832	917
D	3.06	1.26	922	1,035
E	1.94	1.05	549	619
G	5.24	1.14	912	1,027
H	4.88	1.25	901	1,046
J	3.80	1.10	1,011	1,131
Coconut shell charcoal	3.56	0.58	372	436

Source: Reproduced from Puri [10], courtesy of the American Chemical Society.
[a]Before any treatment.
[b]After deashing treatment.

Another point in the surface area determination of activated carbon is the molecule employed. Table 5-13 shows the surface areas of several carbons where $N_{2(g)}$ and CCl_4 vapor were utilized [10]. Nitrogen surface areas were larger, the usual case, than the CCl_4 areas.

Table 5-12 Effect of Different Burnoffs on Carbon for Their Surface Areas

Carbon	*Burnoff (%)*	*Surface Area (m^2/g)*
A	Nil	1,122
	2.5	1,258
	5.9	1,380
	10.4	1,472
	12.4	1,488
B	Nil	1,046
	2.5	1,109
	4.8	1,198
	9.6	1,285
	13.2	1,296
D	Nil	922
	6.5	1,057
	10.4	1,172
Coconut shell charcoal	Nil	320
outgassed at 1,000°C	10.8	750
	20.2	845
	37.6	1,162

Source: Reproduced from Puri [10], courtesy of the American Chemical Society.

Table 5-13 Nitrogen and Carbon Tetrachloride Specific Surface Areas of Carbons

	Surface Area (m^2/g)		
Carbons	*N_2 (78°K)*	*CCl_4 (298°K)*	*Ratio*
Active Carbons			
A	1,122	718	64
B	1,046	616	59
D	922	701	76
E	549	252	46
G	912	565	62
H	901	677	75
Carbon Blacks			
ELF-O	171	165	96
Spheron-6	120	116	97
Mogul	301	258	86
Mogul-A	221	—	95
Graphon	86	85	99

Source: Reproduced from Puri [10], courtesy of the American Chemical Society.

Porosity

Figure 5-2 shows the structures of graphite (a) and turbostratic carbon (b) [11]. Activated carbons are similar to the second type that have microcrystallites only a few layers thick and less than 100 Å in width. The many structural imperfections in activated carbon's microcrystallites result in many possibilities for the edge carbons to react with their surroundings.

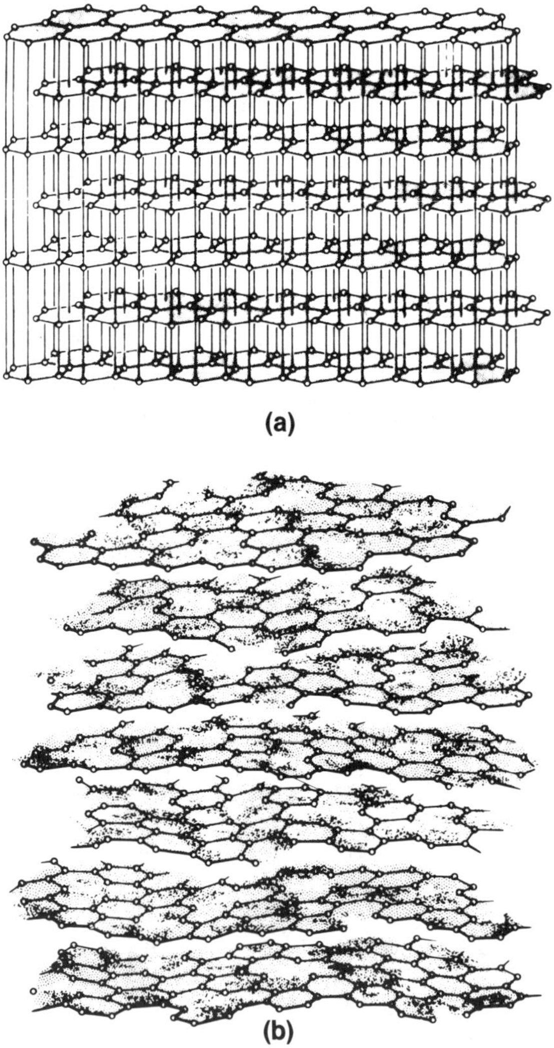

Figure 5-2 Schematic diagram comparing (a), a three-dimensional graphite lattice, with (b), a turbostratic structure. Reproduced from Bokros [11], courtesy of the Marcel Dekker, Inc.

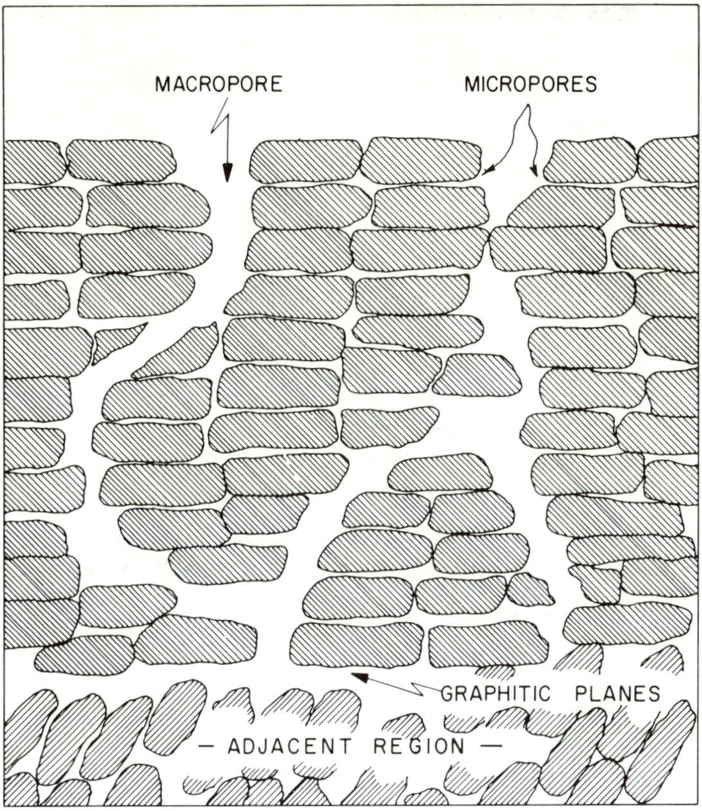

Figure 5–3 Schematic representation of a constituent region of an activated carbon particle. Reproduced from Weber [13], courtesy of John Wiley & Sons, Inc.

However, the porous nature of carbon determines its adsorptive properties. It has been stated that "over 99% of the active sites for adsorption in GAC are located in the interior of the particle" [12]. Figure 5–3 is a schematic of a constituent region of an activated carbon particle where the macropores have diameters of 30 to 100,000 Å and the micropore diameters are in the 10 to 30 Å range [13].

Tables 5–14 and 5–15 summarize the mercury and nitrogen pore volume data for several commercial and experimental carbons (12 × 40 U.S. mesh) [12]. It is obvious that pore surface area does indeed approach 99% of the total area. Also a major part of the total N_2-BET surface area is found in the smallest micropores. In the seventh and eighth columns, the effect of additional activation is noted. GC-25B was steam activated for an hour longer than GC-25A + 26. This increased the total N_2-BET surface area by 83% and opened some of the narrow pores to form wider pores.

Table 5-14 Pore Volumes

	Pore Volume (cm^3/g)		
Carbon	*Hg*	N_2	*Total*
Commercial F	0.500		
Commercial C		0.24	0.60[a]
Commercial A	0.556	0.69	0.99 (0.94[a])
Commercial D		0.65	1[a]
Commercial B		0.70[a]	
GC-16D (1 hr)	0.227		
GC-20 G	0.277	0.44	0.66
GC-21	0.446		
GC-22	0.235	0.33	0.48
GC-25A + 26 (1 hr)	0.465	0.48	0.78
GC-25B (2 hr)	0.958	0.90	1.29
GC-27A	0.885	0.95	1.43
GC-27B	0.984	0.99	1.61

Source: Reproduced from Mattson [12], courtesy of the American Chemical Society.
[a]From the manufacturer's literature.

SURFACE FUNCTIONAL GROUPS

Generalizations

It has been known for a long time that graphite, carbon black, and activated carbon irreversibly chemisorb molecular O_2 [14,15]. This chemically bound O_2 is removable only at elevated temperatures. A study of the chemistry of these carbon surface oxides involved, in early research, indirect methods such as acid-base neutralization and thermal removal of CO and CO_2. More recent studies employed direct investigation of surface functional groups with infrared and ESR (electron spin resonance) spectroscopy [16].

In 1929, several observations were made with sugar-based carbons that were activated at high temperatures in comparison with carbons activated at low temperatures [17]. The former had markedly different surface properties with respect to acid-base properties, electrophoretic mobility, and degree of hydrophilicity. The activation of C at 1,000°C in pure CO_2 or under a vacuum followed by exposure to O_2 at room temperature produces a carbon surface that is capable of raising the pH value of either a neutral or an acidic solution, is hydrophobic, and has a positive electrophoretic mobility [18]. In contrast, the oxidation of C by $O_{2(g)}$ between 200 and 400°C or by adding C to an aqueous oxidizing solution yields a surface characteristic opposite to the previous.

A method of characterizing carbons that are activated and oxidized at different temperatures was developed from their acid-base properties [17]. Those

Table 5–15 Surface Area versus Pore Radius Data and Total N_2 BET Surface Area

	N_2 *BET Surface Area (%)*								
Pore Radius (Å)	*Comml A*	*Comml E*	*GC-18F*	*GC-20G*	*GC-21*	*GC-25A + 26*	*GC-25B*	*GC-27A*	*GC-27B*
300–250	0.0	0.1		0.0	0.0		0.0	0.0	0.0
250–200	0.1	0.0		0.1	0.0		0.0	0.0	0.1
200–150	0.1	0.1		0.1	0.1		0.1	0.1	0.1
150–100	0.2	0.1		0.0	0.0		0.1	0.2	0.3
100–90	0.1	0.1		0.0	0.0	0.0	0.0	0.1	0.1
90–80	0.1	0.1		0.0	0.0		0.1	0.2	0.2
80–70	0.1	0.1	0.0	0.0	0.0		0.1	0.3	0.3
70–60	0.2	0.2		0.0	0.0		0.1	0.5	0.6
60–50	0.2	0.1		0.1	0.1	0.1	0.1	0.5	0.6
50–40	0.4	0.2		0.1	0.0	0.0	0.2	1.0	1.2
40–30	0.7	0.3		0.0	0.0	0.0	0.3	1.5	1.2
30–20	2.1	0.9	0.1	0.2	0.4	0.3	0.8	2.1	2.3
20–15	3.2	2.1	0.4	0.5	0.9	0.8	1.4	2.2	2.4
15–20	13.3	15.0	3.0	4.3	5.4	4.3	13.3	10.7	10.8
10–7	79.1	80.7	96.4	94.4	92.9	94.2	83.4	80.9	79.8
Total N_2 BET area (m^2/g)	1,163.4	404.3	1,013.5	761.0	1,190.2	853.6	1,560.6	1,377.0	1,388.2

Source: Reproduced from Mattson [12], courtesy of the American Chemical Society.

low-temperature oxidized carbons that adsorb $\overline{\text{O}}\text{H}$ ions primarily are called L-carbons. Those that are activated at high temperatures and adsorb H^+ ions are called H-carbons. These classifications have been found to divide above and below an activation-oxidation temperature of 500 to 600°C [16].

Surface Acidity

A considerable amount of research has been conducted on the evolution of CO, CO_2, and H_2O at various temperatures under vacuum [16]. All the chemisorbed oxygen on carbon evolves as these three gases. Furthermore, the quantities of CO_2 and CO evolved from carbon can be determined by increasing the activation temperature from 200 to 800°C.

It has been suggested that surface oxides evolved as CO_2 are responsible for the physicochemical properties of the L-carbons, whereas those oxides evolved as CO are responsible for H-carbon characteristics [19]. Table 5–16 shows data to support these observations that there is a close relationship between the surface acidity of active carbons and the amount of CO_2 evolved on outgassing in vacuum at 1,200°C. By first outgassing sugar char at 300 to 1,000°C and then neutralizing the surface oxides with base and measuring the CO_2 evolved from outgassing, it was observed that, for pretreatment with outgassing above 600°C, CO_2 was no longer removed on additional outgassing at 1,200°C and no bases could be neutralized. These findings support the conclusion that the CO_2-evolving surface functional groups are responsible for the acidic L-carbons [19].

Almost every type of functional group known in organic chemistry has been suggested to be present on activated carbon's surface [16]. The ones suggested most often are carboxyl groups (I), phenolic hydroxyl groups (II), and quinone-type carbonyl groups (III):

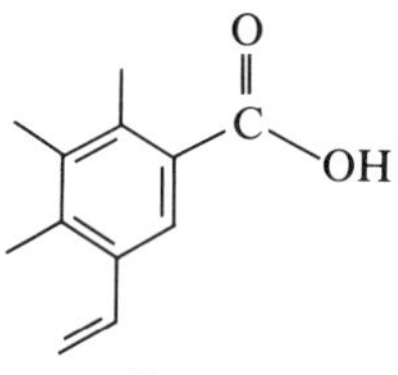

(I)

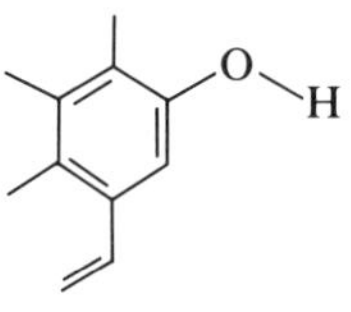

(II)

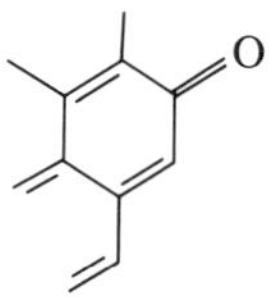

(III)

Other suggested groups are ether, peroxide, and ester groups in the forms of normal (IV) and fluorescein-like lactones (V), carboxylic acid anhydrides (VI), and the cyclic peroxide (VII):

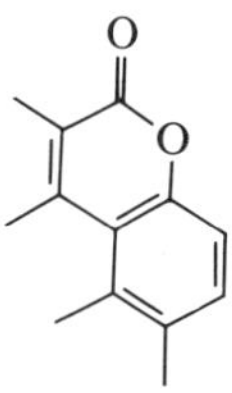

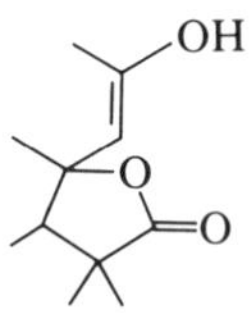

(V)

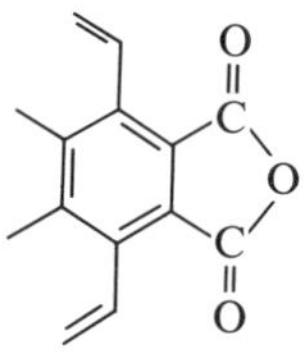

(VI)

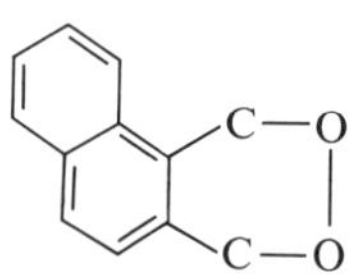

(VII)

Table 5–16 Neutralization of Different Bases by the Various Samples of Charcoal and Carbon Black in Relation to CO_2 Complex

	Amounts of Different Bases Neutralized (meq/100 g)					
Description	*$Ba(OH)_2$*	*NaOH*	n-*butylamine*	*Aqueous Ammonia*	*Gaseous Ammonia*	*CO_2 Evolved on Outgassing (meg/100 g)*
Sugar charcoal, original	655	643	682	354	670	669
Treated with $K_2S_2O_8$	970	985	—	487	898	975
Treated with aqueous chlorine	902	888	879	474	825	881
Treated with aqueous H_2O_2	810	795	802	402	790	819
Outgassed at 300°C	361	366	361	201	347	359
400°C	215	209	202	112	198	206
500°C	166	163	161	88	159	160
600°C	45	43	48	30	38	48
750°C	Nil	Nil	Nil	Nil	Nil	Nil
1000°C	Nil	Nil	Nil	Nil	Nil	Nil
Outgassed 1000°C, treated with HNO_3	1,184	1,166	—	530	955	1,163
Coconut charcoal, original	374	376	381	180	338	384
Treated with $K_2S_2O_8$	749	757	704	362	698	741
Treated with aqueous chlorine	678	668	659	342	624	686
Treated with aqueous H_2O_2	612	618	614	328	—	—

(*cont.*)

Table 5-16 *(continued)*

Description	Amounts of Different Bases Neutralized (meg/100 g)					Co_2 Evolved on Outgassing (meg/100 g)
	$Ba(OH)_2$	NaOH	n-butylamine	Aqueous Ammonia	Gaseous Ammonia	
Outgassed at 300°C	308	316	327	172	302	314
400°C	56	51	57	22	49	59
500°C	3	2	3	Nil	Nil	8
600°C	Nil	Nil	Nil	Nil	Nil	Nil
750°C	Nil	Nil	Nil	Nil	Nil	Nil
1000°C	Nil	Nil	Nil	Nil	Nil	Nil
Outgassed 1000°C, treated with HNO_3	621	612	609	352	584	617
Spheron 6, original	29	28	30	18	—	34
Treated with HNO_3	253	258	250	115	240	257
Mogul, original	139	135	123	68	—	137
Outgassed at 200°C	138	133	—	72	—	137
400°C	82	79	70	44	69	83
700°C	22	22	18	—	—	18
Treated with HNO_3	529	534	524	282	492	524
Mogul A, original	113	115	—	—	—	115
Outgassed at 400°C	59	62	61	—	—	59
500°C	24	18	19	—	—	24
700°C	Nil	Nil	Nil	—	—	Nil

Source: Reproduced from Puri [19], courtesy of Pergamon Press.

Typical identification reactions of organic chemistry were employed to characterize oxygen chemisorbed on carbon as comprising four types of acidic surface groups: a strongly acidic carbonyl group, a more weakly acidic carboxyl group, a phenolic hydroxyl group, and a carbonyl group [20]. These acidic functional groups can be identified by their reaction or lack of reaction with bases of different strengths. For example, group I is neutralized by $NaHCO_3$, Na_2CO_3, and NaOEt, group II is neutralized by Na_2CO_3 or stronger bases but not by $NaHCO_3$. Accordingly, a routine titration with different bases serves to identify the acidic surface oxides present on carbon.

This base titration procedure was utilized to identify acidic surface oxides on Columbia LC325, Darco 551, and Black Pearls 607 and to ascertain the effects of oxidative procedures on these carbon surfaces [21]. Table 5–17 shows the various surface acidities for the three carbons in unoxidized and oxidized states. In the case of the LC325 carbon, oxidation reduced the $N_{2(g)}$ surface area and increased the surface oxides. Vacuum outgassing at 890°C restored the specific surface area and low surface acidity of the original LC325 carbon.

Effects on Adsorption

Subsequent determination of adsorption isotherms for phenol on LC325 showed that the effect of acidic surface oxides was to reduce adsorptive capacity in the 0.0 to 0.3 mmol/1 range [21]. This is seen in Figures 5–4a and b for phenol on two carbons and in Figure 5–4c for sodium benzenesulfonate on Darco 551. The adsorptive power of the oxidized and outgassed Darco 551 for phenol is greater than the original starting material.

Similar effects on the adsorption of phenol and nitrobenzene on Black Pearls 607 and LC325 were observed when the surfaces of these carbons were modified by wet oxidation and reduction [22,23]. Table 5–18 gives the adsorption isotherm plateau values where the equilibrium period was 1 week. In the case of LC325, the adsorptive capacity decreased greatly on oxidation whereupon a reduction step did not restore the surface. This was due to an increase in surface acidities and a decrease in the surface area of the LC325.

There are a couple of explanations for the role of acidic surface oxygen groups in their inhibition of adsorption of phenol and nitrobenzene molecules [22,23]. A major portion of these oxygen groups on carbon is located presumably on the edges of the layer planes where they are not expected to interfere markedly, in a steric sense, with adsorption of organic molecules on the basal planes. Chemically bound oxygen on the edges localizes electrons in surface states and removes them from the π electron system of the basal planes. Depletion of electrons creates an increase in the population of positive holes in the conduction band of the π electron system. This is consistent with the concept that dispersion forces between the phenol's π electron system and the π band of the graphitic planes of the carbon are responsible for adsorption. Removal of electrons from the π band of the carbon by chemisorbed oxygen may interfere with and weaken the adsorptive forces.

Table 5-17 Surface Acidities and Surface Areas of Carbon Adsorbent Samples

Sample	*Base Consumption (Meq/100 g)*				*Surface Groups (meq/100 g)*				*BET (N_2) (m^2/g)*
	NaOEt	*NaOH*	*Na_2CO_3*	*$NaHCO_3$*	*IV*	*III*	*II*	*I*	
LC325	59	48	28	30	11	20	—	30	1,076
LC325 O[a]	310	292	205	150	18	87	55	150	861
LC325 OR[b]	246	210	129	92	36	81	36	92	840
LC325 OD[c]	65	55	15	10	10	40	5	10	1,015
S51	108	71	48	40	37	23	8	40	586
S51 O	421	357	245	185	64	112	60	185	596
S51 OR	215	181	112	80	34	70	32	80	836
S51 OD	60	60	25	20	—	35	5	20	725
P607	265	203	132	109	62	71	23	109	646
P607 O	350	267	185	152	83	82	33	152	613
P607 OR	298	223	142	107	75	81	35	107	640

Source: Reproduced from Coughlin and Tan [21], courtesy of the American Institute of Chemical Engineers.
[a]O = oxidized by $(NH_4)_2 S_2O_8$.
[b]OR = oxidized followed by reduction with hydrogen at room temperature.
[c]OD = oxidized followed by vacuum outgassing at 890°C.

(a)

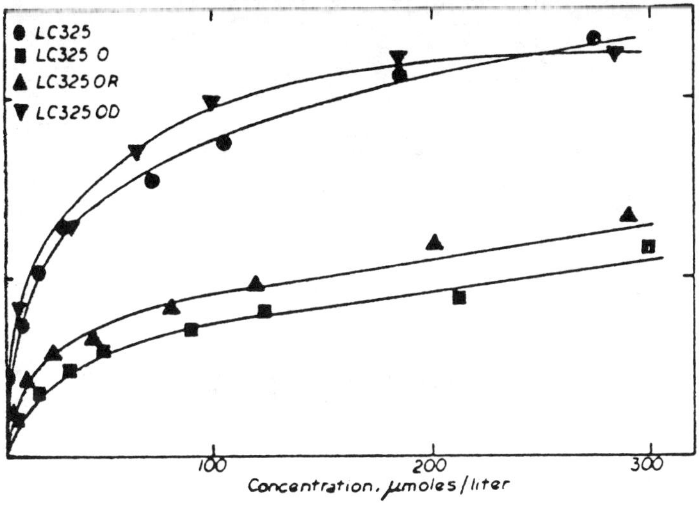

(b)

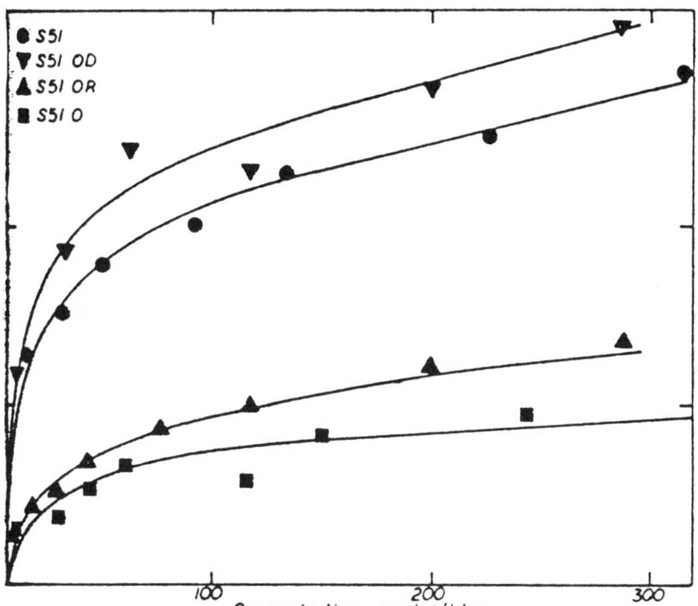

Figure 5–4 (a) Adsorption isotherms of phenol on activated carbon Columbia LC325 in aqueous solution, 30°C, 3 days equilibrium time. (b) Adsorption isotherms of phenol on activated carbon Darco S51 in aqueous solution, 30°C, 3 days equilibrium time. (c) Adsorption isotherms of sodium benzenesulfonate on activated carbon Darco S51 in aqueous solution, 30°C, 3 days equilibrium time. Reproduced from Coughlin and Tan [21], courtesy of the American Institute of Chemical Engineers.

(c)

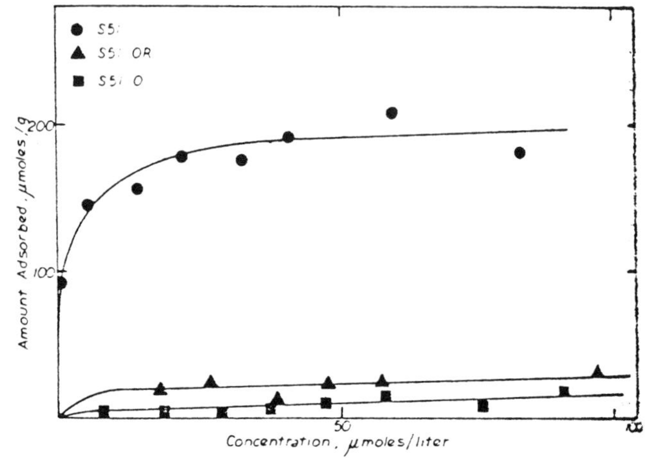

Table 5-18 Plateau Values for the Adsorption of Phenol and Nitrobenzene

	Phenol Adsorption		*Nitrobenzene Adsorption*	
Adsorbent	*μmol/g*	*μmol/m²*	*μmol/g*	*μmol/m²*
P607	290	0.478		
P607, O[a]	142	0.213		
P607, R[b]	330	0.498		
P607, OR[c]	250	0.488		
P607, OROROR	312	0.555		
LC325	1,000	0.833	1,950	1.63
LC325, O	120	0.216	500	0.90
LC325, OR	190	0.475		

Source: Reproduced from Coughlin and Ezra [22], courtesy of the American Chemical Society.
[a]O = oxidized by $(NH_4)_2\ S_2O_8$.
[b]R = reduced with hydrogen at room temperature.
[c]OR = oxidized followed by reduction.

Electrophoretic Characteristics

That the surface of carbon contains negative and positive sites was indicated from the surface acidity studies. Five commercial carbons, in the powdered form, were investigated for their electrophoretic mobility characteristics [24]. This study also determined the intrinsic surface acidity constants of the activated carbons; that is, the acid-base behavior of the surface hydroxyl groups is expressed, for a positively charged surface:

$$\gtrdot COH_2^+ = \gtrdot COH + H^+; \qquad K^1_{int}, \tag{5.6}$$

where

$$K^i_{int} = \frac{[\gtrdot COH][H^+]}{[\gtrdot COH_2^+]}, \tag{5.7}$$

and for a negatively charged surface:

$$\gtrdot COH = \gtrdot CO^- + H^+; \qquad K^2_{int}, \tag{5.8}$$

where

$$K^2_{int} = \frac{[\gtrdot CO^-][H^+]}{[\gtrdot COH]}. \tag{5.9}$$

Table 5–19 shows these acidity constants for three of the activated carbons that are compared with some common organic acids. These pK values would classify the carbons as weak acids and bases, with Nuchar C-190-N as the strongest of the three.

A surface charge, σ_o, in microcoulombs/cm^2, was calculated from the quantities of acid, C_A, and base, C_B, consumed by the carbon's surfaces from titration procedures in reference to a pH_{zpc} (zero point charge) [24]. The equation is:

$$\sigma_o = \frac{F}{S}(C_B - C_A - [O\bar{H}] + [H^+]). \tag{5.10}$$

Table 5–19 Intrinsic Acidity Constants of Activated Carbon and Acidity Constants of Typical Organic Acids

Acid	$pK^{1\,a}_{int}$	$pK^{2\,b}_{int}$	*Source*
Filtrasorb 400	5.70	7.78	[24]
Nuchar N-722	5.54	6.38	[24]
Nuchar C-190-N	3.54	5.49	[24]
o-phthalic	2.89	5.51	[25]
m-phthalic	3.54	4.6	[25]
p-phthalic	3.51	4.82	[25]
Cyclohexane-1:1-dicarboxylic	3.45	6.11	[25]
Malic	3.40	5.11	[25]
Succinic	4.16	5.61	[25]
Oxalic	1.23	4.19	[25]
Malonic	2.83	5.69	[25]

Source: Reproduced from Huang and Ostovic [24], courtesy of the American Society of Civil Engineers.
[a]Or pK_{a1}—for sample, organic acids.
[b]Or pK_{a2}—for sample, organic acids.

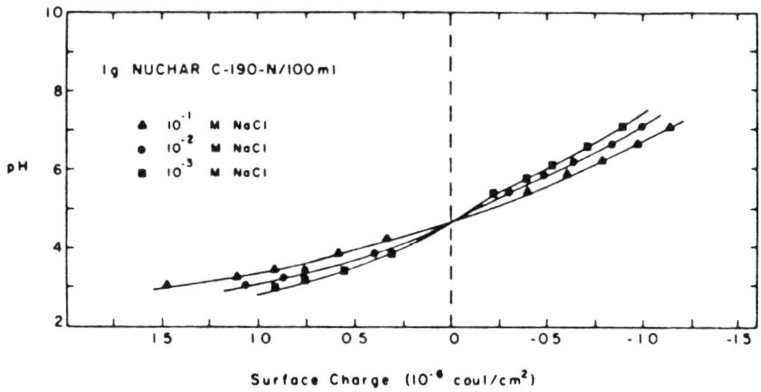

Figure 5–5 Surface charge characteristics of activated carbon Nuchar C-190-N in simple electrolyte (NaCl) solution. Reproduced from Huang and Ostovic [24], courtesy of the American Society of Civil Engineers.

Figure 5–5 shows the surface charge characteristics of Nuchar C-190-N. Table 5–20 gives the pH_{zpc} values for the five carbons. These two surface characteristics are related to the electrophoretic characteristics of a particle. Figure 5–6 shows the measured electrophoretic mobilities of the five carbons with respect to pH.

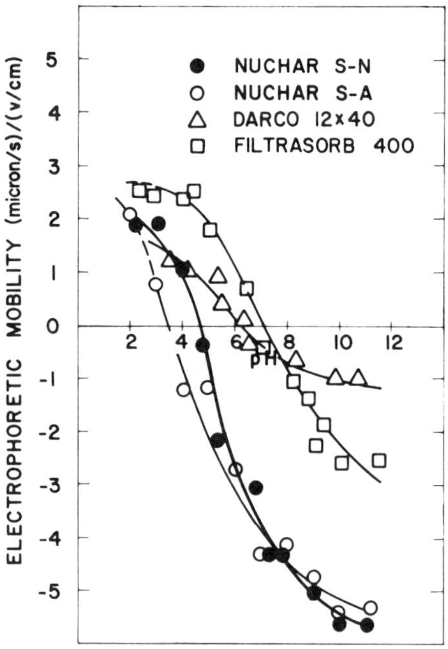

Figure 5–6 Electrophoretic mobility of activated carbon in single electrolyte (NaCl). Reproduced from Huang and Ostovic [24], courtesy of the American Society of Civil Engineers.

Table 5-20 pH_{zpc} for Various Activated Carbons

Carbon	*Method A*[a]	*Method B*[b]
Nuchar C-190-N	4.75	4.52
Nuchar 722	5.70	5.96
Darco	6.15	—
Pittsburgh (PAC)	6.70	—
Filtrasorb 400	7.10	6.78

Source: Reproduced from Huang and Ostovic [24], courtesy of the American Society of Civil Engineers.
[a]Electrophoretic mobility.
[b]From Table 5-19 $pH_{zpc} = 1/2\ (pK^{1}_{\text{int}} + pK^{2}_{\text{int}})$.

The observed positive mobility values at acidic pH values, in turn, become negative as the pH value is increased. This definitely indicates that $O\overline{H}$ and H^+ ions are potential determining species. This finding is consistent with electrophoretic studies of similar surfaces with hydrous oxides of Al and Fe [26].

REFERENCES

1. Hassler, J.W. *Activated Carbon.* Chemical Publishing Co., New York (1974).
2. Blyholder, G., et al. *J. Phys. Chem. 62,* 263 (1958).
3. Blyholder, G., and H. Eyring. *J. Phys. Chem. 61,* 682 (1957).
4. Blyholder, G., and H. Eyring. *J. Phys. Chem. 63,* 1004 (1959).
5. Graham, D. *J. Phys. Chem. 59,* 896 (1955).
6. Wolff, W.F. *J. Phys. Chem. 62,* 829 (1958).
7. Mantell, C.L. *Carbon and Graphite Handbook.* Robert E. Krieger Publishing Co., Huntington, N.Y. (1979).
8. *Process Design Manual for Carbon Adsorption.* U.S. EPA, Technology Transfer, Program 17020 GNR. Washington, D.C. (October 1971).
9. Zogorski, J.S. "The Adsorption of Phenols onto Granular Activated Carbon from Aqueous Solution." Ph.D. dissertation, Department of Environmental Science, Rutgers, The State University, New Brunswick, N.J. (1975).
10. Puri, B.R. "Physicochemical Aspects of Carbon Affecting Adsorption from the Aqueous Phase," Chapter 5 in *Treatment of Water by Granular Activated Carbon.* Advances in Chemistry Series 202. American Chemical Society, Washington, D.C. (1983).
11. Bokros, J.C. In *Chemistry and Physics of Carbon,* Vol. 5. P.C. Walker, ed. Marcel Dekker, New York (1969).
12. Mattson, J.S. *Ind. Eng. Chem. Prod. Res. Devel. 12,* 312 (1973).
13. Weber, W.J. Jr. "Sorption from Solution by Porous Carbon," Chapter 5 in *Principles and Applications of Water Chemistry.* S.D. Faust and J.V. Hunter, eds. John Wiley & Sons, New York (1967).
14. Langmuir, I. *J. Amer. Chem. Soc. 37,* 1139 (1915).
15. Smith, A. *Proc. Royal Soc. (London) A12,* 424 (1863).
16. Mattson, J.S., and H.B. Mark, Jr. *Activated Carbon.* Marcel Dekker, New York (1971).

17. Kruyt, H.R., and G.S. deKadt. *Kolloid Z. 47,* 44 (1929).
18. Steenberg, B. *Adsorption and Exchange of Ions on Activated Charcoal.* Almquist & Wiksells, Uppsala, Sweden (1944).
19. Puri, B.R. *Carbon 4,* 391 (1966).
20. Boehm, H.P., et al. *Angew. Chem. Intern. Ed. Engl. 3*(10), 669 (1964).
21. Coughlin, R.W., and R.N. Tan. *Chem. Eng. Progr. Symp. Ser. 64,* 207 (1968).
22. Coughlin, R.W., and F.S. Ezra. *Environ. Sci. Technol. 2,* 291 (1968).
23. Coughlin, R.W., et al. *J. Colloid Interf. Sci., 28,* 386 (1968).
24. Huang, C.P., and F.B. Ostovic, *J. Environ. Eng. Div. ASCE 104 (EES),* 863 (1978).
25. James, R.O., et al. *Faraday Disc. Chem. Soc. 59,* 142 (1975).
26. Faust, S.D., and O.M. Aly. *Chemistry of Water Treatment.* Butterworths, Stoneham, Mass. (1983).
27. Netzer, A., and D.E. Hughes, *Water Res. 18,* 927 (1984).

6

Adsorption of Organic Compounds

There is a considerable amount of information in the literature concerning adsorption of organic compounds on PACs and GACs. Much of this information has evolved from laboratory studies where the ultimate objective was to apply these data to operational water treatment. One of the difficulties with transfer of laboratory data to plant practice is the wide range of conditions of the adsorption experiments. Consequently, much of the data have limited value in the operational aspects of water treatment. This chapter documents the adsorption of various organic compounds on PAC and GAC as well as the attendant experimental conditions.

NONSPECIFIC ORGANIC CARBON

Analytical Considerations

Several nonspecific analyses bring together a large number of organic compounds under one umbrella of measurement. These include biochemical oxygen demand (BOD), chemical oxygen demand (COD), total organic carbon (TOC), and total organic halogen (TOX). These nonspecific parameters have useful water-quality-monitoring applications. It is easier to measure organic compounds on a routine basis with a nonspecific procedure than to measure individual compounds. There remains, however, the need to correlate the nonspecific analytical methods and compounds of health concern [1]. This should be done on a site-specific basis, according to Suffet [2], because the particular organic compounds that must be controlled will undoubtedly vary among water supplies. Furthermore, many of these compounds of health concern contain one or more of the halogen groups that are measured as TOX. Here again, correlation of TOX contents with organic compounds of health concern has not been studied sufficiently, at least up to 1980 [3]. Table 6–1 is a compilation of several compounds of health concern that have been researched for their adsorption on PAC and GAC.

Table 6–1 Carbon Adsorption of Toxic and Hazardous Organic Compounds in Batch Systems

Compound	*Molecular Weight*	*Type of Carbon*	*Contact Time (min)*	*Equilibrium Range (mg/l)*	*Maximum Adsorbed (mg/g)*	*Freundlich Constants* *K (mg/g)*	*Freundlich Constants* *l/n*	*Langmuir Constants* *b*	*Langmuir Constants* X_m	*Carbon Dosage to Reduce 1 mg/l to 0.1mg/l*	*Reference*
Acenaphthene	154.21	PAC	120.	0.03–2.04	200.	190.	.36	—	—	10.	10
Acenaphthene	154.21	PAC	—	—	—	140.	.43	—	—	—	11
Acenaphthylene	152.21	PAC	120.	0.0028–0.79	79.6	115	.37	—	—	18.	10
Acetophenone	120.14	PAC	120.	0.36–10.3	250.	74.[a]	.44	—	—	34.	10
Acetophenone	120.14	PAC	120.	28.[b]	194.	—	—	—	—	—	12
Acetophenone	120.14	GAC	?	0.1–4[q]	—	3.46[r]	.155	—	—	—	13
2-acetylaminofluorene	222.28	PAC	120.	0.032–10.	431.	318.	.12	—	—	3.7	10
Acridine orange	301.82	PAC	120.	0.01–9.59	388.	180.[c]	.29	—	—	9.9	10
Acridine yellow	273.77	PAC	120.	0.11–9.54	348.	210.[d]	.14	—	—	5.1	10
Acrolein	56.06	PAC	120.	2.76–40.33	10.7	1.2	.65	—	—	3,500.	10
Acrolein	56.06	PAC	120.	694.[b]	61.	—	—	—	—	—	12
Acrolein	56.06	PAC	120.	21.–89.	—	32.	.77	—	—	—	14
Acrylonitrile	53.06	PAC	120.	2.98–48.7	8.47	1.4	.51	—	—	2,200.	10
Acrylonitrile	53.06	PAC	120.	34.–87.5	—	18.	.62	—	—	—	14
Adenine	135.13	PAC	120.	0.34–10.0	200.	71.[d]	.38	—	—	30.	10
Adipic Acid	146.14	PAC	120.	12.2–26.4	120.	20.	.47	—	—	130.	10
Aldrin	365.	PAC	120.	0.0012–0.023	12.	651.	.92	—	—	11.	10
Alizarin	240.32	GAC	?	0.1–0.8[q]	—	1.39[r]	0.95	—	—	—	13
4-aminobiphenyl	169.12	PAC	120.	0.01–9.8	330.	200.	.26	—	—	8.2	10
4-aminobiphenyl	169.12	PAC	120.	0.01–9.8	358.	198.	.26	—	—	8.0	15
Anethole	148.	PAC	120.	0.05–8.55	800.	300.[a]	.42	—	—	8.0	10
o-anisidine	123.15	PAC	120.	0.65–9.8	110.	50.[d]	.34	—	—	39.	10
Anthracene	178.24	PAC	120.	0.00009–0.00521	6.2	376.	.70	—	—	12.	10

Benzene	78.12	PAC	120.	4.0–19.8	58.3	1.	1.6	—	—	35,000.	10
Benzene	78.12	PAC	120.	21[e]	95.	—	—	—	—	—	12
Benzene	78.12	PAC-A	120.	1.4–10.0	—	26.8	1.3	—	—	678.	16
Benzene	78.12	PAC-B	120.	2.0–10.0	—	18.5	1.2	—	—	705.	16
Benzene	78.12	PAC	15.	0.0–25.	—	65.	.42	—	—	—	17
Benzene	78.12	GAC	2 days	0.0–20.	—	36.	.48	—	—	—	18
Benzene	78.12	GAC-A	550.	0.005–0.5	—	.73[g]	.61	—	—	—	19
Benzene	78.12	GAC-B	550.	0.005–0.5	—	1.07[g]	.48	—	—	—	19
Benzene	78.12	GAC-C	550.	0.005–0.5	—	1.12[g]	.39	—	—	—	19
Benzene	78.12	GAC-D	550.	0.005–0.5	—	1.18[g]	.36	—	—	—	19
Benzene	78.12	PAC	120.	5.5–76.6	—	207.	.61	—	—	—	14
Benzene	78.12	GAC	5 days	0.01–0.2	—	1.14[g]	.398	—	—	—	20
Benzene + humic acid (5.0 mg/l)	78.12	GAC	5 days	0.01–0.2	—	.814[g]	.51	—	—	—	21
Benzidine $(HCl)_2$	257.16	PAC	120.	0.03–9.78	490.	220.[d]	.37	—	—	9.4	10
Benzidine	184.24	PAC-A	120.	0.2–10.	—	85.4	.25	—	—	19.	16
Benzidine	184.24	PAC-B	120.	0.003–8.2	—	173.	.29	—	—	10.	16
Benzidine	184.24	PAC	120.	0.01–10.	313.	215.[d]	.16	—	—	6.	15
Benzoic Acid	122.12	GAC	?	0.1–6.[q]	—	3.06[r]	.181	—	—	—	13
Benzoic Acid	122.12	PAC	120.	0.92–10.45	132.	51.[f]	.42	—	—	85,000.	10
Benzoic Acid	122.12	PAC	120.	89.[b]	183.	—	—	—	—	—	12
3,4-benzofluoranthene	252.32	PAC	120.	0.04–0.583 $\times 10^{-3}$	4.41	57.	.37	—	—	37.	10
3,4-benzofluoranthene	252.32	PAC	120.	0.172–0.583 $\times 10^{-3}$	4.29	55.5	.37	—	—	—	15
Benzo(k)fluoranthene	252.32	PAC	120.	0.013–0.194 $\times 10^{-3}$	1.45	181.	.57	—	—	18.	10
Benzo(k)fluoranthene	252.32	PAC	120.	0.065–0.164 $\times 10^{-3}$	683.	183.	.57	—	—	18.4	15
Benzo(ghi)perylene	276.34	PAC	120.	0.015–0.267 $\times 10^{-3}$	0.61	10.7	.37	—	—	200.	10

(cont.)

Table 6–1 *(continued)*

Compound	Molecular Weight	Type of Carbon	Contact Time (min)	Equilibrium Range (mg/l)	Maximum Adsorbed (mg/g)	Freundlich Constants K (mg/g)	Freundlich Constants 1/n	Langmuir Constants b	Langmuir Constants Xm	Carbon Dosage to Reduce 1 mg/l to 0.1mg/l	Reference
Benzo(ghi)perylene	276.34	PAC	120.	0.071–0.27 $\times 10^{-3}$	24.6	10.6	.36	—	—	197.	15
Benz(a)pyrene	252.3	PAC	120.	0.03–0.32 $\times 10^{-3}$	1.07	33.6	.44	—	—	74.	10
Benz(a)pyrene	252.3	PAC	120.	1.35–1.6 $\times 10^{-3}$	1.64	33.6	.44	—	—	—	15
Benzothiazole	135.19	PAC	120.	0.05–9.85	300.	120.[a]	.27	—	—	14.	10
α-BHC	290.83	PAC	120.	0.25–289. $\times 10^{-3}$	122.	303.	.43	—	—	8.	10
β-BHC	290.83	PAC	120.	0.2–104. $\times 10^{-3}$	54.4	220.	.49	—	—	13.	10
γ-BHC (Lindane)	290.83	PAC	120.	0.006–1.02	230.	256.	.49	—	—	11.	10
Bromoform	252.75	PAC	120.	0.013–1.0	12.6	19.6	.52	—	—	150.	10
Bromoform	252.75	GAC	5 days	0.005–0.33	25.3	73.1 176.[p]	.63	—	—	—	19
Bromoform	252.75	GAC	360.	—	—	.185[g]	.32	—	—	—	22
4-bromophenylphenyl ether	249.11	PAC	120.	0.24–10.2	768.	144.	.68	—	—	30.	10
5-bromouracil	190.99	PAC	120.	0.73–9.95	149.	44.[c]	.47	—	—	60.	10
2-butene-1-al	70.09	PAC	120.	11.2–87.6	—	61.7	.63	—	—	—	14
Butyl benzylphthalate	312.36	PAC	120.	0.026–0.44	170.	1,520.	1.26	—	—	11.	10
n-butylphthalate	278.35	PAC	120.	0.42–4.54	342.	220.	.45	—	—	12.	10
Butyric acid	80.11	GAC	?	0.08–7.18[q]	—	.878[r]	.4	—	—	—	13

Carbontetrachloride	153.82	PAC	120.	0.02–1.0	2.85	11.1	.83	—	—	550.	10
Carbontetrachloride	153.82	GAC-A	550.	0.0005–0.1	—	.16[g]	.75	—	—	—	19
Carbontetrachloride	153.82	GAC-B	550.	0.0005–0.1	—	.22[g]	.69	—	—	—	19
Carbontetrachloride	153.82	GAC-C	550.	0.0005–0.1	—	.23[g]	.74	—	—	—	19
Carbontetrachloride	153.82	GAC-D	550.	0.0005–0.1	—	.13[g]	.68	—	—	—	19
Carbontetrachloride	153.82	GAC	2 days	0.0025–0.063	—	11.6	.73	—	—	—	23
Carbontetrachloride	153.82	GAC	?	?	—	38.1	.7	—	—	—	24
Chlordane	409.80	PAC	120.	0.0005–0.22	87.	245.	.38	—	—	8.8	10
Chlorobenzene	112.56	PAC	120.	1.4–25.9	1,700.	91.	.99	—	—	92.	10
Chloroethane	64.52	PAC	120.	0.129–1.06	0.25	.59	.95	—	—	14,000.	10
bis(2-chloroethoxy) methane	173.1	PAC	120.	1.8–54.4	138.4	11.	.65	—	—	350.	10
bis(2-chloroethyl)ether	143.02	PAC	120.	23.5–50.3	173.	.086	1.84	—	—	>100,000.0	10
2-chloroethylvinylether	106.55	PAC	120.	2.28–33.	66.5	3.9	.8	—	—	1,500.	10
Chloroform	119.38	PAC	120.	0.03–1.0	1.87	2.6	.73	—	—	1,900.	10
Chloroform	119.38	GAC	360.	—	—	.016	.6[g]	—	—	—	22
Chloroform	119.38	GAC	5 days (?)	—	—	6.3	.73	—	—	—	25
Chloroform	119.38	GAC	5 days (?)	—	—	30.3[p]	.73	—	—	—	25
bis(2-chloroisopropyl) ether	171.07	PAC	120.	6.2–48.4	273.	24.	.57	—	—	140.	10
p-chloro-*m*-cresol	142.59	PAC	120.	0.27–16.34	242.	124.[h]	.16	—	—	11.	10
2-chloronaphthalene	162.62	PAC	120.	0.02–2.36	350.	280.	.46	—	—	9.3	10
1-chloro-2-nitrobenzene	157.6	PAC	120.	0.11–9.2	386.	130.[a]	.46	—	—	20.	10
o-chlorophenol	125.56	PAC	120.	0.82–19.3	157.	51.[a]	.41	—	—	45.	10
o-chlorophenol	125.56	PAC	14 days	1.0–80.	—	240.	.098	—	—	—	26
o-chlorophenol	125.56	GAC	100 hr	0.0–60.	—	—	—	.405	272.	—	27
o-chlorophenol	125.56	GAC	40 days	1.0–80.	—	240.	.098	—	—	—	26
o-chlorophenol	125.56	GAC	24 hr	1.0–22.5	—	100.[j]	.18	—	—	—	28
o-chlorophenol	125.56	GAC	24 hr	1.0–22.5	—	66.[o]	.24	—	—	—	28

(*cont.*)

Table 6–1 (*continued*)

Compound	*Molecular Weight*	*Type of Carbon*	*Contact Time (min)*	*Equilibrium Range (mg/l)*	*Maximum Adsorbed (mg/g)*	*Freundlich Constants K (mg/g)*	*Freundlich Constants l/n*	*Langmuir Constants b*	*Langmuir Constants Xm*	*Carbon Dosage to Reduce 1 mg/l to 0.1mg/l*	*Reference*
4-chlorophenyl phenyl ether	204.66	PAC	120.	0.26–15.9	262.	111.	.26	—	—	15.	10
5-chlorouracil	146.54	PAC	120.	2.0–9.89	93.	25.[c]	.58	—	—	140.	10
Chrysene	228.28	PAC	?	—	—	6.	.50	—	—	—	11
Crystal violet	407.19	GAC	?	0.1–2.[q]	—	.84[r]	.09	—	—	—	13
Cyclohexanone	98.14	PAC	120.	1.36–10.8	39.5	6.2	.75	—	—	820.	10
DDE	318.03	PAC	120.	0.002–2.07 $\times 10^{-3}$	20.5	232.	.37	—	—	9.	10
DDE	318.03	GAC	?	0.001–0.035	5.1	—	—	—	—	—	29
DDT	354.5	PAC	120.	0.008–8.64 $\times 10^{-3}$	8.4	322.	.50	—	—	8.8	10
DDT	354.5	GAC	?	0.0035–0.02	8.1	—	—	—	—	—	29
dibenzo(a,h)-anthracene	278.33	PAC	120.	0.03–0.126 $\times 10^{-3}$	0.101	69.3	.75	—	—	73.	10
dibenzo(a,h)-anthracene	278.33	PAC	120.	0.07–0.182 $\times 10^{-3}$	399.	70.2	.76	—	—	73.	15
dibromochloromethane	208.29	PAC	120.	0.0009–1.0	5.6	4.8	.34	—	—	410.	10
1,2-dibromo-3-chloropropane	235.34	PAC	120.	0.008–0.996	34.7	53.	.47	—	—	50.	10
1,2-dichlorobenzene	147.0	PAC	120.	0.33–27.5	952.	129.	.43	—	—	19.	10
1,2-dichlorobenzene	147.0	GAC-C	?	?	—	300.	.43	—	—	—	30
1,2-dichlorobenzene	147.0	PAC-A	?	?	—	360.	.56	—	—	—	30

1,2-dichlorobenzene	147.0	PAC-B	?	?	—	140.	.31	—	—	—	30
1,2-dichlorobenzene	147.0	PAC-C	?	?	—	89.	.50	—	—	—	30
1,3-dichlorobenzene	147.0	PAC	?	0.48–24.3	328.	118.	.45	—	—	22.	10
1,4-dichlorobenzene	147.0	PAC	?	0.33–16.7	265.	121.	.47	—	—	22.	10
1,4-dichlorobenzene	147.0	GAC	?	4.0–10.1	206.	—	—	—	—	—	29
1,4-dichlorobenzene	147.0	GAC	550.	0.005–5.0	—	17.1[g]	.37	—	—	—	19
3,3-dichlorobenzidine	253.13	PAC	120.	0.005–2.51	336.	300.[d]	.2	—	—	4.8	10
3,3-dichlorobenzidine	253.13	PAC-A	120.	0.8–2.5	—	126.	.25	—	—	12.8	16
3,3-dichlorobenzidine	253.13	PAC-B	120.	0.6–2.5	—	240.	.95	—	—	5.7	16
3,3-dichlorobenzidine	253.13	PAC	120.	0.008–2.5	310.	304.[d]	.20	—	—	5.	15
dichlorobromomethane	163.83	PAC	120.	0.011–1.0	6.2	7.9	.61	—	—	500.	10
dichlorobromomethane	163.83	GAC	5 days (?)	?	?	24.6 (188.)[p]	.75	—	—	—	25
dichlorobromomethane	163.83	GAC	360.	?	?	.15[g]	.51	—	—	—	22
1,1-dichloroethane	98.96	PAC	120.	0.014–1.0	1.22	1.79	.53	—	—	1.0	10
1,2-dichloroethane	98.96	PAC	120.	0.012–1.0	2.46	3.57	.83	—	—	1,700.	10
1,2-dichloroethane	98.96	PAC	120.	15.8–84.	—	140.5	.75	—	—	—	14
1,2-dichloroethene (trans)	96.94	PAC	120.	0.005–1.0	2.10	3.05	.51	—	—	950.	10
1,1-dichloroethene	96.94	PAC	120.	0.007–1.0	2.83	4.91	.54	—	—	640.	10
2,4-dichlorophenol	163.	PAC	120.	0.01–18.1	229.	157.[h]	.15	—	—	8.0	10
2,4-dichlorophenol	163.	PAC	30.	0.04–0.86	—	122.	.12	—	—	—	31
2,4-dichlorophenol	163.	GAC[i]	2 wk	1.96–220.	—	354.	.095	—	—	—	32
2,4-dichlorophenol	163.	GAC	24 hr	1.0–22.5	—	145.[j]	.13	—	—	—	28
1,2-dichloropropane	112.99	PAC	120.	0.015–1.0	3.6	5.86	.6	—	—	600.	10
1,2-dichloropropene	110.98	PAC	120.	0.012–1.0	5.02	8.2	.46	—	—	320.	10
Dieldrin	380.91	PAC	120.	0.0005–0.21	122.	606.	.51	—	—	4.8	10
Dieldrin	380.91	GAC	?	0.00008–0.0065	6.5	—	—	—	—	—	29
Dieldrin	380.91	GAC	550.	0.0002–0.05	—	2.74[g]	.56	—	—	—	19
Dieldrin	380.91	?	?	?	—	131.	.56	—	—	—	25
diethylphthalate	222.24	PAC	120.	0.86–18.9	283.	110.	.27	—	—	—	10

(cont.)

Table 6–1 (*continued*)

Compound	Molecular Weight	Type of Carbon	Contact Time (min)	Equilibrium Range (mg/l)	Maximum Adsorbed (mg/g)	Freundlich Constants K (mg/g)	Freundlich Constants l/n	Langmuir Constants b	Langmuir Constants Xm	Carbon Dosage to Reduce 1 mg/l to 0.1mg/l	Reference
4-dimethylamino-benzene	225.3	PAC	120.	0.0035–0.28	164.	249.	.24	—	—	6.3	10
4-dimethylamino-benzene	225.3	PAC	120.	0.003–0.28	430.	249.	.24	—	—	6.2	15
N-dimethylnitrosamine	74.08	PAC	120.	6.5–9.0	—	6.8×10^{-5}	6.6	—	—	$>10^5$	10
N-dimethylnitrosamine	74.08	PAC-A	120.	6.5–9.5	—	$4. \times 10^{-5}$	5.64	—	—	$>10^5$	16
N-dimethylnitrosamine	74.08	PAC-B	120.	6.5–9.0	—	$9. \times 10^{-6}$	7.91	—	—	$>10^7$	16
2,4-dimethylphenol	122.17	PAC	120.	0.13–20.4	510.	70.[h]	.44	—	—	36.	10
2,4-dimethylphenol	122.17	PAC	?	?	—	184.	0.9	—	—	—	11
dimethylphenylcarbinol	136.2	PAC	120.	0.02–10.	472.	210.[d]	.34	—	—	9.	10
dimethylphthalate	194.18	PAC	120.	0.1–9.9	250.	97.[a]	.41	—	—	24.	10
4,6-dinitro-*o*-cresol	198.14	PAC	120.	0.2–19.1	560.	169.[h]	.27	—	—	12.	10
2,4-dinitrophenol	184.11	PAC	120.	1.06–9.95	188.	33.[j]	.61	—	—	110.	10
2,4-dinitrophenol	184.11	GAC[i]	2 wk	9.9–142.	—	153.[j]	.12	—	—	—	32
2,4-dinitrotoluene	182.14	PAC	120.	0.09–9.5	400.	146.	.31	—	—	13.	10
2,6-dinitrotoluene	182.14	PAC	120.	0.14–10.6	312.	145.	.32	—	—	13.	10
diphenylamine	169.24	PAC	120.	0.04–10.	246.	120.[a]	.31	—	—	16.	10
1,1-diphenylhydrazine	184.24	PAC	120.	0.39–9.95	191.	135.	.16	—	—	10.	10
1,1-diphenylhydrazine	184.24	PAC-A	120.	0.2–1.5	—	94.8	.28	—	—	18.	16
1,1-diphenylhydrazine	184.24	PAC-B	120.	0.4–9.1	—	149.	.23	—	—	10.	16

1,2-diphenylhydrazine	184.24	PAC	120.	0.054–0.238	710.	16,000.	2.	—	—	5.7	10
α-endosulfan	406.93	PAC	120.	0.000008–0.119	38.	194.	.5	—	—	5.3	10
β-endosulfan	406.93	PAC	120.	0.0002–0.082	30.	615.	.83	—	—	10.	10
Endosulfansulfate	422.93	PAC	120.	0.00003–0.11	51.	686.	.81	—	—	8.5	10
Endrin	381.	PAC	120.	0.0006–0.11	51.	666.	.8	—	—	8.5	10
Endrin	381.	GAC	?	0.000069–0.015	23.6	—	—	—	—	—	29
Ethylbenzene	106.16	PAC	120.	2.1–24.3	720.	53.	.79	—	—	110.	10
Ethylbenzene	106.16	PAC	120.	18.[k]	19.	—	—	—	—	—	12
Ethylbenzene	106.16	PAC	15.	6.0–15.0	—	120.	.29	—	—	—	17
Ethylbenzene	106.16	GAC	2 days	2.0–21.	—	100.	.40	—	—	—	18
bis(2-ethylhexyl)-phthalate	390.56	PAC	120.	1.83–11.0 $\times 10^{-3}$	5.1	11,300.	1.5	—	—	2.5	10
Fluoranthene	202.26	PAC	120.	0.0005–0.06	44.1	664.	.61	—	—	6.0	10
Fluoranthene	202.26	PAC	?	—	—	88.	.38	—	—	—	11
Fluorene	166.26	PAC	120.	0.002–1.21	284.	330.	.28	—	—	5.3	10
Fluorene	166.26	PAC	?	?	—	196.	.57	—	—	—	11
5-fluorouracil	130.08	PAC	120.	4.1–9.8	76.	5.5[c]	1.0	—	—	1,800.	10
Guanine	151.13	PAC	120.	0.25–9.92	296.	120.[d]	.4	—	—	19.	10
Heptachlor	373.5	PAC	120.	0.0058–0.17	132.	1,220.	.95	—	—	6.6	10
Heptachlor	373.5	GAC	?	0.006–0.056	12.3	—	—	—	—	—	29
Heptachlorepoxide	389.32	PAC	120.	0.0004–0.1	70.6	1,038.	.70	—	—	4.3	10
Hexachlorobenzene	284.78	PAC	120.	0.079–0.753 $\times 10^{-3}$	6.24	450.	.6	—	—	8.0	10
Hexachlorobutadiene	260.76	PAC	120.	0.002–1.0	93.8	258.	.45	—	—	48.	10
Hexachloro-cyclopentadiene	272.77	PAC	120.	0.004–0.8	307.	370.	.17	—	—	3.6	10
Hexachloroethane	236.74	PAC	120.	0.001–0.974	50.9	96.5	.38	—	—	20.	10
Hydroquinone	110.12	PAC	120.	0.4–10.	180.	90.	.25	—	—	18.	10
Isophorone	138.21	PAC	120.	0.162–18.1	151.2	32.	.39	—	—	70.	10
Isophorone	138.21	PAC	120.	6.9–48.4	—	223.	.8	—	—	—	14
Methylene chloride	84.94	PAC	120.	0.162–1.0	0.72	1.3	1.16	—	—	10,000.	10

(cont.)

Table 6–1 *(continued)*

Compound	*Molecular Weight*	*Type of Carbon*	*Contact Time (min)*	*Equilibrium Range (mg/l)*	*Maximum Adsorbed (mg/g)*	*Freundlich Constants K (mg/g)*	*Freundlich Constants l/n*	*Langmuir Constants b*	*Langmuir Constants Xm*	*Carbon Dosage to Reduce 1 mg/l to 0.1mg/l*	*Reference*
Methylene chloride	84.94	GAC	5 days	1.4–4.2	4.9	6.6[p]	.73	—	—	—	25
4,4-methylene-bis (2-chloroaniline)	264.28	PAC	120.	0.07–1.38	193.	190.	.64	—	—	21.	10
Naphthalene	128.18	PAC	120.	0.12–8.1	388.	132.	.42	—	—	18.	10
Naphthalene	128.18	PAC-A	120.	0.002–9.3	—	62.8	.30	—	—	29.	16
Naphthalene	128.18	PAC	?	?	—	123.	.41	—	—	—	11
Naphthalene	128.18	PAC-B	120.	0.06–5.3	—	169.	.56	—	—	19.	16
α-naphthol	144.2	PAC	120.	1.91–9.96	400.	180.[a]	.32	—	—	10.	10
β-naphthol	144.2	PAC	120.	0.9–9.73	384.	200.[a]	.26	—	—	8.4	10
β-naphthol	144.2	GAC	100 hr	?	—	—	—	.643	345.	—	27
α-naphthylamine	143.18	PAC	120.	0.04–9.94	358.	160.[d]	.34	—	—	12.	10
β-naphthylamine	143.18	PAC	120.	0.08–1.4	152.	150.	.30	—	—	12.	10
β-naphthylamine	143.18	PAC-A	120.	0.2–1.5	—	77.4	.36	—	—	27.	16
β-naphthylamine	143.18	PAC-B	120.	0.1–1.5	—	166.	.23	—	—	10.	10
p-nitroaniline	138.13	PAC	120.	0.01–9.89	294.	140.[a]	.27	—	—	12.	10
Nitrobenzene	123.11	PAC	120.	0.0009–0.25	42.	68.	.43	—	—	36.	10
Nitrobenzene	123.11	PAC	120.	44.[b]	196.	—	—	—	—	—	12
Nitrobenzene	123.11	GAC	100 hr	2.0–55.	—	—	—	.23	310.	—	27
4-nitrophenyl	199.21	PAC	120.	0.03–1.7	370.	370.	.27	—	—	4.5	10
4-nitrophenyl	199.21	PAC	120.	0.027–1.67	690.	370.	.27	—	—	4.54	15
o-nitrophenol	139.11	PAC	120.	0.12–18.2	246.	99.[h]	.34	—	—	20.	10
o-nitrophenol	139.11	GAC	100 hr	?	—	—	—	.42	240.	—	33
p-nitrophenol	139.11	PAC	120.	0.15–21.3	209.	76.2[h]	.25	—	—	21.	10

p-nitrophenol	139.11	GAC	3–4 wk	5.6–140.	—	1,113.	.17	—	—	—	33
N-nitrosodi-phenylamine	198.07	PAC	120.	0.01–9.2	438.	220.[a]	.37	—	—	9.8	10
N-nitrosodi-*n*-propylamine	130.19	PAC	120.	0.26–20.8	72.	22.4[a]	.26	—	—	67.	10
p-nonylphenol	220.34	PAC	120.	0.02–3.13	348.	250.[j]	.37	—	—	8.3	10
PCB-1221	~200.7	PAC	120.	0.001–0.611	108.	242.	.7	—	—	19.	10
PCB-1232	~232.2	PAC	120.	0.0001–0.353	147.	630.	.73	—	—	7.7	10
PCB-1016	?	GAC	550.	0.0005–0.1	—	3.44[g]	.66	—	—	—	21
PCB-1016	?	GAC-A	550.	0.0005–0.1	—	13.7[g]	.27	—	—	—	21
PCB-1016	?	GAC-B	550.	0.0005–0.1	—	15.4[g]	.25	—	—	—	21
PCB-1016	?	GAC-C	550.	0.0005–0.1	—	6.7[g]	.36	—	—	—	21
PCB-1016	?	GAC-D	550.	0.0005–0.1	—	4.1[g]	.46	—	—	—	21
PCB-1016	?	PAC	48 hr	0.0007–0.02	—	3.44[g]	.644	—	—	—	20
PCB-1016 + 5 mg/l humic acid	?	PAC	48 hr	0.001–0.02	—	μg/g 3.16	.560	—	—	—	20
PCB-1254	?	PAC	48 hr	0.001–0.007	—	μg/g .73	1.14	—	—	—	20
PCB-1254 + 5 mg/l humic acid	?	PAC	48 hr	0.001–0.02	—	μg/g 1.02	.74	—	—	—	20
PCB-1254	?	GAC	550.	0.0005–0.05	—	μg/g .73[g]	1.14	—	—	—	19
Pentachlorophenol	266.4	PAC	120.	0.22–9.23	350.	150.[j]	.42	—	—	16.	10
Phenanthrene	178.24	PAC	120.	0.0022–0.37	89.2	215.	.44	—	—	11.	10
Phenanthrene	178.24	PAC	?	?	—	135.	.45	—	—	—	11
Phenol	94.11	PAC	120.	2.15–9.8	88.	21.[a]	.54	—	—	150.	10
Phenol	94.11	GAC	5 days	?	—	86.	.14	—	—	—	25
Phenol	94.11	PAC	120.	194.[b]	167.[b]	—	—	—	—	—	12
Phenol	94.11	GAC	100 hr	2.0–50.	—	—	—	.176	158.	—	27
Phenol	94.11	GAC	23 days	1.0–95.	—	78.1	.21	—	—	—	26
Phenol	94.11	PAC	4 days	1.0–95.	—	78.1	.21	—	—	—	26

(*cont.*)

Table 6–1 *(continued)*

Compound	Molecular Weight	Type of Carbon	Contact Time (min)	Equilibrium Range (mg/l)	Maximum Adsorbed (mg/g)	Freundlich Constants K (mg/g)	Freundlich Constants 1/n	Langmuir Constants b	Langmuir Constants Xm	Carbon Dosage to Reduce 1 mg/l to 0.1mg/l	Reference
Phenol	94.11	PAC	240.	4.0–32.	—	—	—	.088	234.	—	34
Phenol	94.11	GAC	24 hr	1.0–22.5	—	27.[j]	.48	—	—	—	28
Phenol	94.11	PAC	120.	9.4–85.6	—	140.	.54	—	—	—	14
Phenol	94.11	PAC	240.	0.94–188.	—	4,498.[p]	.41	—	—	—	34
Phenol	94.11	GAC	24 hr	0.3–200.	—	56.	.24	—	—	—	35
Phenol	94.11	GAC	?	0.1–8.[q]	—	2.16[r]	.23	—	—	—	13
Phenylacetic acid	136.15	GAC	?	0.1–10.[q]	—	2.51[r]	.16	—	—	—	13
Phenylacetic acid	136.15	GAC	?	0.05–12.[q]	—	1.0[r]	.165	—	—	—	13
Phenylacetic acid	136.15	GAC	?	0.1–8.[q]	—	.84[r]	.38	—	—	—	13
Phenylacetic acid	136.15	GAC	?	0.1–22.[q]	—	1.14[r]	.15	—	—	—	13
Phenylacetic acid	136.15	GAC	?	0.05–7.[q]	—	1.93[r]	.162	—	—	—	13
Phenylacetic acid	136.15	GAC	?	0.1–8.[q]	—	1.57[r]	.245	—	—	—	13
Phenylacetic acid	136.15	GAC	?	0.1–11.[q]	—	1.[r]	.145	—	—	—	13
Phenylacetic acid	136.15	GAC	?	0.1–7.[q]	—	1.75[r]	.18	—	—	—	13
Phenylacetic acid	136.15	GAC	?	0.1–8.[q]	—	1.82[r]	.03	—	—	—	13
Phenylacetic acid	136.15	GAC	?	0.1–7.[q]	—	1.53[r]	.11	—	—	—	13
Phenylacetic acid	136.15	GAC	?	0.1–6.5[q]	—	1.88[r]	.144	—	—	—	13
Phenylacetic acid	136.15	GAC	?	0.1–8.[q]	—	2.08[r]	.244	—	—	—	13
Phenylmercuric acetate	336.74	PAC	120.	0.02–10.1	503.	270.[c]	.44	—	—	9.2	10
Propionic acid	74.08	GAC	?	0.1–8.3[q]	—	.497[r]	.4	—	—	—	13
Pyrene	202.24	PAC	?	?	—	66.	.24	—	—	—	11
Pyridine	79.10	GAC	?	0.14–7.4[q]	—	1.22[r]	.2	—	—	—	13
Styrene	104.14	PAC	120.	0.07–6.45	316.	120.[a]	.56	—	—	—	10

Styrene	104.14	PAC	120.	18.[n]	28.	—	—	—	—	—	12
1,1,2,2-tetrachloroethane	167.85	PAC	120.	0.004–1.0	6.42	10.6	.37	—	—	360.	10
Tetrachloroethylene	165.83	PAC	120.	0.007–1.0	35.8	50.8	.56	—	—	64.	10
Tetrachloroethylene	165.83	GAC	2 days	0.06–0.5	—	57.7	.48	—	—	—	23
Tetrachloroethylene	165.83	GAC	5 days	?	—	274.[g] 348.[p]	.56	—	—	—	24
Tetrachloroethylene	165.83	GAC	?	?	—	84.1	.41	—	—	—	24
1,2,3,4-tetrahydronaphthalene	132.21	PAC	120.	0.5–10.	440.	74.	.81	—	—	78.	10
1,2,3,4-tetrahydronaphthalene	132.21	PAC	120.	5.3–29.	—	291.	.20	—	—	—	14
Thymine	126.11	PAC	120.	1.77–10.1	104.	27.	.51	—	—	110.	10
Toluene	92.14	PAC	120.	1.71–18.5	106.	26.1	.44	—	—	96.	10
Toluene	92.14	GAC	?	1.6–4.4	8.75	—	—	—	—	—	29
Toluene	92.14	PAC	120.	66.[l]	50.	—	—	—	—	—	12
Toluene	92.14	PAC	15.	6.0–15.0	—	110.	.28	—	—	—	17
Toluene	92.14	GAC	2 days	2.0–20.	—	90.	.30	—	—	—	18
Toluene	92.14	PAC	120.	2.8–49.	—	348.	.48	—	—	—	14
Toluene	92.14	PAC	?	?	—	40.	.35	—	—	—	11
1,2,4-trichlorobenzene	181.45	PAC	120.	0.38–15.	500.	157.	.31	—	—	12.	10
1,1,1-trichloroethane	133.41	PAC	120.	0.0035–1.0	2.4	2.48	.34	—	—	800.	10
1,1,1-trichloroethane	133.41	GAC	2 days	0.09–0.8	—	42.7	.57	—	—	—	23
1,1,2-trichloroethane	133.41	PAC	120.	0.006–1.0	3.8	5.81	.6	—	—	620.	10
Trichloroethylene	131.39	PAC	120.	0.005–1.0	8.7	28.0	.62	—	—	130.	10
Trichloroethylene	131.39	GAC-C	?	?	—	32.0	.45	—	—	—	30
Trichloroethylene	131.39	PAC-A	?	?	—	16.5	.54	—	—	—	30
Trichloroethylene	131.39	PAC-B	?	?	—	28.	.52	—	—	—	30
Trichloroethylene	131.39	PAC-C	?	?	—	11.	.67	—	—	—	30
Trichloroethylene	131.39	GAC	?	?	—	26.2	.5	—	—	—	24
Trichlorofluoromethane	137.4	PAC	120.	0.056–5.0	8.98	5.6	.24	—	—	280.	10

(*cont.*)

Table 6–1 *(continued)*

Compound	*Molecular Weight*	*Type of Carbon*	*Contact Time (min)*	*Equilibrium Range (mg/l)*	*Maximum Adsorbed (mg/g)*	*Freundlich Constants*		*Langmuir Constants*		*Carbon Dosage to Reduce 1 mg/l to 0.1mg/l*	*Reference*
						K (mg/g)	*1/n*	*b*	*Xm*		
2,4,6-trichlorophenol	197.45	PAC	120.	0.58–19.1	692.	155.1[m]	.40	—	—	15.	10
2,4,6-trichlorophenol	197.45	GAC	24 hr	1.0–22.5	—	200.[j]	.15	—	—	—	28
2,4,6-trichlorophenol	197.45	GAC	24 hr	1.0–22.5	—	88.[o]	.57	—	—	—	28
Uracil	112.09	PAC	120.	4.3–9.95	66.	11.[a]	.63	—	—	350.	10
p-xylene	106.2	PAC	120.	1.5–18.	130.	85.	.19	—	—	15.	10
o-xylene	106.2	PAC	15.	5.0–15.0	—	130.	.27	—	—	—	17
o-xylene	106.2	GAC	2 days	1.5–20.0	—	120.	.22	—	—	—	18

[a]Data pooled from three isotherms, pH = 3.0, 7.0, and 9.0.
[b]From an initial concentration of 1,000. mg/l.
[c]Data pooled from two isotherms, pH = 3.0 and 7.0.
[d]Data pooled from two isotherms, pH = 7.0 and 9.0.
[e]From an initial concentration of 416. mg/l.
[f]At pH = 3.0.
[g]When C_e = 1.0 μg/l.
[h]At pH = 5.5.
[i]12 × 16 mesh.
[j]At pH = 7.0.
[k]From an initial concentration of 115. mg/l.
[l]From an initial concentration of 317. mg/l.
[m]At pH = 6.0.
[n]From an initial concentration of 180 mg/l.
[o]At pH = 9.0.
[p]When C_e = 1.0 mol/l.
[q]mm/l.
[r]mm/g.

Equilibrium Studies

Ultraviolet light adsorption at 240 nm was used to evaluate 10 commercial carbons for their ability to remove total organic matter from the Rhine River [4]. Typical log graphs were obtained (not shown) when the quantity of organic chlorine adsorbed (mg) per (g) carbon was plotted against equilibrium concentration (based on ultraviolet absorbance). The relative positions and slopes of the isotherms were influenced by variable organic contents as reflected by sampling location on the lower Rhine River and on the time of year and rate of flow of the river. Apparently much useful operational data were derived from the ultraviolet plots.

Pilot Plant Studies

At Düsseldorf, West Germany, the efficiencies of these carbons for removal of TOC and total chlorinated organic compounds (TOCl) were evaluated [5,6] (see also Chapters 8 and 9). Specific chlorinated organics were studied also, including hexachlorocyclohexane, bis (2-chloropropyl) ether, and hexachlorobutadiene. Their removal paralleled the removal of TOCl. Distribution of TOCl and specific chlorinated organics was determined with respect to depth in several carbon beds operating in parallel for as long as 6 months. Chlorinated aromatic compounds such as 2-dichlorobenzene were adsorbed strongly on all carbons and apparently were not displaced significantly by other competing adsorbates.

Pretreatment

Removal of ultraviolet-absorbing organics from preozonated Rhine River bank filtrate was accomplished by employing chemical coagulation with $FeCl_3$ and $Al_2(SO_4)_3$ prior to carbon adsorption [7]. Adsorption isotherms were generated in systems without coagulation, with 1 mg/l $FeCl_3$ and with 1 mg/l $Al_2(SO_4)_3$. The best carbon adsorption was obtained when alum was the pretreatment coagulant. $FeCl_3$ had very little effect on subsequent carbon adsorption.

Ozonation of organic matter in a raw water supply and prior to carbon adsorption has been investigated [4]. In this case, the dissolved organic matter content (DOC), COD, and ultraviolet absorbance were employed as parameters of treatment efficiency. Some reduction of DOC was effected by ozone. For example, the greatest reduction of 11 to 38% was recorded for ultraviolet-absorbing organics. Subsequent experiments evaluated the effect of alum on organic content. Here, the carbon adsorption isotherm following ozonation differed very little from those using nonozonated water that had been treated with alum. However, when ozonation and coagulation were used together prior to carbon adsorption, removals of organic matter were significantly better. These observations were confirmed more or less by Rook [8] in a pilot-plant study using Rhine River water. A 15% reduction of TOC was effected by breakpoint chlorination and coagulation. Subsequent ozonation did not increase TOC removal, but an additional 10% of TOC was removed by double layer filtration following ozonation. Another 10% was achieved by carbon adsorption.

Breakthrough

The performance of GAC filters has been evaluated for TOC removal [9]. Some operators have argued that complete removal of TOC results in complete removal of organic compounds. This is simply not the case. In actual plant practice, partial breakthrough of TOC has occurred when fresh GAC is employed. This is probably due to elution of soluble organics from the carbon. Under normal operation of a GAC column, the TOC (C_e) of the effluent steadily increases until breakthrough occurs (see Chapter 8). If the operation were permitted to continue, the TOC of the effluent would increase until it equals the influent TOC. Apparently this is not the situation in actual water treatment plant operation [9]. The effluent's TOC seldom reached the influent TOC but reaches a steady-state value lower than that of the influent. This suggests that the GAC continues to remove TOC on an indefinite basis that has been attributed to biodegradation or to slow mass transfer into micropores of the GAC. All of this is well and good if this steady-state TOC content meets a particular water quality standard.

In the operation of a GAC adsorber, the TOC of immediate breakthrough is an important parameter that has been reported from several studies [9] (see Chapter 8). As recorded from 47 datasets, there is an immediate breakthrough of TOC with most of the points in the C_e:C_i range of 0.1 to 0.5 after start-up. This immediate breakthrough of organics is one of the operational limitations of GAC filters where many factors contribute to this problem, one of which is the EBCT, or bed depth. This is seen in Figure 6–1a where the C_e:C_i ratios are higher in the shallower beds. Figure 6–1b shows that a plot of C_e:C_i versus EBCT results in a considerable scatter of points. Here again, lower contact times yield higher C_e:C_i ratios.

Another important relationship in the operation of a GAC filter is the time to steady state (t_{ss}) [9]: How long can a GAC filter operate and produce an acceptable water quality or meet a given standard? The answer depends on many factors, but one of the most important is the EBCT. Figure 6–1c shows a t_{ss} versus EBCT plot that was composited from many studies [9]. It is an empirical log-log plot with the linear equation: $t_{ss} = 15 \times \text{EBCT}^{0.62}$. Under typical conditions of water treatment, there is a positive correlation between t_{ss} and EBCT. However, this relationship is unexpected [9], that is, an increase in EBCT should increase the efficiency of utilization of the carbon bed, and the value of the exponent in the preceding empirical equation should be equal to or greater than unity.

It is not clear why the statistical plot seen in Figure 6–1c was derived. Some observers have attributed it to a high molecular weight faction of the TOC characterized by strong sorption with slow intraparticle transport. Another expla-

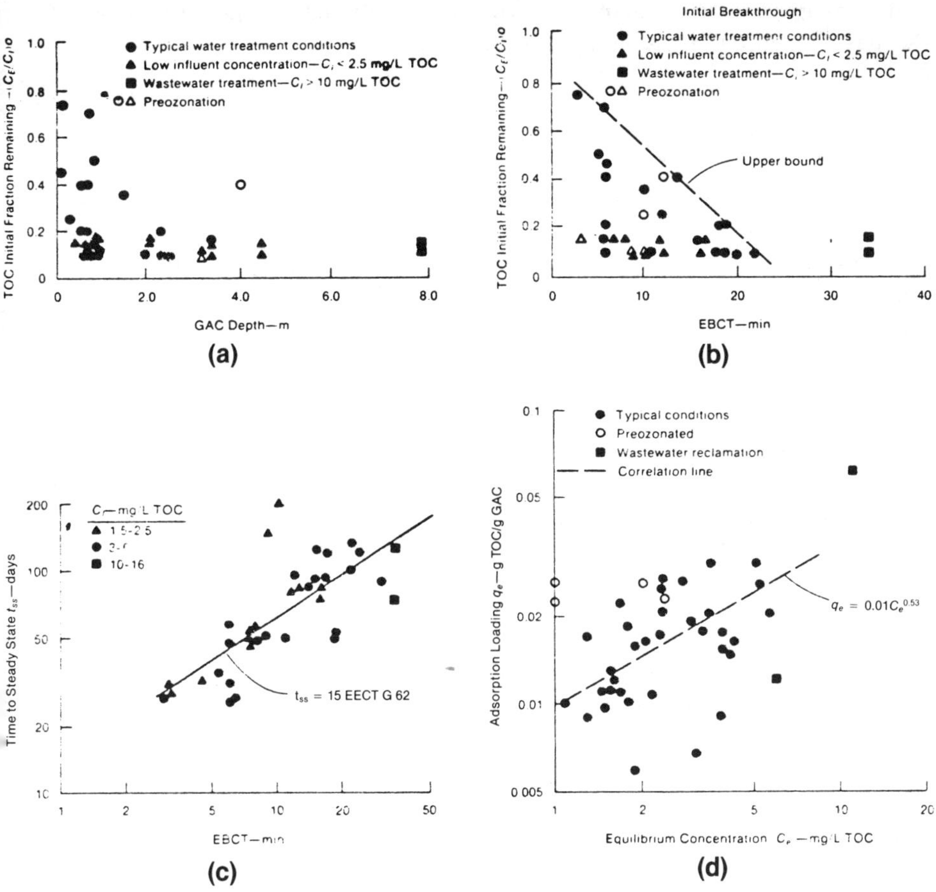

Figure 6–1 (a) Dependence of initial fraction remaining $(C_e/C_1)_o$ on GAC bed depth. Typical conditions are taken to be $2.5 \leq C_1 \leq 10$ mg/lTOC, without preozonation. (b) Dependence of initial fraction remaining on EBCT. (c) Observed run length at which steady state is reached. (d) Equilibrium adsorption capacity correlation for all data: Adsorption loading q_e versus equilibrium solution concentration C_e. Reproduced from Roberts and Summers [9], courtesy of the American Water Works Association.

nation is that the residual TOC reflects the net adsorption of organic compounds, of which some are weakly adsorbed and some are strongly adsorbed. Consequently, the exponent represents a net adsorption of these compounds. Evidence to support this explanation is seen in the scatter of points in Figures 6–1c and 6–1d. The latter figure is a typical Freundlich isothermal plot where q_e (quantity adsorbed) is graphed against C_{ess} (equilibrium concentration at steady state). This is a statistical plot of 40 datasets from many studies. There is considerable scatter, with the line having a correlation coefficient, *r*, of .52. The 95% confidence interval for the exponent $(1/n)$ of .53 is $-.02$ to $+1.08$. This correlation was based on many investigations, many types of carbons, and many types of organics. The scatter in Figure 6–1d is entirely expected.

ORGANIC PRIORITY POLLUTANTS

Most of the compounds on the Environmental Protection Agency's (EPA) priority pollutant list are organic. They are considered to be toxic and hazardous to various target organisms, including humans. Consequently, it is necessary to remove or to reduce their concentrations to a safe level [1]. Table 6–1 is a rather long list of PAC and GAC carbon adsorption data derived from many laboratory studies of batch systems. Most of the data came from Dobbs and Cohen [10]. Other sources are included also for comparative purposes. The Freundlich equation constants are applicable only to the experimental systems from which they were calculated. These constants are also employed to calculate carbon dosages for the reduction of any initial concentration to any desired residual. The Freundlich *K* values for various compounds can be compared for their affinities of adsorption. Most of the *K* values in Table 6–1 are those calculated when $C_e = 1.0$ mg/l. Some were calculated for $C_e = 1.0$ μg/l. These *K* values are footnoted. Some Langmuir constants are given also.

PAC was the predominant carbon employed for the laboratory studies in Table 6–1. Consequently, the data have a limited value when applied to treatment plant operation. However, adsorption of organics on GAC usually follows from PAC. See other tables below for additional adsorption data of organics on GAC. Nevertheless, the PAC data are useful for individual adsorptive characteristics.

Many compounds in Table 6–1 have been studied in detail. For example, benzene has been adsorbed on at least five PACs and five GACs. *K* values ranged from 1.0 to 26.8 mg/g for PAC and from 0.73 to 36. for GAC. This range indicates a low affinity for benzene on carbon. Dosages of 678. and 705. mg/l were calculated for a reduction of 1.0 mg/l of benzene to 0.1 mg/l. These quantities of carbon are high for routine water treatment.

Another poorly adsorbed compound is CCl_4. The *K* value is 11.1 mg/g for PAC and ranges from 0.13 to 11.6 mg/g ($C_e = 1.0$ μg/l). Apparently, CCl_4 is too polar for carbon adsorption. Chloroform, bromoform, dichlorobromomethane, and dibromochloromethane (the trihalogenated hydrocarbons) show similar adsorptive characteristics. Consequently, carbon is not an effective water treatment for these compounds.

In contrast, 1,2-dichlorobenzene is adsorbed quite well by PAC and GAC. *K* values range from 89. to 360. mg/g for PAC, and 300. mg/g was reported for one GAC. The 19 mg/l carbon dosage for 1.0 to 0.1 mg/l is well within a normal water treatment operation.

PCBs (polychlorinated biphenyls) are readily adsorbed on PAC and poorly on GAC. Apparently there is little penetration of PCB molecules into the porous GAC.

Trichloroethanes and ethylenes are moderately adsorbed on PAC and GAC.

Consequently, carbon treatment would be slightly effective for their removal from water.

More information is given in the next sections for selected compounds from Table 6–1, phenols, pesticides, humic and fulvic acids, taste and odor compounds, hydrocarbons, and other organics.

CARBON TETRACHLORIDE

Several in-depth laboratory studies have been conducted for the removal of CCl_4 from water by PAC and GAC [19,23,36]. A summary of adsorption equilibrium (500 to 600 min) studies for CCl_4 on Filtrasorb 400 is seen in Table 6–2 [36]. The Freundlich equation "was found to provide the most straightforward and generally satisfactory method for fitting and describing the experimental data" [36]. Very little effect was observed from $[H_3^+O]$, carbon particle size, and background solution effect (5 ppm humic acid). When the carbon was presaturated with humic acid, however, lesser amounts of CCl_4 were adsorbed.

Rates for CCl_4 adsorption onto GAC were conducted in the presence and absence of humic acid [36]. In the absence of humic acid, rate coefficients $k_f = 4.1 \times 10^{-3}$ cm/sec and $D_s = 3 \times 10^{-10}$ cm^2/sec were obtained. The k_f value is the film transfer coefficient, whereas the D_s is an internal diffusion coefficient. In the presence of 5 mg/l humic acid, k_f and D_s values were 3.8×10^{-3} cm/sec and 1.9×10^{-10} cm^2/sec, respectively. This suggested that humic acid molecules retard the diffusion of CCl_4 within the carbon pores [36], provided the humic and molecules can enter the internal pores. Also there is the question of the statistical significance of the difference between the two sets of data.

TRIHALOMETHANES

Adsorption of trichloroethylene (TCE), tetrachloroethylene (PCE), and carbon tetrachloride on Nuchar WV-6 GAC was studied extensively in equilibrium and column modes [23]. Individual Freundlich isotherm constants are seen in Table 6–3 for the three compounds. These values are somewhat similar to those from other studies. Quantitative differences are due to type of carbon and adsorption time. K and $1/n$ values are given for PCE and TCE in a competitive situation with each other. The lower K values of 44.7 and 25.1 mg/g indicate that there is a considerable competitive effect. This effect undoubtedly would occur in an actual water treatment operation.

Four trihalogenated hydrocarbons—chloroform, bromoform, dibromochloromethane, and dichlorobromomethane—were examined for their adsorption on Nuchar WV-6 GAC also in equilibrium and column modes. Table 6–4 shows Freundlich constants for their adsorption on GAC, whereas their removal in a column operation is seen in Table 6–5. The order of appearance of the four compounds coming off the column is predicted quite well from isothermal data.

Table 6–2 Summary of Adsorption Equilibrium Studies for Carbon Tetrachloride

Background Solution	*Carbon Size, U.S. Standard Mesh Size*[d]	*Initial Concentration (μg/l)*	*Buffer KH_2PO_4*	*Freundlich Isotherm Constants*[a]			*Remarks*
				K_f	1/n	r^b	
OFW[c]	30/40	2,160	—	.344	.551	.975	
OFW	50/60	—	—	.497	.585	.942	Minicolumn
OFW (pH 5)	220/325	100	$1 \times 10^{-3} M$	.209	.603	.998	
OFW (pH 7)	200/325	100	$1 \times 10^{-3} M$	.224	.630	.993	pH effect
OFW (pH 9)	200/325	100	$1 \times 10^{-3} M$	.208	.601	.995	
OFW	35/50	100	$1 \times 10^{-3} M$	.243	.818	.954	Particle size effect
OFW	80/100	100	$1 \times 10^{-3} M$	.234	.700	.995	
OFW	16/20	125	—	.209	.678	.965	
OFW + 5 ppm humic acid	16/20	125	—	.262	.674	.987	Background solution effect
Tap water	30/40	178	—	.195	.594	.970	
OFW	30/40	125	—	.215	.610	.985	
OFW	30/40	142	—	.169	.517	.992	Presaturated carbon
OFW	30/40	143	—	.202	.633	.996	Virgin carbon

Source: Reproduced from Weber and Pirbazari [36], courtesy of the American Chemical Society.
Note: All studies were conducted at pH 7 unless otherwise noted.
[a]Based on C_e and q_e expressed in μg and mg/g, respectively.
[b]r = Correlation coefficient for fit of the isotherm to the experimental data.
[c]Organic-free water.
[d]Filtrasorb 400.

Table 6–3 Freundlich Isotherm Constants for Light Halogenated Hydrocarbons Used in this Study

Compound	*K (mg/g)*	*l/n*
Single compounds		
CCl_4	11.6	.71
PCE	57.7	.48
TCE	42.7	.57
Mixed compounds		
PCE	44.7	.48
TCE	25.1	.49

Source: Benedict [23].

Table 6–4 Freundlich Isotherm Constants for Light Halogenated Hydrocarbons on GAC

Compound	*K[a] (mg/g)*	*l/n*	*PAC[b] Dosage (mg/l)*
$CHCl_3$	.0165	.60	1,900.
$CHClBr_2$	.130	.62	410.
$CHCl_2Br$	.150	.51	500.
$CHBr_3$	.185	.32	150.

Source: Youssefi [37].
Note: Phosphate buffer, pH 7. Temperature = 24°C.
[a]Contact time = 360 min.
[b]To reduce initial content of 1 mg/l to 0.1 mg/l (from Table 6–1).

Table 6–5 GAC Performance for Removal of Light Halogenated Hydrocarbons on Columns

Compound	*C_o (mg/l)*	*Breakthrough Point*	*Amount Adsorbed from a Mixture*	*Amount Adsorbed as Single Compounds*
$CHCl_3$	1.20	0.10	0.593	0.829
CCl_4	1.89	0.094	1.88	—
$CHCl_2Br$	3.69	0.10	3.14	3.22
$CHClBr_2$	2.99	0.10	3.64	—
$CHBr_3$	11.14	0.05	11.12	20.45

Source: Youssefi [37].
Note: Phosphate buffer, pH 7. Temperature = 24°C.

$CHClBr_2$ and $CHCl_2Br$ are transposed in the column study perhaps because of similar adsorptive characteristics. However, quantities of the four compounds adsorbed in the carbon column suggest this mode of treatment is somewhat inefficient. This inefficiency is seen (Table 6–4) also in the quantities of PAC needed to reduce an initial 1.0 mg/l to 0.1 mg/l. All those PAC dosages in Table 6–4 are much too high for normal water treatment plant operation [37].

A laboratory study confirmed the inefficiency of PAC to remove total trihalomethanes (TTHM) from drinking water [38]. Various quantities of PAC were added to tap water that contained some free chlorine residual, 0.12 mg/l (Table 6–6). Among other things, at least 50 mg/l PAC are needed to effect a 50% removal of TTHM within a treatment time of 60 minutes. This study was also consistent with others where the bromine-containing compounds were removed more efficiently than chloroform.

Much laboratory research has been conducted to simulate plant conditions whereby TTHMs are formed from precursors in the presence of chlorine and PAC. In one such study, the combination of chlorine after 30 min and 3 mg/l humic acid caused a 55.5% increase in the [$CHCl_3$] in water that was not dechlorinated [39]. Conversely, dechlorinated water (after 30 min) required less carbon to reduce the [$CHCl_3$]. Undoubtedly, this occurs in actual plant operation if precursors and/or chlorine are not removed prior to the addition of PAC.

Pilot plant studies of Philadelphia's drinking water were conducted for removal of several organics including $CHCl_3$ [40]. The pilot plant consisted of two 7.5 cm (ID) × 1 m columns of GAC (F-400) (one column received chlorinated water but the second did not). The Delaware River was the raw source of the drinking water. Influent chloroform contents ranged from 71. to 176. μg/l over 14 weeks (average 102. μg/l). Chloroform was 79 to 94% of the TTHM. Also evaluated was a rechlorination of effluent waters to a chlorine content of 1.5 mg/l. This simulates formation of TTHM in distribution systems. However, it is not necessarily the maximum TTHM formation.

Breakthrough curves for instantaneous $CHCl_{3(i)}$ and $CHCl_{3(1.5)}$ concentrations are seen in Figures 6–2a, b, c, and d. $CHCl_{3(i)}$ did not exceed 50 μg/l in the effluent until the fifth week of operation (Figure 6–2a). $CHCl_{3(1.5)}$ effluent contents exceeded 50. μg/l in the third week of operation and 100. μg/l during the fifth week. Influent precursors for $CHCl_3$ (as $CHCl_3$) were determined as the difference between $CHCl_{3(i)}$ and $CHCl_{3(1.5)}$. These values ranged from 23. to 155. μg/l over 14 weeks with an average of 85. μg/l. Figure 6–2c shows the precursor breakthrough reaching 45. μg/l within 2 weeks of operation. After 4 weeks, the precursor contents of the influent and effluent were identical. On numerous occasions in the two columns, precursor effluent contents exceeded the influent. In this study, this occurred during the fifth week for the dechlorinated column that was 2 weeks prior to 100% breakthrough of the chlorinated system. Desorption of $CHCl_3$ precursors apparently was responsible for their occasional increase in the effluent. Cumulative mass loadings of $CHCl_{3(i)}$ reached 1.6 mg/g in the two columns after 100 days, which is higher than can be achieved in a short-term laboratory study [22]. Cumulative mass loadings of $CHCl_3$ precursors reached

Table 6–6 THM Concentrations in PAC Experiments, Louisville, Ky., Tap Water

		THMs[a]							
PAC Dose (mg/l)	*Exposure Time (minutes)*	*$CHCl_3$ (μg/l)*	*Percent Removal*	*$CHBrCl_2$ (μg/l)*	*Percent Removal*	*$CHBr_2Cl$ (μg/l)*	*Percent Removal*	*TTHM (μg/l)*	*Percent TTHM Removal*
0	0	35.2	—	12.3	—	0.7	—	48.2	—
(Control)	5	30.3	—	10.3	—	1.1	—	41.7	—
	15	31.8	—	10.2	—	0.7	—	42.7	—
	30	30.2	—	10.5	—	0.7	—	41.4	—
	60	29.4	—	8.7	—	0.5	—	38.6	—
25	0	35.2	0	12.3	0	0.7	0	48.2	0
	5	26.3	13	7.2	30	0.3	73	33.8	19
	15	24.1	24	5.9	42	0.3	57	30.3	29
	30	23.2	23	5.3	50	0.2	71	28.7	31
	60	19.2	35	3.9	55	0.1	80	23.2	40
50	0	35.2	0	12.3	0	0.7	0	48.2	0
	5	25.0	17	4.9	52	0.2	82	30.1	28
	15	18.1	43	3.1	70	0.1	86	21.3	50
	30	17.7	41	2.7	74	0.1	86	20.5	50
	60	15.5	47	2.1	76	0.1	80	17.7	54
100	0	35.2	0	12.3	0	0.7	0	48.2	0
	5	17.4	43	2.8	73	0.1	91	20.3	51
	15	13.2	58	1.4	86	0.0	100	14.6	66
	30	10.1	67	1.0	90	0.0	100	11.1	73
	60	9.7	67	0.8	91	0.0	100	10.5	73

Source: Weil [38].
Note: Free chlorine residual = 0.12 mg/l.
[a]No bromoform found.

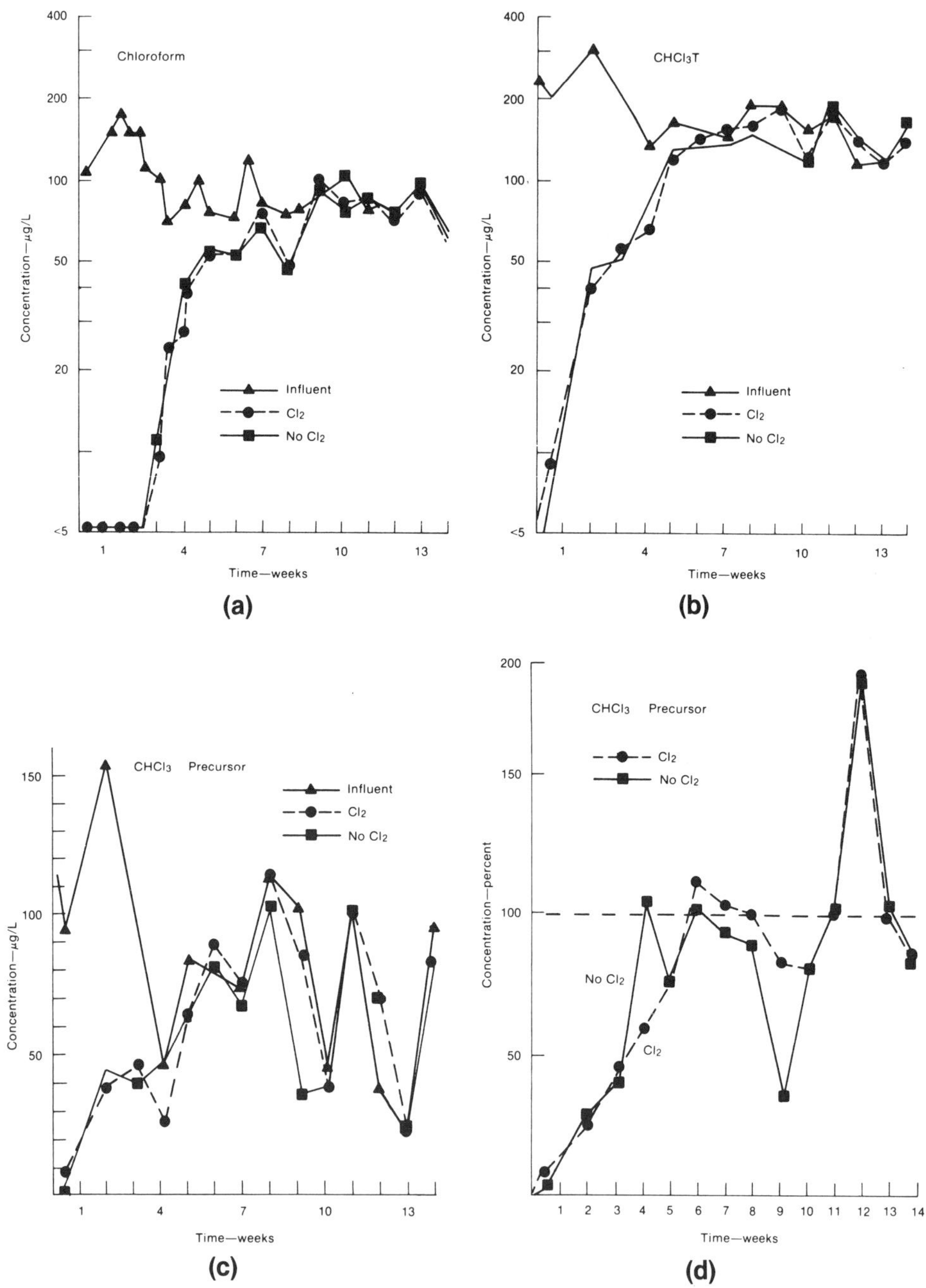

Figure 6–2 (a) Instantaneous influent-effluent $CHCl_3$ concentrations. (b) Terminal influent-effluent $CHCl_{3(1.5)}$ concentrations. (c) $CHCl_3$ precursor influent-effluent concentrations-mg/1. $CHCl_3$ precursor is the difference between $CHCl_{3(1.5)}$ and instantaneous $CHCl_3$. (d) $CHCl_3$ precursor influent-effluent concentrations-percent. $CHCl_3$ precursor is the difference between $CHCl_{3(1.5)}$ concentrations. Reproduced from Yohe et al. [40], courtesy of the American Water Works Association.

the range of 0.8 to 1.1 mg/g after 100 days. These findings indicate that a steady-state condition is reached in carbon columns with effluents containing significant quantities of $CHCl_3$ and its precursors.

FULVIC AND HUMIC ACIDS

Equilibrium Studies

PAC has been used for many years to remove naturally occurring color in surface waters, especially lakes, ponds, and reservoirs. That fulvic and humic acids are major components of the organic structure of naturally colored colloids has been well established by many researchers [1]. However, it has been shown recently that removal by carbon adsorption was researched quantitatively in batch and column operations. Table 6–7 is a summary of experimental conditions with the attendant Freundlich constants for humic and fulvic acid adsorption on PAC and GAC in batch systems. Contact times ranged from 550 min to 5.17 day, which are beyond the design of a conventional water treatment plant. The intent of the researchers was to reach equilibrium conditions.

An indication of the adsorptive capacities of various carbons is seen in the K values in Table 6–7. These values are the milligrams of fulvic and/or humic acids adsorbed per gram of carbon when the C_e is 1.0 mg/l. PACs normally adsorbed more color than the GACs. In contrast, some carbons exhibited extremely low capacities for the humic and fulvic acids. Also, there was a tendency for greater adsorption at slightly acid pH values.

An extensive study was conducted on adsorption of humic acids from various sources including Michigan Finch soil and extracts of leaves from a hardwood forest [45]. Since the term *humic acids* is defined operationally [1], it is expected that the physicochemical properties would vary; hence, adsorptive characteristics would be different (Table 6–8). Adsorption isotherm studies were conducted on 50 × 60 mesh F-400 and WV-G over 7 days at room temperature (Figures 6–3a, b, c, and d).

That $[H_3^+O]$ is a factor affecting the adsorption of fulvic acid is seen in Figure 6–3a. The extent of adsorption increases as the pH value is lowered. This is undoubtedly due to association of the oxygen-containing functional groups. A slight advantage may be gained in actual operation by lowering the pH value to 7.0 and below. However, the exclusive use of carbon would have to be color removal to justify lowering the pH value.

Figure 6–3b shows the adsorption of various types of humic and fulvic acids. There appears to be a significant difference among the four isotherms. These compounds are adsorbed in the reverse order of their oxygen-containing functional groups; that is, the lower the concentration of these groups, the greater is the extent of adsorption. This is consistent with the fact that carbon has many oxygen-containing groups on its internal and external surfaces.

The molecular weight of soil fulvic acid definitely affects adsorption. This is seen in Figures 6–3c and 6–3d where lower molecular weights are adsorbed to

Table 6–7 Adsorption of Fulvic and Humic Acids by Carbon in Batch Systems

Constitutent	*Type of Carbon*	*Contact Time*	*pH*	*Freundlich Constants* K *(mg/g)*	1/n	*Reference*
Fulvic acids	GAC	> 48 hr	—	1.26[a]	.26	41
Fulvic acids	Bars	> 48 hr	—	1.10	.54	41
Fulvic acids	PAC	> 48 hr	—	2.49	.15	41
Humic acids	GAC-A	550. min	—	12.4	.44	19
Humic acids	GAC-B	550. min	—	6.9	.62	19
Humic acids	GAC-C	550. min	—	13.4	.42	19
Humic acids	GAC-D	550. min	—	9.8	.49	19
Humic acids	GAC	5.17 days	4.5	4.0[c]	~1.	22
Humic acids	GAC	5.17 days	7.0 and 8.7	6.0[c]	1.06	22
Humic acids	GAC-A	4–5 days	8.5	1.8	—	42
Humic acids	GAC-B	4–5 days	5.5	33.	—	42
Commercial humic acid	PAC-A[b]	5 days	6.0	1.095	1.818	43
	B	5 days	6.0	6.617	1.022	43
	F	5 days	6.0	1.982	1.374	43
	I	5 days	6.0	1.098	1.524	43
Peat fulvic acid	A	5 days	6.0	5.282	0.519	43
	B	5 days	6.0	7.365	0.516	43
	F	5 days	6.0	3.726	0.565	43
	I	5 days	6.0	1.913	0.532	43
Peat fulvic acid of < 1,000 mol wt	A	5 days	6.0	31.146	0.610	43
	B	5 days	6.0	18.535	0.553	43
	F	5 days	6.0	21.184	0.621	43
	I	5 days	6.0	11.018	0.733	43
Peat fulvic acid of > 50,000 mol wt	A	5 days	6.0	4.163	0.555	43
	B	5 days	6.0	6.546	0.545	43
	F	5 days	6.0	1.493	1.285	43
	I	5 days	6.0	0.725	1.465	43
Peat fulvic acid remaining after alum coagulation	A	5 days	6.0	20.224	0.592	43
	B	5 days	6.0	20.007	0.451	43
	F	5 days	6.0	15.123	0.624	43
	I	5 days	6.0	5.913	1.006	43
Humic acids	GAC	550 min	7.0	6.44[c]	1.61	44
Humic acids	GAC	550 min	7.0	1.02[d]	1.21	44

[a]Expressed as gC/m^3.
[b]Pulverized GAC.
[c]When $C_e = 1$ μg/l.
[d]Tap water.

Table 6–8 Oxygen-containing Functional Groups in the Humic Substances

Humic Substance	*Total Acidity (meq/g TOC)*	*Carboxyl (meq/g TOC)*	*Phenolic-OH (meq/g TOC)*	*Carbonyl (mmol/g TOC)*
Commercial humic	10.4	6.9	3.5	4.7
Soil humic	29.9	13.1	16.8	10.1
Soil fulvic	39.5	25.0	14.5	—
Leaf fulvic	11.6	6.7	4.9	—

Source: Reproduced from McCreary and Snoeyink [45], courtesy of Pergamon Press.

a greater extent than the higher ones. The latter are too large to penetrate into the internal pore structure of carbon. This is also the reason why humic and fulvic acids show an almost immediate breakthrough from carbon columns.

Pretreatment of water containing fulvic acids by coagulation greatly enhances carbon (F-400, 14 × 20 mesh) adsorption of these organics [46]. Figures 6–4a and b show several adsorption isotherms with and without alum pretreatment. It appears that coagulation removes a substantial portion of the fulvic acids (as measured by TOC-ultraviolet light). This coagulation leaves a fraction that is strongly adsorbed or the presence of Al^{3+} ions enhances complex formation between the organics and carbon. In all probability, coagulation removes the larger fulvic acid molecules that cannot penetrate the porous carbon, leaving smaller molecules that can penetrate.

An extensive study was conducted on the effect of Ca, Mg, and Na on the adsorptive capacity of activated carbon (F-400, 14 × 20 mesh for batch tests and 60 × 80 mesh for column tests) for various sources of fulvic acid [47]. Ca, Mg, and Na salts enhanced the adsorption of fulvic acids onto F-400. The order of effectiveness is $Ca^{2+} > Mg^{2+} > Na^{+}$. An example is an increase of peat fulvic acid adsorption from 5 mg/g in the absence of calcium to 30 mg/g in the presence of 10^{-3} M Ca^{2+} (40 mg/l). These observations are explained by charge reduction of the negative sites on the carbon's surface and on the fulvic acids. There is also the possibility of complex formation between the carbon and fulvic acid by Ca^{2+} and Mg^{2+}. Na^{+} would have very little complexation with fulvic acid.

Column Studies

Breakthrough curves for humic acid from columns of granular carbon (F-400) are seen in Figures 6–5a and b [22,44]. It appears that breakthrough occurs rapidly; consequently, carbon filters would have little effectiveness for humic acid removal under treatment plant conditions. Figure 6–5b is an interesting breakthrough profile of humic acids with two backgrounds (DDW = deionized distilled water). Apparently, the inorganic ions, Ca^{2+}, Mg^{2+}, and Fe^{3+}, can complex

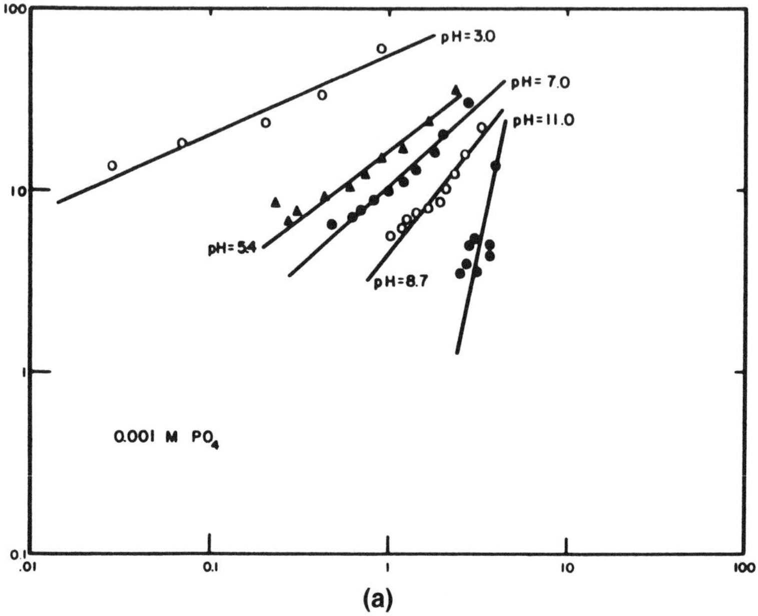

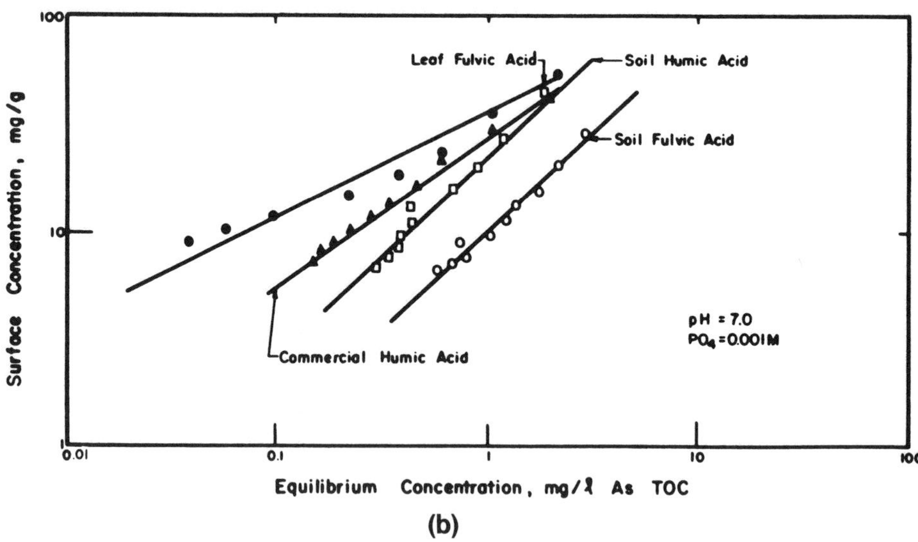

Figure 6–3 (a) Effect of pH on the adsorption of soil fulvic acid. (b) Adsorption of various types of humic substances. (c) Adsorption of molecular weight fractions of soil fulvic acid. (d) Adsorption of molecular weight fractions of soil humic acid. Reproduced from McCreary and Snoeyink [45], courtesy of Pergamon Press.

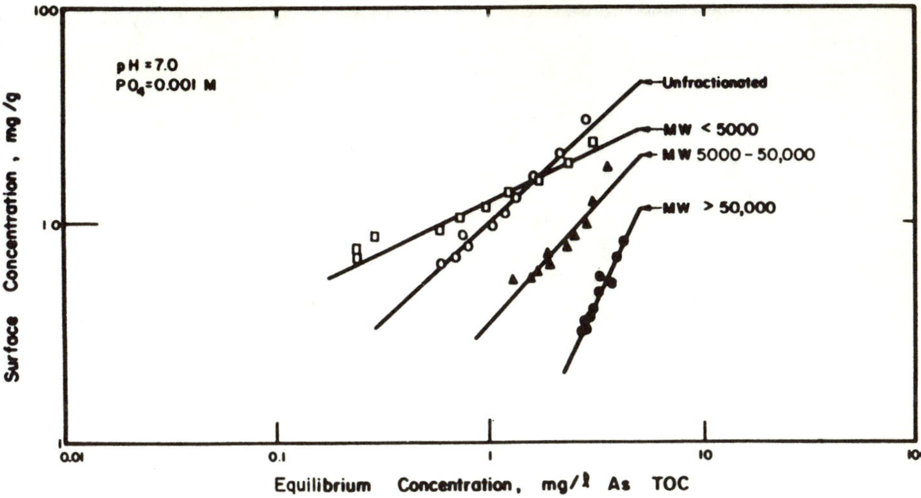

(c)

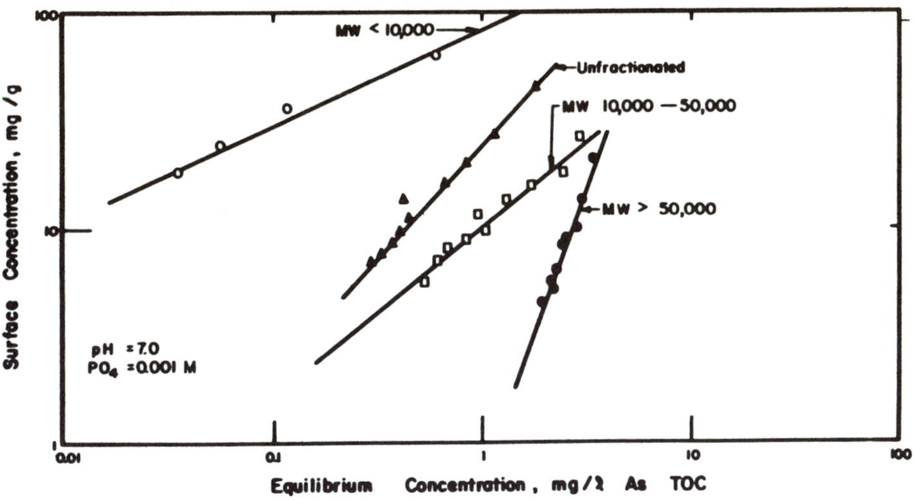

(d)

the humic acids with a subsequent increase in adsorption. This observation would be similar to actual treatment plant conditions.

Several GACs (20 × 40 mesh) in columns were evaluated for their removal of fulvic acids [43]. A mathematical model was constructed from equilibrium parameters, the surface diffusion coefficient, and the column mass transfer coefficient for prediction of column performance. Adsorption breakthrough curves for peat fulvic acid before and after alum coagulation were calculated for four carbons (Figures 6–6a and b). In this case, the treatment objective was 100 μg/l

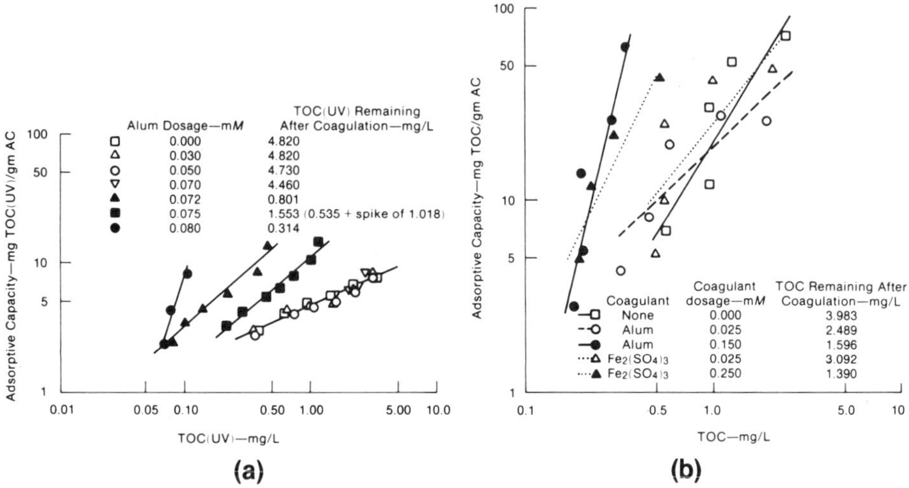

Figure 6–4 (a) Adsorption isotherms for untreated, pretreated, and spiked fulvic acid solutions. Coagulation at pH 6.0 with [Na_2SO_4] held constant at 0.6 m*M*; adsorption at pH 8.3 with 1.0 m*M* $NaHCO_3$ added. (b) Adsorption isotherms for untreated and pretreated groundwater, coagulation at pH 6.0; adsorption at pH 6.0 with 10.0 m*M* NaH_2PO_4 added. Reproduced from Randtke and Jepsen [46], courtesy of the American Water Works Association.

THMFP (trihalomethane formation potential) or less from fulvic acids. Previous studies of TOC correlation with THMFP indicated that the column's effluent should not exceed 2.20 mg/l TOC. For this objective, carbon bed service times for carbons I, F, A, and B were 1.69, 2.88, 3.38, and 3.72 days, respectively. These service times are relatively short when compared to other organics. Alum coagulation prior to carbon adsorption greatly extends these service times for carbons I, B, F, and A to 38, 40, 59.4, and 61.4 days, respectively. These times are significantly higher than for noncoagulated systems. Also, this demonstrates the tremendous advantage of coagulation prior to carbon adsorption. This advantage is reflected also in the rate of adsorption of fulvic acid as seen in Figures 6–6c and d [43].

THM formation from chlorinated precursors such as humic and fulvic acids, as well as their subsequent removal, has been a concern in water treatment plants. This problem was addressed in a study of the GAC (30 × 40 mesh F-400) column treatment of chlorinated humic and fulvic acids [48]. In this case, the organics were determined by TOX and TOC. The test solutions consisted of humic acid (5.5 mg/l as TOC), Cl_2 (10. mg/l), and a phosphate buffer (0.001 M), at pH 6.0 and 21°C. The columns were run for 5 days. Figures 6–7a and b show breakthrough curves for chlorinated humic acid expressed as TOX and TOC, respectively. Here again, humic acid broke through rapidly. No $CHCl_3$ was found in the effluent of this column. Essentially, the same results were observed for chlorinated fulvic acid (Figures 6–7c and d) [48].

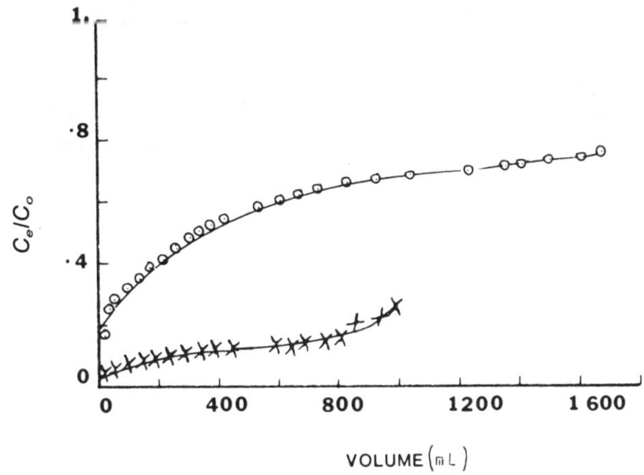

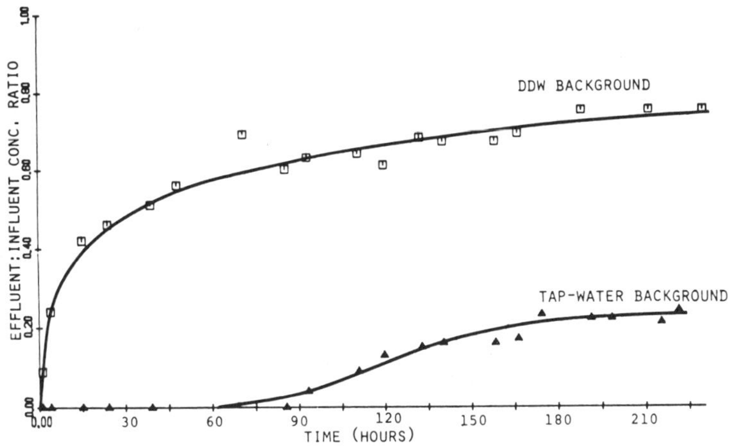

Figure 6–5 (a) Breakthrough curves for humic acid on GAC: ○ = humic acid/g carbon = 5.03; x = humic acid/g carbon = 2.57; pH = 7.00, temperature = 24°C. Reproduced from Youssefi and Faust [22], courtesy of Ann Arbor Science Publishers, Inc. (b) Breakthrough profiles for humic acids in GAC adsorption columns. Reproduced from Weber et al. [44], courtesy of Ann Arbor Science Publishers, Inc.

In another column experiment, the test solution was fulvic acid (10. mg/l as TOC), Cl_2 (10. mg/l), and a 0.001 M phosphate buffer at pH 6.3 and 21°C. This column was shorter and contained less carbon than the one described earlier. The intent was to obtain a higher percentage of saturation of carbon by the chlorinated fulvic acid. That this was accomplished is seen in Figures 6–8a and b [48]. Again, the fulvic acid broke through very quickly, and influent and effluent TOX

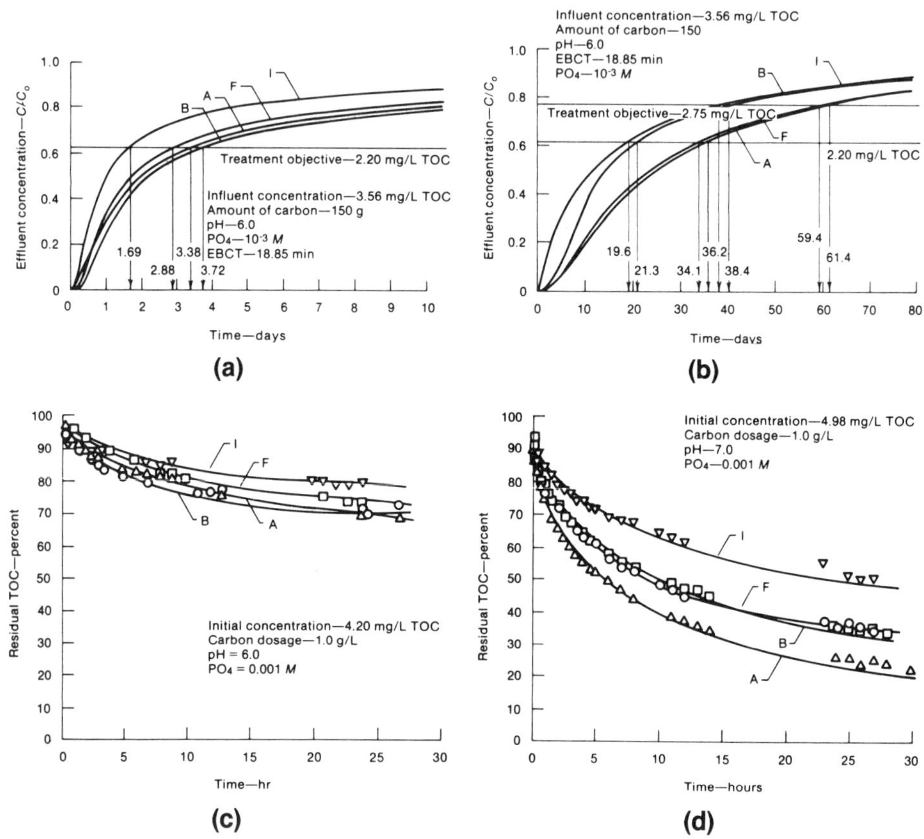

Figure 6–6 (a) Predicted breakthrough curves for peat fulvic acid. (b) Predicted breakthrough curves for peat fulvic acid remaining after coagulation with alum. (c) Rate of adsorption of peat fulvic acid. (d) Rate of adsorption of fulvic acid remaining after coagulation with alum. Reproduced from Lee et al. [43], courtesy of the American Water Works Association.

reached a steady-state condition (Figure 6–8a). Chloroform was formed apparently in the influent water that was removed subsequently. However, $CHCl_3$ broke through after 160. liters of water, and the effluent concentration rose rapidly to a steady-state condition with the influent.

PHENOLIC COMPOUNDS

Phenolic compounds, especially the chlorinated species, impart medicinal iodine tastes to drinking water [1]. Most of this taste arises from the chlorination treatment for disinfection and taste and odor control. PAC and GAC have been used in the water treatment industry for many years in a rather empirical manner for

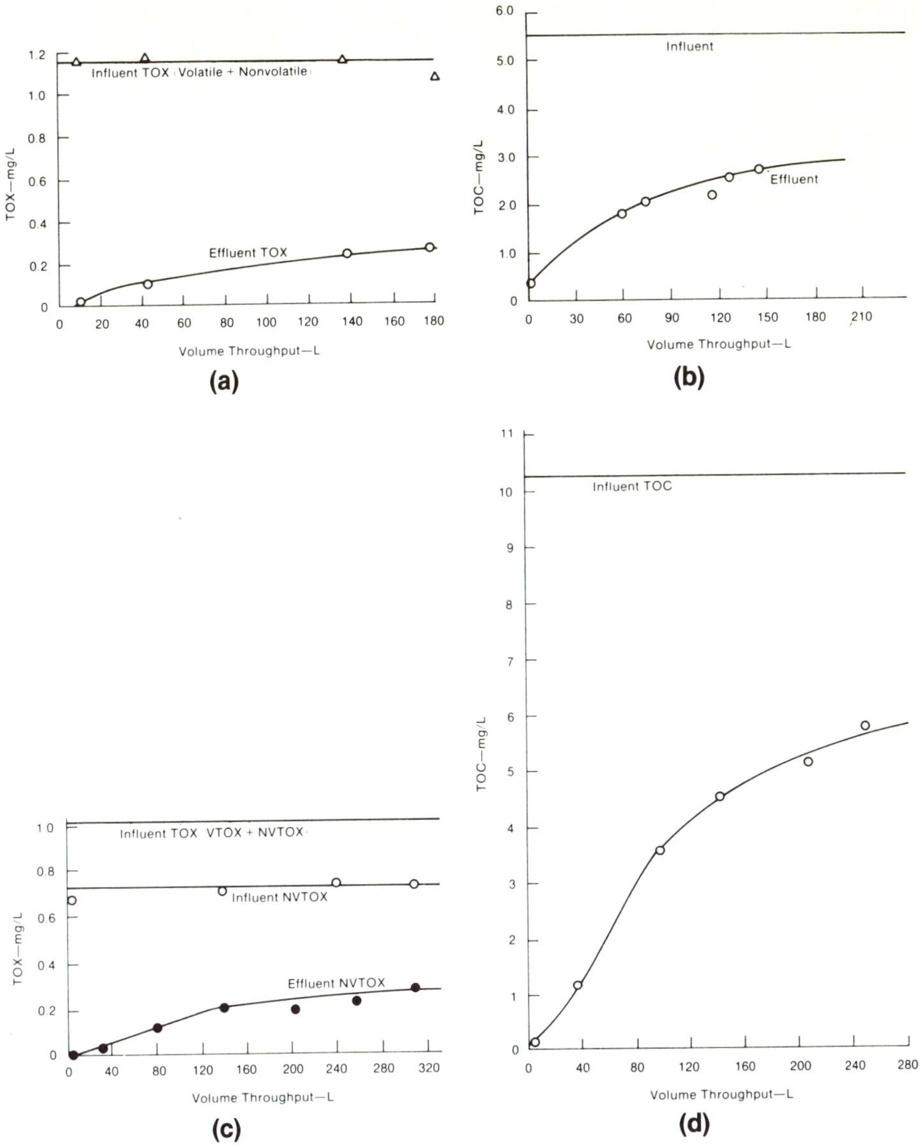

Figure 6–7 (a) TOX breakthrough for chlorinated humic acid. (b) TOC breakthrough for chlorinated humic acid. (c) TOX breakthrough for chlorinated fulvic acid, Column 1. (d) TOC breakthrough for chlorinated fulvic acid, Column 1. Reproduced from Quinn and Snoeyink [48], courtesy of the American Water Works Association.

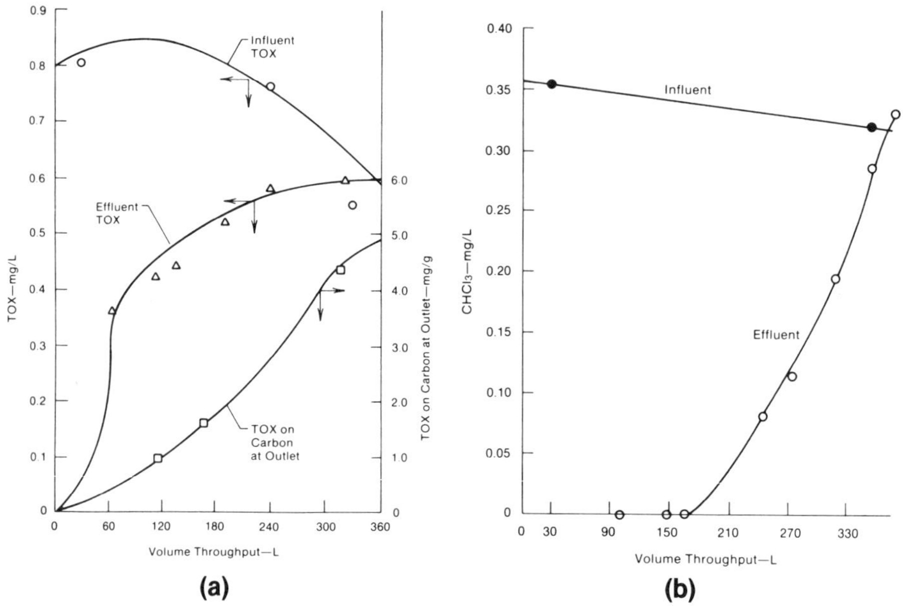

Figure 6-8 (a) TOX breakthrough for chlorinated fulvic acid, Column 2. (b) $CHCl_3$ breakthrough for chlorinated fulvic acid, Column 2. Reproduced from Quinn and Snoeyink [48], courtesy of the American Water Works Association.

taste and odor removal [49]. Only since the late 1950s and early 1960s has PAC and GAC treatment found quantitative application in laboratory and pilot plant operation.

Equilibrium Studies

Tables 6-1 and 6-9 contain Freundlich and Langmuir adsorption constants for several phenols on PAC and GAC under a wide range of experimental conditions. Adsorption capacities (K and X_m) ranged from 25. to 138. mg/g (when $C_e = 1.0$ mg/l) and from 2,162. to 2,942. mg/g when $C_e = 1$ mol/l (Table 6-9). These capacities are modest for carbon adsorption of organic compounds. However, it is feasible to use carbon for phenol removal at water treatment plants where contact times (15 to 60 minutes) are somewhat less than those reported in Tables 6-1 and 6-9. Langmuir monolayer coverage constants, X_m, ranged from 103. to 671. mg/g. In general, PAC exhibited greater capacities than GAC. Some studies included the effect of pH on adsorption; that is, the un-ionized form of a phenol shows greater adsorption than its ionized form. It is doubtful that the $[H_3^+O]$ would be a major factor under treatment plant conditions because most phenols are weak acids in water. Substituted phenols were adsorbed to a greater extent than phenol.

Many experimental and operational factors affect the adsorption of organics on carbon. Phenols are no exception. Many of these factors assist in an understanding of the mechanism of adsorption. Rarely would there be a change in the mode of operation at a water treatment plant to effect optimum conditions. Nonetheless, much treatment information is obtained from these studies [52].

The type and origin of a specific activated carbon do affect the extent of adsorption of phenolic compounds [32]. This point is well documented in Tables 6–1 and 6–9. In contrast, however, there is the issue of the effect of carbon particle size on adsorption of organics, in general, and phenols, in particular. Figure 6–9 shows no influence of particle size on the adsorption of 2,4-DCP and 2,4-DNP on GAC (Columbia LCK) [32]. This observation is at variance with previous studies with detergents on the same carbon [50]. It is possible that both effects are correct (see Chapter 3). Phenol is a small molecule that can penetrate

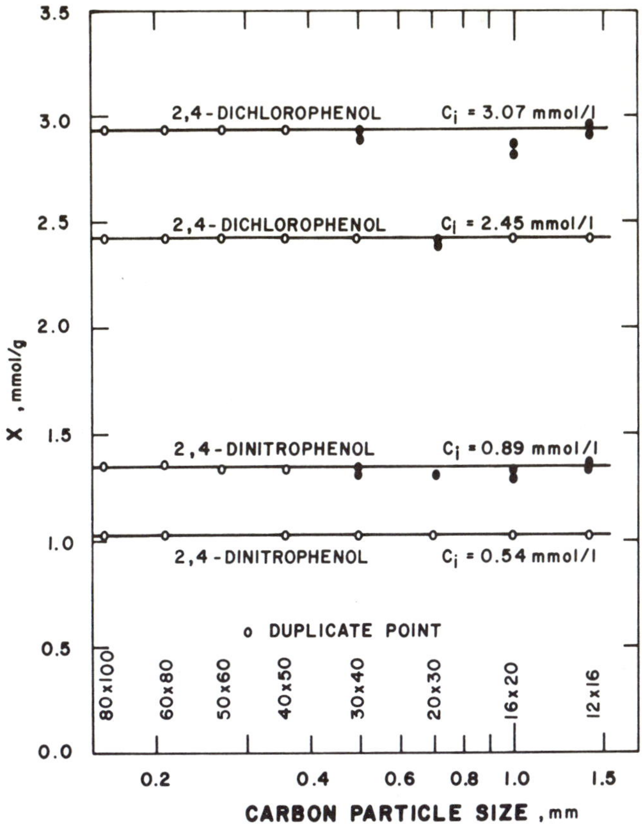

Figure 6–9 Influence of carbon particle size on the adsorption of 2,4-dinitrophenol and 2,4-dichlorophenol at 20°C. Reproduced from Zogorski and Faust [32], courtesy of Ann Arbor Science Publishers, Inc.

Table 6–9 Adsorption of Phenolic Compounds by Carbon in Batch Systems

				Freundlich Constants		*Langmuir Constants*		
Phenol	*Type of Carbon*	*Contact Time*	*pH*	K *(mg/g)*	l/n	X_m *(mg/g)*	b^{-1}	*References*
Phenol	GAC	24 hr	9.0	25.	.37	—	—	28
4-chloro	GAC	24 hr	7.0	82.	.29	—	—	28
4-chloro	GAC	24 hr	9.0	78.	.20	—	—	28
2,4-dichloro	GAC	24 hr	9.0	70.	.27	—	—	28
2,6-dichloro	GAC	24 hr	7.0	138.	.13	—	—	28
2,6-dichloro	GAC	24 hr	9.0	25.	.57	—	—	28
Phenol	GAC	1–2 wk	?	—	—	103.	9.26[a]	50
2-CH_3O-	GAC	100 hr	6.5	—	—	296.	20.[a]	27
2-CH_3-	PAC	240 min	?	2,162.[c]	.24	383.	18.6[b]	34
3-CH_3-	PAC	240 min	?	—	—	383.	18.6[b]	—
4-CH_3-	PAC	240 min	?	—	—	383.	18.6	—
2-C_2H_5-	PAC	240 min	?	—	—	532.	27.1[b]	34
3-C_2H_5-	PAC	240 min	?	—	—	532.	27.1[b]	—
4-C_2H_5-	PAC	240 min	?	—	—	532.	27.1[b]	—
2,3-$(CH_3)_2$	PAC	240 min	?	—	—	552.	30.3[b]	34
2,6-$(CH_3)_2$	PAC	240 min	?	2,942.[c]	.22	552.	30.3[b]	—
3,4-$(CH_3)_2$	PAC	240 min	?	—	—	552.	30.3[b]	—
2-isopropyl	PAC	240 min	?	—	—	671.	33.6[b]	34
DNOSBP[d]	GAC	2 wk	?	—	—	444.	5.8[b]	51
DNOCHP[e]	GAC	2 wk	?	—	—	500.	6.8[b]	51
Phenol	GAC	4 wk	4.0	—	—	123.	13.[a]	35
Phenol	GAC	4 wk	12.0	—	—	34.45	45.[a]	35
p-nitrophenol	GAC	4 wk	4.0	—	—	292.	4.2[a]	35
p-nitrophenol	GAC	4 wk	10.0	—	—	63.5	7.7[a]	35

2,4-dinitro-	GAC	4 wk	2.0	—	—	460.	2.5[a]	35
2,4-dinitro-	GAC	4 wk	9.5–10.	—	—	95.	5.3[a]	35
Phenol	GAC	10 days	7.0	—	—	99.	7.7[a]	53
Phenol	GAC	10 days	9.5	—	—	85.	12.5[a]	53
2,4-dichloro-	GAC	10 days	7.0	—	—	380.	.93[a]	53
2,4-dichloro-	GAC	10 days	9.5	—	—	233.	33.3	53
2-CH_3-	GAC	100 hr	6.5	—	—	240.	22.[a]	27
3-CH_3-	GAC	100 hr	6.5	—	—	240.	22.[a]	27
4-CH_3-	GAC	100 hr	6.5	—	—	240.	22.[a]	27
4-Cl-	GAC	100 hr	6.5	—	—	272.	19.[a]	27
p-nitrophenol	GAC-C[f], 6 × 10	5 days	3.0	81.76	.167	—	—	54
p-nitrophenol	GAC-C, 10 × 14	5 days	3.0	90.32	.175	—	—	54
p-nitrophenol	GAC-C, 14 × 20	5 days	3.0	91.34	.183	—	—	54
p-nitrophenol	GAC-C, 20 × 25	5 days	3.0	93.04	.169	—	—	54
p-nitrophenol	GAC-C, 25 × 30	5 days	3.0	91.63	.176	—	—	54
p-nitrophenol	GAC-C, > 30	5 days	3.0	93.04	.178	—	—	54
p-nitrophenol	GAC-B[g], 10 × 14	5 days	3.0	59.08	.161	—	—	54
p-nitrophenol	GAC-B, 14 × 20	5 days	3.0	61.00	.156	—	—	54
p-nitrophenol	GAC-B, 20 × 25	5 days	3.0	59.24	.163	—	—	54
p-nitrophenol	GAC-B, 25 × 30	5 days	3.0	61.95	.147	—	—	54
p-nitrophenol	GAC-B, > 30	5 days	3.0	57.96	.154	—	—	54
Phenol	GAC	2 days	?	50.	.26	213.	—	95
p-Cl-	GAC	4 wk	?	126.	.25	434.	—	95

[a]μmol/l.
[b]mmol/l.
[c]When $C_e = 1$ mole/l.
[d]2,4-dinitro-*o*-sec-butyl phenol.
[e]2,4-dinitro-*o*-cyclohexyl phenol.
[f]F-400.
[g]WV-L

carbon's pores regardless of particle size. In contrast, dodecylbenzenesulfonate is a large molecule and cannot penetrate the pores.

Temperature affects the exothermic process of adsorption. Isotherms were developed for phenol at temperatures of 8, 20, and 29°C. These are seen in Figure 6–10 where the quantity of phenol adsorbed was decreased as the temperature was increased [32]. However, there was a lesser influence on the adsorption process when the phenol content was 200 μmol/l and higher. Temperature cannot be controlled under a treatment plant situation; rather, carbon dosage and/or filter size can be manipulated to compensate for warmer temperatures. The lesser efficiency of adsorption is partially offset by faster rates of adsorption at higher temperatures.

Hydronium ion concentration affects the associated-dissociated state of the adsorbate, and this, in turn, affects the extent of adsorption. Figure 6–11 shows that adsorption of 2,4-DCP is decreased as the pH value is lowered from 6.0 or is raised from 6.0. Two effects of $[H_3^+O]$ are observed here: (1) lesser adsorption occurs when the anion of 2,4-DCP predominates above pH 6.0 and (2) as the pH value is lowered from 6.0, the H_3^+O competes with the un-ionized 2,4-DCP for adsorption sites. Similar pH effects on the adsorption of 2,4-DNP were observed [32]. Most of the influence of $[H_3^+O]$ on the extent of adsorption is exerted on the adsorbate. However, weak acidic functional groups on the carbon's surface are affected also, in turn influencing adsorption to a lesser degree than the adsorbate.

Table 6–10 gives some Freundlich isotherm constants for 2,4-DCP and 2,4-DNP at pH values ranging from 2. to 10.9 [32]. Comparison of the K values

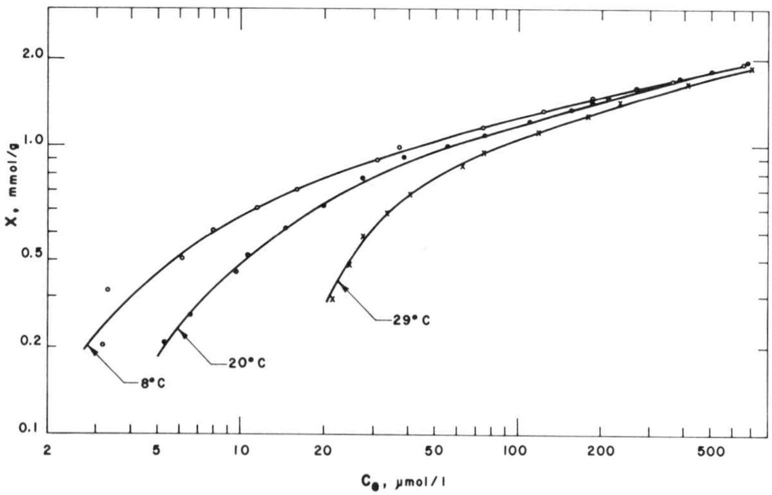

Figure 6–10 Influence of temperature on the adsorption of phenol at pH 6.3, 16 × 20 mesh adsorbent. Reproduced from Zogorski and Faust [32], courtesy of Ann Arbor Science Publishers, Inc.

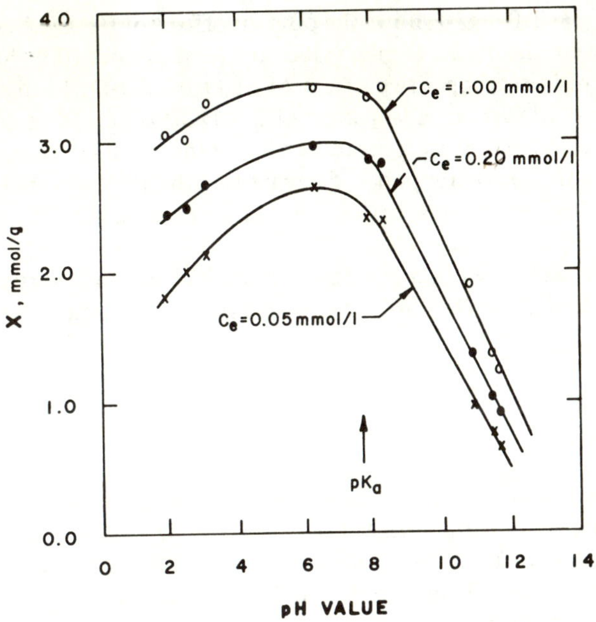

Figure 6–11 Variation in the adsorption of 2,4-dichlorophenol with pH. Reproduced from Zogorski and Faust [32], courtesy of Ann Arbor Science Publishers, Inc.

Table 6–10 Freundlich Constants for 2,4-DCP and 2,4-DNP at Several pH Values

		Freundlich Constants	
Compound	*pH*	K *(mg/g)*	1/n
2,4-DCP[a]	2.0	275.	.114
	6.3	322.	.100
	8.3	306.	.117
	10.9	109.	.207
2,4-DNP[b]	2.1	321.	.145
	6.0	166.	.156
	8.0	129.	.128
	10.4	116.	.147

Source: Adapted from Zogorski and Faust (32), courtesy of Ann Arbor Science Publishers, Inc.
Note: GAC, 2 weeks contact time.
[a]pK_a = 7.8; 25°C.
[b]pK_a = 4.0; 25°C.

(mg/g when C_e = 1.0 mg/l) shows the obvious effects of [H_3^+O]. Lesser quantities of 2,4-DCP were adsorbed at pH values of 2., 8.3, and 10.9 than at 6.3. As indicated, there is a competitive effect of H_3^+O on 2,4-DCP adsorption under acid conditions. Maximum adsorption of 2,4-DCP occurred at pH values just below its *pKa* value of 7.8. In contrast, 2,4-DNP showed its maximum adsorption at a pH value of 2.0 whereupon it decreased as the [H_3^+O] was decreased. Apparently, 2,4-DNP competes better with H_3^+O than 2,4-DCP because it is a stronger acid (*pKa* = 4.0).

Similar results were observed from the effect of pH on adsorption of phenol, PNP, 2,4-DNP, and 2,4-DCP on GAC where pH values ranged from 2.0 to 10.6 [33,53,55].

Column Studies

Laboratory studies of GAC in a column mode are designed to simulate treatment plant operation. Also, these studies bridge the gap between equilibrium data and pilot plant and/or treatment plant design. Column studies take isotherm and rate constants and translate them into design factors (see Chapter 8).

Typical breakthrough curves for phenol on four carbon columns are seen in Figure 6–12 [53]. From these curves, volumes to C_e/C_o = 0.1 and to saturation and the quantities of adsorbed phenol were taken. These values are given in Table 6–11 where the quantities adsorbed, 130. to 228. mg/g, at saturation or bed exhaustion compare very well to Langmuir X_m values in Table 6–1, 158. to 234. mg/g. Also seen in Figure 6–12 is the difference in adsorptive capacities of four carbons. Pitt Cal carbon had the highest capacity of the four in this laboratory study [53].

Adsorption of 2,4-DCP was determined also on a GAC column [53]. Breakthrough curves were similar to those in Figure 6–12 except that 2,4-DCP had a greater adsorptive capacity than phenol. For example, 2,4-DCP was adsorbed to the extent of 513. mg/g, or three times the capacity of phenol (Table 6–12), at pH 7.0. However, responses of 2,4-DCP and phenol to the effect of [H_3^+O] were quite different. There was an 84% reduction in the quantity of 2,4-DCP when the pH value was increased from 7.0 to 9.5, whereas there was only a 25% reduction for phenol. This can undoubtedly be explained by the stronger acidity of 2,4-DCP (*pKa* = 7.85) over phenol (*pKa* = 9.98).

A mixture of phenol and 2,4-DCP, at pH 7.0, was applied to a GAC column. Subsequent breakthrough curves showed that the two phenols were adsorbed separately. There were some competitive effects, however. Breakthroughs to C_e/C_o = 0.1 were more rapid (Table 6–12), and the adsorbed quantities of either phenol were reduced considerably. For phenol, the reduction was from 159. to 114. mg/g (28%), and for 2,4-DCP it was from 513. to 177. mg/g (65%). The presence of phenol apparently adversely affected the adsorption of 2,4-DCP.

The effects of flow rate, pH, bed depth, phenol concentration, and spent and unused carbon are seen also in Table 6–11. These are very important factors

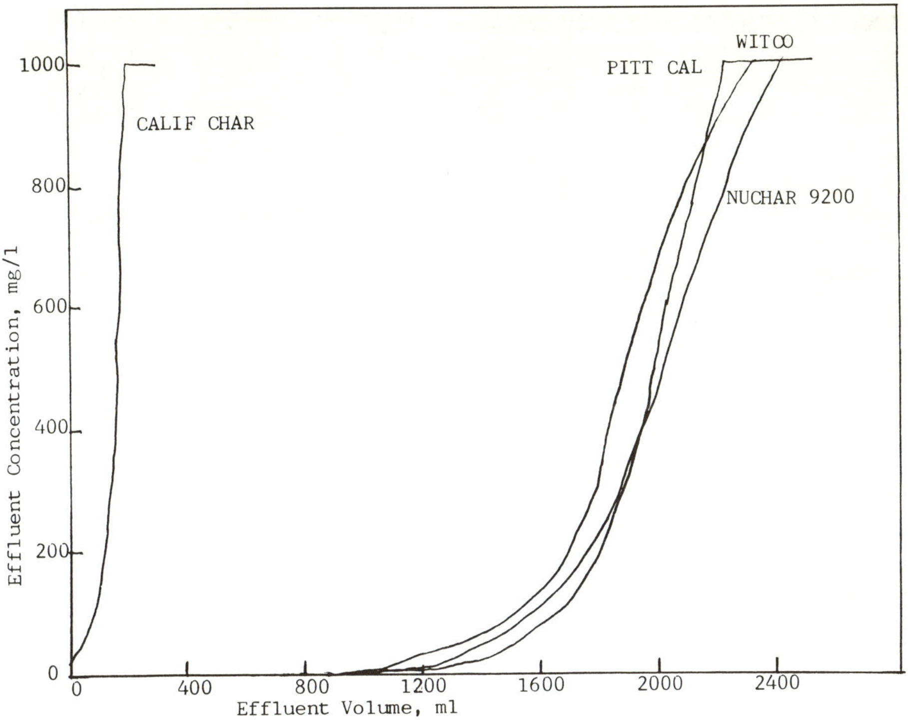

Figure 6–12 Breakthrough curves for phenol on four carbons [53]. C_i = 1,000 mg/l, flow rate = 1.8 gal/min/ft^3, pH = 7.0.

in the design of carbon columns, although it is doubtful that phenol contents of 100. to 1,000. mg/l will be found in raw water supplies. Flow rate and bed depth affect the contact time within the bed. Greater amounts of phenol are adsorbed as the contact time becomes longer (see Chapter 8).

Operational parameters for the optimum removal of phenols from water by a fixed bed of GAC were investigated in considerable detail under laboratory conditions [55]. In addition to the usual data from breakthrough curves, the height of mass transfer zone (*HMTZ*), rate of movement of mass transfer zone (*RMTZ*), and the fractional capacity of the adsorbent within the mass transfer zone, ϕ, were obtained. These three characteristics are related to either kinetics of adsorption of the adsorbate or adsorptive capacity of the adsorbent (see Chapter 3).

The *HMTZ* is a measure of the rate of removal of adsorbate by the adsorbent. The more rapid the kinetics of adsorption, the shallower is the *HMTZ*. It is calculated from:

$$HMTZ = H\left[\frac{V.9 - V.02}{V.02 + \phi(V.9 - V.02)}\right], \qquad (6.1)$$

Table 6-11 Capacities of Carbon Columns for Phenol Adsorption

	Volume of Effluent at C_e/ C_o = 0.1 (ml)	*Amount Adsorbed (mg/g)*	*Volume to Saturation (ml)*	*Amount Adsorbed (mg/g)*
Flow rate (gal/min/ft³)				
3.37	1400	136	2200	169.7
2.10	1620	159	2200	187.5
1.04	2000	180	2800	228.0
Effect of pH (2.1 gal/min/ft³)				
pH 9.5	1120	119	1700	131
pH 8.0	1400	137	2100	171
pH 7.0	1620	159	2200	187
Effect of bed depth (2.1 gal/min/ft³)				
21 cm	1620	159	2200	187.5
10 cm	760	145	1400	185
Effect of concentration (1.04 gal/min/ft³)				
1000 mg/l	2000	180.0	2800	228.0
500 mg/l	4000	180.0	—	—
100 mg/l	4000	—	—	—
Spent carbon (1.04 gal/min/ft³)				
	1000	96.0	1900	130
Unused carbon (1.04 gal/min/ft³)				
	2000	180.0	2800	228
Effect of phenol type (2.1 gal/min/ft³)				
phenol	1620	159	2200	187.5
2,4-dichloro-phenol	2600	460	—	—

Source: Aly and Faust [53].
Note: Pitt Cal carbon was used.

where ϕ = the fractional capacity (see Equation 6.2), $V.02$ and $V.9$ = the volumes (in liters) of solution treated to the .02 and .9 breakthrough points, respectively, and H = depth of the adsorbent [55].

The fractional capacity, ϕ, is a measure of the adsorptive efficiency of the carbon located in the mass transfer zone. It is defined as the amount of adsorbate actually adsorbed by carbon within the mass transfer zone relative to the total potential capacity of adsorbent within the zone [55]. Rapid, sharp breakthrough curves usually have high values of ϕ, whereas lower values are observed for gradual, flat breakthrough curves. The fractional capacity is calculated from:

$$\phi = \frac{\int_{V.02}^{V.9} (C.9 - C.02)dV}{(V.95 - V.02)\, C_i}, \qquad (6.2)$$

Table 6–12 Capacitics of Carbon Columns for 2,4-Dichlorophcnol Adsorption

pH	*Flow Rate (gal/min/ft³)*	*Volume of Effluent at C_e/C_o = 0.1 (ml)*	*Amount Adsorbed (2,4-DCP) (mg/g)*	*Comparative Amount of Phenol Adsorbed (mg/g)*
7.0	2.1	2,600	513 (3.147 mmol)	159 (1.69 mmol)
8.0	2.1	1,800	354 (2.17 mmol)	137 (1.46 mmol)
9.5	2.5	410	80 (0.49 mmol)	119 (1.26 mmol)
Mixture of phenol and 2,4-dichlorophenol (500 mg/l each)				
Phenol	2.4	1,190	114 (1.213 mmol)	
2,4-dichloro-phenol	—	1,800	177 (1.086 mmol)	

Source: Aly and Faust [53].
Note: Pitt Cal carbon was used.

where $C.9$ and $C.02$ represent the concentrations at 90% and 2% of exhaustion, respectively. C_i is the initial concentration. The fractional capacity is the amount of adsorbate actually adsorbed by carbon in the primary adsorption zone from the breakpoint ($V.02$) to exhaustion ($V.95$) (arbitrarily selected) [55]. This quantity may be calculated by integration of the quantity ($C.9 - C.02$) over V between the limits of $V.02 - V.9$. This is the numerator in Equation 6.2. The total capacity is the denominator, ($V.95 - V.02$) C_i (again, arbitrarily selected).

Rate of movement of the MTZ down the column depends on the affinity of the adsorbate for the adsorbent. This term is important because it indicates the rate at which the adsorbent will be exhausted. The *RMTZ* is calculated from [55]:

$$RMTZ = \frac{HMTZ \times Q}{\mathrm{V.9} - \mathrm{V.02}}, \tag{6.3}$$

where Q is the volumetric flow rate of the adsorbate solution.

Variations of the *HMTZ, RMTZ,* and ϕ are seen in Tables 6–13a, b, c, and d for several phenols [55]. From adsorption isotherms, the order of increasing affinity for Columbia LCK carbon is phenol, 4-nitrophenol and 4-methoxyphenol, and 2,4-DCP [32]. That the same sequence occurred in a column operation is seen in the *RMTZ* values in Table 6–13a. Phenol broke through first, followed by the others.

Bed depth is a variable frequently considered in column operation. Increasing bed depth and maintaining the flow rate constant increases the contact time

Table 6–13a Variation in the Characteristics of the Mass Transfer Zone of Several Adsorbates

Adsorbate	ϕ	*HMTZ (ft)*	*RMTZ (ft/day)*
Phenol	.61	0.24	2.6
4-methoxyphenol	.55	0.31	2.2
4-nitrophenol	.42	0.52	2.0
2,4-dichlorophenol	.40	0.49	1.8
2,4-dichlorophenol	.41	0.52	1.8

Source: Reproduced from Zogorski and Faust [55], courtesy of the American Institute of Chemical Engineers.
Note: 25°C, pH = 6.3, 2.5 gal/min/ft^2, C_i = 5.5 mmol/l, and 16 g of 20×25 mesh Columbia LCK Carbon.

Table 6–13b Variation in the Characteristics of the Mass Transfer Zone of 2,4-DCP with Depth of Adsorbent

Depth of Adsorbent (ft)	ϕ	*HMTZ (ft)*	*RMTZ (ft/day)*	*Contact Time (min)*
0.34	.39	0.55	1.8	0.7
0.69	.40	0.49	1.8	1.4
0.73	.41	0.52	1.8	1.5
1.05[a]	.36	0.66	1.7	2.1
1.73	.40	0.61	2.1	3.6
2.21	.36	0.68	1.7	4.5

Source: Reproduced from Zogorski and Faust [55], courtesy of the American Institute of Chemical Engineers.
Note: 27°C, 2.5 gal/min/ft^2, pH = 6.3, C_i = 5.6 mmol/l, and 20×25 mesh Columbia LCK Carbon.
[a]Average of two duplicate experiments.

within the column. Table 6–13b shows that the three efficiency characteristics are independent of bed depth for 2,4-DCP. One exception would be the *HMTZ* for the 0.34 foot deep bed.

Table 6–13c gives the variation in characteristics of the mass transfer zone (MTZ) for four 2,4-DCP concentrations. Increasing the initial concentration of 2,4-DCP from 0.18 to 5.6 mmol/l decreased the fractional capacity of carbon within the mass transfer zone by 0.14 feet. This observation is due to a greater mass of 2,4-DCP being transported to the carbon's surface per unit of time [56]. An increase in *HMTZ* would be expected to increase for the same reason. There was an increase also in *RMTZ,* with an increase in 2,4-DCP concentration. Pre-

Table 6–13c Variation in the Characteristics of the Mass Transfer Zone with the Concentration of Adsorbate

Concentration of 2,4-DCP (mmol/l)	ϕ	*HMTZ (ft)*	*RMTZ (ft/day)*
0.177	.55	0.36	0.078
1.06	.50	0.36	0.39
5.50	.40	0.49	1.8
5.60	.41	0.52	1.8

Source: Reproduced from Zogorski and Faust [55], courtesy of the American Institute of Chemical Engineers.
Note: 25°C, pH = 6.3, 2.5 gal/min/ft^2, and 16 g of 20×25 mesh Columbia LCK Carbon.

Table 6–13d Variation in the Characteristics of the Mass Transfer Zone of 2,4-DCP with Flow Rates

Rate of Flow (gal/min/ft^2)	ϕ	*HMTZ (ft)*	*RMTZ (ft/day)*
2.0	.41	0.49	1.4
3.8	.40	0.77	2.9
7.0	.38	1.2	5.3
9.7	.37	1.6	7.3

Source: Reproduced from Zogorski and Faust [55], courtesy of the American Institute of Chemical Engineers.
Note: 27°C, pH = 6.3, C_i = 5.5 mmol/l, and 20×25 mesh Columbia LCK Carbon.

vious studies with adsorption isotherms showed that an increase of [2,4-DCP] from 0.18 to 5.6 mmol/l increased the capacity of carbon by approximately 46% [32]. Thus, the rate of movement of the mass transfer zone should decrease from capacity considerations alone. However, this increase in concentration also increased the mass flow rate of adsorbate by a factor of 31. Thus, the effect of mass flow rate far exceeds the change in adsorptive capacity of the carbon. The net result is an increase in *RMTZ*.

Table 6–13d shows the effect of flow rate on the characteristics of the mass transfer zone. Increasing the flow rate decreases the contact time within the column, but at the same time, the mass of adsorbate applied per unit of time is increased. The net effect is to increase *HMTZ* and *RMTZ*.

Results similar to those of Zogorski et al. [55, 56] were reported by Rosene [57] (see also Chapter 4). All this evidence indicates that phenolic compounds are removed easily from water in GAC columns.

PETROCHEMICALS

Equilibrium Studies

On occasion it is necessary to treat a water for specific removal of organic compounds from petroleum and/or petrochemical sources [1]. A laboratory study was conducted to determine the amenability of some 93 organic compounds to PAC adsorption [12]. Single dosage experiments were conducted by using 100 ml aliquots of 1,000 mg/l stock solutions of each compound with 5 g/l of PAC (Nuchar WV-G). Representative compounds from alcohols, aldehydes, amines, pyridines, morpholines, aromatics such as benzene and toluene, esters, ethers, glycols and glycol ethers, ketones, organic acids, and oxides were investigated. Adsorbability was quite variable, ranging from 3.6% for methanol to 100% for butyl ether. Some adsorption isotherms were drawn for five compounds at pH values of 2., 7., and 11. and for two hours of shaking. Freundlich constants are seen in Table 6–14 for five compounds on four carbons. Qualitatively, these compounds show low adsorptive capacities through the K values, mg/g, compared to many other compounds in Table 6–1.

Several adsorption isotherms were conducted for 15 minutes on PAC (type BDH) for five petrochemicals in an effort to simulate treatment plant conditions [17]. Table 6–15a gives the Freundlich isothermal constants for benzene, toluene, *o*-xylene, ethylbenzene, and cyclohexane. Relative adsorbabilities may be ob-

Table 6–14 Freundlich Constants for Adsorption of Some Petrochemicals on GAC

		Butanol		*Butyraldehyde*		*Ethyl Acetate*		*Methyl Ethyl Ketone*		*Butyric Acid*	
Carbon[a]	*pH*	K[b]	1/n	K[b]	1/n	K[b]	1/n	K[b]	1/n	K[b]	1/n
A	2	7.	.44	2.2	.60	3.6	.58	6.6	.41	6.	.19
	7	—	—	8.9	.46	2.4	.68	23.	.24	—	—
	11	5.5	.51	6.0	.62	6.6	.85	14.	.48	—	—
B	2	4.4	.45	5.8	.75	1.3	.67	2.4	.48	3.1	.53
	7	—	—	3.3	.57	.6	.83	24.	.18	—	—
	11	2.2	.60	2.4	.78	.16	1.0	5.5	.44	—	—
C	2	3.6	.47	1.6	.95	1.3	.65	.22	.85	1.4	.65
	7	—	—	—	—	.92	.75	36.	.16	—	—
	11	5.5	.44	8.3	.99	.3	.88	2.9	.51	—	—
D	2	.1	.92	—	—	.04	1.18	—	—	.63	.68
	7	—	—	—	—	—	—	—	—	—	—
	11	.1	.93	—	—	—	—	—	—	—	—

Source: Giusti [12].

[a]Carbon A = Witco Grade 517 (12×30), Carbon B = Filtrasorb 400 (12×40), Carbon C = Nuchar WV-G (12×40), Carbon D = Nacar G107 (12×30).

[b]mg/g when C_e = 1.0 mg/l.

Table 6–15a Adsorption Parameters for PAC

	Single		*System Bisolute*[a]		*Multisolute*	
Compound	K[b]	l/n	K	l/n	K	l/n
Benzene	.065	.42	.026	.63	.017	.67
Toluene	.110	.28	.046	.50	.030	.73
o-xylene	.130	.27	.065	.67	.044	.71
Ethylbenzene	.120	.29	.060	.58	.042	.68
Cyclohexane	.095	.20	—	—	.050	.90

Source: Reproduced from El-Dib et al. [7], courtesy of Pergamon Press.
[a]Bisolute systems are (benzene/toluene) and (*o*-xylene/ethyl benzene).
[b]mg/mg.

Table 6–15b Freundlich Adsorption Parameters for GAC

	Single-Solute System		*Mixed-Solute System*	
Compound	K[a]	l/n	K	l/n
Benzene	.036	.48	.008	.60
Toluene	.090	.30	.023	.50
o-xylene	.120	.22	.075	.27
Ethylbenzene	.100	.40	.065	.30

Source: Reproduced from El-Dib and Bradawy [8], courtesy of Pergamon Press.
[a]mg/mg.

tained from a comparison of K values (mg/mg) in the single-solute column. The order of decreasing adsorption is *o*-xylene > ethylbenzene > toluene > cyclohexane > benzene. Effects of bi- and multisolute systems are seen also through a comparison of the K values.

Table 6–15b contains Freundlich isothermal constants for four petrochemicals on GAC (F-400) shaken for 2 days [18]. The quantities of adsorbed compounds are slightly less on GAC than on PAC, which is expected. Data from the mixed-solute system show effects of competition for adsorption sites. This would be the case in actual treatment plant operation.

Column Studies

Two continuous column operations were conducted for single-component systems of methyl ethyl ketone and butyraldehyde on GAC (Witco Grade 517, 12 × 30 mesh) [12]. The breakthrough curves (not shown) exhibited steep slopes that are typical of single-component systems. The loading was 132. mg for methyl

ethyl ketone/g of carbon that was 83% of the 160. mg/g loading from the isotherm study. For butyraldehyde, the continuous column loading was 192. mg/g for an 87% of the isotherm value of 220. mg/g.

Several studies were performed for two- and four-component systems in continuous column operation [12]. A mixture of 500 mg/l each of butanol and ethyl acetate was applied until the carbon was exhausted. These two compounds were adsorbed to a lesser extent (237. mg/g) than a predicted quantity (270. mg/g) from single-component equilibrium tests. The observation of less total adsorption from a mixture was at variance with previous studies with phenol and a benzenesulfonate detergent [50]. In contrast, a two-component system of methyl ethyl ketone and butyraldehyde showed a slightly higher, 275. mg/g, total adsorption in column operation than individually, 270. mg/g. When a four-component system of butyraldehyde, ethyl acetate, butanol, and butyric acid was examined, there was considerably less total adsorption, 290. mg/g, than the sum of individual compounds, 575. mg/g. This competitive situation is indicative of actual treatment plant operation.

Toluene and a mixture of toluene, ethylbenzene, and *o*-xylene were applied to a column of F-400 GAC in a laboratory study [18]. More or less typical breakthrough curves were observed (not shown). The number of bed volumes free from toluene in the single-solute solution was 27.4. This volume was reduced to 19.5 in the mixed system. Again, this is a case of competitive adsorption. The order of breakthrough (Table 6–16) was toluene, ethylbenzene, and *o*-xylene that was predicted by the Freundlich *K* values seen in Table 6–15b.

ORGANIC PESTICIDES

Organic pesticides have been in raw water supplies since about 1940 [1]. However, it was not until the early 1960s that research was conducted for their removal from water by various treatment processes [58]. The lack of sensitive analytical

Table 6–16 Breakthrough and Column Characteristics

Compound	*Contact Time (min)*	*Influent Concentration (mg/l)* Single-Solute	*Mixed-Solute*	*Bed Volume*[a]	*Carbon Dose* x/M, *(mg/g)*
Toluene	2	72	—	27.4	42.9
Toluene	2	—	35	19.5	14.3
o-xylene	2	—	35	23.5	21.4
Ethylbenzene	2	—	35	25.5	19.5

Source: Reproduced from El-Dib and Bradawy [18], courtesy of Pergamon Press.
[a]Before breakthrough at a concentration of 0.1 mg/l^{-1}.

methodology retarded research efforts in this problem area. With the introduction of gas-liquid chromatography and other techniques in the middle 1960s, more studies were initiated and reported. This was continued for about 10 to 15 years whereupon research efforts decreased because many chlorinated hydrocarbons were banned. Organic phosphates, carbamates, ureas, and so forth then were employed instead of the chlorinated compounds. However, many of these compounds disappear rapidly from water because of hydrolytic reactions. Thus, the presence of organic pesticides in water has become less a problem in the 1980s. There are, however, drinking water standards for endrin, lindane, toxaphene, 2,4-D, 2,4,5-TP, and methoxychlor [1].

Equilibrium Studies

One of the first reports concerning the removal of pesticides from water by adsorptive techniques was given in 1945 [59]. This laboratory study reported on removing DDT from water supplies with various combinations of carbon and chemical coagulants. Removal efficiencies of 80 to 98% of 0.1 to 10. mg/l of DDT were recorded. Complete removal was observed when activated carbon was provided for a 15 minute contact period after coagulation and sedimentation but before sand filtration. DDT is extremely insoluble in water and is in the form of a suspension rather than being dissolved. This and other chlorinated hydrocarbons are removed rather easily from water by carbon.

Another early study was reported by Cohen et al. [60] for the laboratory removal of fish poisons such as rotenone and toxaphene and their commercial formulations from water. Chemical coagulation with alum was ineffective. Subsequent experimentation showed that 8 to 10 mg/l of PAC (Aqua Nuchar A) were able to reduce within 30 minutes 0.3 mg/l of toxaphene and rotenone to one-half the TLm for fathead minnows. In a later report, PAC was employed in a field study at Dickinson, North Dakota, to reduce the odor of organic solvents used in formulations of fish poisons [61]. It was concluded that removal of odor also assured removal of the pesticide residue.

In the 1960s, there was considerable research on use of 2,4-D and its many derivatives as an aquatic herbicide in lakes, reservoirs, and other bodies of water. Consequently, there were some studies [31,62] on the removal of these compounds by various treatment processes. One such study involved carbon removal of 2,4-D and 17 of its chlorinated derivatives from water [62]. Efficiencies ranged from 8 to 70% removal of a 10^{-4} M concentration of the herbicides. An inverse correlation between adsorption and the compound's solubility was observed. Phenoxyacetic acid, the most soluble, was adsorbed the least (8%). As solubility decreased, adsorptive efficiency was increased.

Effects of various conventional water treatment processes on removal of the sodium salt and several esters of 2,4-D were studied in the laboratory [31]. Coagulation with ferrous sulfate and alum at contents up to 100 mg/l failed to remove 2,4-D. PAC treatment was the most effective method for removal of 2,4-

D compounds, 2,4-DCP, and the odor-producing substances in commercial formulations from natural waters. Esters of 2,4-D required less carbon dosage than the sodium salt of 2,4-D, as seen in Table 6–17. These dosages were calculated from Freundlich isothermal constants (30 min of contact).

The effectiveness of PAC (Aqua Nuchar A) on the removal of several pesticides was examined again in a laboratory study using distilled water and also raw Little Miami River water [63]. A 1 hour contact time was used. Table 6–18 gives the carbon dosages needed to reduce the initial pesticide contents of 10 ppb to 1.0 and 0.1 ppb and of 1.0 ppb to 0.1 and 0.05 ppb, respectively. Because of organic matter in the raw river water, a larger dosage of carbon was required than with distilled water to obtain the same removal. Subsequently, it was determined that a two-step treatment with PAC was more economical than attempting to obtain the required removal in one dosage.

In the early 1960s, laboratory research shifted from PAC to GAC for the removal of organic compounds, especially the pesticides. One of these early efforts is seen in Table 6–19, where 0.274 mm Columbia LCK carbon was employed in an equilibrium study for a period of 2 weeks [51]. The adsorption data were fitted into the Langmuir isothermal equation from which the X_m and b^{-1} constants were calculated. The b^{-1} values, parathion excepted, indicate that the approach to monolayer coverage occurs at relatively high residual concentrations. These b^{-1} values also provide an insight into breakthrough from carbon columns. Normally, larger b^{-1} values call for greater column depths.

The herbicides, Diquat and Paraquat, were researched in the 1960s for their effectiveness as aquatic herbicides. Consequently, their removal from water was a concern at that time. These compounds were not effective aquatic herbicides, but in the 1980s, they were found to be useful for controlling other vegetation.

Table 6–17 Carbon Dosages Required to Reduce the Concentration of 2,4-D Compounds

	Carbon Dosage (mg/l)			
Initial Concentration[b]	*Sodium Salt*[a]	*Isopropyl Ester*	*Butyl Ester*	*Isooctyl Ester*
10	306	150	165	179
5	153	74	82	89
3	92	44	49	53
1	31	14	15	16
		Freundlich Isotherms		
K (mg/g)	118.	283.	200.	117.
l/n	.56	.63	.52	.33

Source: Reproduced from Aly and Faust [31], courtesy of the American Water Works Association.
[a]Final concentration 0.1 mg/l.
[b]Expressed as the acid equivalent.

Table 6–18 Activated Carbon Dosages Required to Reduce the Pesticide Level in Distilled Water and in Little Miami River Water

		10 ppb[a]		*1.0 ppb*[a]	
Pesticide	*Method*	*1.0*[b]	*0.1*[b]	*0.1*[b]	*0.05*[b]
Parathion	JT[c]	2.5	5	0.5	0.6
	p[d]	5	10	0.9	1.1
2,4,5-T ester	JT	2.5	17	1.5	3
	p	14	44	3	5
Endrin	JT	1.8	14	1.3	2.5
	p	11	126	11	2.3
Lindane	JT	0.2	12	1.1	2
	p	29	70	6	9
Dieldrin	JT	3	12	1.1	1.7
	p	18	85	7	12

Source: Reproduced from Roebeck et al. [63], courtesy of the American Water Works Association.
[a]Initial level of pesticide.
[b]Pesticide level after treatment.
[c]Jar test, in which pesticide is removed from distilled water by activated carbon alone, with a contact time of an hour.
[d]Plant treatment, in which pesticide is removed from river water by conventional treatment and activated carbon.

Table 6–19 Langmuir Equilibrium Constants for Organic Pesticides

	X_m		b^{-1}	
Compound	*mmol/g*	*mg/g*	*μmol/l*	*mg/l*
2,4-D	1.75	387	10.5	2.32
2,4,5-T[a]	1.75	448	6.7	1.71
Silvex	1.72	464	6.9	1.86
Parathion	1.82	530	0.82	0.24

Source: Reproduced from Weber and Gould [51], courtesy of the American Chemical Society.
[a]2-(2,4,5-trichlorophenoxy) propionic acid.

Several water treatment methods were examined for the removal of Diquat and Paraquat: activated carbon, chemical coagulation, and adsorption on clay minerals with subsequent coagulation [64]. Activated carbon (PAC) was more effective for Paraquat than for Diquat. For example, 44 mg/l PAC was able to reduce 1.0 mg/l Paraquat to 0.1 mg/l within 30 minutes whereas 82 mg/l PAC would be needed for Diquat (Table 6–20). However, the PAC dosages in Table 6–20 may be too impractical and uneconomical for the specific removal of these two herbicides from water. They would be effectively removed in an indirect manner because they are strongly adsorbed on particulate matter that is coagulated easily.

Table 6–20 Effect of Contact Time on Amounts of Carbon Required to Reduce Diquat and Paraquat Concentrations in Water

Initial Herbicide Concentration (mg/l)	*Carbon Dosage[a] (mg/l)*			
	Diquat		*Paraquat*	
	30 min	*60 min*	*30 min*	*60 min*
1	82	57	44	41
2	173	121	92	87
3	264	184	140	133
5	445	311	238	225
7	627	438	335	317
10	900	629	480	456

Source: Reproduced from Zarins and Faust [64], courtesy of Springer-Verlag.
[a]To residuals of 0.1 mg/l, at 20°C and pH 7.2.

The carbamates, Sevin and Baygon, were studied in the laboratory for their adsorption on GAC (Darco, 0.6 and 1.2 mm particle sizes) [65]. Equilibrium systems were shaken for 3 days at room temperature (25 ± 2°C). This study was unique in that the adsorption data could be fitted to both the Freundlich and Langmuir equations. The adsorption parameters (constants) for these two equations are seen in Tables 6–21a and b. It appears that the smaller-sized particle (0.6 mm) adsorbed more of the two pesticides than the larger particle (1.2 mm) according to the Langmuir X_m value. Also, Sevin was adsorbed to a greater extent than Baygon.

PAC (BDH) was employed for adsorption of 10 phenylamide pesticides from water within a 15 minute period of shaking [66]. The Freundlich equation was able to describe the adsorption data from which the appropriate constants are given in Table 6–22. Comparison of *K* values gives a qualitative indication of adsorbability of the 10 pesticides. Carbon dosages to remove 1 mg/l of com-

Table 6–21a Adsorption Parameters of Langmuir Isotherms for Sevin and Baygon

Compound	*Particle Size of Carbon (mm)*	*Limiting Values of X_m Monolayer Adsorption*		*Reciprocal of Half-Coverage Concentration* b *(l/μmol)*	*Half-Coverage Concentration (μmol/l)*
		(μmol/g)	*(mg/g)*		
Sevin	0.6	1,250	251.0	0.08	12.0
	1.2	800	160.8	0.25	4.0
Baygon	0.6	500	104.5	0.05	20.0
	1.2	530	117.0	0.07	14.5

Source: Reproduced from El-Dib et al. [65], courtesy of Pergamon Press.

Table 6-21b Adsorption Parameters of Freundlich Isotherms for Sevin and Baygon

Compound	*Particle Size of Carbon (mm)*	K[a] *(mg/g)*	*Slope* (l/n)
Sevin	0.6	.14	.25
	1.2	.11	.16
Baygon	0.6	.04	.26
	1.2	.04	.25

Source: Reproduced from El-Dib et al. [65], courtesy of Pergamon Press.
[a]When C_e = 1.0 mg/l.

pound to 0.1 mg/l are within the normal range (except Vitavax) of water treatment plants. The 5 mg/l requirements are somewhat high.

Removal of methoxychlor from water by various water treatment processes was reported [67]. Neither chemical coagulation and filtration nor lime soda softening was able to remove significant amounts of methoxychlor from a 5.0 mg/l solution. Treatment by PAC (Nuchar Special) was able to reduce 1.0 and 5.0 mg/l of methoxychlor to the MCL value of 0.1 mg/l (no contact time was given). Freundlich isotherms were drawn for the two pesticidal concentrations. For the 1.0 mg/l solution, the isothermal constants were K = 136. mg/g and $1/n$ = .456, and for the 5 mg/l solution they were K = 92. mg/g and $1/n$ = .172.

Table 6–22 Adsorption Parameters and Carbon Dosages to Remove Phenylamides Studied

		Adsorption Parameters		*Carbon Dose to Remove*	
Compound	*Solubility (mg/l)*	K[a] *(mg/mg)*	l/n	*5 mg/l*	*1 mg/l*
IPC	250	.045	.467	312	57
CIPC	88	.050	.187	151	27
Fenuron	2900	.028	.456	500	92
Monuron	230	.066	.333	160	29
Diuron	42	.095	.109	66	12
Linuron	75	.080	.157	88	16
Neburon	6	.080	.301	122	22
Stam	50	.060	.222	136	25
Karsil	8	.100	.238	84	15
Vitavax	20	.023	.451	602	110

Source: Reproduced from El-Dib and Aly [66], courtesy of Pergamon Press.
Note: Final concentration, 0.1 mg/l.
[a]When C_e = 1.0 mg/l.

Table 6–23 Langmuir Constants for Adsorption Isotherms of Pesticides on Activated Carbon

	X_m		b
Compound	*(mM/g)*	*(mg/g)*	*(1/μmol)*
2,4-dichlorophenxyacetic acid	1.75	387.	0.095
2,4,5-trichlorophenxyacetic acid	1.75	448.	0.150
2-(2,4,5-trichlorophenoxy) propionic acid	1.72	464.	0.145
0,0-diethyl-0-*p*-nitrophenyl phosphorothioate (Parathion)	1.82	530.	1.22

Source: Morris and Weber, Jr. [68].

Three phenoxy herbicides and parathion were examined for their adsorbability on GAC (0.273 mm Columbia LCK carbon) over a 2 wk period in a laboratory study [68]. Langmuir constants are given in Table 6–23 for these four pesticides (25°C).

Column Studies

Carbon column studies have been surprisingly conspicuous by their absence in the literature. Here again, this may be due to reduced use of chlorinated hydrocarbons and the relative instability of phosphates, carbamates, and ureas in water. Limited data are available for the GAC (Darco, 0.6 and 1.2 mm) column removal of Sevin and Baygon from water [65]. Column characteristics and adsorption data are seen in Tables 6–24a and b for Sevin and Baygon, respectively. For the latter, Baygon was adsorbed to a lesser extent than Sevin, which was predicted

Table 6–24a Carbon Column Characteristics and Adsorption Data for Sevin

Column	*Influent Concentration (mg/l)*	*Particle Size (mm)*	*Bed Depth (mm)*	*Contact Time (min)*	*Throughput Volume (l)*[a]	*Bed Volumes*[a]	*x/M (mg/g)*
A	5	1.2	10	3.75	118.0	1,800	22.4
B	20	1.2	10	3.75	28.0	425	21.5
C	20	1.2	10	1.0	11.0	170	8.5
D	20	1.2	10	0.5	1.0	15	0.8
E	20	0.6	10	0.5	<1.0	—	—
F	20	0.6	10	3.75	27.0	410	20.5
G	20	1.2	30.5	3.75	80.0	400	20.5
H	20	0.6	30.5	3.75	75.0	375	19.0

Source: Reproduced from El-Dib et al. [65], courtesy of Pergamon Press.
[a]Volume of effluent passed before breakthrough at 0.1 mg/l Sevin.

Table 6–24b Carbon Column Characteristics and Adsorption Data for Baygon

Column	*Influent Concentration (mg/l)*	*Particle Size (mm)*	*Bed Depth (cm)*	*Contact Time (min)*	*Throughput Volume (l)*[a]	*Bed Volumes*[a]	x/M *(mg/g)*
M	10	1.2	10	0.5	<1.0	—	—
N	5	1.2	10	0.5	<1.0	—	—
O	5	1.2	10	1.0	<1.0	—	—
P	5	0.6	10	1.0	<1.0	—	—
Q	5	0.6	10	3.75	17.0	255	3.2
R	5	1.2	10	3.75	18.0	273	3.4
S	5	0.6	30.5	3.75	47.5	235	3.0
T	5	1.2	30.5	3.75	50.0	250	3.2

Source: Reproduced from El-Dib et al. [65], courtesy of Pergamon Press.
[a]Volume of effluent passed before breakthrough at 0.1 mg/l Baygon.

from equilibrium studies (Tables 6–21a and b). The effect of contact time (in minutes) on breakthrough of the two pesticides is obvious. However, the 30.5 cm bed depth with 3.75 minutes of contact is suspect data. No mention was made of how the contact time was held constant despite an increase of bed depth from 10 to 30.5 cm. The rate of flow should have been increased to keep the contact time constant in the 10 cm and 30.5 cm columns [55].

A brief examination of GAC (F-400) column removal of methoxychlor was reported [67, p. 285]. "After water containing 1., 5., and 10. mg/l methoxychlor was passed through the GAC column, there was no detectable methoxychlor in any sample."

COMPOUNDS CAUSING TASTES AND ODORS

Despite the fact that tastes and odors are an endemic water quality problem, very few specific adsorption studies have been conducted for organoleptic organic compounds. Most reports in the literature are empirical and operational in nature [1]. Some of the compounds reported previously, especially the halogenated phenols, may cause tastes and odors in drinking water. See Chapter 8 for operating experiences in taste and odor removal.

A limited study of the adsorption on n-butylmercaptan and butyl disulfide on GAC (F-400, 20 × 30 mesh) was conducted in a 48 hour equilibrium test [69]. Adsorption data were plotted in accord with the Langmuir equation. The appropriate constants were as follows:

Compound	X_m *mg/g*	*b (l/mol)* $\times 10^{-4}$
n-butylmercaptan	307	.71
Butyl disulfide	428	7.7

The X_m values suggest a reasonable quantity of adsorption, and according to the authors, equilibrium occurred in 15 minutes, but the compounds were allowed to continue to react for 48 hours [69].

Two naturally occurring odorous compounds, 2-methylisoborneal (MIB) and geosmin were investigated for their adsorption on GAC (F-200 and F-400, 40 × 50 mesh) in equilibrium studies (4 to 5 days) [42]. Geosmin is produced by some actinomycetes and blue-green algae, whereas MIB is a product of actinomycetes [1]. Figure 6–13a is a log-log plot of the data for MIB in the absence and presence of humic acid. The latter was included to interfere with adsorption of MIB and geosmin. At 1 μg/l MIB, 0.9 mg/g was adsorbed on the GAC from distilled water solutions. The presence of humic acid did inhibit adsorption of MIB. That geosmin was adsorbed to a greater extent is seen in Figure 6–13b where 4. mg/g was the surface coverage when C_e = 1 μg/l. Again, humic acid interfered with adsorption of geosmin.

Column breakthrough curves were constructed for MIB and geosmin in the absence of humic acid [42]. These are seen in Figure 6–13c, about which the authors state: "An obvious characteristic of these curves is the manner in which

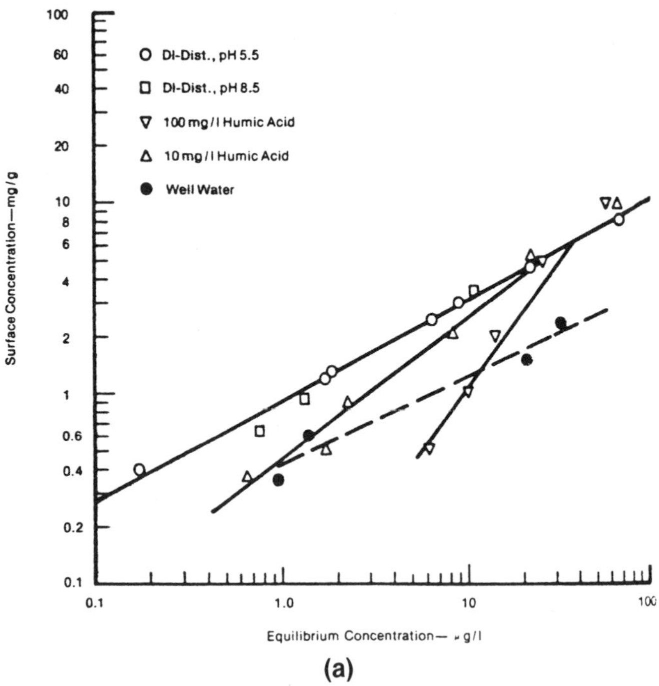

Figure 6–13 (a) Adsorption of MIB. (b) Adsorption of geosmin. (c) Column breakthrough curves for MIB and geosmin. Reproduced from Herzing et al. [42], courtesy of the American Water Works Association.

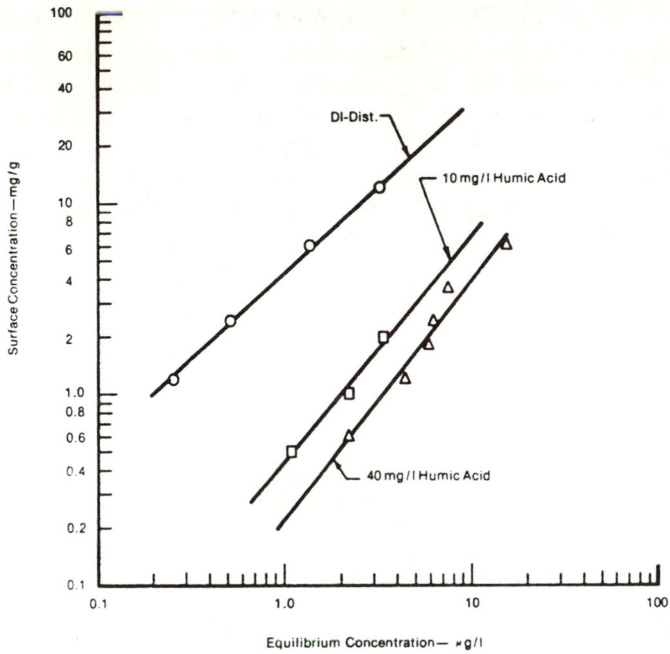

(b)

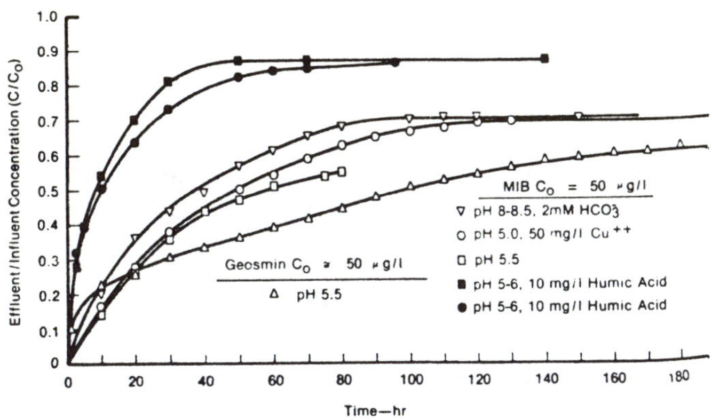

(c)

they level off rather than converge to $C_e/C_i = 1$. This suggests that a portion of the capacity is being utilized very slowly and that a long operational time would be required for complete saturation." This may be true, but it does not guarantee that the earthy, musty odor would be eliminated from the water within the same time frame.

Additional studies were conducted for PAC (100 × 200 mesh) adsorption of MIB [70]. In this case, four bituminous coal-based carbons and one lignite-based carbon were employed in a comparative study. Adsorption isotherms (7 days of shaking) are seen in Figure 6–14a for three carbons with MIB in distilled water alone and in the presence of 10 mg/l humic acid. There was very little difference between these three carbons in their ability to adsorb MIB (compare the surface coverages at 1. μg/l C_e). Again, humic acid inhibited MIB adsorption on the three PACs. Figure 6–14b shows that the capacities of carbons B-3 and

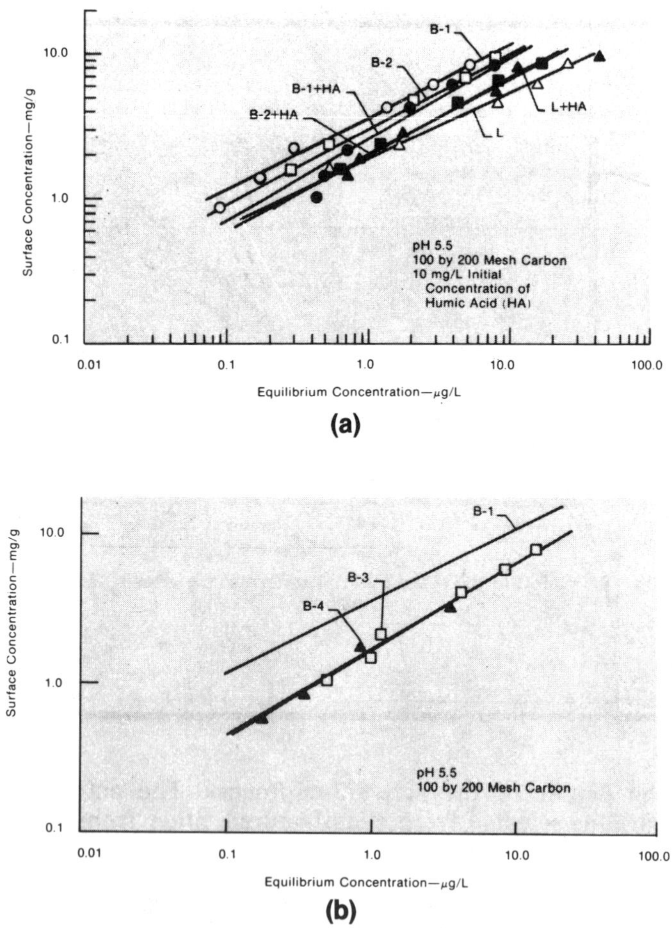

Figure 6–14 (a) MIB adsorption isotherms on carbons B-1, B-2, and L. (b) MIB adsorption isotherms on carbons B-3 and B-4. Reproduced from Chudyk et al. [70], courtesy of the American Water Works Association.

B-4 were somewhat less than B-1, B-2, and L carbons. These isotherms are representative of treatment because MIB and other naturally occurring odorous substances are usually in the μg/l concentration range in natural waters [1].

POLYCYCLIC AROMATIC HYDROCARBONS

Polycyclic aromatic hydrocarbons (PAHs) are considered to be naturally occurring because their origin is usually in petroleum [71]. They are released into the environment through a pyrolytic industrial operation of some sort. Examples are preparation of acetylene from natural gas; hydrolysis of kerosene to benzene, toluene, and other organic solvents; and pyrolysis of wood. There is a legitimate concern about the occurrence of these compounds in human's environment because of the carcinogenic properties exhibited in laboratory animals by some PAHs. For example, 3,4-benzpyrene (BP) is, perhaps, the most carcinogenic of all PAHs [71].

One of the ironic aspects of activated carbon treatment of potable water for organics is that it is a source of PAH and other organics through leaching processes. In screening tests, 22 varieties of activated carbons (all European) were examined for their aromatic content by a two-dimensional thin-layer chromatographic technique [72]. Ten PAHs were found in various contents (Table 6–25). It is not clear how these PAHs occur in carbon: do they occur naturally or do they result from the activation process?

Eight PAHs were subjected to adsorption by ten activated carbons in a laboratory study [72]. Experimental conditions were 3 μg/l PAH, 1,000 mg/l activated carbon, 30 min contact time, and 15°C in 1 l synthetic sewage. The median adsorption was 99.78%, with 99% as the minimum and 100% as the maximum. A PAC, Norit NK, was perhaps the most efficient for PAH. Additional experiments were conducted that used 10. and 100. mg/l activated carbon (Norit NK), and the contact time was varied from 0.5 to 45 hours. Results are seen in Table 6–26 where adsorption was somewhat time dependent. The author concluded that "30 minutes of contact time is not long enough using 100. mg/l of AC. Good results with such a dose require very long and, therefore unrealistic contact times" [72, p. 145]. This may be true, but it does not take into consideration the residual PAH after a given contact time. If a drinking water standard is established, it may not be necessary to achieve a zero PAH content.

Adsorption isothermal data for 11 PAHs from water onto F-400 PAC (200 mesh) at 25°C were generated by batch shake testing for 1. to 2.5 days [73]. Five equations were evaluated for their fit to the adsorption data, whereupon it was concluded that the Langmuir, two linearized forms, was the most useful. Table 6–27 is a summary of the appropriate isothermal equation parameters that are necessary for treatment design. PAHs are adsorbed readily due to their hydrophobic nature and low solubility in water.

Table 6–25 Concentrations of Polycyclic Aromatic Hydrocarbons in Some Types of GACs and PACs

		Detected Amounts (μg/kg)											
Activated Carbon	*Type*	*I*	*II*	*III*	*IV*	*V*	*VI*	*VII*	*VIII*	*IX*	*X*	*S*	S^{++}
Carboraffin C[a]	Powdered	80.0			1.2		0.8		0.8			83	1
Brillonit Norm[a]	Powdered	100										100	
EPN-A-Kohle[b]	Powdered	28.0										28	
Gerdit[c]	Powdered	53.6			5.2	1.2	2.8		4.0	1.2		68	8
Gerdit[c]	0–0.5 mm	110.0	12.4		12.4	9.6			10.0	2.4		157	37
Gerdit[c]	1–2.0 mm	216.0			21.2	3.6	8.4		5.2	2.4		257	27
Gerdit[c]	2–4 mm	81.6			3.2	2.0			2.0	0.8		90	6
Hydraffin 118 ff[a]	Powdered	1.2										1	
Hydraffin BD[a]	0.5–2.5 mm	60.0	3.6	3.6	8.0	1.2	0.8		2.4	0.8	20.0	100	16
Hydraffin BD Extra[a]	0.5–1.5 mm	3.6										4	
Hydraffin LW[a]	0.5–3.0 mm	40.0		8.0	3.2	0.8	0.8		1.2	0.8	20.0	75	13
Hydraffin LW Extra[a]	0.5–3.0 mm	80.0			2.4		1.6		0.8	0.4		85	3
Hydraffin TC spez, I/II[a]	0.5–2.5 mm	28.0			1.2	0.8	0.8		0.8			32	2
KO-A-Kohle[b]	Powdered	80.0										80	
Norit FND[d]	25% < 5 μm 6% 5–10 μm	28.0										28	

Norit P[d]	12% 10–20 μm	1.6										2	
	26% 20–50 μm												
Norit NK[d]	13% 50–75 μm												
	3% 75–90 μm												
Norit FNA[d]	15% > 90 μm												
Norit RBWII[d]	Rods Ø = 1.6 mm	40.0		4.0	1.6	0.4	0.8		0.8	0.4		48	6
Olit Normal[a]	< 0.2 mm	60.0			2.0		0.8		0.8			64	2
R_2-A-Kohle[b]	Granular	60.0	8.0	12.0	4.0	1.6	3.2		2.8	1.6		93	27
A-Kohle Nr. 2186[e]	< 0.2 mm	2.8										3	

Source: Reproduced from Borneff [72], courtesy of Ann Arbor Science Publishers, Inc.
Polycyclics:
I: fluoranthene
II: benzo[a]anthracene
III: benzo[j]fluoranthene
IV: benzo[b]fluoranthene
V: benzo[a]pyrene
VI: benzo[g,h,i]perylene
VII: perylene
VIII: benzo[k]fluroanthene
IX: indeno[1,2,3-cd]pyrene
X: pyrene
S: sum of polycyclics I-X
S^{++}: sum of carcinogenic polycyclics
[a]Lurgi GmbH, Frankfurt/Main.
[b]VEB-Chemiefaserwerk, F. Engels.
[c]Jägersberg, Hamburg.
[d]Norit Corp, Amsterdam.
[e]Merck AG, Darmstadt.

Table 6–26 Adsorption of Polycyclic Aromatic Hydrocarbons by PAC (Norit NK)

Hydrocarbon	*Carbon Dose (mg/l)*	*Residue of the Hydrocarbon after a Contact Time of 0.5 to 45 hr (%)*				
		0.5 hr	*4 hr*	*22 hr*	*28 hr*	*45 hr*
Fluoranthene	100	1.2	0.6	0.5	0.8	2.5
	10	18.2	7.1	1.6	1.2	0.7
Benzo[a]anthracene	100	2.8	0.9	2.2	0.6	1.1
	10	20.7	15.5	2.4	1.6	0.8
Benzo[b]fluoranthene	100	19.1	2.1	0.8	0.9	1.7
	10	50.1	33.9	5.0	2.6	1.1
Benzo[a]pyrene	100	19.1	3.2	0.5	1.1	1.3
	10	31.2	26.8	3.7	2.0	1.0
Benzo[g,h,i]perylene	100	42.2	18.2	2.3	2.2	1.3
	10	62.2	51.0	6.3	3.2	1.8
Benzo[j]fluoranthene	100	14.5	3.1	0.5	0.9	1.5
	10	41.0	25.1	4.0	2.0	1.1
Indeno[1,2,3-c,d]pyrene	100	37.0	10.1	1.4	1.2	0.7
	10	37.2	31.0	3.4	2.0	1.3
Pyrene	100	1.7	1.1	1.1	0.4	1.2
	10	11.2	6.2	1.3	0.6	0.3
Sum of the hydrocarbons	100	11.1	3.2	1.0	0.9	1.6
	10	27.6	18.1	2.8	1.6	0.9
Sum of carcinogenic	100	18.7	3.6	1.3	1.0	1.3
hydrocarbons	10	36.1	27.1	3.8	2.1	1.1

Source: Reproduced from Borneff [72], courtesy of Ann Arbor Science Publishers, Inc.

MISCELLANEOUS ORGANIC COMPOUNDS

Equilibrium Studies

In the 1960s synthetic detergents were the predominant water quality problem. Much research was conducted on their removal from raw drinking water supplies by PAC, GAC, and other treatments. An extensive study was conducted for the GAC (Columbia LC carbon) adsorption of several alkylbenzenesulfonates, nitrochlorobenzenes, octylphenoxynonylethoxy ethanol, and phenol [74]. Table 6–28 gives the Langmuir parameters for various compounds at 30°C. These systems were shaken for periods of 1 to 4 weeks. All compounds were adsorbed quite well with exception of the first three compounds. Most detergents listed in Table 6–28 are probably not in use in the 1980s. However, compounds of a similar nature and structure may occur in raw water supplies. Also, the data in Table 6–28 demonstrate the effect of molecular structure on the adsorption of complex compounds by porous GAC.

Langmuir and Freundlich isothermal constants were obtained for several industrial organics on PAC (F-400) in an effort to predict the preferentially ad-

Table 6–27 Summary of Isotherm Equation Parameters

Compound	Henry's Law		Langmuir: High C_e Linear Form			Langmuir: Low C_e Linear Form			BET			Freundlich			Redlich-Peterson[b]			
	K_h	r^2	q^o	b	r^2	q^o	b	r^2	q^o	B	r^2	K_f	1/n	r^2	K_{hr}	K_{fr}	m	r^2
NA	2,390	.59	580	2.26	.96	475	5.67	.85	283	−516	.93	277	.430	.87	1,310	580	0.0	.96
ACY	5,270	.22	500	3.41	.97	275	25.4	.70	100	−54	.87	266	.302	.88	70,000	359	.392	.74
ACE	16,800	.72	465	21.7	.98	332	52.1	.79	286	1,220	.98	624	.457	.87	17,600	390	.0643	.80
FLE	5,640	.28	440	13.8	.93	52.1	248	.66	256	64.2	.97	674	.604	.90	[c]	674	.604	.90
PH	≥13,000[a]	[c]	231	12.4	.75	101	115	.62	129	40.7	.84	273	.406	.84	[c]	273	.406	.84
AN	6,140	.54	23.1	244	.72	21.4	290	.91	17.8	24.5	.80	330	.620	.81	[c]	330	.620	.81
FLA	≥59,000[a]	[c]	104	181	.88	60.2	983	.70	77.0	87.2	.93	242	.314	.81	249,000	215	.285	.77
PY	30,500	.29	82.8	522	.97	64.0	802	.68	48.8	1,340	.96	389	.386	.78	[c]	389	.386	.78
TRP	≥5,900[a]	[c]	36.8	121	.16	17.0	3,860	.38	22.8	8.80	.26	532	.685	.55	[c]	532	.685	.55
BaA	≥20,400[a]	[c]	14.7	933	.56	8.77	2,660	.58	1 0.8	19.4	.64	216	.500	.67	[c]	216	.500	.67
CHR	≥136,000[a]	[c]	38.4	3,220	.57	16.7	19,500	.28	20.9	14.4	.77	716	.458	.62	[c]	716	.458	.62

Source: Reproduced from Walters and Luthy [73], courtesy of the American Chemical Society.
[a]Tabulated value was estimated from inspection of log-log isotherm plots and represents a minimum value.
[b]Nonlinear regression analysis indicated that the Redlich-Peterson equation reduced to the Langmuir equation for NA and to the Freundlich equation for FLE, PH, AN, PY, TRP, BaA, and CHR. Tabulated Redlich-Peterson parameters for these compounds correspond to those tabulated for the Langmuir or Freundlich equations.
[c]Nonrealistic values obtained.

Table 6–28 Langmuir Parameters for Adsorption Isotherms

Adsorbate	*Diameter of Adsorbent Particles (mm)*	*Limiting Value of X for Monolayer Adsorption, X_m* (μM/g)	(mg/g)	*Reciprocal of Half-coverage Concentration,* b *(l/μM)*
Alkylbenzenesulfonates:				
benzenesulfonate	0.273	161	29	0.28
p-toluenesulfonate	0.273	222	43	0.33
2-hexyl-	0.273	313	83	2.56
2-octyl-	0.273	345	101	2.90
2-decyl-	0.273	370	119	3.60
2-dodecyl-	0.273	400	139	4.55
3-dodecyl-	0.273	400	139	4.63
6-dodecyl-	0.273	400	139	4.55
2-tetradecyl-	0.273	435	164	6.76
Nitrochlorobenzenes:				
o-	0.503	1,400	220	1.1
p-	0.503	1,500	240	1.8
Octylphenoxynonyl-ethoxyethanol	0.503	330	215	6.7
Phenol (C_{eq} = 5–21 μmol/l)	0.273	1,090	103	0.11

Source: Reproduced from Weber and Morris [74], courtesy of the American Society of Civil Engineers.
Note: All experiments were conducted at solution temperatures of 30°C.

sorbed compound in 22 bisolute activated carbon column systems [75]. Table 6–29 shows the adsorption parameters that represent 2 hours of shaking at room temperature. The correlation coefficients indicate the degree of fit to the Langmuir and Freundlich equations. A good fit is suggested by r values $> .9$, and a bad fit would be $< .9$. In this case, the Freundlich equation provides extremely good fits, and its K values were a successful predictor of preferential adsorption in the 22 bisolute systems on carbon columns [75]. That is, the K value correctly predicted the preferred compound in all 22 bisolute systems. By performing single-solute isotherms, the order of bisolute breakthrough is predicted [75].

In a study on the effect of molecular structure on adsorption of several organic compounds, isotherms were developed on 12 × 40 mesh CAL activated carbon [27]. Continuous agitation was supplied for 100 hours. Table 6–30 gives the Langmuir constants for 17 organic compounds. Several phenols were investigated also (see Table 6–9 for their isothermal constants). There was a general tendency for adsorption capacity (X_m values) to increase with molecular weight. This relationship has been reported frequently [50,68,74].

Adsorption of aniline, phenol, benzyl alcohol, benzaldehyde, and nitrobenzene was examined to compare effects of the following substituents on adsorbability: NH_2, OH, CH_2OH, CHO, and NO_2 [27]. Adsorption decreased in

Table 6–29 Adsorption Parameters for Organics on PAC

Compound	Number of Points	Langmuir X_m (mmol/g)	Langmuir b (l/mmol)	Langmuir r[a]	Freundlich K (mmol/g)	Freundlich n	Freundlich r
i-propanol	7	.79	.123	.993	.092	1.37	.996
n-propanol	7	.92	.158	.984	.127	1.37	.991
i-butanol	6	1.04	.482	.995	.310	1.70	.980
n-butanol	6	1.06	.954	.989	.472	1.84	.983
n-pentanol	6	2.85	.858	.992	1.18	1.97	.985
n-hexanol	6	3.21	1.35	.997	1.76	2.60	.977
Propionaldehyde	6	1.21	.107	.870	.157	1.51	.969
Butyraldehyde	6	1.19	.745	.989	.487	1.59	.992
Ethyl acetate	7	1.54	.531	.863	.511	1.82	.937
i-propyl acetate	7	1.66	.954	.973	.856	3.01	.918
Propyl acetate	9	7.13	.153	.773	.959	1.33	.990
Butyl acetate	9	10.7	.166	.814	1.46	1.21	.994
Methyl ethyl ketone	4	1.00	.526	.994	.348	1.83	.994
Diethyl ketone	4	1.67	.911	.998	.766	2.10	.997
Methyl *i*-butyl ketone	7	1.95	1.44	1.000	1.12	3.01	.989

Source: Reproduced from Arbuckle and Romagnoli [75], courtesy of the American Institute of Chemical Engineers.
[a]Correlation coefficient.

this order: NO_2 > CHO > OH and CH_2OH > NH_2 when the monolayer capacities were compared on a mass (mg/mg carbon) or molar (mol/g carbon) basis. The order of decrease in adsorption is in accord with a decrease in molecular weight, with benzyl alcohol being the exception. In all probability, the molecular weight controls adsorption in this case rather than polarity, hydrogen bonding, solubility, and other such factors as suggested by the authors [27].

The effects of several experimental factors on the adsorption of some of the compounds in Table 6–30 were investigated subsequently in batch and column systems [76]. These were [H_3^+O], carbon particle size, carbon bed depth, and flow rate as factors affecting adsorption of 2-methylpyridine and *o*-cresol on 12 × 40 mesh CAL carbon (100 hours agitation). Three particle sizes of carbon—0.43, 0.9 to 1.1, and 1.2 mm diameter—were investigated for their adsorptive capacity for 2-methylpyridine. The data from adsorption isotherms (not shown) showed that a change in particle size had no effect on the adsorptive capacity of CAL carbon. This finding is in agreement with those of other investigators [32]. Here, again, the 2-methylpyridine molecule was small enough to enter the internal pore structure of the carbon regardless of particle size.

Adsorption isotherms for 2-methylpyridine and *o*-cresol at three pH values were developed [76]. An increase in pH value from 6.8 to 9.5 did not have an appreciable effect on the adsorptive capacities for these two compounds. How-

Table 6–30 Some Adsorption Characteristics of Organic Pollutants

Compound	*Molecular Weight*	*Solubility in Water at 20°C Approximately (g/100g)*	*Langmuir Equation Constants* X_m *(mg/mg Carbon)*	*Langmuir Equation Constants* b *(l/mg)*
Aniline	93.13	3.70	.108	.177
Benzyl alcohol	108.15	3.80	.158	.176
Benzaldehyde	106.13	0.30	.210	.170
2-naphthol	144.19	0.068	.345	.643
o-toluidine	107.16	1.7	.186	.256
p-toluidine	107.16	0.7	.186	.256
o-anisidine	123.16	sl.sol.	.240	.417
m-anisidine	123.16	sl.sol.	.185	.338
p-anisidine	123.16	v.sl.sol.	.248	.505
Anisaldehyde	136.16	0.2	.305	.468
Salicylaldehyde	122.13	1.7	.240	.417
Vanillin	152.15	1.0	.375	1.110
Pyridine	79.10	miscible	.055	.154
2-methylpyridine	93.13	v.sol.	.106	.262
Quinaldine	143.19	sl.sol.	.296	.400
Pyrrole	67.09	sl.sol.	.030	.152
Indole	117.15	0.02	.240	.417

Source: Reproduced from Al-Bahrani and Martin [27], courtesy of Pergamon Press.
sl.sol. = slightly soluble
v.sol. = very soluble

ever, an increase in $[H_3^+O]$ from $10^{-6.8}$ to $10^{-3.5}$ M resulted in a significant decrease in the capacity to adsorb 2-methylpyridine. A similar effect was observed for *o*-cresol, the extent of which was less. The absence of an effect at pH values of 6.8 and 9.5 is in disagreement with other researches (see under phenols). The explanation is, however, relatively simple. Both *o*-cresol (pKa = 10.28) and 2-methylpyridine (pKa = 9.57) are weak acids. Consequently, the associated forms for both compounds predominate at pH 6.8 and 9.5 where their adsorptive capacities would be similar.

The influence of molecular structure, solubility, substituent groups, and $[H_3^+O]$ has been the objective of many researches of carbon adsorption. At this point, some generalizations are offered about these effects on the adsorbability of organics by activated carbons. This should not be confused with effects of physical parameters on adsorption and the development of adsorption models therefrom (see Chapter 2). Table 6–31 is a summary of general observations about the influence of molecular structure and so forth [77]. This summary was offered as a preamble to a study of the effect of functional group, molecular size, and $[H_3^+O]$ on the adsorption of 15 organics on PAC (Westvaco Nuchar, 4 hours, 25°C). For functional groups, carboxyl, hydroxyl, and amine were selected on aliphatic chains of 3, 5, 7, and 9 carbon atoms. For molecular size, aromatic

Table 6 31 Influence of Molecular Structure and Other Factors on Adsorbability of Organics by Activated Carbon

1. Increasing solubility decreases its adsorbability.
2. Branched chains are usually more adsorbable than straight chain compounds. Increasing length of the chain decreases solubility.
3. Substituent groups affect adsorbability:

Substituent Group	*Effects*
Hydroxyl	Generally reduces adsorbability depending on structure.
Amino	Similar to that of hydroxyl but somewhat greater. Many amino acids are not adsorbed to any appreciable extent.
Carbonyl	Varies with molecule; glyoxylic acids are more adsorbable than acetic acid, but a similar increase does not occur when higher-molecular-weight fatty acids are introduced.
Double bonds	Variable as with carbonyl.
Halogens	Variable.
Sulfonic	Usually decreases adsorbability.
Nitro	Often increases adsorbability.
Aromatic rings	Greatly increases adsorbability.

4. Generally, strongly ionized solutes are not as adsorbable as weakly ionized solutes; that is, undissociated molecules are in general preferentially adsorbed.
5. The extent of adsorption depends on the ability of hydrolysis to form an adsorbable acid or base.
6. Unless the screening action of the carbon pores intervenes, large molecules are more sorbable than small molecules of similar chemical nature. This is attributed to more solute carbon chemical bonds being formed, making desorption more difficult.
7. Molecules with low polarity are more sorbable than highly polar solutes.

Source: Reproduced from Reimers et al. [77], courtesy of Ann Arbor Science Publishers.

rings with acidic, hydroxyl, and amine functional groups were chosen. For the effect of $[H_3^+O]$, adsorption of each compound was run at two pH values. Freundlich isotherms were used to describe the adsorption, with K and $1/n$ values tabulated. Nothing unusual was discovered in this study. The major factor influencing adsorbability of the 15 organics was the $[H_3^+O]$. In general, the associated form of the compound was adsorbed to a greater extent than the dissociated form. There were a couple of exceptions: in polyethyleneimine, aniline, and phenol, the charged specie was adsorbed more than the uncharged specie.

Column Studies

Table 6–32 is a summary of the appropriate experimental conditions and results from adsorption of six low molecular weight organic solvents on GAC [24]. This table is a compilation of several researches.

Table 6–32 Summary of Adsorption of Volatile Organic Compounds by GAC

Compound	*Average Concentration (μg/l)*	*Bed Depth*		*EBCT*[a] *(min)*	*Loading (m³/m³)*		*Exhaustion Capacity*[b] *(m³/m³)*
		(m)	*(ft)*		*Breakthrough of 0.1 μg/l (except as noted)*	*Exhaustion (influent = effluent)*	
Trichloroethylene	177	0.8	2.5	9	>20,160	>20,160	21,500
	4	0.8	2.5	8.5	>60,900 but <123,340	>123,340	99,900
	3[c]	0.8	2.5	18	>32,500	>32,500	100,560
	0.4[d]	0.8	2.5	18	>32,500	>32,500	199,800
Tetrachloroethylene	1,400	0.8	2.5	9	12,300	33,100	17,500
	94[c]	0.8	2.5	18	>32,500	>32,500	57,400
	9[d]	0.8	2.5	18	>32,500	>32,500	162,800
	4	0.8	2.5	8.5	>60,900 but <123,340	>123,340	237,600
	1	0.8	2.5	9	>20,160	>20,160	475,200
1,1,1-trichloroethane	100	0.6	2	7.5	1,300[e]		3,800
		1.2	4	15	2,700[e]		3,800
		1.8	6	22.5	3,900[e]		3,800
	237[c]	0.8	2.5	18	15,700	30,800	2,600
	23[d]	0.8	2.5	18	>32,500	>32,500	12,000
	38	0.8	2.5	8.5	11,800	26,000	9,400
	1	0.8	2.5	9	16,400	22,500	94,400

Carbontetrachloride	12	0.8	2.5	10	14,000	25,000	9,400
		0.8	2.5	5	6,050		9,400
cis-1,2-dichloroethylene[f]	18	0.8	2.5	6	4,100	15,800	9,400
		1.5	5	12	7,100	14,300	9,400
		2.3	7.5	18	8,100	13,700	9,400
	6	0.8	2.5	9	14,200	19,000	15,300
	2	0.8	2.5	8.5	29,600	48,600	26,000
2-dichloroethane	8	0.8	2.5	20	1,700	8,640	3,500
	1.4[c]	0.8	2.5	18	17,400	45,900	5,700
	0.8[d]	0.8	2.5	18	>32,500	>32,500	5,000
	2	0.9	3	11	3,300	9,160	4,000
		1.8	6	22	3,400	7,850	4,000
		2.7	9	33	4,150	>7,000	4,000
		3.6	12	44	>7,000	>7,000	4,000
		0.8	2.5	17.5	2,500	7,450	4,000
		0.7	2.4	17	2,500	7,450	4,000

Source: Reproduced from Love and Eilers [24], courtesy of the American Water Works Association.
[a]Empty bed contact time.
[b]Estimated by Freundlich isotherms.
[c]Adsorption of nonaerated water.
[d]Adsorption after 10 min aeration; air-to-water ratio 4:1 (vol:vol).
[e]5 μg/l in effluent.
[f]Isotherm capacities based on trans-1,2-dichloroethylene.

ORGANICS IN MIXED SYSTEMS

A few examples were given earlier for adsorption of various organic compounds in bisolute and/or multisolute systems. It is the competitive nature of the adsorption process that has more pragmatic application to water and wastewater treatment. Competition arises from synthetic organic compounds, hydronium ions, naturally occurring compounds such as humic and fulvic acids, and inorganic ions in water. It is difficult to duplicate experimentally a natural water with all its constituents and to be able to sort out all competitive effects in the adsorptive process. Nevertheless, several studies have been designed to understand these effects.

Equilibrium Studies

A quantitative recognition of the competitive nature of adsorption on carbon probably began in the late 1950s and early 1960s when synthetic detergents were in vogue [50,68,78]. Many studies have been conducted, in the laboratory, with mixed systems—that is, two or more compounds in competition for adsorption by PAC or GAC. Most experiments have been conducted with two solutes, which is not representative of a real water treatment situation.

An early study was conducted in an adsorptive equilibrium system for combinations of phenol-DBS and *p*-nitrochlorobenzene (*p*-NCB)-DBS on a GAC (Columbia LC carbon, 20 × 40 mesh, 3 weeks) [78]. In this situation, a Langmuir equation was employed to describe the competitive adsorptive equilibria:

$$X_i = \frac{X_i' \, b_i \, C_i}{1 + \sum_{j=1}^{m} b_j \, C_j}, \tag{6.4}$$

where X_i = amount of solute, *i*, adsorbed per unit weight of carbon at concentration C_i; X_i' = limiting value of X_i for complete monolayer formation (X_m is also used frequently); C_i = the concentration of solute remaining in bulk solution phase at equilibrium; and b_i = the energy-related constant of solute *i*. The *j* terms in Equation 6.4 refer to additional solutes in the mixture.

Figure 6–15 shows adsorptive equilibria for the two pairs of organic compounds cited earlier. Experimentally determined values for C_i and X_i were used to calculate the functions, $C_a/C_n/X_a$ and $C_a/C_p/X_a$, from Equation 6.4 for the equilibrium adsorption of DBS in solutions containing NCB and phenol. The circles in Figure 6–15 represent experimental data, and the solid lines are calculated values. "A more stringent test results, however, if values of X_a', b_p, and b_a are taken from Langmuir adsorption isotherms for individual solutes in pure solution, and fit of the data for the mixed systems with the straight line, thus obtained is observed" [78, p. 448]. This conclusion applies when systems are at equilibrium and maximum adsorptive capacity.

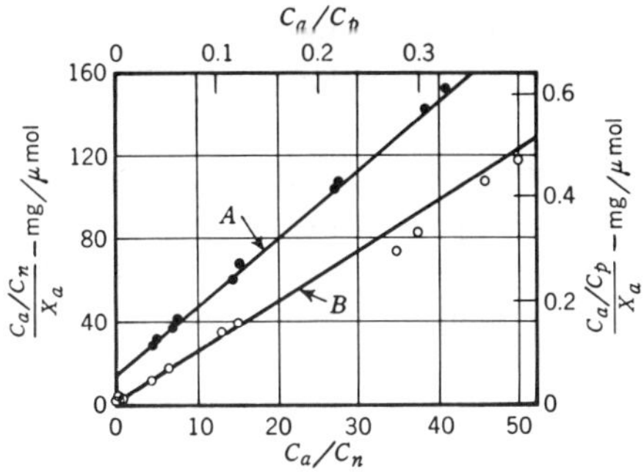

Figure 6–15 Competitive adsorption equilibria for DBS. The solid circles are for adsorption of DBS on 0.126 mm carbon from solutions containing phenol, and the open circles for adsorption on 0.273 mm carbon from solutions containing *p*-NCB. The lines drawn through the data have been calculated from Equations 6–5 and 6–6 with values of X' and b determined experimentally for equilibrium adsorption from pure solutions of each adsorbate. Curve A refers to the right ordinate and upper abscissa; Curve B, to the left ordinate and lower abscissa. Reproduced from Weber and Morris [78], courtesy of the American Water Works Association.

Another aspect of this study was the inhibitive effect of phenol on the adsorption of DBS and vice versa [78]. In a solution containing equal contents of phenol and DBS by weight, the equilibrium adsorption of phenol was 570 μmol/g,or 76% of the 750 μmol/g adsorbed in a single-solute system. The quantity of adsorbed DBS was 530 μmol/g, or 88% of 600 μmol/g adsorbed in a single-solute system. The combined quantity of 1,100 μmol/g was more than the individual compound acting alone. This suggests that total adsorptive capacities may be increased with mixtures of solutes. This is plausible where solutes are seeking different adsorption sites on the carbon's surface.

Additional studies were conducted on the competitive nature of carbon adsorption during the phenol-detergent era. Various combinations of PNP, PBP, and sodium benzenesulfonate (BS) were examined on 40 × 50 mesh F-400 carbon with an agitation period of 3 days for batch systems [79]. At that time, there was no mathematical model to predict the extent of adsorption of individual compounds from a bisolute solution where a portion of the adsorption occurred without competition. Such a model was developed based on the hypothesis that adsorption without competition occurs when $X_{m,1} \neq X_{m,2}$. "Further, it was assumed that the number of sites for which there was no competition was equal to the quantity $(X_{m,1} - X_{m,2})$, where $X_{m,1} > X_{m,2}$. On this basis, the following equations are proposed:

$$X_1 = \frac{(X_{m,1} - X_{m,2})b_1C_1}{1 + b_1C_1} + \frac{X_{m,2}b_1C_1}{1 + b_1C_1 + b_2C_2} \tag{6.5}$$

and

$$X_2 = \frac{X_{m,2} b_2 C_2}{1 + b_1 C_1 + b_2 C_2}, \tag{6.6}$$

where $X_{m,1} > X_{m,2}$. X_1 and X_2 = amount of solutes 1 and 2 adsorbed per unit weight, or per unit surface area, of adsorbent at equilibrium concentrations, C_1 and C_2, respectively; $X_{m,1}$ and $X_{m,2}$ = maximum values of X_1 and X_2, respectively, that are obtained from single-solute systems; and b_1 and b_2 = constants that are a function of the energy of adsorption of components 1 and 2, respectively, and that are derived from single-solute systems. The first term on the right side of Equation 6.5 is the Langmuir expression for the number of molecules of species 1 that adsorbs without competition on the surface area proportional to $(X_{m,1} - X_{m,2})$. The second term represents the number of species 1 adsorbed on the surface area proportional to $X_{m,2}$ under competition with species 2 and is based on the Langmuir model for competitive adsorption. The number of molecules of species 2 adsorbed on the surface area proportional to $X_{m,2}$ under competition with species 1 can be calculated from Equation 6.6'' [79]. When $X_{m,1} = X_{m,2}$, this model is the same as Equation 6.4. The single and bisolute equilibrium data for this study were fitted to the Langmuir equation.

Table 6-33a gives X_m and b values for five ionic species as well as for pH values of systems that determined the neutral or anionic nature of the com-

Table 6-33a Langmuir Parameters for Different Adsorbates

Adsorbate	X_m *(10^{-4} mol/g)*	b *(10^4 l/mole)*	*Equilibrium Concentration Range (10^{-3} mol/l)*	*pH*
Neutral PNP	19.0	6.75	0.04 to 0.30	4.1 ± 0.1
	24.1	1.00	0.30 to 1.58	4.1 ± 0.1
Anionic PNP	3.03	35.8	0.001 to 0.03	11.3 ± 0.2
	5.90	1.06	0.20 to 1.59	11.3 ± 0.2
	9.10	0.27	0.05 to 4.0	10.5 ± 0.1
	2.70	20.1	0.005 to 0.01	12.5 ± 0.1
	5.90	0.83	0.24 to 1.65	12.5 ± 0.1
Neutral PBP	19.0	6.75	0.04 to 0.30	3.2 ± 0.1
	24.1	1.00	0.30 to 1.58	3.2 ± 0.1
Anionic PBP	5.26	16.25	0.004 to 0.04	11.8 ± 0.2
	15.0	0.79	0.2 to 1.19	11.8 ± 0.2
	5.40	5.68	0.005 to 0.04	12.5 ± 0.1
	12.2	0.91	0.04 to 1.31	12.5 ± 0.1
Anionic BS	1.92	61.6	0.004 to 0.65	3.7 ± 0.3 and 6.1 ± 0.6
	2.94	0.17	0.05 to 4.43	10.5 ± 0.1

Source: Reproduced from Jain and Snoeyink [79], courtesy of the Water Pollution Control Federation.

pounds. Analysis of these and other isotherm data (not shown) indicated that neutral PNP/neutral PBP was described accurately by the Langmuir model for competitive adsorption that presumes competition among species for available sites. Isothermal data from mixtures of neutral PNP/anionic BS, anionic PNP/anionic BS, anionic PNP/anionic PBP, and anionic PNP/anionic BS were evaluated as the result of adsorption of one compound on a portion of the surface (or at different adsorption sites) without competition. Thus, these data (Tables 6–33b to 6–33e) were fitted better to Equations 6.5 and 6.6 than to Equation 6.4, the Langmuir competitive adsorption model. Whatever the mechanism, this study provided additional evidence of competitive adsorption and/or evidence for alternative adsorption sites for the competing compounds. The latter is contrary to Langmuir's uniform energy surface.

Several organic compounds were admixed in a laboratory study of competition on GAC (CAL 12, 12 × 40 mesh, 100 hours of shaking) [80]. The Langmuir equation was employed to describe the adsorption isotherms for solutes as

Table 6–33b Adsorption of Anionic PBP in the Presence of Anionic PNP

	PBP Adsorbed (C = 5 × 10⁻⁴M)		
PNP (M)	*Experimental (mol/g)*	*Predicted by the Proposed Model (mol/g)*	*Predicted by the Langmuir Competitive Model (mol/g)*
5×10^{-5}	9.5×10^{-4}	9.0×10^{-4}	8.2×10^{-4}
5×10^{-4}	7.2×10^{-4}	7.2×10^{-4}	5.3×10^{-4}
5×10^{-3}	6.0×10^{-4}	5.3×10^{-4}	1.2×10^{-4}

Source: Reproduced from Jain and Snoeyink [79], courtesy of the Water Pollution Control Federation.

Table 6–33c Adsorption of Anionic *p*-nitrophenol in the Presence of Anionic Benzenesulfonate

	PNP Adsorbed (C = 5 × 10⁻⁴M)		
BS (M)	*Experimental (mol/g)*	*Predicted by the Proposed Model (mol/g)*	*Predicted by the Langmuir Competitive Model (mol/g)*
5×10^{-5}	6.5×10^{-4}	5.2×10^{-4}	5.2×10^{-4}
5×10^{-4}	6.0×10^{-4}	4.6×10^{-4}	4.0×10^{-4}
5×10^{-3}	5.6×10^{-4}	4.0×10^{-4}	1.2×10^{-4}

Source: Reproduced from Jain and Snoeyink [79], courtesy of the Water Pollution Control Federation.

Table 6-33d Analysis of Isotherms

Adsorbate	X *(mol/g) Single Solute*	X *(mol/g) in Presence of 5 × 10⁻⁴M of Competing Species*
PNP (5 × 10^{-4}*M*)	4.5 × 10^{-4}	2 × 10^{-4}
PBP (5 × 10^{-4}*M*)	1.0 × 10^{-3}	8 × 10^{-4}
PNP + PBP (5 × 10^{-4}*M* for each species)	—	1 × 10^{-3}

Source: Reproduced from Jain and Snoeyink [79], courtesy of the Water Pollution Control Federation.

Table 6-33e Analysis of Isotherms

Adsorbate	X *(mol/g) Single Solute*	X *(mol/g) in Presence of 5 × 10⁻⁴M of Competing Species*
BS (5 × 10^{-4}*M*)	1.5 × 10^{-4}	3 × 10^{-5}
PNP (5 × 10^{-4}*M*)	6 × 10^{-4}	6 × 10^{-4}
BS + PNP (5 × 10^{-4}*M* for each species)	—	6.3 × 10^{-4}

Source: Reproduced from Jain and Snoeyink [79], courtesy of the Water Pollution Control Federation.

individuals or in a mixture. Table 6–34 gives the appropriate constants, X_m and *b,* for seven compounds from individual systems. Figures 6–16a and b show the individual adsorption isotherms, whereas Figures 6–16c and d are the isotherms for the compounds in admixtures. Comparison of these isotherms readily shows the competitive effects; the adsorption capacity for each compound was reduced by approximately 50% for admixtures versus individual systems. Figure 6–16d shows a significant point in competitive adsorption; namely, there is a decrease in the concentration of compounds 2, 3, 4, and 5 as the C_{eq} is increased. That is, there is an apparent decrease in the adsorbed quantities of compounds 2, 3, 4, and 5. It appears that nitrobenzene is interfering with the adsorption of the other four compounds. Data to support this statement are given in Table 6–34 where nitrobenzene has the highest adsorptive capacity.

Competitive adsorption of 2,4-DCP and 2,4,6-trichlorophenol (TCP) with humic substances on GAC (60 × 80 mesh, bituminous-based activated carbon) was investigated in equilibrium studies (12 days) [81]. Here, again, the Langmuir model (Equation 6.4) was able to describe and to predict the competitive effects

Table 6–34 Some Adsorptive Characteristics of Organic Pollutants

Compound	Molecular Weight	Langmuir Equation Constants X_m (mg/mg)	b (l/mg)
Pyridine	79.10	.055	.154
2-methylpyridine	93.13	.106	.262
o-cresol	108.14	.240	.417
2-chlorophenol	128.56	.272	.405
Nitrobenzene	123.11	.310	.230
o-methoxyphenol	124.15	.296	.400
Quinaldine	143.19	.296	.400

Source: Reproduced from Martin and Al-Bahrani [80], courtesy of Pergamon Press.

with a reasonable degree of accuracy. Three X values were developed: X_{obsd} = the observed quantity of solute adsorbed per gram of activated carbon at a particular equilibrium concentration (C_{eq}), X_{ss} = the single-solute X_m value taken from the appropriate isotherm, and X_c = surface concentrations expected for each adsorbate as a function of the C_{eq} of each adsorbate. When $X_{ss} < X_c$, competition has occurred. When $X_{obsd} \rightarrow X_c$, the Langmuir model is predicting the competitive adsorption with accuracy.

Table 6–35 summarizes the chlorophenol competitive studies [81]. At pH 5.2, DCP and TCP are primarily undissociated. The reduction in capacity for DCP (X_{obsd}/X_{ss}) ranged from 34 to 77%, whereas for TCP the reduction was 12 to 27%. At pH 7, these reductions were 15 to 31% for DCP and 4 to 47% for TCP. At this pH value, DCP is 88% associated and TCP is 91% anionic. At pH 9.1, the reductions are 6 to 24% for DCP and 37 to 73% for TCP. Here, the anionic DCP has a greater effect on anionic TCP than the reverse. At pH 5.2, the humic substances caused significant reductions in the carbon's capacity to adsorb TCP. Overall, reductions ranged from 32 to 62%. Leaf fulvic acid was the most effective inhibitor. At pH 9.0, reductions were 21 to 64%, and commercial humic acids and leaf fulvic acids were greater inhibitors than soil fulvic acid. This reversal of inhibition was due to the effect of $[H_3^+O]$ on adsorption of the humic substances. For example, at pH 5.2, 48% of the soil fulvic acid was adsorbed, whereas at pH 9.1 only 14% was removed. In any event, humic substances do interfere with adsorption of chlorophenols and certainly other organic contaminants.

Much research has been conducted on the virtues of the Freundlich and Langmuir models to describe a particular set of adsorption data. Many researchers attempt to force their data into one or the other equation. Rarely does the data fit both equations, but there are exceptions [65]. A Freundlich-type multi-

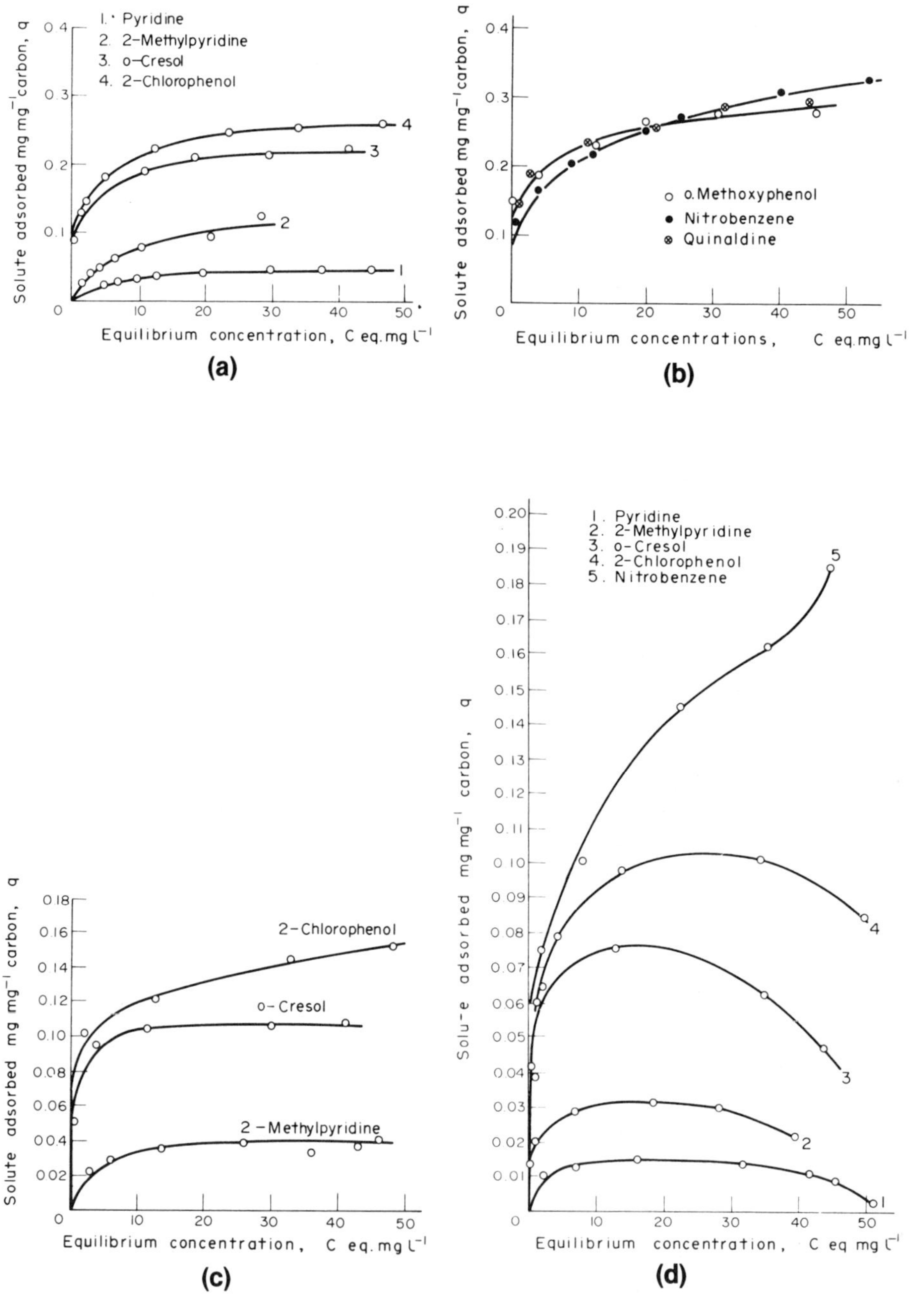

Figure 6–16 (a) and (b) Adsorption isotherms. (c) and (d) Adsorption isotherms of individual solutes in admixture. Reproduced from Martin and Al-Bahrani [80], courtesy of Pergamon Press.

component isotherm equation has been developed to describe adsorption data of various bicomponent systems [82,83]. The isothermal equation for component i in a k-component system is:

$$q_i = K_i C_i \left(\sum_{j=1}^{k} a_{ij} C_j \right)^{(n_i - 1)} \tag{6.7}$$

This equation applies to systems in which each component obeys the Freundlich isotherm:

$$q_i = K_i C_i^{n_i}. \tag{6.8}$$

K_i and n_i are determined from the monocomponent systems. The competitive coefficients, a_{ij}, describe inhibition to adsorption of component i by component j. They can be determined from thermodynamic data or experimentally from bicomponent systems. "The proposed isotherm is clearly superior to the commonly employed competitive Langmuir isotherm when the components follow the Freundlich isotherm since fewer parameters are required for the Freundlich isotherm than for the multiscction Langmuir isotherm" [82]. This statement is interpreted to mean the Freundlich equation was fitted to their adsorption data better than the Langmuir equation. So it is with carbon adsorption.

The proposed isotherm was employed to describe the adsorption data of phenol, PBP, and BS on GAC (F-400, mesh unknown) from the respective bicomponent and tricomponent aqueous solutions (1 week for equilibrium) [82]. The bicomponent equation can be written:

$$\frac{C_1}{C_2} = \frac{1}{C_2}\beta_1 - a_{12}\,; \qquad \frac{C_2}{C_1} = \frac{1}{C_1}\beta_2 - a_{21}\,, \tag{6.9}$$

with

$$\beta_i = \left(\frac{K_i C_i}{q_i} \right)^{1/(1 - n_i)}, \tag{6.10}$$

where the C terms are equilibrium concentrations and β and a are competition coefficients that are determined experimentally.

Isotherm constants for these three compounds (mg/l contents) are:

i	*Solute*	K_i	n_i	e_{i1}	a_{i2}	a_{i3}
1	Phenol	.037	.371	1.0	9.7	0.26
2	PBP	.130	.225	0.1	1.0	0.025
3	BS	.013	.352	3.85	42	1.0

Table 6–35 Summary of Chlorophenol Competitive Adsorption

	2,4-Dichlorophenol			2,4,6-Trichlorophenol		
	Bottle	X_{obsd}/X_{ss}	X_{obsd}/X_c	*Bottle*	X_{obsd}/X_{ss}	X_{obsd}/X_c
Chlorophenol competition at pH 5.2[a]	1	0.66	1.07	1	0.84	1.35
($C_{eq,DCP} \approxeq 0.35\ C_{eq,TCP}$, $C_{0,DCP} = 2.70$ mg/l,	2	0.46	0.89	2	0.73	1.18
$C_{0,TCP} = 20.80$ mg/l)	3	0.44	0.96	3	0.78	1.19
	4	0.36	0.84	4	0.85	1.16
	5	0.30	1.05	5	0.88	1.03
	6	0.23	0.94	6	0.82	1.06
Chlorophenol competition at pH 7.0[a]	1	0.85	1.16	1	0.65	1.52
($C_{eq,DCP} \approxeq 0.14C_{eq,TCP}$, $C_{0,DCP} = 6.58$ mg/l,	2	0.84	1.25	2	0.63	1.28
$C_{0,TCP}$, = 10.76 mg/l)	3	0.82	1.37	3	0.71	1.13
	4	0.77	1.47	4	0.53	0.97
	5	0.74	1.89	5	0.96	0.78
	6	0.69	4.65	6	0.85	0.61
Chlorophenol competition at pH 9.1[a]	1	0.76	1.37	1	0.63	0.96
($C_{eq,DCP} \approxeq 0.21\ C_{eq,TCP}$, $C_{0,DCP} = 7.67$ mg/l,	2	0.78	1.31	2	0.57	0.99
$C_{0,TCP} = 2.89$ mg/l)	3	0.89	1.27	3	0.46	0.93
	4	0.89	1.08	4	0.35	0.91
	5	0.94	0.98	5	0.27	1.32

	Humic Substances		2,4,6-Trichlorophenol
	Type	Initial Concn, mg/l TOC	X_{obsd}/X_{ss}
2,4,6-trichlorophenol competition at pH 5.2 ($C_{0,TCP}$ = 23.80 mg/l, 10^{-2} M phosphate buffer)	Commercial humic acid	10	0.57
			0.68
		50	0.45
			0.55
	Leaf fulvic acid	10	0.48
			0.61
		50	0.38
			0.48
	Soil fulvic acid	10	0.53
			0.67
		50	0.42
			0.54
2,4,6-trichlorophenol competition at pH 9.1 ($C_{0,TCP}$ = 23.80 mg/l, 10^{-2} M phosphate buffer)	Commercial humic acid	10	0.64
			0.61
		50	0.60
			0.36
	Leaf fulvic acid	10	0.64
			0.58
		50	0.44
			0.39
	Soil fulvic acid	10	0.79
			0.74
		50	0.62
			0.56

Source: Reproduced from Murin and Snoeyink [81], courtesy of the American Chemical Society.

[a]C_{eq} = equilibrium concentration, C_0 = initial concentration, X_{obsd} = observed, X_c = calculated, X_{ss} = single-solute X.

The extent of reduction due to adsorptive competition for each component in a multicomponent system is exhibited by a plot of curves of C_{eq} values versus carbon load, *m,* for the respective monocomponent systems along with the multicomponent systems [82]. Figure 6–17 shows such plots. For this situation, BS and phenol were inhibited in the multicomponent system, whereas the strongly adsorbed PBP was almost unaffected by the presence of the other two constituents.

An important operational aspect of carbon adsorption is the influence of natural organics in raw water on the removal of specific compounds. The influence of humic substances was discussed. Another part of this problem is the adsorption of individual organics in the presence of ligin sulfonic acids (LSA) that occur in the Rhine River [84]. Using flocculated LSA (pretreated with alum) as a model mixture instead of a given raw water, competitive adsorption experiments were conducted with PNP as the second constituent (GAC was presumably used). Figure 6–18a shows several adsorption isotherms for LSA in the presence of PNP, and a displacement of the LSA isotherms as the PNP content is increased is seen. Figure 6–18b shows the influence of LSA on PNP adsorption. Here, the effect is slightly different; that is, there is a clear divergence from linearity at the highest [PNP] on the log-log plot. This divergence apparently is related somehow to the strength of the adsorption of the bisolutes. Weaker adsorbates will bend in the presence of stronger adsorbates. This is seen in Figure

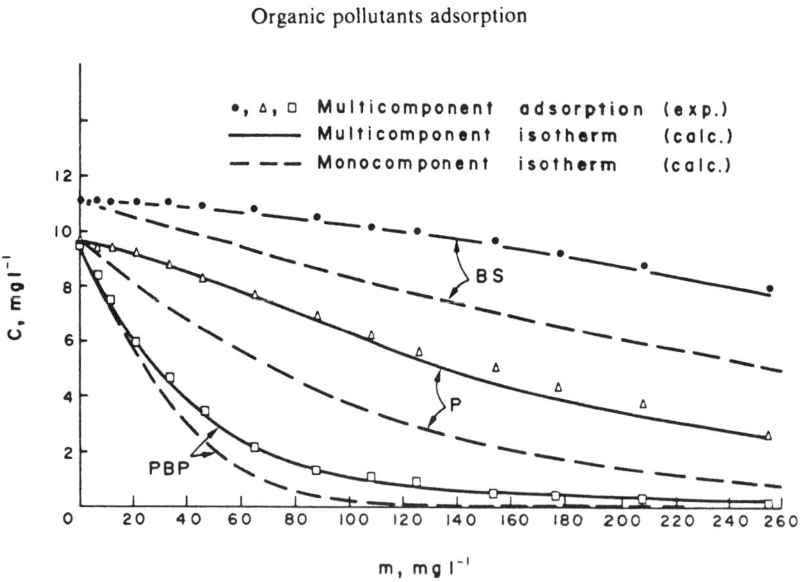

Figure 6–17 Description of adsorption in a three-component system by the proposed isotherm employing the bicomponent competition coefficients: correlation between experimental (exp.) and calculated values (calc.). Reproduced from Sheindorf et al. [82], courtesy of Pergamon Press.

6–18b where [PNP] = 10. mg/l and in Figure 6–18c where the tetrachloroethane content is 0.5 mg/l in the presence of PNP. The latter system did not contain any LSA but is representative of competition for adsorption sites that tetrachloroethane faces. The competitive effects were quantitated and interpreted with a Freundlich model [84].

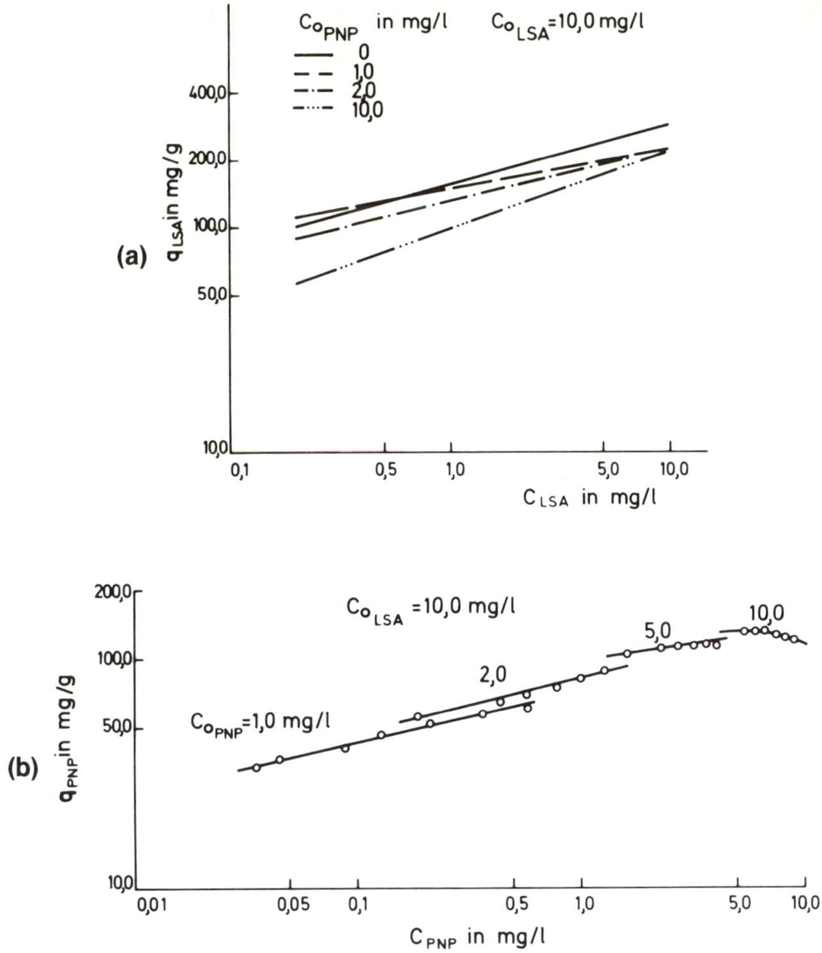

Figure 6–18 (a) Adsorption isotherms of LSA with different initial concentrations of PNP. $C_{o_{LSA}}$ of 10 mg/l is on a total weight basis and not on a DOC basis. (b) Adsorption isotherms of PNP in the presence of LSA. $C_{o_{LSA}}$ and $C_{o_{PNP}}$ are the initial concentrations of flocculated LSA and PNP, respectively, used to obtain these isotherms. (c) Adsorption isotherms for PNP and tetrachloroethane as single solutes and in mixture. $C_{o_{PNP}}$ and $C_{o_{Tetra}}$ are the initial concentrations of PNP and tetrachloroethane, respectively, used to obtain these isotherms. Reproduced from Frick et al. [84], courtesy of Ann Arbor Science Publishers, Inc.

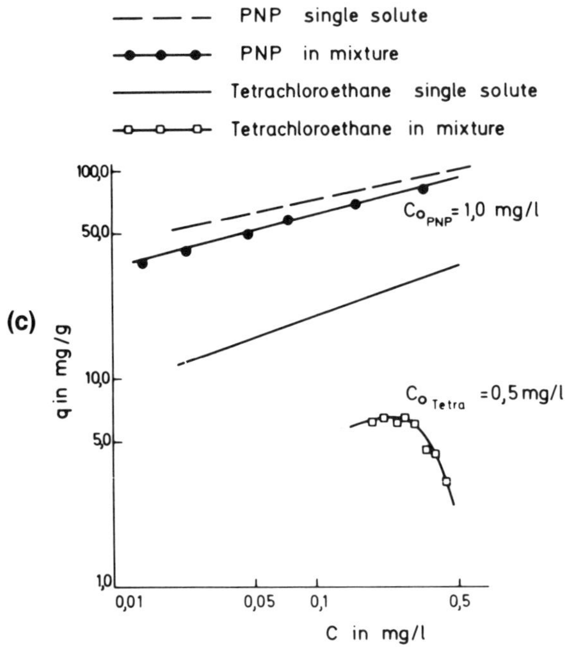

Figure 6-18 *(continued)*

Displacement effects and other phenomena due to competitive adsorption are predicted only when the unknown mixtures are treated as a multicomponent mixture [85]. It is necessary to have single-solute data (for example, Freundlich or Langmuir parameters) and the initial concentrations. Furthermore, the multicomponent organic mixture is divided into three groups that have single-solute behavior (assumed). In this way, the adsorption equilibrium should be predicted for the sum of organics and their effect on the adsorption of individual compounds in water.

The single-solute adsorbability of three components of organic matter is defined as a nonadsorbable fraction, a weakly adsorbable fraction, and a strongly adsorbable fraction [85]. Figure 6–19a gives the experimental work that is required to obtain the necessary data. Point A represents the nonadsorbable fraction, point B is the weakly adsorbed substance(s), and point C is the distribution ratio between strongly and weakly adsorbed substances. It is necessary to add a tracer compound to the experimental system as a reference for the strength of adsorption.

Figure 6–19b is an example of the application of this type of analysis to multisolute mixtures of unknown composition. Here, the measured isotherm for humic acid (as dissolved organic carbon, DOC) was calculated as a response to the simultaneous adsorption in a three-component mixture. The Freundlich con-

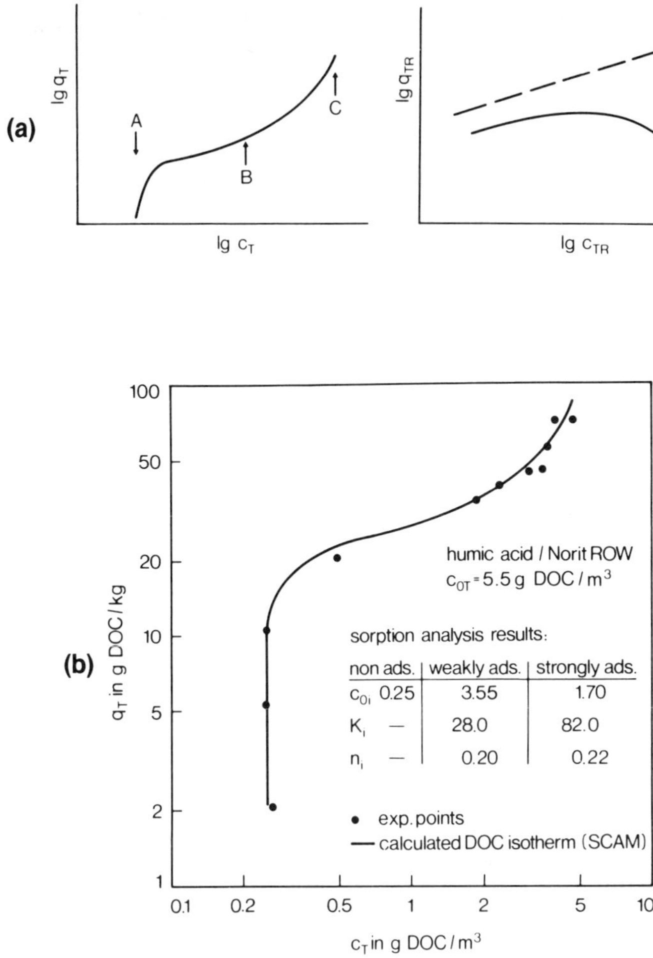

Figure 6–19 (a) Adsorption experiments as basis for the sorption analysis of multisolute mixtures of unknown composition. The plot on the top represents an overall isotherm with equal volume of the unknown mixture and varying carbon dosage (~10–1,000 mg carbon/l solution), determination of the total equilibrium concentration by DOC measurement. The plot on the bottom represents a tracer substance isotherm with equal volume of the unknown mixture + amount of the tracer substance + varying carbon dosage (~10–50 mg carbon/l solution); determination of the tracer equilibrium concentration by ultraviolet or GAC measurements. (b) Result of the sorption analysis applied to the humic acid solution. (c) *p*-nitrophenol adsorption isotherm in mixture with the humic acid referred to in Figure 6–19b. Conditions: C_oPNPh = 5.2 g DOC/m^3, and C_oHA = 5.5 g DOC/m^3. Key: --, PNPh single solute isotherm; -O-, PNPh in mixture with humic acid; and -··- PNPh isotherm calculated with SCAM. Reproduced from Frick and Sontheimer [85], courtesy of the American Chemical Society.

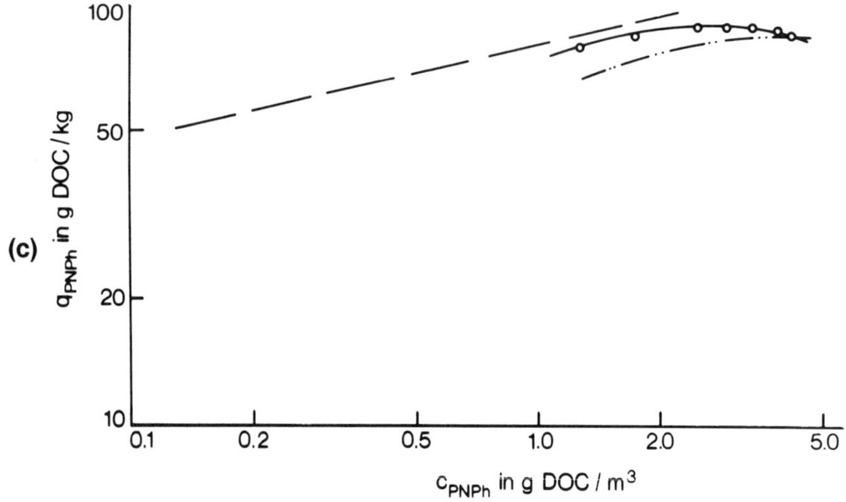

Figure 6–19 (*continued*)

stants and initial concentrations, C_o, seen in the figure were used in the predicted isotherm by a simplified competitive adsorption model (SCAM) [86,87]. Of course, one objective of this exercise is to predict competition between the organic background and the individual compounds that must be removed from water.

A comparison was made between the overall measured PNP isotherm in a mixture with humic acid (Figure 6–19b) and the calculated isotherm (Figure 6–19c). This mixture of PNP was treated mathematically as a four-component system with the three fictive components of humic acid and the single-solute isotherm (Freudlich) parameters (not given here) of the PNP. The calculated isotherm shows a strong influence of the humic acid components on the PNP adsorption and "agrees approximately with the experimental points" (Figure 6–19c) [80].

Column Studies

Granular carbon columns are predominantly employed at treatment plants for the removal of a variety of organic compounds in a competitive situation. It is desirable to have a predictive capability of these columns' (beds, filter, fixed bed) performance prior to their installation. Several mathematical models have been developed for this purpose and tested in laboratory situations [87]. One example is the competitive model based on the Langmuir equation, which is cited earlier. According to some researchers [79], a disadvantage of the Langmuir competitive model is its inability to describe all competitive equilibrium data. An alternative model is the Ideal Adsorbed Solution (IAS) model that was developed originally by Myers and Prausnitz [88] and by Radke and Prausnitz [89]. However, the IAS model had to be tested in mixtures more complex than two-component systems. A simplified model was developed empirically for multicomponent systems [90] and was compared to the IAS model for equilibrium situations [91].

In the IAS model, the spreading pressure, π_i, is calculated from single-solute adsorption equilibria [87]:

$$\pi_i = \frac{RT}{A} \int_0^{c_o} q_i^o \frac{dc_i^o}{c_i^o} , \tag{6.11}$$

where R = ideal gas law constant, T = absolute temperature, A = surface area per unit weight, q_i^o= solid-phase loading, and c_i^o= the fluid-phase concentration. Adsorption isotherms that are described by the same $q_i^o = f_1 c_i^o$, are described also by the same $\pi_i = f_2 c_i^o$ and $\pi_i = f_3 q_i^o$. From the IAS model:

$$c_i\,(\pi, T) = z_i\, c_i^o(\pi, T) \; . \tag{6.12}$$

Equation 6.12 states that "the liquid-phase concentration in a mixture, c_i, is given by multiplication of the corresponding single-solute concentration at the same spreading pressure and temperature as the mixture, by c_i, the mole fraction surface coverage of component i in the mixture system" [87]. Where there are identical isotherms, the spreading pressures in the mixture and single-solute systems are equal when all values of $c_i^o$are identical and if:

$$q_T = q_i^o , \tag{6.13}$$

where q_T = the total amount that is adsorbed in the mixture. The identity relationship:

$$\Sigma z_i = 1 \tag{6.14}$$

is combined with Equation 6.12 to give:

$$c_i^o = \Sigma c_i \; . \tag{6.15}$$

The single-solute equilibrium data are described (assumed) by the Freundlich equation:

$$q_i^o = K c_i^{o\,1/n} \; . \tag{6.16}$$

Substitution of q_T for q_i and Σc_i for c_i yields:

$$q_T = K(\Sigma c_i)^{1/n} \; . \tag{6.17}$$

This equation relates the total load in the sorbed-phase to liquid-phase concentrations of each component in the mixture. Individual loadings can be computed also since the ratio of the solid-phase loadings can be expressed from Equation 6.12:

$$\frac{q_i}{q_{i+1}} = \frac{c_i}{c_{i+1}} \qquad (6.18)$$

$$\vdots$$

$$\frac{q_i}{q_N} = \frac{c_i}{c_N}$$

and

$$q_i = Kc_i\,(\Sigma c_i)^{(1/n\,-\,1)}. \qquad (6.19)$$

Another simplification of Baldauf's [86] model is made at this point, which permits handling of the more realistic case in which single-solute isotherms are not identical. First, hypothetical solutes are defined in a manner that they all have the same isotherm equation and, therefore, the same dependence on spreading pressure. The Freundlich equation is used again:

$$\bar{q}_i^o = K'\,\bar{c}_i^{o\,1/n'}, \qquad (6.20)$$

where $\bar{c}_i^o$ = the newly defined single-solute concentration, $\bar{q}_i^o$ = the corresponding load, and $1/n'$ [author note: orignal reference [87] uses n'] and K' = the Freundlich constants that are obtained by averaging the values obtained for each solute. Redefining the solutes in this manner permits simplification of the IAS model:

$$\bar{q}_T = K'(\Sigma\bar{c}_i)^{1/n'}. \qquad (6.21)$$

The hypothetical concentrations in Equation 6.21 must be expressed in terms of actual concentrations. A beginning point is when the $1/n'$ values are identical [90], whereupon Equation 6.12 becomes:

$$\frac{q_1}{q_2} = \left[\frac{K_1}{K_2}\right]^{1/n} \frac{c_1}{c_2} = \frac{q_1^{o\,1/n}}{q_2^{o\,1/n}}, \qquad (6.22)$$

where q_1^o and q_2^o must be evaluated at c_1 and c_2 of the mixture. Assuming similar behavior of the hypothetical solutes:

$$\frac{\bar{q}_1}{\bar{q}_2} = \frac{q_1}{q_2}, \qquad (6.23)$$

it follows that:

$$\frac{\bar{q}_1}{\bar{q}_2} = \frac{\bar{q}_1^{o\ 1/n}}{\bar{q}_2^{o\ 1/n}} = \frac{q_1^{o\ 1/n}}{q_2^{o\ 1/n}} . \tag{6.24}$$

Substitution of the appropriate Freundlich equations for the q terms in Equation 6.24 leads to:

$$\bar{c}_i = \left(\frac{K_i}{K'} c_i^{1/n_i} \right)^{1/n'} . \tag{6.25}$$

This equation relates the hypothetical solute concentrations to the actual solute concentrations. An analogous approach can be used for individual loads similar to Equation 6.19:

$$q_i = K'^{\left(\frac{1/n' - 1}{1/n'}\right)} (K_i c_i^{1/n_i})^{1/n'} \left[\sum_N \left(\frac{K_i}{K'} c_i^{1/n_i} \right)^{1/n'} \right]^{1/n' - 1} . \tag{6.26}$$

This simplified model agrees exactly with the IAS model when the adsorption of solutes is described by either identical Freundlich equations or the same equations with equal n values. In addition, this model "also allows for application of concepts of the IAS theory in describing competitive adsorption when the n values are unequal" [87].

The simplified model shown here has been compared with the IAS model for these systems: PNP/PCP, PNP/benzoic acid (BA), and PNP/phenol [87,90]. Previous work demonstrated that single-solute isotherms were not described by a single set of Freundlich constants. Agreement between the two models is seen by examination of the high (10^{-4} to 10^{-3} M) and low (10^{-5} and 10^{-4} M) concentration ranges for the case of PNP/phenol competition in Figures 6–20a and b, respectively. Simplified model results lie above values from the IAS model in the high content range and below at low concentrations. According to the researchers, "A point of exact agreement between the two models is implied. Within reasonable proximity to this point, the simplified model gives a good approximation" [87].

The IAS model was employed to describe the competitive equilibrium cited earlier [84]. Several mathematical models of fixed bed adsorption were tested wherein the only difference between them was the assumption of the rate-limiting step(s) of mass transfer. The simplest situation was controlled by external film transfer (see also Chapter 3). Here, the driving force for adsorption is determined from:

$$\frac{dq_i}{dt} = \frac{ka}{\rho_p} (C_i - C_i^*), \tag{6.27}$$

where k is the mass transfer coefficient (cm/sec), a = the external surface area of the particle per unit volume (cm^2/cm^3), and ρ_p = the density of the particle

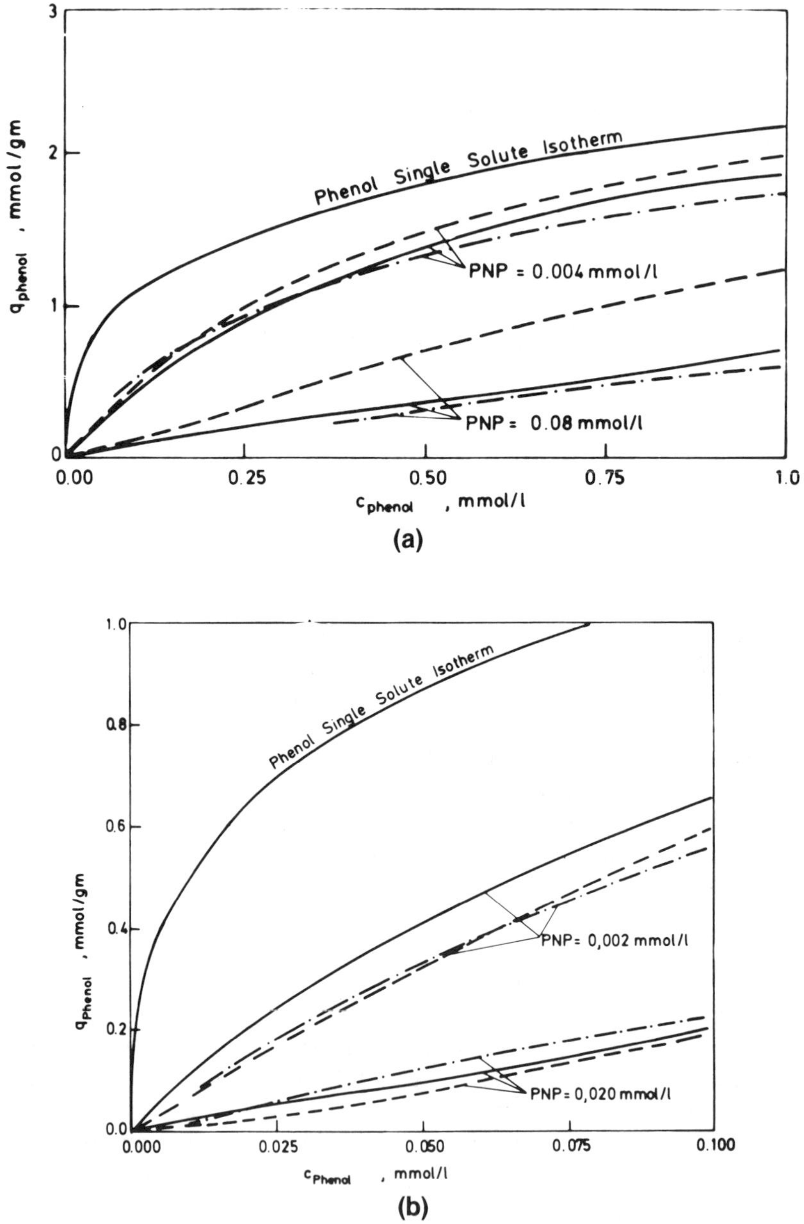

Figure 6–20 (a) Adsorption equilibria of phenol in the presence of PNP; -·-, interpolated line of constant PNP concentration from experimental data;——, IAS model; - - -, simplified model. (b) Adsorption equilibria of phenol in the presence of PNP; -·-, interpolated line of constant PNP concentration from experimental data; ——, IAS model; ---, simplified model. Reporduced from DiGiano et al. [87], courtesy of Ann Arbor Science Publishers, Inc.

(g/cm^3). C_i is the fluid-phase concentration internal to the film (mol/cm^3), and C_i^* is the fluid-phase concentration (mol/cm^3), which would be in equilibrium with the amount adsorbed on the solid phase, q_i (g/g) [84]. The value of C_i^* is found from the competitive equilibrium model. Within the sorber bed at any given point, there is a wide range of equilibrium. Consequently, the simplified IAS model was given a demanding test [87].

That external film transfer was rate limiting in a low concentration range for several bisolute mixtures was reported by Frick et al. [84, p. 214]. The three major characteristics of breakthrough behavior are:

1. peak concentration of the more weakly adsorbed component, which is caused by the displacement effect,
2. breakthrough time of the more weakly adsorbed component, and
3. the separation factor, which is measured arbitrarily as the time between appearance of 20% of the feed concentration of each component in the effluent. [87]

These characteristics are a function of dimensionless bed length that relates to bed contact time. The system constants are seen in Table 6–36.

Figure 6–21a shows the breakthrough curves at a specific bed length that were obtained from two equilibrium models for PNP/BA systems [87]. This bisolute system had produced the greatest difference between the two models in

Table 6–36 System Constants for Fixed Bed Sorber Calculations

Flowrate (Q)	= 5 l/hr
Bed diameter	= 3.4 cm
Application rate	= 5.5 m/hr
Particle diameter (D)	= 0.132 cm
Particle density (ρ_s)	= 0.59 g/cm^3
Weight of carbon/length of bed (W/L)	= 3.5 g/cm
Packed bed density	= 0.385 g/cm^3
Bed porosity	= 0.35
Free liquid diffusivity	= 0.9×10^{-5} cm^2/sec
Mass transfer coefficient (k)	= 2.5×10^{-3} cm/sec
Dimensionless bed length (ξ)	$= \dfrac{WK}{Q\rho_s \left(\frac{D}{2}\right) L} z$

Source: Reproduced from DiGiano et al. [87], courtesy of Ann Arbor Science Publishers, Inc.
Note: Empty bed contact time corresponding to $\xi = 1$ is 0.68 min.

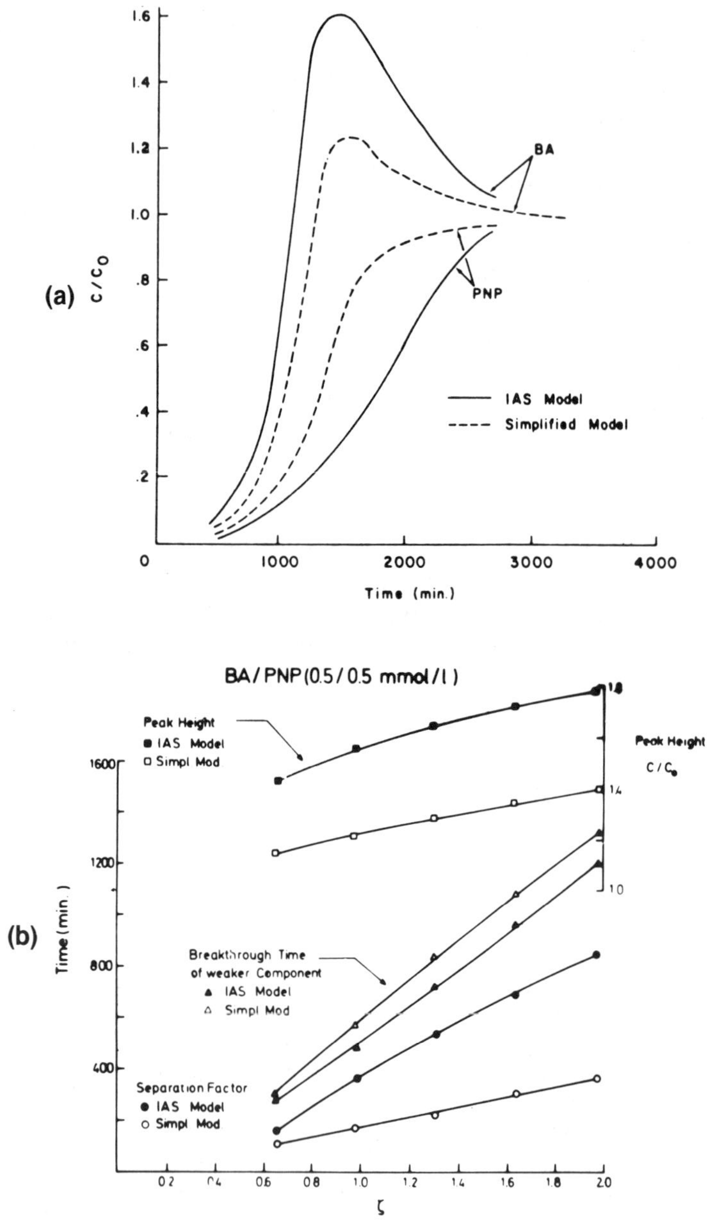

Figure 6–21 (a) Comparison of calculated breakthrough curves for PNP and BA using the IAS and simplified competitive equilibrium models as input to the fixed bed adsorber model with external film diffusion-controlled mass transport; $\xi = 1.3$. (b) Characterization of the calculated breakthrough behavior of BA (weaker component) and PNP in a fixed bed adsorber with external film diffusion-controlled mass transport and competitive adsorption equilibria described by the simplified and IAS models. Reproduced from Di-Giano et al. [87], courtesy of Ann Arbor Science Publishers, Inc.

equilibrium calculations [90]. Comparison of the breakthrough characteristics in Figure 6–21b suggests that the simplified model:

1. underestimates concentration of the displaced BA;
2. gives a reasonable estimate of breakthrough time, although longer than that predicted by the IAS model; and
3. underestimates the separation factor. [87]

The simplified model apparently reduces the difference in adsorbability of the two compounds; that is, PNP becomes a less strongly adsorbed compound whereas BA becomes more strongly adsorbed. Similar results were observed for other bisolutes: PNP/PCP, PNP/phenol, and DCP/TCP [87]. In any event, these results represent the state-of-the art in predicting breakthrough curves from equilibrium models. (Figures 6–21a and b).

Other mathematical models are available for multicomponents on fixed beds [91–94]. Their pragmatic usefulness is, however, questionable.

REFERENCES

1. Faust, S.D., and O.M. Aly. *Chemistry of Water Treatment.* Butterworths, Stoneham, Mass. (1983).
2. Suffet, I.H. *J. Amer. Water Works Assoc. 72,* 41 (1980).
3. *Drinking Water and Health,* Vol. 2. National Academy Press, Washington, D.C. (1980).
4. Sontheimer, H., and D. Maier. *Gas Wasserfach, Wasser/Abwasser 113,* 187 (1972).
5. Sontheimer, H. *Lehrstuhl fur Wasserchemic der Universitat Karlsruhe,* West Germany (1974).
6. Kölle, W., et al. *Vom Wasser 44,* 203 (1975).
7. Scheidtmann, H., et al. *Gas Wasserfach, Wasser/Abwasser 114,* 267 (1973).
8. Rook, J.J. *Water Treat. Exam. 21,* 259 (1972).
9. Roberts, P.V., and R.S. Summers. *J. Amer. Water Works Assoc. 74,* 113 (1982).
10. Dobbs, R.A., and J.M. Cohen. "Carbon Adsorption for Toxic Organics." EPA 600/8-80-023. Municipal Environmental Research Laboratory, U.S. EPA, Cincinnati, Ohio, 45268 (April, 1980).
11. Matthews, J.E. "Evaluation of the Effectiveness of Granular Activated Carbon Adsorption." EPA 600/52-81-067. Robert Kerr Environmental Research Laboratory, U.S. EPA, Ada, Okla. (1981).
12. Giusti, D.M., et al. *J. Water Pollut. Contr. Fed. 46,* 947 (1974).
13. Spahn, H., and E.U. Schünder. *Chem. Eng. Sci. 30,* 529 (1975).
14. Arbuckle, W.B. *AICHE Symp.,* no. 197, *76,* 61 (1980).
15. Fochtman, E.G. "Biodegradation and Carbon Adsorption of Carcinogenic and Hazardous Organic Compounds." EPA 600/2-81-032. Municipal Environmental Research Laboratory, U.S. EPA, Cincinnati, Ohio (1981).
16. Fochtman, E.G., and R.A. Dobbs. "Adsorption of Carcinogenic Compounds by Activated Carbon," Chapter 7 in *Activated Carbon Adsorption of Organics from the Aqueous Phase,* Vol. 1. I.H. Suffet and M.J. McGuire, eds. Ann Arbor Science Publishers, Ann Arbor, Mich. (1980), p. 157.
17. El-Dib, M.A., et al. *Water Res. 12,* 1131 (1978).
18. El-Dib, M.A., and M.I. Bradawy. *Water Res. 13,* 255 (1979).

19. Weber, W.J. Jr., and M. Pirbazari. *J. Amer. Water Works Assoc. 74,* 203 (1982).
20. Pibazari, M., and W.J. Weber, Jr. "Adsorption of Polychlorinated Biphenyls from Water by Activated Carbon," Chapter 15 in *Chemistry in Water Reuse,* Vol. 2. W.J. Cooper, ed. Ann Arbor Science Publishers, Ann Arbor, Mich. (1981), p. 309.
21. Pibazari, M., and W.J. Weber, Jr. "Adsorption of Benzene from Water by Activated Carbon," Chapter 14 in *Chemistry in Water Reuse,* Vol. 2. W.J. Cooper, ed. Ann Arbor Science Publishers, Ann Arbor, Mich. (1981), p. 285.
22. Youssefi, M., and S.D. Faust. "Adsorption and Formation of Light Halogenated Hydrocarbons and Humic Acid in Water by Granular Activated Carbon," Chapter 5 in *Activated Carbon Adsorption of Organics from the Aqueous Phase,* Vol. 1. I.H. Suffet and M.J. McGuire, eds. Ann Arbor Science Publishers, Ann Arbor, Mich. (1980), p. 133.
23. Benedict, A.L. "The Adsorption and Desorption of Trichloroethylene, Tetrachloroethylene, and Carbon Tetrachloride from Granular Activated Carbon." M.S. thesis, Department of Environmental Science, Rutgers University, New Brunswick, N.J. (1982).
24. Love, O.T. Jr., and R.G. Eilers. *J. Amer. Water Works Assoc. 74,* 413 (1982).
25. Weber, W.J. Jr., et al. Chapter 8 in *Viruses and Trace Contaminants in Water and Wastewater.* W. Redman et al., eds. Ann Arbor Science Publishers, Ann Arbor, Mich. (1977).
26. Peel, R.G., and A. Benedek. *Environ. Sci. Technol. 14,* 66 (1980).
27. Al-Bahrani, K.S., and R.J. Martin. *Water Res. 10,* 731 (1976).
28. Aly, O.M., and S.D. Faust. "The Adsorption of Chlorinated Phenols by Activated Carbon." Presented at the Kendall Award Symposium, American Chemical Society, 163rd National Meeting, Boston (April 9, 1972).
29. Suffet, I.H. "An Evaluation of Activated Carbon for Drinking Water Treatment," Chapter 4 in *Drinking Water & Health,* Vol. 2. National Academy Press, Washington, D.C. (1980), p. 251.
30. Weber, W. J. Jr. *Ind. Water Eng.* July/August, 19 (1981).
31. Aly, O.M., and S.D. Faust. *J. Amer. Water Works Assoc. 57,* 221 (1965).
32. Zogorski, J.S., and S.D. Faust. "Equilibria of Adsorption of Phenols by Granular Activated Carbon," Chapter 9 in *Chemistry of Wastewater Technology.* A.J. Rubin, ed. Ann Arbor Science Publishers, Ann Arbor, Mich. (1978), p. 143.
33. Snoeyink, V.L., et al. *Environ. Sci. Technol. 3,* 918 (1969).
34. Singer, P.C., and C. Yen. "Adsorption of Alkylphenols by Activated Carbon," Chapter 8 in *Activated Carbon Adsorption of Organics from the Aqueous Phase,* vol. 1. I.H. Suffet and M.J. McGuire, eds. Ann Arbor Science Publishers, Ann Arbor, Mich. (1980), p. 167.
35. DiGiano, F.A., and W.J. Weber, Jr. "Mathematical Modeling of Sorption Kinetics in Finite and Infinite-Bath Systems." Technical Report, WP-00706. Federal Water Pollution Control Administration, U.S. Department of the Interior, Washington, D.C. (1969).
36. Weber, W.J. Jr., and M. Pirbazari. "Removal of Carbon Tetrachloride from Water by Activated Carbon," Chapter 7 in *Treatment of Water by Granular Activated Carbon.* M.J. McGuire and I.H. Suffet, eds. Advances in Chemistry Series 202. American Chemical Society, Washington, D.C. (1983), p. 121.
37. Youssefi, M. "The Determination, Formation, and Removal of Halogenated Pollutants with Emphasis on Those Derived from Chlorination of Naturally Occurring Organic Compounds in Water." Ph.D. dissertation, Department of Environmental Science, Rutgers University, New Brunswick, N.J. (1978).

38. Weil, J.D. "Aeration and Powdered Activated Carbon Adsorption for the Removal of Trihalomethanes from Drinking Water." M.E. thesis, University of Louisville, Louisville, Ky. (1979).
39. Hoehn, R.C., et al. "Chlorination and Water Treatment for Minimizing Trihalomethanes in Drinking Water," Chapter 39 in *Water Chlorination: Environmental Impact and Health,* Vol. 2. R.L. Jolley et al., eds. Ann Arbor Science Publishers, Ann Arbor, Mich. (1978), p. 519.
40. Yohe, T.L., et al. *J. Amer. Water Works Assoc. 73,* 402 (1981).
41. van Breemen, A.N., et al. *Water Res. 13,* 771 (1979).
42. Herzing, D.R., et al. *J. Amer. Water Works Assoc. 69,* 223 (1977).
43. Lee, M.C., et al. *J. Amer. Water Works Assoc. 73,* 440 (1981).
44. Weber, W.J. Jr., et al. "Potential Mechanisms for Removal of Humic Acids from Water by Activated Carbon," Chapter 16 in *Activated Carbon Adsorption of Organics from the Aqueous Phase,* Vol. 1. I.H. Suffet and M.J. McGuire, eds. Ann Arbor Science Publishers, Ann Arbor, Mich. (1980), p. 317.
45. McCreary, J.J., and V.L. Snoeyink. *Water Res. 14,* 151 (1980).
46. Randtke, S.J., and C.P. Jepsen. *J. Amer. Water Works Assoc. 73,* 411 (1981).
47. Randtke, S.J., and C.P. Jepsen. *J. Amer. Water Works Assoc. 74,* 84 (1982).
48. Quinn, J.E., and V.L. Snoeyink. *J. Amer. Water Works Assoc. 72,* 483 (1980).
49. *Water Quality and Treatment,* 3rd ed. American Water Works Association, Denver (1971).
50. Morris, J.C., and W.J. Weber, Jr. "Adsorption of Biochemically Resistant Materials from Solution." 999-WP-11. U.S. Public Health Service, Cincinnati, Ohio (1964).
51. Weber, W.J. Jr., and J.P. Gould. "Sorption of Organic Pesticides from Aqueous Solution," Chapter 23 in *Organic Pesticides in the Environment.* Advances in Chemistry Series 60. A.A. Rosen and H.F. Kraybill, eds. American Chemical Society, Washington, D.C. (1966), p. 280.
52. Huang, J.-C., and J.T. Garrett. *Water & Sew. Works 124,* 64 (1977).
53. Aly, O.M., and S.D. Faust. "Adsorption of Phenolic Compounds on Activated Carbon." Technical Report. Department of Environmental Sciences, Rutgers University, New Brunswick, N.J. (August 11, 1970).
54. Randtke, S.J., and V.L. Snoeyink. *J. Amer. Water Works Assoc. 75,* 406 (1983).
55. Zogorski, J.S., and S.D. Faust. "Water—1976: I. Physical, Chemical, Wastewater Treatment." Symp. Series 166. *73,* 54. American Institute of Chemical Engineering, New York (1977).
56. Zogorski, J.S., et al. *J. Colloid Interf. Sci. 55,* 329 (1976).
57. Rosene, M.R. "Controlling Mechanisms for Granular Activated Carbon Adsorption Columns in the Liquid Phase," Chapter 9 in *Treatment of Water by Granular Activated Carbon.* M.J. McGuire and I.H. Suffet, eds. Advances in Chemistry Series 202. American Chemical Society, Washington, D.C. (1983), p. 201.
58. Gomaa, H.M., and S.D. Faust. "Removal of Organic Pesticides from Water to Improve Quality," Chapter 15 in *Pesticides in Soil and Water.* W.D. Guenzi et al., eds. Soil Science Society of America, Madison, Wis. (1974), p. 413.
59. Carollo, J.A. *J. Amer. Water Works Assoc. 37,* 1310 (1945).
60. Cohen, J.M., et al. *J. Amer. Water Works Assoc. 52,* 1551 (1960).
61. Cohen, J.M., et al. *J. Amer. Water Works Assoc. 53,* 233 (1961).
62. Leopold, A.C., et al. *Weeds 8,* 48 (1960).
63. Roebeck, G.G., et al. *J. Amer. Water Works Assoc. 57,* 181 (1965).
64. Faust, S.D., and A. Zarins. *Residue Reviews 29,* 151 (1969).
65. El-Dib, M.A., et al. *Water Res. 9,* 795 (1975).

66. El-Dib, M.A., and O.A. Aly. *Water Res. 11,* 617 (1977).
67. Steiner J. IV, and J.E. Singley. *J. Amer. Water Works Assoc. 71,* 284 (1979).
68. Morris, J.C., and W.J. Weber, Jr. "Adsorption of Biochemically Resistant Materials from Solution." 999-WP-33. U.S. Public Health Service, Cincinnati, Ohio (1966).
69. Ishizaki, C., and J.T. Cookson, Jr. *J. Water Pollut. Cont. Fed. 45,* 515 (1973).
70. Chudyk, W.A., et al. *J. Amer. Water Works Assoc. 71,* 529 (1979).
71. Faust, S.D., and O.M. Aly. *Chemistry of Natural Waters.* Butterworths, Stoneham, Mass. (1981).
72. Borneff, J. "Elimination of Carcinogens (Excluding Haloforms) by Activated Carbon," Chapter 6 in *Activated Carbon Adsorption of Organics from the Aqueous Phase,* Vol. 1. I.H. Suffet and M.J. McGuire, eds. Ann Arbor Science Publishers, Ann Arbor, Mich. (1980), p. 145.
73. Walters, R.W., and R.G. Luthy. *Environ. Sci. Technol. 18,* 395 (1984).
74. Weber, W.J. Jr., and J.C. Morris. *J. Sanit. Eng. Div. ASCE 90(SA3),* 79 (1964).
75. Arbuckle, W.B., and R.J. Romagnoli. *Water—1979,* AICHE Symp. Series *76(197)* (1980).
76. Martin, R.J., and K.S. Al-Bahrani. *Water Res. 12,* 879 (1978).
77. Reimers, R.S., et al. "Evaluation of Organics Adsorption by Activated Carbon," Chapter 8 in *Chemistry of Wastewater Technology.* A.J. Rubin, ed. Ann Arbor Science Publishers, Ann Arbor, Mich. (1978), p. 123.
78. Weber, W.J. Jr., and J.C. Morris. *J. Amer. Water Works Assoc. 56,* 447 (1964).
79. Jain, J.S., and V.L. Snoeyink. *J. Water Pollut. Cont. Fed. 45,* 2463 (1973).
80. Martin, R.J., and K.S. Al-Bahrani. *Water Res. 11,* 991 (1977).
81. Murin, C.J., and V.L. Snoeyink. *Environ. Sci. Technol. 13,* 305 (1979).
82. Sheindorf, C., et al. *Water Res. 16,* 357 (1982).
83. Sheindorf, C., et al. *J. Colloid Interf. Sci. 79,* 136 (1981).
84. Frick, B., et al. "Predicting Competitive Adsorption Effects in Granular Activated Carbon Filters," Chapter 11 in *Activated Carbon Adsorption of Organics from the Aqueous Phase,* Vol. 1. I.H. Suffet and M.J. McGuire, eds. Ann Arbor Science Publishers, Ann Arbor, Mich. (1980), p. 229.
85. Frick, B.R., and H. Sontheimer. "Adsorption Equilibria in Multisolute Mixtures of Known and Unknown Composition," Chapter 11 in *Treatment of Water by Granular Activated Carbon.* M.J. McGuire and I.H. Suffet, eds. Advances in Chemistry Series 202. American Chemical Society, Washington, D.C. (1983), p. 247.
86. Baldauf, G., et al. *Vom Wasser 49,* 315 (1977).
87. DiGiano, F.A., et al. "Simplifying the Description of Competitive Adsorption for Practical Application in Water Treatment," Chapter 10 in *Activated Carbon Adsorption of Organics from the Aqueous Phase,* Vol. 1. I.H. Suffet and M.J. McGuire, eds. Ann Arbor Science Publishers, Ann Arbor, Mich. (1980), p. 213.
88. Myers, A.L., and J.M. Prausnitz. *Amer. Inst. Chem. Eng. J. 11,* 121 (1965).
89. Radke, C.J., and J.M. Prausnitz. *Amer. Inst. Chem. Eng. J. 18,* 761 (1972).
90. DiGiano, F.A., et al. *Chem. Eng. Sci. 33,* 1667 (1978).
91. Tien, C., et al. *Amer. Inst. Chem. Eng. J. 22,* 498 (1976).
92. Hsieh, J.S.C., et al. *Amer. Inst. Chem. Eng. J. 23,* 263 (1977).
93. Wang, S.C., and C. Tien. *Amer. Inst. Chem. Eng. J. 28,* 565 (1982).
94. Balzli, M.W., et al. *Trans. Inst. Chem. Eng. 56,* 145 (1978).
95. McKay, G., et al. *Water Res. 19,* 491 (1985).

7

Removal of Inorganic Compounds

The nearly ubiquitous distribution in natural waters of inorganic constituents such as Ag, As, Ba, Cd, Cr, Cu, F, Hg, H_2S, NO_3^-, Pb, radionuclides, Se, and Zn has created concern about their physiological and other effects on humans [1]. Additional sources of these inorganics would be wastewaters of various types. Chlorine and chlorine-containing compounds are included in this chapter for treatment situations where their contents must be lowered before entry into a drinking water distribution system. Most of these inorganics have maximum contaminant levels (MCL) for potable waters whereupon treatment is necessary when these levels are exceeded [1]. Many treatment processes are reported elsewhere [2]. For a description of the geologic sources, typical contents in natural waters, and details of their aqueous chemistries, see Faust and Aly [2,3].

ARSENIC

Aqueous Chemistry

Arsenical compounds are among the most widely distributed elements in the earth's crust and the biosphere. The concern about arsenic in the environment is given in Faust and Aly [2]. For the most part, arsenic occurs in an inorganic form in aquatic environments and results from the dissolution of solid phases such as $As_2O_{3(s)}$ (arsenolite), $As_2O_{5(s)}$ (arsenic anhydride), and $AsS_{2(s)}$ (realgar). Two weak acids may occur in the water phase: $HAsO_2$ (arsenious acid, (III) oxidation state) and H_3AsO_4 (arsenic acid, (V) oxidation state). Since the solubilities of the oxides are high, arsenic occurs in the anionic form, which is dependent on the pH value of the water. There are environments of pH values and redox values where reduced oxidation states and methylated forms of arsenic occur: $CH_3H_2AsO_3$ (monomethylarsonic acid), $(CH_3)_2HAsO_2$ (dimethylarsinic acid), $(CH_3)_2HAs$ (dimethylarsine), $(CH_3)_3As$ (trimethylarsine), and AsH_3 (arsine). These arsenical compounds, however, are not stable in the oxidizing conditions of water treatment plants. Faust and Aly [3] give more information about the environmental chemistry of arsenic.

Removal

Inasmuch as most arsenical compounds occur in nature in anionic forms, removal by activated carbon is not expected to be effective. For example, jar-test studies indicated that PAC was ineffective for removal of either As(III) or (V). Less than 3% reductions were obtained for solutions of 0.5 mg/l of As(V) using 100. mg/l PAC (Nuchar). For As(III), 0.5 mg/l, the removals were less than 8% with PAC doses up to 300. mg/l. Also, removals were not influenced by $[H_3^+O]$ in the 10^{-6} to $10^{-8.2}$ M range [4].

Pilot plant studies were performed for removal of As(III,V), using virgin GAC (Filtrasorb 200) filters in parallel with a dual-media filter [5]. Chemical coagulation preceded the filtration portion of the pilot plant. Initial As(III) contents ranged from 0.16 to 0.54 mg/l, whereas initial As(V) contents ranged from 0.16 to 0.81 mg/l. Experimental runs of 70 to 102 hours were conducted. Percentage removals were 1. to 82. for As(III) and 14. to 96. for As(V) for the dual-media filter. The virgin GAC effected 29 to 84% removals of As(III) and 11 to 97% of As(V). Raw water sources were tap, well, Ohio River, and gravel pit. These pilot plant studies were somewhat limited; consequently, no consistent arsenic removals were obtained.

As(III) and As(V) have been removed from fresh and seawater by adsorption on alumina, bauxite, and GAC (8 × 30 mesh) [6]. Batch systems, 25°C, were shaken continuously for 48 hours for the GAC. Table 7-1 shows the efficiency of As(V) removal by GAC only. Apparently, As(III) removal was negligible and was not reported. A Langmuir adsorption isotherm was drawn for As(V) on the carbon whose parameters were q_o (monolayer coverage) = 0.01 mM/g and b = 1,580. l/mM for a concentration range of 0.01 to 0.028 mM. The removals in Table 7-1 are extremely high for an inorganic compound on carbon, but the optimum pH value of 3.1 is unlikely for a water treatment process.

Table 7-1 Efficiency of As(V) Removal by GAC

Activated Carbon (3 mg/l)	q_{100}[a]	q[b]	*Percent*	*pH*	*Arsenic Removed (mg/g of solids)*[c]
As(V)					0.34
12.8 μM	4.26	4.11	96.5	3.1	
25.4 μM	8.46	7.82	92.4	3.1	
34.7 μM	11.6	9.69	85.5	3.2	

Source: Reproduced from Gupta and Chen [6], courtesy of the Water Pollution Control Federation.
Note: Solvent matrix: water.
[a] q_{100} = adsorption density at 100% removal.
[b] q = adsorption density (μM/g of solids).
[c] At optimum pH, 50 μg/l arsenic concentration.

It would appear from these studies that activated carbon, powdered or granular, is not a particularly efficient adsorbent for arsenical compounds. No data were given for desorption of arsenic from carbon; consequently, there could be only a once-through use of this adsorbent. Other treatments are available for arsenic removal from water [2].

BARIUM

Aqueous Chemistry

This element is one inorganic whose content in potable water is limited to 1.0 mg/l [1]. Trace concentrations, 9.0 to 152. μg/l, were found in many surface waters in North America [3]. Barium contents in some groundwaters of northern Illinois have been as high as 19 mg/l, with 2. to 7.0 mg/l being a more common range [7]. Since Ba is one of the alkaline earth elements, its chemical properties are similar to Ca and Mg; that is, $BaCO_3$ and $BaSO_4$ are very insoluble in water, and $Ba(OH)_2$ is somewhat more soluble. Consequently, the occurrence of barium in natural waters would be mostly in the cationic form, Ba^{2+}.

Removal

Since Ba is removed from water much more efficiently by lime soda softening treatment [2], activated carbon studies are few. The effectiveness of three PACs (Amoco PX-21, Darco HDP, and Aqua Nuchar) for removal of 5 mg/l Ba at pH values of 7., 8., and 9 was reported [7]. With PAC quantities up to 100 mg/l, only a 7% reduction was observed at a pH value of 9. A pilot plant study suggested that no barium was removed by GAC adsorption [7].

CADMIUM

Aqueous Chemistry

This element, likewise, has an MCL for potable water of 0.01 mg/l [1]. Surface water contents of Cd in North America ranged from 21. to 130. μg/l [1]. It is not especially widespread in natural geologic sources. However, Cd may occur in substantial concentrations due to discharge of wastewaters, especially plating wastes. The aqueous chemistry of cadmium is, for the most part, dominated by Cd^{2+}, $CdCO_{3(s)}$ (otavite), and $Cd(OH)_{2(s)}$ [2,3]. Otavite has a solubility of approximately 300 μg/l for pH values greater than 10. The solubility of $Cd(OH)_{2(s)}$ is 44. μg/l at pH 10 for an aged form of this solid.

Removal

Because the occurrence of Cd in natural waters is infrequent, few studies have been conducted for its specific removal by various treatments. PAC and GAC have some capacity to remove Cd from water [8]. For example, 100 mg/l of PAC-1 (Amoco PX-21) and PAC-2 (Darco HDP) effected 29 and 36% removals, respectively, of a 0.05 mg/l Cd solution at a pH value of 7.0 [8]. At the higher pH value of 9.0, Cd removal was increased from 26 to 53% with PAC-2, whereas removal by PAC-1 was increased to 43%. Another PAC-3 (Aqua Nuchar) was ineffective with only 7 and 27% removals at pH values of 7.0 and 9.0, respectively.

Similar results were reported by the U.S. EPA's Drinking Water Research Division, with 200 mg/l quantities of PAC-3 applied to 0.02 and 0.10 mg/l solutions of Cd at pH values of 8.1 to 8.3 [8]. Removals were in the 20 to 30% range. Slightly higher reductions were observed with 0.03 mg/l Cd in Ohio River water in the pH 7.6 to 8.4 range. Pilot plant tests were conducted also with fresh and exhausted GAC as a filter medium in parallel with a conventional dual-media filter. As seen in Table 7-2, GAC does adsorb Cd. It is interesting to note that the exhausted GAC filters removed greater quantities of Cd than the fresh GAC. The data in Table 7-2 were collected from an 8 week series at the pilot plant. There was no chemical pretreatment of the water ahead of the carbon filters.

Four activated carbons, Nuchar WV-L (8 × 30 mesh) and Aqua Nuchar, Nuchar S-A, and Darco HDC (50 × 200 mesh) were examined for their capacity to adsorb Cd, Zn, and Pb in laboratory batch systems (24 hr) [9]. Freundlich and Langmuir isotherms were employed to describe the adsorption data for free Cd^{2+} and Cd^{2+} complexed with ethylenediaminetetraacetic acid (EDTA) and

Table 7-2 Results of Pilot Tests on Removal of Cadmium by GAC

Test Number	*Length of Test Run (hr)*	*Average Raw Water Concentration (mg/l)*[a]	*Average Percentage Removed by GAC Filters*[c]	
			Fresh GAC[b]	*Exhausted GAC*[c]
1	103	0.028	22	54
2	104	0.027	18	48
3	104	0.032	19	43
4	81	0.031	8	26
5	101	0.027	30	40
6	102	0.029	29	34
7	102	0.029	10	34
8	101	0.029	7	18

Source: Reproduced from Sorg et al. [8], courtesy of the American Water Works Association.
[a]Spiked Cincinnati tap water.
[b]Filtrasorb 100, Calgon Corp., Pittsburgh, PA.
[c]Filtrasorb 200, Calgon Corp., Pittsburgh, PA.

Table 7-3 Summary of Experimental Adsorption Isotherm Parameters for Nuchar WV-L PAC

		Langmuir		*Freundlich*		
Adsorbate	*pH*	X_m	K^a	k^b	1/n	*Ligand: Cd Ratio*
Cadmium	5.7	6	5	1.6	.34	0
Cadmium	7.1	—	—	3.7	.54	0
Cadmium	8.1	247	37	8.4	.78	0
EDTA	7.1	14	5	3.3	.41	—
Cd-EDTA	7.1	—	—	3.2	.50	0.1
Cd-EDTA	7.1	14	7	3.7	.33	0.5
Cd-EDTA	7.1	5	4	1.8	.24	1.0
1,10-phenanthroline	7.1	1,131	12	57.1	.83	—
Cd-phenanthroline	7.1	544	12	29.8	.85	1.0
Cd-phenanthroline	8.1	684	16	18.7	1.03	1.0

Source: Reproduced from Rubin and Mercer [9], courtesy of Ann Arbor Science Publishers, Inc.
Note: Range of initial concentrations: 50–500 mg/l activated carbon; 5–50 μM cadmium; 1–50 μM EDTA; and 30–90 μM phenanthroline. X_m is in μmol/g, and K is in μM.
[a] $K = 1/b$ (see Chapter 1).
[b] $k = \mu M/g$ when $C_e = 1\ \mu M$.

1,10-phenanthroline. The appropriate isothermal parameters from the two equations are given in Table 7–3. Typical isotherms are seen in Figures 7–1a and b for free Cd^{2+} at three pH values and EDTA-complexed Cd.

The effect of pH value is obvious from Figure 7–1a for adsorption of free Cd^{2+} on the GAC, Nuchar WV-L. This effect may be explained, perhaps, by the formation of $Cd(OH)_{2(s)}$ at the highest pH value of 8.1 or by a greater negative charge on the surface of the carbon at pH 8.1. Some isoelectric point pH data were offered to support the latter explanation. The two complexing agents apparently did not enhance Cd adsorption to a large extent. This finding was in disagreement with previous reports [10]. These experiments do, indeed, demonstrate Cd adsorption by PAC and GAC. However, the quantities, 500 mg/l, required for significant removal may be beyond practical application in a treatment situation.

Three commercial GACs (presumably, no mesh sizes were given), Filtrasorb 400, Nuchar C-190-N, and Nuchar 722, were evaluated in laboratory batch systems (20 hr) for their capacity to adsorb Cd(II). Here again, the experimental quantity of carbon, 1 g/l, is unrealistic for practical operation. However, the adsorbability of Cd on these carbons was demonstrated. This adsorption study was supplemented with extensive surface charge measurements of several carbons (see Chapter 5 for these results). Figure 7–2a shows the effects of carbon type and $[H_3^+O]$ on Cd removal from water [11]. Here, again, the 10. g/l activated carbon quantity is somewhat too high for routine water treatment. In any event, Cd is adsorbed to a small extent.

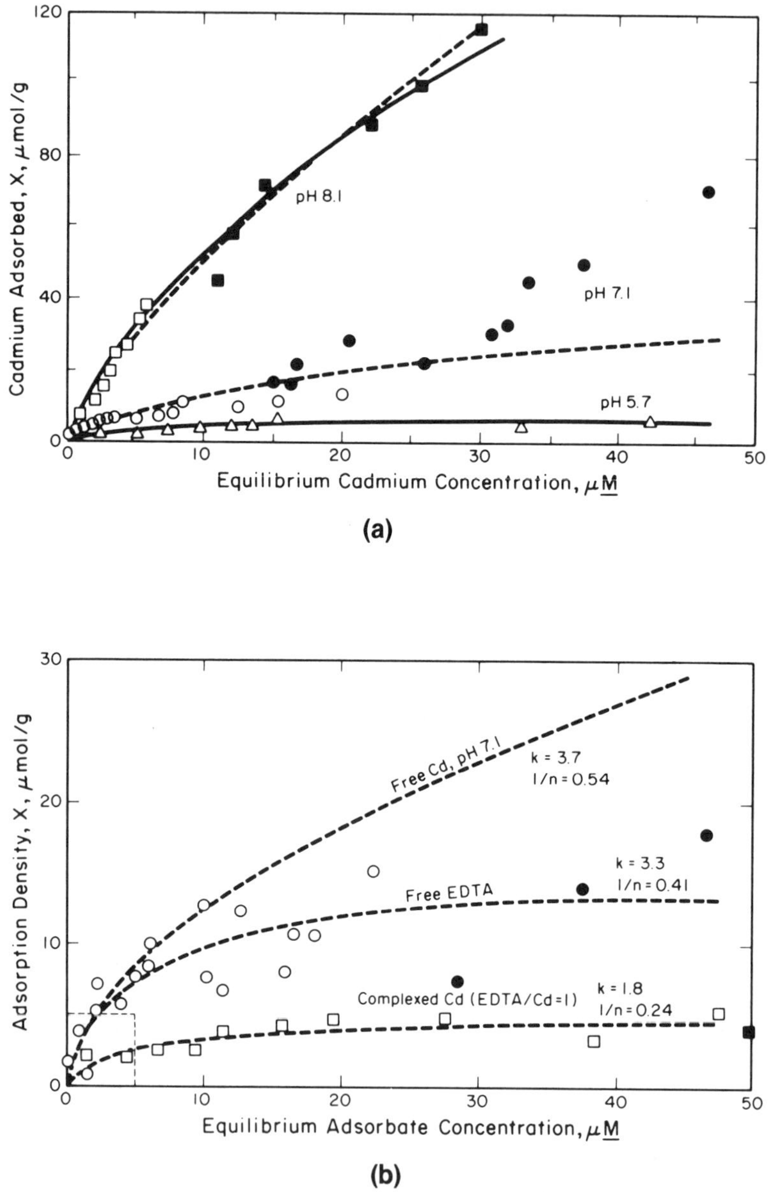

Figure 7-1 (a) Adsorption of cadmium on Nuchar WV-L at pH 5.7, 7.1, and 8.1. Solid line is calculated Langmuir isotherm; broken lines are calculated Freundlich isotherms. Carbon dose: 500 mg/1 (open symbols) and <500 mg/1 (solid symbols). (b) Adsorption of free EDTA and complexed cadmium by Nuchar WV-L at pH 7.1. Circles are EDTA; squares are complexed cadmium. Carbon dose: 500 mg/1 (open symbols) and <500 mg/1 (solid symbols). Reproduced from Rubin and Mercer [9], courtesy of Ann Arbor Science Publishers, Inc.

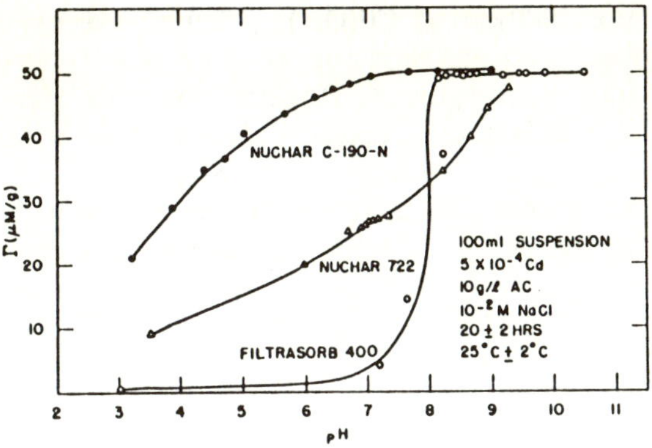

(a)

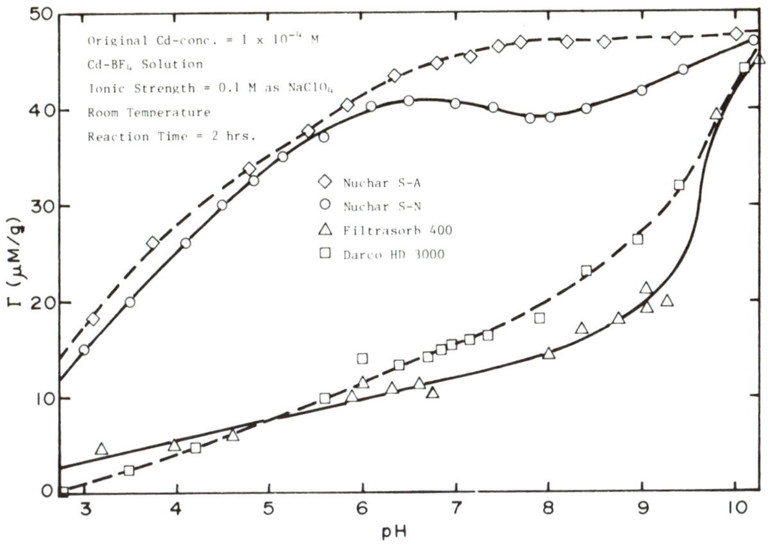

(b)

Figure 7–2 (a) Effect of carbon type on cadmium(II) removal. Reproduced from Huang and Ostovic [11], courtesy of the American Society of Civil Engineers. (b) Comparison of Cd(II) adsorption capacity by GAC and PAC, as affected by pH. Reproduced from Huang and Smith [12], courtesy of Ann Arbor Science Publishers, Inc.

Two PACs (Nuchar S-N and Nuchar S-A) and two GACs (Darco HD 3000 and Filtrasorb 400) were examined in the laboratory for Cd removal from plating wastes [12]. Figure 7–2b shows the adsorptive capacity of four carbons, 1. g/l, over the pH range of 3. to 10. These data were obtained from batch experiments with a reaction time of 2 hours. The figure shows that PAC is more effective than GAC. That the $[H_3^+O]$ affects the adsorption of Cd is seen also. The convergence of the four curves to the maximum adsorption at pH 10. is explained by "substantial precipitation of $Cd(OH)_{2(s)}$ and/or $CdCO_{3(s)}$" [12]. Typical Langmuir adsorption isotherms were obtained (not shown) for two carbons and for Cd-BF_4 and $Cd(CN)_2$ components of plating solutions. Here again, the PAC was more effective. In this study, Langmuir adsorption isotherms and pH effects were also presented [12]. This study tends to confirm the data presented by Sorg et al. [8]. These studies were oriented toward treatment of plating wastes, strongly suggesting that GAC may not be a practical drinking water treatment for Cd removal.

CHROMIUM

Aqueous Chemistry

Total chromium has an MCL of 0.05 mg/l for drinking water [1,2]. In natural waters, Cr contents are limited by the low solubility of Cr(III) oxides. This is seen in the Cr contents of 15 rivers of North America that ranged from 0.7 μg/l (Sacramento River) to 84 μg/l (Mississippi River) [3]. There is also the possibility of contamination from discharge of industrial wastewaters. The essential aqueous chemistry of Cr is its occurrence in either the +3 or +6 oxidation state. Cr(III) occurs as a cation with $Cr(OH)_{3(s)}$ being very insoluble in water. At pH 7., its solubility has been reported to be $10^{-14.3}$ M as Cr (3.8×10^{-11} mg Cr_2O_3/l) [13].

Removal

Of the several studies of removal of Cr by PAC and GAC [14–18], some are oriented to treatment of industrial wastewaters. For example, a 10 g/l batchwise quantity of a GAC (a calcined coke) was used for Cr(VI) removal over the pH range of 2 to 10 [16]. The greatest removals occurred at pH 2, whereupon there was an exponential decrease of effectiveness to a pH value of about 6, where there was virtually no reduction in [Cr]. In another batch study, a 10,000 mg/l quantity of Filtrasorb 400 was utilized for Cr(VI) and Cr(III) removal [17]. This GAC responded somewhat differently to pH effects. Virtually no adsorption of Cr (either valence) occurred below pH 2 or above pH 8, with the maximum occurring at pH values of 5 to 6. Loosely packed columns of various quantities of Filtrasorb 400 were operated continuously at 2 gpm/ft^2 at pH 2.5 and 10^{-3} M Cr(VI) [18]. No Cr was removed within the first 50 or so bed volumes because of the number of H^+ ions required to hydrolyze the carbon surface. All this work was directed to treating industrial wastewaters and has limited application to drinking waters. Sorg [15] reported that several jar-test studies with PAC (Filtrasorb 400) dosages of 10 to 200 mg/l showed very low removals, about 10% of 0.15 mg/l Cr(VI), from pH 6.8 to 8.3. Sorg concluded [15, p. 455] "that GAC treatment probably is not a very effective method for Cr(VI) removal from drinking water with low organic concentrations." The same conclusion may be drawn for PAC.

Freundlich and Langmuir parameters have been reported for adsorption of Cr(VI) on calcined coke [16] and Filtrasorb 400 [17]. Table 7-4 gives the Freundlich values for coke systems with 20/40 and 100/200 mesh sizes. These were obtained from a continuously mixed (24 hours) batch system with 10 g/l coke and chromate concentrations ranging from 10^{-4} to 6×10^{-3} M (5.2 to 32 mg/l Cr). The effect of $[H_3^+O]$ is obvious in Table 7-4a where pH values of 6 and 7

Table 7-4a Freundlich Parameters for Adsorption of Cr(VI) on Calcined Coke

	No. 20/40		*No. 100/200*	
pH	K_F[a]	1/n	K_F	1/n
2	2.730×10^{-4}	0.342	3.213×10^{-4}	0.230
3	1.063	0.330	1.353	0.197
4	0.759	0.356	0.723	0.180
5	0.520	0.372	0.455	0.175
6	0.467	0.432	0.249	0.168
7	0.254	0.426	0.109	0.165

Source: Reproduced from Huang and Wu [16], courtesy of the Water Pollution Control Federation.

[a] $\frac{x}{m} = K_F C^{1/n}$, T in mol/g, C in mol/l, at 10^{-1} M NaCl, and 10 g/l coke.

Table 7-4b Langmuir Constants for Cr(VI) and CN Adsorption on Filtrasorb 400

		Langmuir Constants	
Adsorbate	*pH*	X_m *(mmol/g)*	K_L[a]
Cr(VI)	6.0	11.1	4.6
	6.5	8.33	2.3
	7.0	5.00	1.9
	7.5	4.17	0.4
	8.0	0.91	1.7
	3.0	0.54	0.5
CN	4.0	0.49	0.3
	8.0	0.95	1.6
	11.0	0.42	0.5

Source: Reproduced from Huang and Wu [17], courtesy of Pergamon Press Ltd.

[a]$K_L = 1/b$ (see Chapter 1).

Table 7–4c Adsorption Constants for Cr^{3+} on F-400

Langmuir		*Freundlich*	
a *(dm³/mg)*	K *(dm³/g)*	K_F *(mg/g)*	1/n
.015	.52	2.3	.44

Source: Reproduced from McKay, et al. [47], courtesy of Pergamon Press Ltd.
Note: Contact time = 2 days.

are more appropriate for drinking water treatment. In the F-400 (~18 mesh) study, the experimental conditions were 10. g/l carbon, 24 hours of continuous shaking, and 23 ± 1°C. Table 7–4b shows the Langmuir constants for Cr(VI) and CN calculated from the isotherms in Figures 7–3a and b. In this situation, F-400 was more effective for the removal of Cr(VI) from solution. For example, at pH 7.0, the X_m value was 5.0×10^{-3} M/g of F-400, whereas the K_F value was 2.54×10^{-5} M/g of coke. This is a qualitative contrast only inasmuch as the Freundlich (K_F) and Langmuir (X_m) constants cannot be compared directly. Table 7–4b gives Langmuir constants also for adsorption of CN on F-400. These data are included because this study was directed toward treatment of chromium plating wastewaters where CN is always present [17,18]. Table 7–4c gives adsorption constants for Cr^{3+} on F-400 [47].

A carbon column study was conducted for the adsorption of Cr(VI)-containing industrial wastewaters on granular F-400 at a pH value of 2.5. This, obviously, is not a practical pH value for drinking water treatment.

Several jar-test studies were conducted for the removal of Cr(VI) (0.15 mg/l) with PAC (Aqua Nuchar A) doses of 10. to 200 mg/l. Figure 7–4 shows that extremely low results were obtained in the pH range of 6.8 to 8.3. For example, a 200 mg/l dose of PAC effected a 13% removal of Cr(VI).

Pilot plant studies were examined for GAC removal of low contents of Cr(VI) from wastewaters [19,20]. Sand filtration preceded the GAC columns. Cumulative removals of Cr following GAC treatment were 96.6 and 98% in the two studies. Another pilot plant study by the U.S. EPA yielded similar results: > 96% and > 98% removal of Cr(III), 0.13 mg/l, and 10 to > 98% removal of Cr(VI), 0.14 to 1.29 mg/l [15]. For the latter, removals were, for the most part, less than 93%. A conventional chemical coagulation pilot plant was employed to treat Ohio River water and a well water spiked with various contents of Cr(III,VI). GAC filters followed a dual-media filter that, in turn, followed chemical coagulation by either alum, ferrous sulfate, or ferric sulfate [15]. It was speculated that the high removals were achieved by Cr reacting with organics in the raw water that, in turn, were removed by chemical coagulation, filtration, and GAC adsorption [15,19].

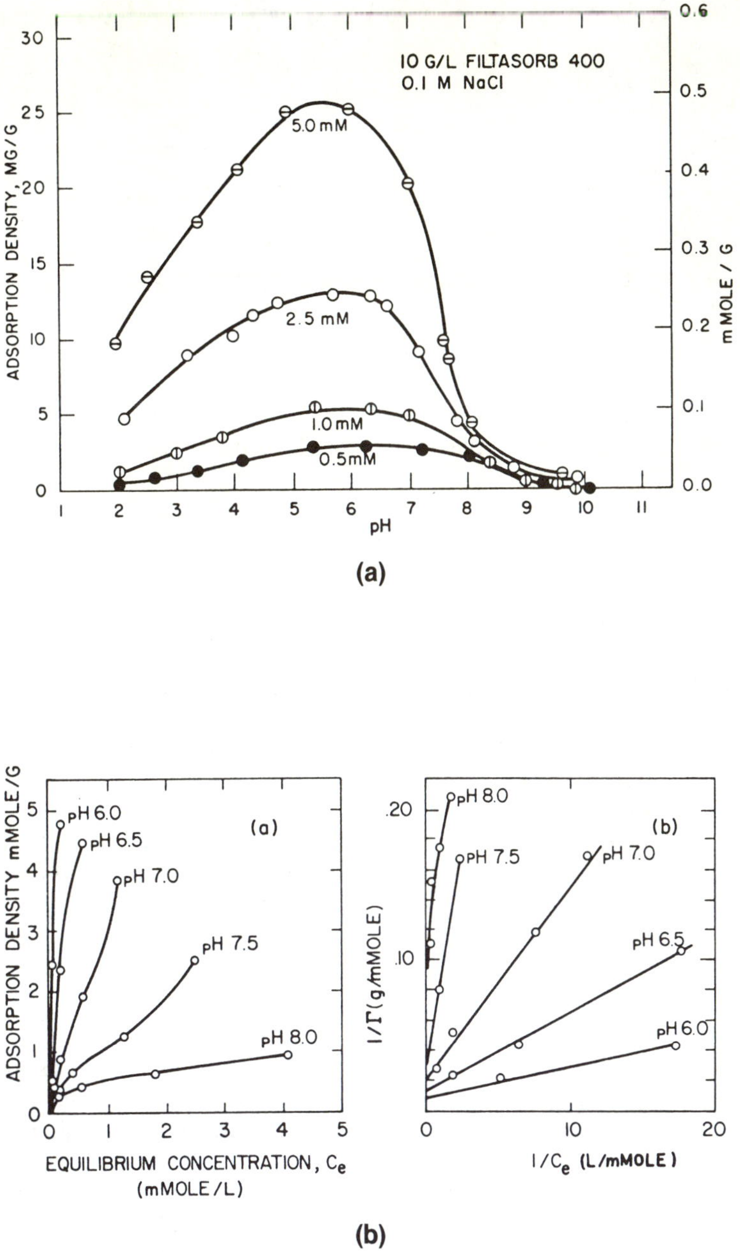

Figure 7-3 (a) The effect of pH and total Cr(VI) on the adsorption of Cr(VI). (b) The adsorption isotherm of Cr(VI). (left) A plot of adsorption density versus equilibrium Cr(VI) concentration. (right) A linearized Langmuir plot. Reproduced from Huang and Wu [17], courtesy of Pergamon Press.

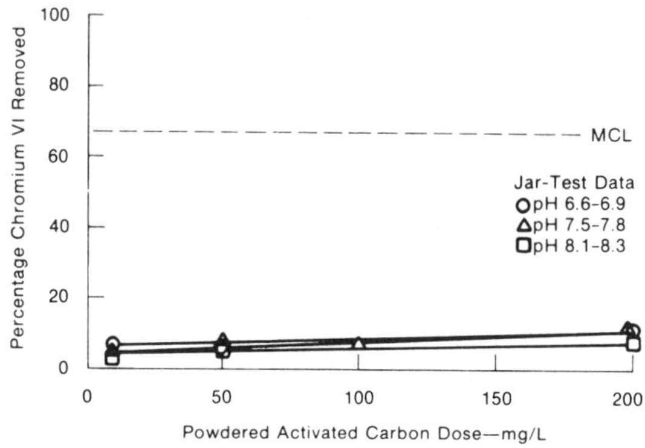

Figure 7-4 Removal of chromium(VI) (0.15 mg/l) from river water by PAC. Reproduced from Sorg [15], courtesy of the American Water Works Association.

COPPER

Aqueous Chemistry

The MCL for copper is 1.0 mg/l in drinking water, which is a secondary standard. Copper is not distributed widely in natural waters. In the famous study of 15 North American rivers [3], copper contents ranged from 0.83 to 105. μg/l. Discharge of industrial wastewaters may increase copper contents to greater than 1.0 mg/l. As cited in Faust and Aly [2], copper ions may originate from acidic corrosion of copper pipes. As high as 3.0 mg/l has been measured in household well water [2]. In these cases, neutralization of the water with calcite beds will eliminate the corrosive condition and the copper ions.

Removal

Since contamination of ground- and surface waters is rare and the drinking water standard for copper is on the secondary list, it is not surprising that very little research has been reported for its removal by activated carbon. A literature review of the chemical interactions of many inorganics and activated carbon [14] reported several Cu-activated carbon studies that have little relevance to drinking water treatment. Some were concerned with copper removal from seawater (1.65 M NaCl). Others employed chelating agents for copper in seawater followed by adsorption on activated carbon [14]. Another investigator reported adsorption of Cu(II)-amino acid complexes on granular F-400 [21]. Perhaps the lack of research on carbon adsorption for copper removal is a result of the availability of other treatments such as calcite for the neutralization of corrosive acidity and for the adsorption of copper ions [2].

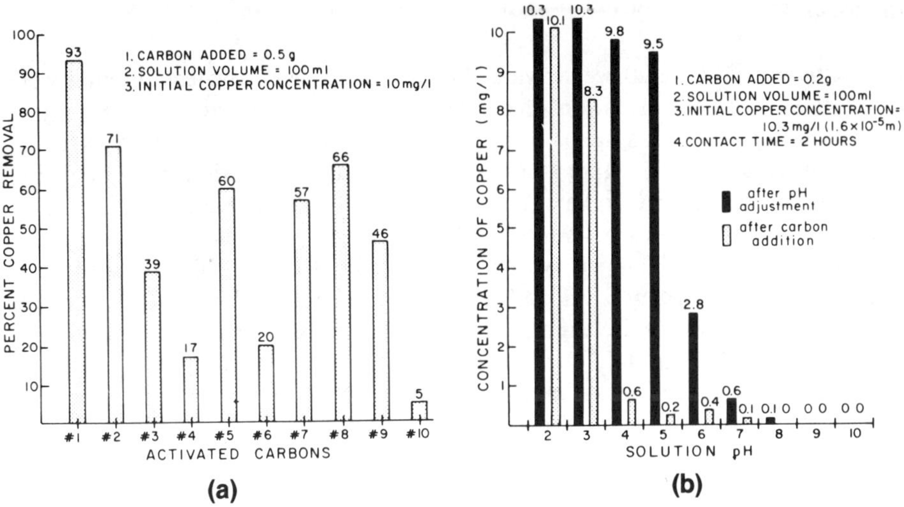

Figure 7–5 (a) Activated charcoal comparison for copper removal at pH 3.3 and contact time of 2 hours. (b) Copper concentration after pH adjustment [1] and carbon adsorption [2] versus solution pH. Reproduced from Netzer and Hughes [48], courtesy of Pergamon Press.

The adsorption of copper by several carbons was reported for nine pH values and for several contact times up to 120 min [48]. Figure 7–5a shows the results for 10 carbons (see Table 5–9). The Barney Cheney NL 1266 apparently adsorbed the most copper. Figure 7–5b shows the effect of pH on copper adsorption, where precipitation was significant above 6.0. At a pH value of 5.0, there was an approximate 98% removal by adsorption.

FLUORIDE

Health Concerns

The MCL for F^- in drinking water ranges from 1.4 to 2.4 mg/l. It is considered an essential nutrient in human metabolism [2]. Fluoride is added to many drinking waters in small quantities to prevent dental caries. Above the optimum value for this prevention, the mottling of teeth—that is, dental fluorosis—occurs. Intake of excessively large amounts of fluoride over prolonged periods of time may produce other health effects including bone changes, crippling fluorosis, and death from a single dose of 2,250 to 4,500 mg [2]. In most natural waters, fluoride contents range from μg/l to 1 to 2 mg/l. It was discovered that natural groundwaters with about 1.0 mg/l prevented dental caries. There are occasional natural waters with unusually high $[F^-]$—for example, 32 mg/l in southeastern Arizona and 67 mg/l in the Union of South Africa [3].

Removal

Four activated carbons, three activated bauxites, and two activated aluminas were investigated for defluoridation of distilled water, seawater, and simulated geothermal water [22]. All systems were continuously mixed batches with 25 g/l of the adsorbents at 25°C. Rate studies indicated that adsorption was completed within 12 hours. The researchers, however, selected a 48 hour contact time. Table 7-5 shows a summary of pH studies in which percentage removals are reported. The initial $[F^-]$ was 20 mg/l. Adsorbent 1 was Hydrodarco 3000; adsorbent 2, Filtrasorb 300; adsorbent 6, Alcoa F-1; and adsorbent 9, Milwhite. Optimum pH values in the acid range are expected for the alumina and bauxite because of the competition of OH^- and F^- for the positively charged sites. It appears that these two adsorbents can, at least, remove F^- to the MCL from a 20 mg/l content and are superior to the two GACs.

LEAD

Aqueous Chemistry

The primary drinking water standard for lead is 0.05 mg/l [1]. Natural sources of lead include $PbCO_{3(s)}$ (cerussite), $PbS_{(s)}$ (galena), and $PbSO_{4(s)}$ (anglesite). Lead contents in natural waters may range from < 1.0 to 890. μg/l [3]. The principal problem with lead in drinking water comes from corrosion in water distribution systems [2]. Other sources of lead would come from weathering of lead paints, metallic products, and industrial wastewaters [3]. The aqueous chemistry of lead is controlled by the solubilities of $Pb(OH)_{2(s)}$ (plumbous hydroxide) and $PbCO_{3(s)}$. For example, at a pH value of 10., Pb contents would be 33. μg/l from $Pb(OH)_{2(s)}$ and 26. μg/l from $PbCO_{3(s)}$. These values are below the MCL value.

Removal

Since lead is not a widespread problem in drinking water, few studies have been conducted with the removal of Pb by PAC or GAC. In laboratory jar tests it was reported that 10 mg/l PAC effected an 89% removal of 0.15 mg/l Pb and 200 mg/l PAC effected a 98% reduction at pH 7.3 to 7.4 [8]. In a parallel study using GAC (Filtrasorb 200) in a pilot plant, Pb removals were 95% and greater from 0.11 to 0.20 mg/l Pb in the raw water.

Other attempts were made to remove Pb from water by activated carbon. Several trace metals including the Pb at various pH values were removed with large quantities of PAC, 10,000 mg/l [23]. In the range of 3.0 to 5.0 pH, removals were substantial: initial Pb contents of 100. and 10. mg/l were reduced to less than 0.1 mg/l. At higher pH values, 5.0 to 10.0, the combination of pH adjust-

Table 7–5 Optimum pH_{48} for the Maximum Removal of Fluoride

	Activated Carbon				Activated Alumina (Adsorbent 6)		Activated Bauxite (Adsorbent 9)	
	Adsorbent 1		Adsorbent 2					
Background Solution	Optimum pH_{48}	Percentage Maximum Removal Efficiency	Optimum pH_{48}	Percentage Maximum Removal Efficiency	Optimum pH_{48}	Percentage Maximum Removal Efficiency	Optimum pH_{48}	Percentage Maximum Removal Efficiency
DDW[a]	6.2	84	6.0	45	4.9–8.0	95	5.5–6.8	95
Diluted (1:3) SW[b]	6.2	72	6.0	24	4.9–8.0	95	5.5–6.8	95
Diluted (1:3) SGW[c]	6.2	72	5.9	24	4.9–8.0	95	5.5–6.8	95

Source: Reproduced from Choi and Chen [22], courtesy of the American Water Works Association.
Note: pH_{48} is the pH after 48 hours of contact between adsorbent and solution.
[a]Deionized-distilled water.
[b]Fourfold dilution of seawater.
[c]Fourfold dilution of simulated geothermal water.

ment and PAC also produced very high removals (>99%). However, most of these removals were due to precipitation of $Pb(OH)_{2(s)}$ from the pH adjustment.

Table 7-6 shows the Langmuir X_m values for adsorption of Pb, Cd, and Zn on four PACs at pH values of 6.5 and 8.0 [9]. It is clear that adsorption was better at the higher pH value. Here, again, this may be due to precipitation of metallic hydroxides as well as adsorption.

The adsorption of lead by several GACs and PACs was reported for nine pH values and for contact times up to 120 minutes [48]. Figure 7-6a shows the results for lead on 10 carbons where the Barney Cheney NL 1266 GAC was the best adsorbent. Figure 7-6b shows the effect of pH on lead adsorption where precipitation is a significant factor also above 5.0. At a pH value of 4.0, there was a 95% removal of lead, apparently by adsorption.

MERCURY

Aqueous Chemistry

The MCL for mercury in drinking water is 0.002 mg/l [1]. There are very little historical data on the occurrence of Hg in ground-, surface, and finished waters in the United States. This is mainly because of the cumbersome analytical techniques for Hg in water. Mercury contents of < 0.5 to 6.8 μg/l were reported for several surface waters in the United States [3]. Of the several mercury-bearing

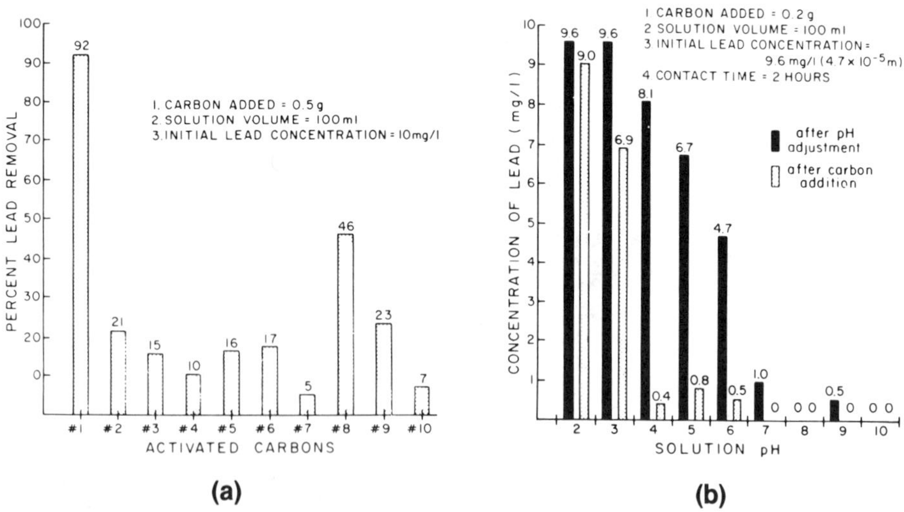

Figure 7-6 (a) Activated charcoal comparison for lead removal at pH 2.7 and contact time of 2 hours. (b) Lead concentration after pH adjustment [1] and carbon adsorption [2] versus solution pH. Reproduced from Netzer and Hughes [48], courtesy of Pergamon Press.

Table 7–6 Comparison of Metals and Powdered Carbons

Activated Carbon[a]	*Specific Surface Area (m^2/g)*	*Isoelectric Point*[b]	X_m *at pH 6.5*[c]			X_m *at pH 8.0*		
			Cd	*Zn*	*Pb*	*Cd*	*Zn*	*Pb*
Darco HDC	650	3.8	3.2	5.5	9.2	178	340	870
Nuchar WV-L	1,000	4.3	3.2	5.5	9.2	160	310	821
Aqua-Nuchar	1,000	6.2	2.0	2.7	5.6	125	220	620
Nuchar S-A	1,500	8.3	<0.3	<0.3	<0.3	<10	<10	<10

Source: Reproduced from Rubin and Mercer [9], courtesy of Ann Arbor Science Publishers, Inc.
[a]50–200 mesh, 24 hours of continuous shaking.
[b]Isoelectric pH of carbons by the titration procedure described by Parks and de Bruyn [24].
[c]X_m is in μmol/g.

minerals in nature, only a few exist abundantly. The most common are sulfides [cinnabar and metacinnabar (HgS)] and native mercury. In aqueous systems, mercury can exist in one of three oxidation states: as the free metal, Hg^0; as the mercurous ion, Hg_2^{2+}; or as the mercuric ion, Hg^{2+}. Mercury readily forms complexes with many inorganic anions and organic constituents for which methylation reactions have been documented [3]. An important characteristic of Hg is its tendency to adsorb and adhere to various types of surfaces, which has immediate application to water treatment.

Removal

Several investigations have been conducted on the utilization of PAC and GAC for Hg removal. Figure 7–7a shows the log-log plot of the adsorption data for inorganic Hg and CH_3Hg^+ [25]. These data indicate that each milligram per liter of PAC (Nuchar Aqua A) removes 0.0001 mg of inorganic Hg and CH_3Hg^+. Pilot plant studies with PACs (Darco-B and Darco-M) and GACs (Hydrodarco 1030 and Filtrasorb 100) were conducted [15]. When PAC was used in conjunction with ferric sulfate, Hg removals were greater than by either treatment alone. Cumulative removals for inorganic mercury by GAC filtration ranged from 49 to 97%, whereas organic Hg removals ranged from 80 to nearly 100% [15]. Exhausted GAC was as effective as fresh GAC. Figure 7–7b shows the effect of pH on removal of inorganic mercury by PAC (Darco HDB) [10]. Figure 7–7c shows the adsorption of CH_3Hg^+ by three different types of PACs: PAC-1, lignite base (Darco HDB); PAC-2, petroleum base (Amoco PX-21); and PAC-3, bituminous base (Nuchar Aqua) [10]. With the exception of PAC-3, methyl mercury removals were decreased as the pH value was increased from 7 to 9. PAC-1 apparently was the most effective of the three powdered carbons. At pH 7., the MCL for Hg was reached by 80 and 100 mg/l PAC-1 [10]. Batch and column tests were conducted for Hg removal by PAC (Filtrasorb 300, 270 mesh) and GAC (Filtrasorb 300, 8 × 30 mesh) [26]. Typical log-log plots of the adsorption data from batch tests for pH values of 4 and 10 are seen in Figure 7–8. Experimental conditions were 2.0 g carbon, 200 ml of 10.0 mg/l Hg(II), and 6 hours of agitation at 20°C. Greater adsorption occurred at the lower pH value of 4. For example, at 1 mg/l Hg, the quantity adsorbed was 0.14 mg Hg/g carbon at pH 10, whereas the quantity at pH 4 was 1 mg Hg/g. Breakthrough curves for the carbon columns are seen in Figure 7–9 for pH values 4 and 10. For the latter system, Hg broke through within the first day, whereas no breakthrough occurred within 5 days at pH 4. PAC has been employed with conventional coagulation treatment to enhance the removal of mercury. For example, a PAC dosage of 65. mg/l and an alum dosage of 30. mg/l effected a reduction in excess of 70% in the inorganic mercury content. This was greater than the 40% removal with only alum in Ohio River water [25]. Similar results were obtained with another PAC (Nuchar W-C1000N) used in conjunction with either alum or iron coagulants [27]. A 30.

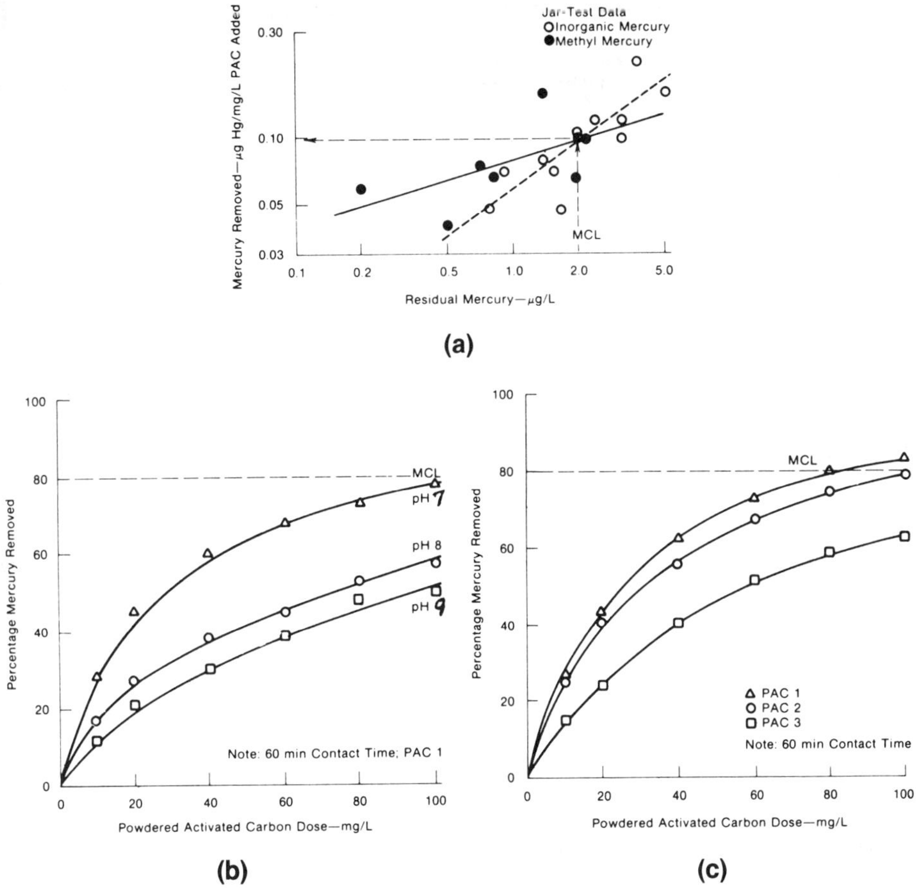

Figure 7–7 (a) Freundlich isotherms for mercury adsorption with PAC. Reproduced from Logsdon and Symonds [25], courtesy of the American Water Works Association. (b) Effect of pH on removal of inorganic mercury (0.010 mg/l) with PAC treatment. Reproduced from Thiem [10], courtesy of the American Water Works Association. (c) Removal of methyl mercury (0.010 mg/l as Hg) by PAC at pH 7. Reproduced from Sorg [15], courtesy of the American Water Works Association.

mg/l quantity of this PAC increased mercury (0.5 mg/l) removals from 35 to 73% for alum systems and from 67 to 85% for iron systems.

Several pilot plant experiments have evaluated the additional benefits of PAC with conventional water treatment for organic (methyl) mercury removal [15]. Table 7–7a shows results of these experiments, and Table 7–7b shows results from studies on the removal of inorganic mercury. The presence of PAC apparently increases mercury removal when used in conjunction with conventional treatments [15]. (Some researchers do not consider ferric sulfate to be a conven-

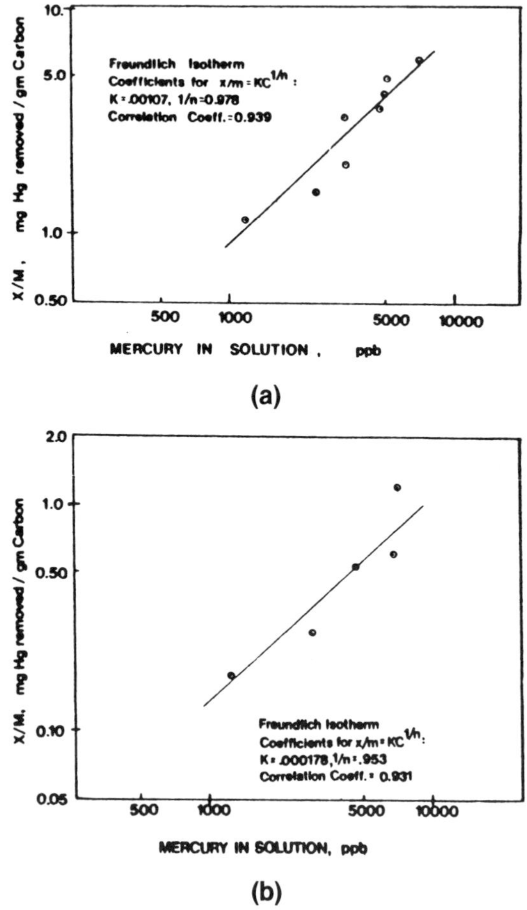

Figure 7–8 Adsorption isotherm for Hg(II) and PAC at 20°C and (a) pH 4 and (b) pH 10. Reproduced from Humenick and Schnoor [26], courtesy of the American Society of Civil Engineers.

tional coagulant.) The increase in mercury removal is attributed to PAC and GAC adsorption. Also, the reduction data for virgin and exhausted activated carbon columns suggest continued mercury removal after the carbon has been exhausted for organics removal. This may or may not be due to mercury adsorption. Chelation of mercury may be significant here. Some experimental evidence exists to support this [10]. Table 7–7c shows some adsorption constants for Hg^{2+} [47].

Eleven different brands of commercial activated carbon were evaluated for their removal of Hg(II) in a laboratory study [28]. A very high percentage (99 to 100%) of total Hg removal was obtained by all types of activated carbons, especially at pH values of 4 to 5 (Figure 7–10) with dosages of 1,000 mg/l. Above and below these pH values, removal of Hg decreased considerably. Freundlich isotherms were used to describe the adsorption, with *K* values ranging from 6.3

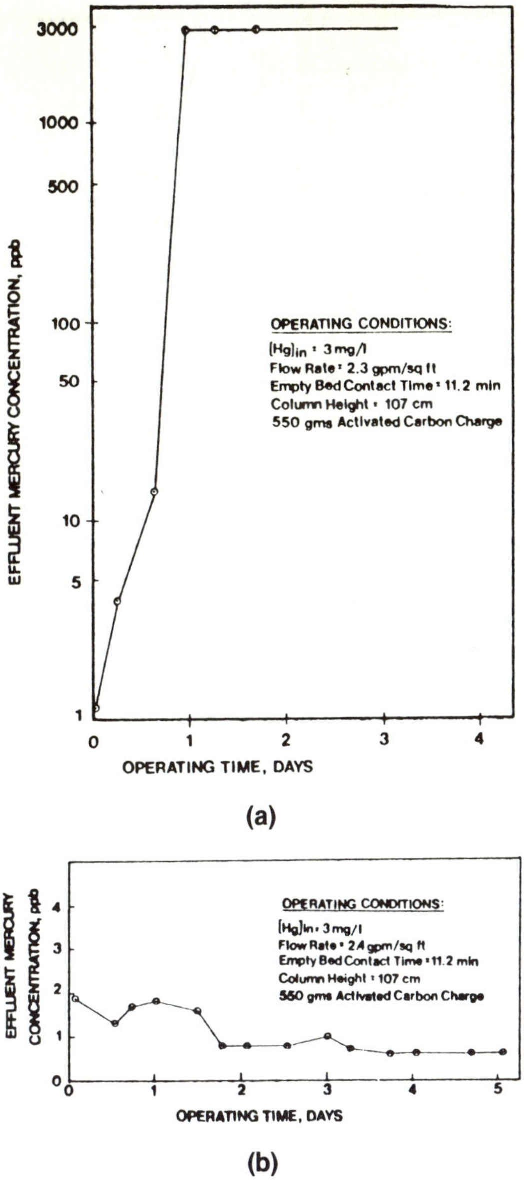

Figure 7–9 Carbon-only system: column run at 25°C and (a) pH 10 and (b) pH 4. Reproduced from Humenick and Schnoor [26], courtesy of the American Society of Civil Engineers.

(pH 11) to 62.9 (pH 4) μmol/g. Adsorption and reduction were thought to be the mechanisms of removal.

The adsorption of methylmercuric chloride from aqueous solutions on Filtrasorb 200 (0.9 mm diameter) at 25°C was examined in batch experiments (10

Table 7-7a Results of Conventional Coagulation Pilot Tests for Inorganic Mercury Removal

Length of Test (hr)	*Test Water*	*Coagulant*		*pH of Treated Water*	*Turbidity of Raw Water (NTU)*[a]	*Average Raw Water Mercury Concentration (mg/l)*[b]	*Percentage Mercury Removed*				
								Filtered Water			
									GAC Filters		
		Type	*Dose (mg/l)*				*Settled Water*	*Dual-Media Filter*	*Virgin GAC*[c]	*Exhausted GAC*[c]	*Virgin GAC*[d]
93	Ohio River	Ferric sulfate	28	8.0	5	0.0055	13	20	55	49	—
99	Ohio River	Alum	30	8.0	31–76	0.007	63	74	—	87	>93
91	Ohio River	Ferric sulfate	29	8.1	10–20	0.008	73	73	—	80	88

Source: Reproduced from Sorg [15], courtesy of the American Water Works Association.
[a]Turbidity of tap water adjusted with bentonite clay.
[b]Raw water spiked with $HgCl_2$.
[c]Filtrasorb 200, Calgon Corp. Pittsburgh, PA.
[d]Hydrocarco 1030, ICI Americas, Inc., Wilmington, DE.

Table 7–7b Results of Conventional Coagulation and PAC Pilot Plant Tests for Organic Mercury Removal

PAC (mg/l)	Length of Test (hr)	Test Water	Coagulant		pH of Treated Water	Turbidity of Raw Water (NTU)	Average Raw Water Mercury Concentration (mg/l)	Percentage Mercury Removed				
									Filtered Water			
										GAC Filters		
			Type	Dose (mg/l)				Settled Water	Dual-Media Filter	Virgin GAC	Exhausted GAC	Virgin GAC
9[a]	54	Cincinnati tap[c]	Ferric sulfate	31	7.0	<1.0	0.0054[d]	24	31	100[e]	100[f]	86[g]
17[a]	54	Cincinnati tap	Ferric sulfate	31	7.0	<1.0	0.0045	32	38	99	100	99
7[b]	57	Cincinnati tap	Ferric sulfate	35	7.0	<1.0	0.0056	27	40	100	100	98
7[b]	40	Cincinnati tap	Ferric sulfate	28	7.0	<1.0	0.0055	5	8	100	100	99
0	61	Cincinnati tap	Ferric sulfate	28	6.8	<1.0	0.0037	0	19	100	97	84

Source: Reproduced from Sorg [15], courtesy of the American Water Works Association.
[a]Darco-B, ICI Americas, Inc., Wilmington, Del.
[b]Darco-M, ICI Americas, Inc., Wilmington, Del.
[c]Water spiked with CH_3 HgCl
[d]Detectable limit of 0.0005 mg/l was used to compute percentage where effluent was at or below detectable limit; value reported with > symbol.
[e]Hydrodarco 1030, ICI Americas, Inc., Wilmington, Del.
[f]Filtrasorb 100, Calgon Corp., Pittsburgh, PA.
[g]Filtrasorb 200, Calgon Corp., Pittsburgh, PA.

Table 7–7c Adsorption Constants for Hg^{2+} on F-400

Langmuir		Freundlich	
a (dm³/mg)	K (dm³/g)	K_F (mg/g)	1/n
.034	4.68	20.	.46

Source: Reproduced from McKay, et al. [47], courtesy of Pergamon Press, Ltd.
Note: Contact time = 2 days.

days) and in packed beds with continuous liquid flow [29]. An isotherm was obtained for which neither the Freundlich nor Langmuir equation fit the data. However, a Temkin equation, $Ca = 0.054\ln(0.174\ C_L)$ (where Ca = adsorbed quantity of CH_3HgCl, gmol/cm³, and C_L = concentration in solution, gmol/cm³), described the adsorption reasonably well. Breakthrough curves were measured in packed beds as a function of particle size (0.057 to 0.263 mm diameter), bed length (0.79 to 4.7 cm), and liquid velocity (1.6 to 4.7 cm/sec). Channeling effects distorted the breakthrough curves.

RADIONUCLIDES

Health Concerns

The primary basis for the U.S. EPA regulation of radionuclides in drinking water is the carcinogenic potential of these substances [30]. These regulations are more complex than the MCLs established for other inorganic contaminants. For example, the MCL for alpha emitters is a combination of radium 226 and 228 and is 5 picocuries per liter (pCi/l). Gross alpha activity (excluding radon and uranium) must not exceed 15 pCi/l. Another regulation concerned with fabricated radionuclides states: "the average annual concentration of beta particles and photon radioactivity from man-made radionuclides in drinking water shall not produce an annual dose equivalent to the total body or any internal organ greater than 4 millirem/year" [30].

Of several radionuclides found in the environment, radium is perhaps the most significant because it is naturally occurring and is found often in groundwaters in several locations throughout the United States [3]. Radium has similar chemistry to calcium, so it is a bone-seeking element, with 80 to 85% of what is retained in the body being deposited in the bone [1].

Figure 7–10 The total Hg(II) removal capacity of representative activated carbons: (a) Nuchar SA, (b) Darco S-51, (c) HD3000 (fresh). The reaction time was 24 hours; carbon doses were 1 g/l. Reproduced from Huang and Blankenship [28], courtesy of Pergamon Press.

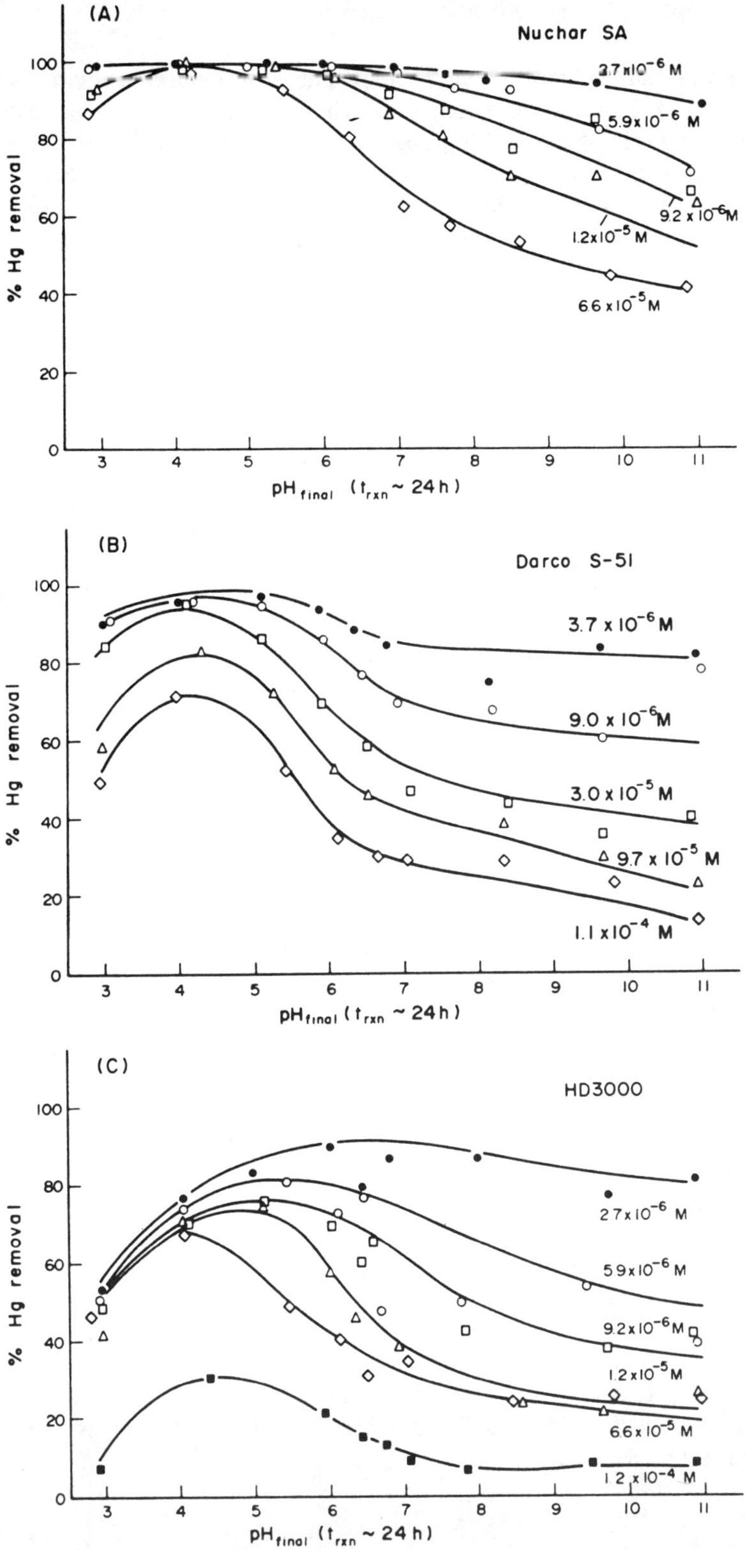
(A)
Nuchar SA
3.7x10^{-6} M
5.9x10^{-6} M
9.2 x 10^{-6}M
1.2x10^{-5} M
6.6 x 10^{-5}M
% Hg removal
pH$_{final}$ (t$_{rxn}$ ~ 24 h)
(B)
Darco S-51
3.7 x 10^{-6} M
9.0 x 10^{-6}M
3.0 x 10^{-5} M
9.7 x 10^{-5} M
1.1 x10^{-4} M
% Hg removal
pH$_{final}$ (t$_{rxn}$ ~ 24 h)
(C)
HD3000
2.7x10^{-6} M
5.9x10^{-6} M
9.2x10^{-6}M
1.2x10^{-5}M
6.6x10^{-5} M
1.2 x 10^{-4} M
% Hg removal
pH$_{final}$ (t$_{rxn}$ ~ 24 h)

Another naturally occurring radionuclide is uranium that occurs as the uranyl ion (UO_2^{2+}) in natural waters [31]. This ion occurs mostly in wastewaters from uranium mining and milling operations and nuclear fuel-processing plants. The average content of uranium in drinking water in the United States is estimated to be 2 pCi/l [32]. Thus, an average person would ingest 1,440 pCi in 1 year by consuming 2 l water per day. Perhaps some attention should be given to this element.

Removal

Radium removal by PAC and GAC is not expected since Ca and Ba are poorly adsorbed, also. Pilot plant studies support this statement [7]. Percentage removals for dual-media filters and GAC filters were essentially identical, which indicated that adsorption of Ra^{226} by GAC did not occur. Any removal of Ra^{226} by GAC was attributed entirely to filtration.

Some 20 municipal water treatment plants in the western part of the United States were surveyed for uranium removal [31]. Analyses of raw, intermediate-stage, and treated water samples indicated that conventional treatments, including activated carbon, were not effective for removing uranium from raw waters when its concentration was in the range of 0.1 to 16. μg/l [31].

SELENIUM

Aqueous Chemistry

The MCL for selenium in drinking water is 0.01 mg/l for several human health reasons [5]. The occurrence of soluble Se compounds in some soils and other geologic materials suggests that natural waters would be effective in leaching and transporting this element from drainage areas to floodplains and other large bodies of water. However, levels encountered in surface waters are usually low—for example, averaging 0.2 μg/l in nine rivers throughout the world [33]. Rivers receiving drainage from seleniferous soils may contain high concentrations of Se under the proper pH conditions. For example, tributaries of the Colorado River receiving such drainage contained up to 400 μg/l [34]. There are, of course, anthropogenic sources of Se and the attendant wastes and wastewaters [2].

Selenium exists in four oxidation states in aqueous systems: −2, 0, +4, and +6. The reduced state, −2, is represented by the selenide species, and the elemental state, Se^0, exists in different allotropic forms, whereas the two higher oxidation states are represented by selenites, +4, and selenate, +6. Selenious acid, H_2SeO_3, is a weak acid that dissociates into two selenite species, $HSeO_3^-$ and SeO_3^{2-}. Selenic acid, H_2SeO_4, is a stronger acid and forms, essentially, one anion in water, SeO_4^{2-}.

Removal

The WSRD of the U.S. EPA employed jar-test studies to ascertain removal of Se(IV, VI) from water by PAC [5]. Less than 4% of either Se species was removed with PAC dosages up to 100 mg/l on groundwater spiked with Se contents of 0.03 and 0.1 mg/l. There were no effects of $[H^+]$ on this Se removal. Furthermore, pilot plant studies incorporating GAC with chemical coagulation indicated that carbon did not adsorb any Se [4]. These findings are not surprising inasmuch as Se is anionic and inorganic in nature.

ZINC

Aqueous Chemistry

Zinc has a secondary drinking water standard of 5.0 mg/l, principally because of its bitter metallic taste and milky turbidity when $Zn(OH)_{2(s)}$ precipitates [2]. Zinc appears in aquatic environments mostly from the corrosion of galvanized metals. Then are the usual anthropogenic sources. Corrosion of zinc is thermodynamically feasible, and the product is Zn^{2+} ions. $ZnO_{2(s)}$, $ZnCO_{3(s)}$, and $Zn(OH)_{2(s)}$ are important solid phases and occur whenever their solubility product constants are exceeded [2].

Removal

Since zinc, alone, has not been a widespread contamination problem, few studies have been conducted for its removal. Table 7-6 gives some Langmuir X_m values for four powdered carbons [9]. Experimental conditions were 5,000 mg/l PAC and 24 hours of continuous shaking. That there was a definite effect of $[H^+]$ is seen where greater quantities of Zn were adsorbed at a pH value of 8.0 than at 6.5. This may be due to a greater negative charge on the surface of carbon [14] or to precipitation of $Zn(OH)_{2(s)}$ with its subsequent adsorption.

CHLORINE COMPOUNDS

Chlorine

Background

Carbon has been employed for a long time for the incidental or purposeful removal of chlorine compounds from water. Baylis was perhaps the first person to report the production of chlorides and carbon dioxide when aqueous chlorine was adsorbed on activated carbon [35]. It is necessary to remove excessive quantities of Cl_2 following treatment for taste and odor-causing compounds [2]. Carbon is utilized for chlorine removal from finished waters for food processing and other industries.

According to Magee [36] and Kovach [37], the reaction between HOCl and carbon is:

$$\mathrm{HOCl} + \mathrm{C}^* = \mathrm{HCl} + \underset{\text{(surface oxide).}}{\mathrm{CO}} \tag{7.1}$$

This is an oxidation-reduction reaction. Another reaction proposed by Baylis [35]:

$$2\mathrm{Cl}_2 + \mathrm{C} + 2\mathrm{H}_2\mathrm{O} = 4\mathrm{HCl} + \mathrm{CO}_2. \tag{7.2}$$

In any event, carbon catalyzes or reduces the valence state of Cl_2 and apparently does not adsorb it.

Kinetic studies have been conducted for these reactions. In a packed bed reactor, a two-stage reaction was observed [36]. The first stage involved a very rapid rate of HOCl uptake that was attributed to diffusion. During this stage there was a net increase of chlorine-containing species on the carbon's surface. Continued application of HOCl developed a second stage and slower reaction. Here the HOCl removal was equal to the Cl^- released in accord with Equation 7.1. It was hypothesized that the carbon's surface became saturated with HOCl, and a subsequent reduction reaction became the rate-limiting step [36]. Also, it was observed that the reaction between OCl^- and C was approximately 25% of that between HOCl and C.

Kinetics

Several laboratory studies have been conducted on the kinetics of Cl_2-C reaction. Two PACs, Darco G-60 and Nuchar C-1000N, were employed in one of these studies conducted at various temperatures [38]. Figure 7–11 shows the uptake of chlorine (as OCl^- at pH 10.) by Nuchar C-1000N as a function of time under uncontrolled temperature conditions. There is an initial very rapid rate of reaction followed by a slower, linear rate that was labeled a first-order process [38]. This two-step removal is similar to the kinetics of adsorption of organic compounds on GAC (see Chapter 3). In addition, there appears to be a relation between the initial, nonlinear portion of the curve and the linear portion; that is, the greater the amount of chlorine removed in the first portion of the removal process, the smaller the first-order rate constant (taken from the slope of the linear portion).

In a more definitive study [39,40,41], an assumption was made that the Cl_2-C reaction proceeds by a reversible adsorption step:

$$\mathrm{C}^* + \mathrm{HOCl} \underset{k_{2\mathrm{H}}}{\overset{k_{1\mathrm{H}}}{\rightleftharpoons}} \mathrm{CHOCl}^*, \tag{7.3}$$

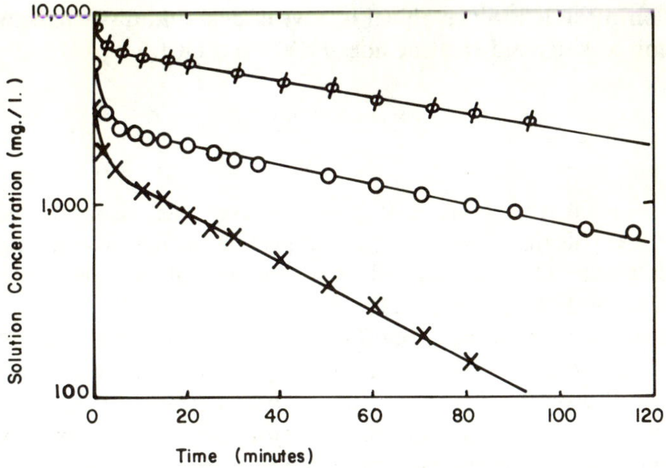

Figure 7–11 Aqueous chlorine concentration as a function of time. 3.0 g/l Nuchar C-1000N at pH 10.0. Reproduced from Olson and Binning [38], courtesy of Ann Arbor Science Publishers, Inc.

followed by an irreversible dissociation:

$$CHOCl^* \xrightarrow[k_{3H}]{} CO^* + H^+ + Cl^- , \tag{7.4}$$

where k_{1H}, k_{2H}, and k_{3H} are HOCl reaction rate constants. In these reactions, C* represents an activated carbon site, CO* is a surface oxide on the carbon, and CHOCl* is an adsorbed HOCl molecule. The reaction rates are controlled by a number of active C sites and surface oxide sites.

The number of surface oxide sites on carbon have been determined experimentally [42]. The fraction of active sites occupied by the oxides, f_o, is [41]:

$$f_o = \frac{Q}{k_{5H} + Q} , \tag{7.5}$$

where Q = the amount of HOCl that has reacted per unit surface area of the carbon, and k_{5H} is a constant. An assumption is made whereby the reaction in Equation 7.3 is in equilibrium and that the total number of active sites, S, on fresh carbon is given by the sum of C*, CO*, and CHOCl*. Subsequently, a surface reaction equation was derived for the rate of conversion of HOCl in terms of C_H and f_o:

$$R_H\,(C_H,f_o) = \frac{k_{3H}\,S\,C_H}{k_{4H} + C_H}\,(1 - f_o) , \tag{7.6}$$

where $k_{4H} = k_{2H}/k_{1H}$ (adsorption-desorption equilibrium constant) and C_H =

concentration of HOCl. Since HOCl is a weak acid and dissociates with $[H^+]$, an expression was derived that includes HOCl and $O\bar{C}l$:

$$R_{FC}\,(C_{FC}, f_o) = \frac{SC_{FC}(1 - f_o)}{k_{4FC} + C_{FC}}\,(k_{3H}f_H + k_{3\bar{O}}f_{\bar{O}})\,, \tag{7.7}$$

where C_{FC} = total free chlorine, k_{4FC} = the reversible adsorption equilibrium constant of free chlorine, k_{3H} and $k_{3\bar{O}}$ = respective surface dissociation rate constants of HOCl and OCl, and f_H and $f_{\bar{O}}$ = fractions of total free chlorine in the form of HOCl and $O\bar{C}l$, respectively.

Physical processes responsible for the transport of reactants and products between the bulk solution phase and the carbon surfaces within the pore structure are similar to those affecting adsorption of organic molecules (see Chapter 6). Intraparticle diffusional resistance plays a major role [39–42]. A material balance on a chlorine residual species within a carbon's pore yields:

$$\frac{\partial C}{\partial t} = \left[\frac{D_c}{L_p^2}\right]\frac{\partial^2 C}{\partial \eta^2} - \frac{4}{d_p}R_C(C,Q)\,, \tag{7.8}$$

where C = concentration of the chlorine residual species, t = real time, D_c = diffusivity of the chlorine species, η = a reduced position variable measured from the mouth of the pore, d_p = diameter of a straight cylindrical pore, and L_p = pore half-length.

The total quantity of a chlorine residual species that has been converted locally is an integral over time of the local reaction [41]. It is, in differential form:

$$\frac{\partial Q}{\partial t} = R_C(C,Q)\,. \tag{7.9}$$

Equations 7.8 and 7.9 are in a general format that can be specified for a type of chlorine.

In the reaction rate studies of chlorine and carbon, the type of reactor affects the time and space variation of the bulk fluid concentration, C_B, of a given chlorine species [39–42]. For example, in the completely mixed, closed batch reactor, the bulk concentration of a given chlorine species is allowed to vary with time. Thus, C_B can be obtained from a material balance on chlorine residual in the bulk fluid phase [41]:

$$\frac{dC_B}{dt} = \left(\frac{A_p d_p}{4}\right)\left(\frac{m}{\epsilon}\right)\left(\frac{D_c}{L_p^2}\right)\left(\frac{\partial C}{\partial \eta}\right)_{\eta=0} - k_d C_B\,, \tag{7.10}$$

where A_p = pore surface area per unit weight of carbon, m = mass of carbon per unit volume, ϵ = reactor porosity, k_d = a first-order self-decay rate constant for the chlorine species (experimentally determined), and d_p = diameter of a straight cylindrical pore.

Another type of reactor is the semibatch where concentration of a specific chlorine residual species remains constant at a predetermined level, C_o. In this case [41]:

$$C_B = C_o. \tag{7.11}$$

Both semibatch and closed batch mathematical models gave results in which the observed reaction rate was insensitive to the independent time variable, t, and showed considerable variation only with C_B and the total chlorine residual conversion, $\overline{Q}$ [40–43]. It was concluded that the reaction rate for a single particle can be correlated in terms of C_B and $\overline{Q}$ where the latter is the total conversion that has occurred within that particle. The generalized equation is [41]:

$$R(C_B, \overline{Q}) = \left(\frac{J_1 C_B}{1 + J_2 C_B}\right)^{J_3} \left(\frac{1}{1 + J_4\overline{Q}\exp(J_5 C_B)}\right)^{J_6} \left(\frac{1}{1 + [J_7\overline{Q}\exp(J_8 C_B)]} J_9\right)^{J_{10}}, \tag{7.12}$$

where $R(C_B,\overline{Q})$ is the overall reaction rate per unit weight of uniform-size GAC and the Js are constants.

The constants in Equations 7.5 to 7.12 can be evaluated from available literature data, manufacturer's bulletins, or experiments for individual chlorine species [39–42]. Table 7–8 gives values for several of these constants for this and other models [39]. Also there is some trial and error fitting of data from experiments to that of the predictive model [39–42]. Figures 7–12a to d show the results of semibatch experiments with 60 × 80 mesh carbon particles. These curves represent the predicted rates of HOCl conversion computed from the semibatch reactor model (points represent experimental data). Two effects are noted from the curves in Figures 7–12a to d. First, the rate decreases sharply as the quantity of HOCl reacted per unit weight of carbon, $\overline{Q}$, increases. Second, the rate increases with HOCl concentration but is not proportional to it; that is, the effect of C_{FC} in the denominator of Equation 7.7 is apparent. These observations were confirmed by two closed batch experiments [40].

Particle size is a factor affecting the kinetics of dechlorination inasmuch as the major portion of the reactive surface area of activated carbon lies within the granule. Consequently, diffusion of the Cl_2 is required for a reaction to occur. For larger particles, this diffusion and the associated diffusional path length may be the rate-limiting step and decrease the overall reaction rate. In Equation 7.8, the term (D_c/L_p^{z2}) is included, where L_p is taken as one-sixth of the arithmetic mean of the particle size range. This term represents an effective diffusional path.

Table 7–8 Values of Constants in Mathematical Models for Chlorine-Carbon Reactions

	pH 4		*pH 7.6*	*pH 10*		
Constant	T = 23°C	T = 35°C	T = 23°C	T = 2°C	T = 23°C	T = 35°C
4 k_3S/d_p, g/cm³-sec	6.08×10^{-3}	1.22×10^{-2}	3.82×10^{-3}	4.06×10^{-4}	1.56×10^{-3}	3.13×10^{-3}
k_3S/k_5	2.17	4.34	1.36	0.145	0.557	1.12
k_4, g/cm³	1.81×10^{-5}	1.81×10^{-5}	1.81×10^{-5}	1.81×10^{-5}	1.81×10^{-5}	1.81×10^{-5}
k_d, sec^{-1}	2.23×10^{-7}	5×10^{-7}	2.23×10^{-7}	2.23×10^{-7}	2.23×10^{-7}	2.23×10^{-7}
D_c, cm²/sec	1×10^{-5}	1.34×10^{-5}	1×10^{-5}	0.52×10^{-5}	1×10^{-5}	1.34×10^{-5}
L_p (60 × 80), cm	0.0035					
V_p, cm³/g	0.94					

Source: Reproduced from Suidan et al. [39], courtesy of the American Chemical Society.

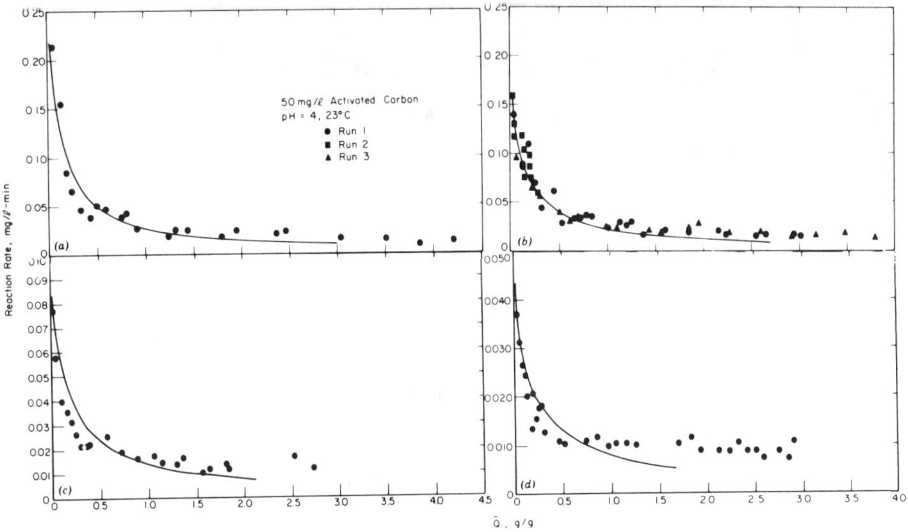

Figure 7–12 Chlorine-carbon (60 × 80 mesh) reaction in semibatch reactor: (a) C_o = 30 mg/l; (b) C_o = 20 mg/l; (c) C_o = 10 mg/l; (d) C_o = 5 mg/l. Reproduced from Suidan et al. [40], courtesy of the American Society of Civil Engineers.

As seen in Figure 7–13, the closed batch reactor model predicted the effect of particle size reasonably well (again, points represent experimental data).

That $[H^+]$ is a factor affecting the kinetics of dechlorination is seen in Figure 7–14. At pH 4, the chlorine is predominantly HOCl; at pH 10, $\overline{O}Cl$ pre-

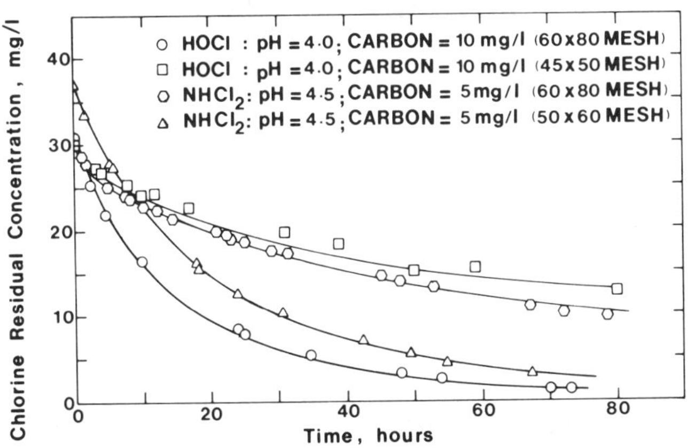

Figure 7–13 Effect of particle size on chlorine residual-activated carbon reaction in a closed batch, 23–24°C. Reproduced from Suidan et al. [41], courtesy of the Ann Arbor Science Publishers, Inc.

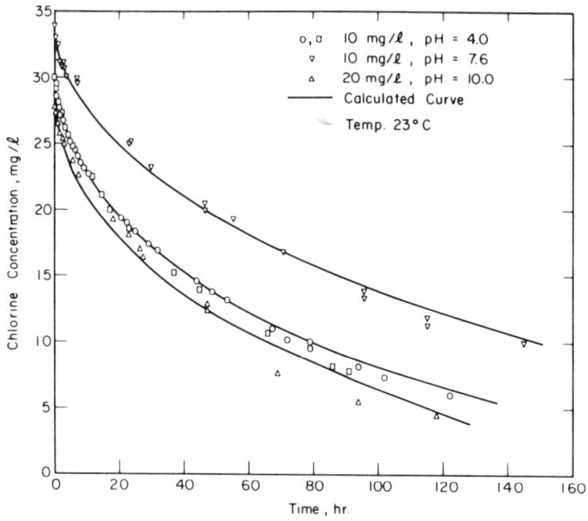

Figure 7-14 Effect of pH on carbon-chlorine reaction in closed batch reactor. Reproduced from Suidan et al. [39], courtesy of the American Chemical Society.

dominates; and at pH 7.6, the concentrations of each are equal. It appears that the model predicts the kinetics at the three pH values rather well.

Temperature is expected to affect diffusivity and reaction rate constants. Experiments were conducted at 2° and 35°C for which diffusion constants were calculated from [41]:

$$D_c\overline{T}_2 = D_c\overline{T}_1 \left(\frac{\mu_1}{\mu_2} \frac{\overline{T}_2}{\overline{T}_1} \right), \tag{7.13}$$

where μ = viscosity and $\overline{T}$ = absolute temperature [43]. The expected effect of temperature is seen in Figure 7-15 where faster reactions occur at the higher temperature. Also it is noted that the HOCl-C reaction is quite a bit faster than the $\overline{O}$Cl-C reaction.

Chloramines

Background

That the reaction between the ammonium ion (NH_4^+) and chlorine (HOCl and $\overline{O}$Cl) yields principally mono-(NH_2Cl) and dichloramines ($NHCl_2$) at pH values normal to water and wastewater treatment is well established [2]. Nitrogen trichloride (NCl_3) may also form, but the pH value must be below 4.

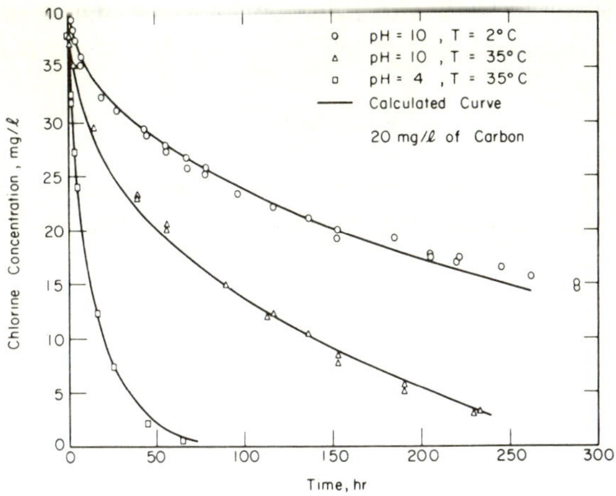

Figure 7–15 Effect of pH and temperature on carbon-chlorine reaction in closed batch reactor. Reproduced from Suidan et al. [39], courtesy of the American Chemical Society.

Studies have been conducted on the reaction of chloramines with activated carbon [41,44,45] that led to the following stoichiometry:

$$C^* + 2NHCl_2 + H_2O = CO^* + 4H^+ + 4\overline{Cl} + N_{2(g)}\,, \qquad (7.14)$$

$$C^* + NH_2Cl + H_2O = CO^* + H^+ + \overline{Cl} + NH_3\,, \qquad (7.15)$$

and

$$CO^* + 2NH_2Cl = C^* + 2H^+ + 2\overline{Cl} + H_2O + N_{2(g)}. \qquad (7.16)$$

Reaction 7.16 may occur after reaction 7.15 has proceeded for some 20 hours [44].

Kinetics

Surface reaction rate expressions for the reduction of dichloramine by activated carbon have been formulated [45]. A reaction mechanism was postulated:

$$C(H_2O)^* + NHCl_2 \underset{k_{2DC}}{\overset{k_{1DC}}{\rightleftharpoons}} C\,(H_2O)(NHCl_2)^*, \qquad (7.17)$$

where k_{1DC} and k_{2DC} are dichloramine reaction rate constants. This adsorption-desorption step is followed by an irreversible decomposition step:

$$2C(H_2O)(NHCl_2)^* \xrightarrow{k_{3DC}} CO^* + 4H^+ + 4\overline{Cl} + C(H_2O)^* + N_{2(g)}. \qquad (7.18)$$

Comparison of Reaction 7.18 with Reaction 7.4 shows that four times as much $NHCl_2$ (as Cl_2) is required than when free Cl_2 is the activated carbon surface oxidant.

Accumulation of oxides on the carbon's surface from the $NHCl_2$-C reaction should be similar to the Cl_2-C system. However, this accumulation of surface oxides as a function of the extent of reaction, Q, would be slower because of the greater quantity of dichloramine in Reaction 7.18. Consequently, the fraction of active sites occupied by oxides is [45]:

$$f_o = \frac{Q^{\frac{1}{2}}}{k_{5DC} + Q^{\frac{1}{2}}}, \qquad (7.19)$$

where Q = the quantity of $NHCl_2$ that has reacted per unit surface area of the carbon, and k_{5DC} is a constant. In a manner similar to that for free chlorine cited earlier, a surface reaction rate equation was derived for conversion of $NHCl_2$:

$$R_{DC}(C_{DC}, f_o) = 2k_{3DC}\left(\frac{C_{DC}}{C_{DC} + k_{4DC}}\right)^2 S^2(1-f_o)^2, \qquad (7.20)$$

where C_{DC} = concentration of $NHCl_2$, and $k_{4DC} = k_{2DC}/k_{1DC}$.

For the NH_2Cl reaction with the surface oxide site, an empirical equation was derived [41]:

$$R_{MC}(C_{MC}, Q) = \left(\frac{q_2 C_{MC}}{q_1 + C_{MC}}\right)^2 \left(1 + \frac{q_3 Q^3}{1 - q_4 Q^3} + \frac{q_5}{q_6} - \frac{q_5 Q^3}{1 + q_6 Q^3}\right), \qquad (7.21)$$

where C_{MC} = the $[NH_2Cl]$ and q_i are constants. As the value of Q increases, Equation 7.21 reaches a steady-state form:

$$R_{MC}C_{MC} = q_7\left(\frac{C_{MC}}{q_1 + C_{MC}}\right)^2, \qquad (7.22)$$

where $q_7 = (1 - q_3/q_4)$.

As seen in Figures 7-13 and 7-16, the model reaction rate equation fits the data reasonably well. Also, the chloramine reactions proceed at rates considerably slower than those of hypochlorites.

Packed Bed Reactors

Mathematical models were developed for packed or fixed GAC beds in removal of chlorine compounds [41,42,46]. They consisted of three nonlinear partial dif-

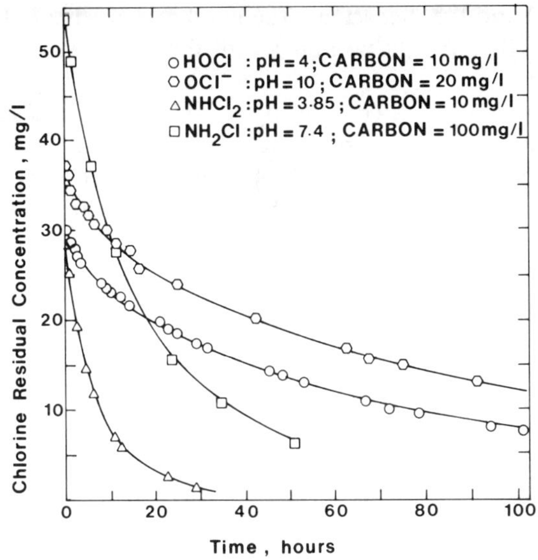

Figure 7–16 Chlorine residual-activated carbon (60 × 80 mesh) reaction in a closed batch, 23–24°C. Reproduced from Suidan et al. [41], courtesy of Ann Arbor Science Publishers, Inc.

ferential equations whose numerical solution "is rather complicated and costly in terms of computer time" [40]. An axial dispersion term was included in this model, and simplifications led to eventual solution of the models [41]. In any event, fixed bed experiments were conducted with two different carbon particle sizes, as seen in Figure 7–17. Here, again, the agreement between the experi-

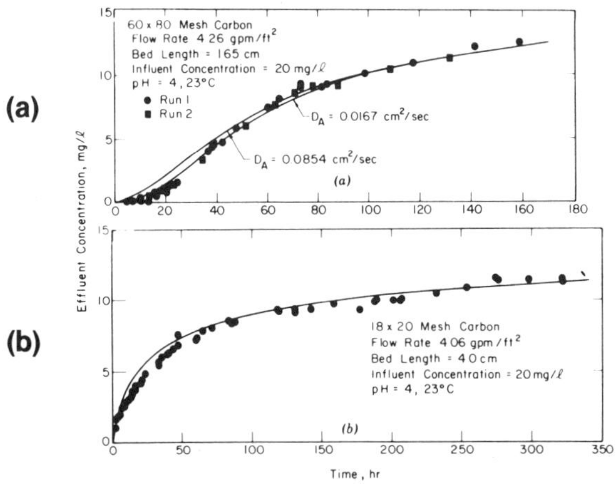

Figure 7–17 Chlorine breakthrough curve: (a) 60 × 80 mesh; (b) 18 × 20 mesh. Reproduced from Suidan et al. [40], courtesy of the American Society of Civil Engineers.

mental (the points) and the theoretical (the lines) was excellent. The rather sharp breakthrough of HOCl should be noted with the larger carbon particle size in Figure 7-17.

Additional modeling was conducted for NH_2Cl-GAC because this reaction is very slow [46]. Figures 7-18, 7-19, and 7-20 show the predicted performance curves for the effects of influent [NH_2Cl], flow rate, contact time, GAC bed depth, and particle size. These curves should be useful for the design and op-

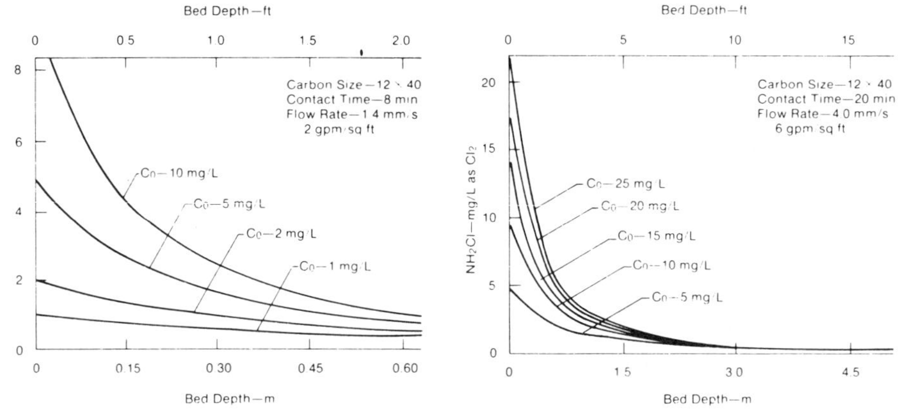

Figure 7-18 Effect of influent concentration on the profile of NH_2Cl concentration in a packed GAC adsorber at steady state. Reproduced from Kim and Snoeyink [46], courtesy of the American Water Works Association.

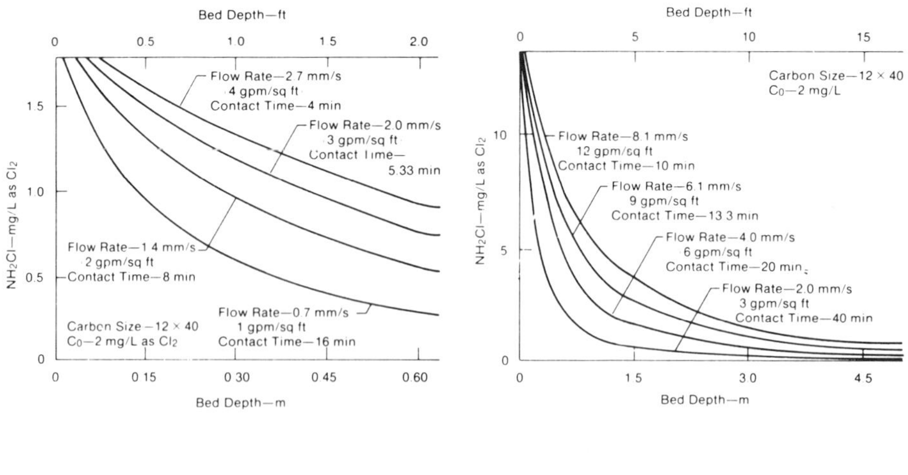

Figure 7-19 Effect of flow rate and contact time on the profile of NH_2Cl concentration in a packed GAC adsorber at steady state. Reproduced from Kim and Snoeyink [46], courtesy of the American Water Works Association.

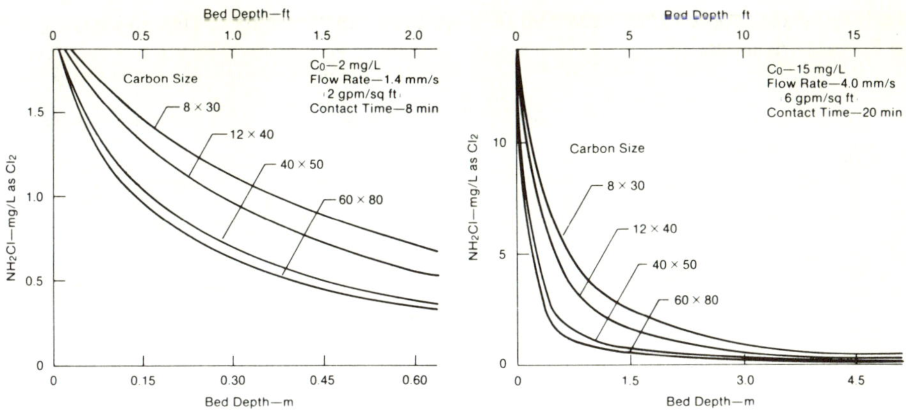

Figure 7–20 Effect of particle size on the profile of NH_2Cl concentration in a packed GAC adsorber at steady state. Reproduced from Kim and Snoeyink [46], courtesy of the American Water Works Association.

eration of GAC beds for NH_2Cl removal. However, they were generated for only one carbon (F-400) and one pH value. It would be desirable to have some actual plant practice for confirmation.

HYDROGEN SULFIDE

Aqueous Chemistry

Hydrogen sulfide is listed in the National Secondary Drinking Water Regulations as a contributor to odorous water; that is, it is responsible for a rotten egg smell [2]. Also, H_2S reacts with many of the metals cited here to produce, for example, black stains or black deposits of iron sulfide. H_2S cannot be considered a normal constituent of natural waters but is associated with waters high in sulfate and organic matter [2]. The microbially mediated oxidation of organic matter utilizes SO_4^{2-} as an electron acceptor, which results in reduced sulfur as H_2S. Also, this odorous substance is associated frequently with Fe and Mn from groundwaters in or near former swampy areas. In either case, H_2S can be removed easily from water by aeration, chemical oxidation, or adsorption onto carbon.

$$H_2S = H^+ + HS^-; \qquad \log Ka_1 = 10^{-6.88} \tag{7.23}$$

and

$$HS^- = H^+ + S^{2-}; \qquad \log Ka_2 = 10^{-12.2} \text{ to } 10^{-17.1}. \tag{7.24}$$

Table 7-9 Fixation of Hydrogen Sulfide from Aqueous Phase by Carbons in Relation to Surface Unsaturation

Carbon	*Surface Unsaturation (mEq/g)*		*H_2S Fixed (mEq/g)*	
	Before[a]	*After*[b]	*Before*[a]	*After*[b]
A	1.23	4.35	1.47	4.74
B	1.55	3.65	1.72	3.42
C	1.39	4.01	1.32	4.18
D	0.83	3.08	1.05	3.38
E	1.25	2.43	1.62	2.84
Coconut charcoal outgassed at 750°C	3.50	—	3.37	—

Source: Reproduced from Puri [49], courtesy of the American Chemical Society.
[a]Carbons without any treatment.
[b]Carbons after surface oxidation following by outgassing.

Thus, H_2S is a weak acid, and the secondary dissociation is not realized in a practical sense. This gaseous substance can be quite soluble in water: at 760 mm Hg and 25°C, the solubility of H_2S is 3,380 ppm [2]. There are 30 or more ionic and molecular sulfur species in existence, of which 5 are thermodynamically stable in water at 25°C and 1 atm: HSO_4^-, SO_4^{2-}, S^0, H_2S, and HS^- [2]. Other species such as thiosulfate, polysulfides, and polythionates are unstable in water. Sulfate is stable under oxidizing conditions and over the entire pH range.

Removal

H_2S is removed easily from water by either PAC or GAC. Carbon treatment is recommended for domestic water supplies that contain H_2S [2]. Despite the widespread use of carbon for removal of this gas, very little quantitative data are available in the literature. Table 7-9 gives some data where H_2S adsorption was measured before and after chemical treatment of the carbon surfaces [49]. Significant quantities of H_2S were adsorbed.

REFERENCES

1. "National Interim Primary Drinking Water Regulations." *Federal Register 40(248)*, 59566 (1975).
2. Faust, S.D., and O.M. Aly. *Chemistry of Water Treatment.* Butterworths, Stoneham, Mass. (1983).
3. Faust, S.D., and O.M. Aly. *Chemistry of Natural Waters.* Butterworths, Stoneham, Mass. (1981).
4. Logsdon, G.S., et al. "Trace Metals in Water Supplies." *Proc. 16th Water Quality Conf.* Univ. Bull. 71, Univ. Illinois, Urbana-Champaign, Ill. (1974), p. 111.

5. Sorg, T.J., and G.S. Logsdon. *J. Amer. Water Works Assoc. 70,* 379 (1978).
6. Gupta, S.K., and K.Y. Chen. *J. Water Pollut. Cont. Fed. 50,* 493 (1978).
7. Sorg, T.J., and G.S. Logsdon. *J. Amer. Water Works Assoc. 72,* 411 (1980).
8. Sorg, T.J., et al. *J. Amer. Water Works Assoc. 70,* 680 (1978).
9. Rubin, A.J., and D.L. Mercer. "Adsorption of Free and Complexed Metals from Solution by Activated Carbon," Chapter 8 in *Adsorption of Inorganics at Solid-Liquid Interfaces.* M.H. Anderson and A.J. Rubin, eds. Ann Arbor Science Publishers, Ann Arbor, Mich. (1981), p. 295.
10. Thiem, L., et al. *J. Amer. Water Works Assoc. 68,* 447 (1976).
11. Huang, C.P., and F.B. Ostovic. *J. Environ. Eng. Div. ASCE 104 (EE5),* 863 (1978).
12. Huang, C.P., and E.H. Smith. "Removal of Cd(II) from Plating Waste Water by an Activated Carbon Process," in *Chemistry in Water Reuse.* W.J. Cooper, ed. Ann Arbor Science Publishers, Ann Arbor, Mich. (1981), p. 355.
13. Pourbaix, M. *Atlas of Electrochemical Equilibria in Aqueous Solutions.* Pergamon Press, New York (1966).
14. Huang, C.P. "Chemical Interactions Between Inorganics and Activated Carbon," in *Carbon Adsorption Handbook.* P.N. Cheremisinoff and F. Ellerbusch, eds. Ann Arbor Science Publishers, Ann Arbor, Mich. (1978), p. 281.
15. Sorg, T. J. *J. Amer. Water Works Assoc. 71,* 454 (1979).
16. Huang, C.P., and M.H. Wu. *J. Water Pollut. Cont. Fed. 47,* 2437 (1975).
17. Huang, C.P., and M.H. Wu. *Water Res. 11,* 673 (1977).
18. Bowers, A.R., and C.P. Huang. *Prog. Water Tech. 12,* 629 (1980).
19. Linstedt, K.D., et al. *J. Water Pollut. Cont. Fed. 43,* 1507 (1971).
20. Maruyama, T., et al. *J. Water Pollut. Cont. Fed. 47,* 962 (1975).
21. Elliott, H.A., and C.P. Huang. *Environ. Sci. Technol. 14,* 87 (1980).
22. Choi, W.W., and K.Y. Chen. *J. Amer. Water Works Assoc. 71,* 562 (1979).
23. Netzer, A., and J.D. Norman. *Water Pollut. Res. in Canada 8,* 110 (1973).
24. Parks, G.A., and P.L. deBruyn. *J. Phys. Chem. 66,* 967 (1962).
25. Logsdon, G.S., and J.M. Symonds. *J. Amer. Water Works Assoc. 65,* 554 (1973).
26. Humenick, M.J. Jr., and J.L. Schnoor. *J. Environ. Eng. Div., ASCE* 100EE6, 1249 (1974).
27. Ebersole, G., and J.T. O'Connor. "Removal of Mercury from Water by Conventional Treatment Processes." Presented at American Water Works Association Universities Forum Program, Chicago (1972).
28. Huang, C.P., and D.W. Blankenship. *Water Res. 18,* 37 (1984).
29. Ammons, R.D., et al. *Ind. Eng. Chem. 16,* 263 (1977).
30. "Drinking Water Regulations—Radionuclides." *Federal Register 4,* 133 (July 9, 1976).
31. White, S.K., and E.A. Bondutti. *J. Amer. Water Works Assoc. 75,* 354 (1983).
32. Drury, J.S., et al. *Uranium in U.S. Surface, Ground, and Domestic Waters,* Vol. II. ORNL/EIS-192/V2. Oak Ridge National Laboratory, Oak Ridge, Tenn. (April 1981).
33. Klarkar, D.P., et al. *Geochem. Cosmochim. Acta 32,* 285 (1968).
34. Scott, R.C., and P.T. Voegeli, Sr. "Radiochemical Analyses of Ground and Surface Water in Colorado, 1954–1961." Colorado Water Conservation Board Basic Data Report No. 7. Denver, Colo. (1961).
35. Baylis, J. *J. Amer. Water Works Assoc. 21,* 787 (1929).
36. Magee, V., *Proc. Soc. Water Treat. Exam. 5,* 17 (1956).
37. Kovach, J.L., *Ind. Water Eng. 8,* 30 (1971).
38. Olson, L.L., and C.D. Binning. "Interactions of Aqueous Chlorine with Activated

Carbon,'' Chapter 12 in *Chemistry of Water Supply Treatment and Distribution.* A.J. Rubin, ed. Ann Arbor Science Publishers, Ann Arbor, Mich. (1974), p. 253.

39. Suidan, M.T., et al. *Environ. Sci. Technol. 11*, 785 (1977).
40. Suidan, M.T., et al. *J. Environ. Eng. Div. ASCE, 103(EE4),* 677 (1977).
41. Suidan, M.T., et al. ''Reduction of Free and Combined Chlorine with Granular Activated Carbon,'' Chapter 19 in *Activated Carbon Adsorption of Organics from the Aqueous Phase,* Vol. 1. I.H. Suffet and M.J. McGuire, eds. Ann Arbor Science Publishers, Ann Arbor, Mich. (1980), p. 397.
42. Snoeyink, V.L., et al. ''Active Carbon Dechlorination and the Adsorption of Organic Compounds,'' Chapter 11 in *Chemistry of Water Supply Treatment and Distribution.''* A.J. Rubin, ed. Ann Arbor Science Publishers, Ann Arbor, Mich. (1974), p. 233.
43. Perry, J.H. *Chemical Engineering Handbook.* McGraw-Hill, New York (1950).
44. Bauer, R.C., and V.L. Snoeyink. *J. Water Pollut. Cont. Fed. 45,* 2290 (1973).
45. Kim, B.R., et al. *J. Water Pollut. Cont. Fed. 50,* 122 (1978).
46. Kim, B.R., and V.L. Snoeyink. *J. Amer. Water Works Assoc. 72,* 488 (1980).
47. McKay, G., et al. *Water Res. 19,* 491 (1985).
48. Netzer, A., and D.E. Hughes. *Water Res. 18,* 927 (1984).
49. Puri, B.R. ''Physicochemical Aspects of Carbon Affecting Adsorption from the Aqueous Phase,'' Chapter 5 in *Treatment of Water by Granular Activated Carbon.* M.J. McGuire and I.H. Suffet, eds. Advances in Chemistry Series 202. American Chemical Society, Washington, D.C. (1983), p. 77.

8

Operational, Pilot Plant, and Case Studies

Certainly the current water treatment technology is oriented toward the removal of contaminants, mostly organic compounds, by activated carbon. In recent years, GAC has been the predominant mode of treatment despite its many limitations. PAC had been employed in a successful manner for many years for removal of taste and odor constituents from natural surface waters. Much of this PAC was applied empirically and depended on the operator's experience to control the carbon dosage. This, plus a sludge disposal problem, led to the eventual replacement of PAC by GAC. It was thought that if taste and odor organics could be removed by PAC, then organics should be removed by GAC. Consequently, the change in technology of carbon treatment began in the early 1960s when the Kanawha River at Nitro, West Virginia, became highly contaminated with petrochemicals that imparted tremendous tastes and odors to the drinking water. This chapter traces the historical development of PAC and GAC as it relates to the operational and design aspects of drinking water treatment.

TASTE AND ODOR REMOVAL

Nitro, West Virginia

The first full-scale GAC water treatment plant was placed on-stream at Nitro, West Virginia, in 1966 [1]. One of the principal features of this system was that filter sand was replaced with 14 × 40 mesh GAC to remove simultaneously turbidity as well as to adsorb organic compounds and to reactivate the carbon on-site. This step was preceded by 3 years of small-scale experiments and plant-scale tests. Figure 8–1 shows the flow diagram of the original system, the experimental system, and the new system (this system has been replaced since by a regional system; see Anderson et al. [22]). It should be noted that the pretreatment of double aeration was retained in the newest system. Figure 8–2 shows a summary of the operating results for odor removal. The threshold odor value of 3 was the drinking water standard in effect that was met, at least during 1966 and 1967 [2].

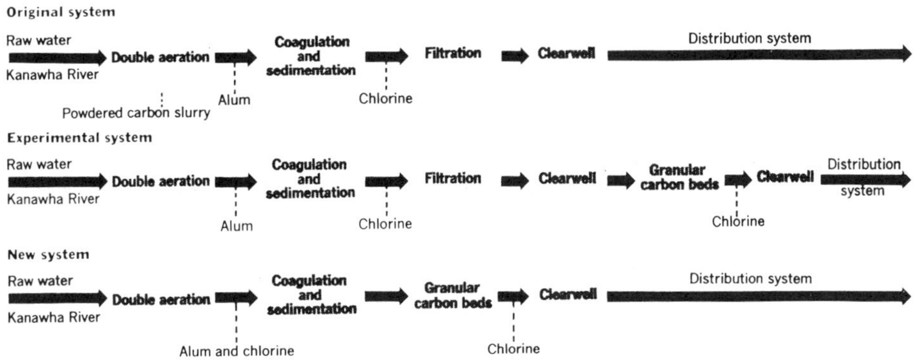

Figure 8–1 Flow diagram of Nitro, West Virginia, water purification plant. Reproduced from Hager [1], courtesy of the American Chemical Society.

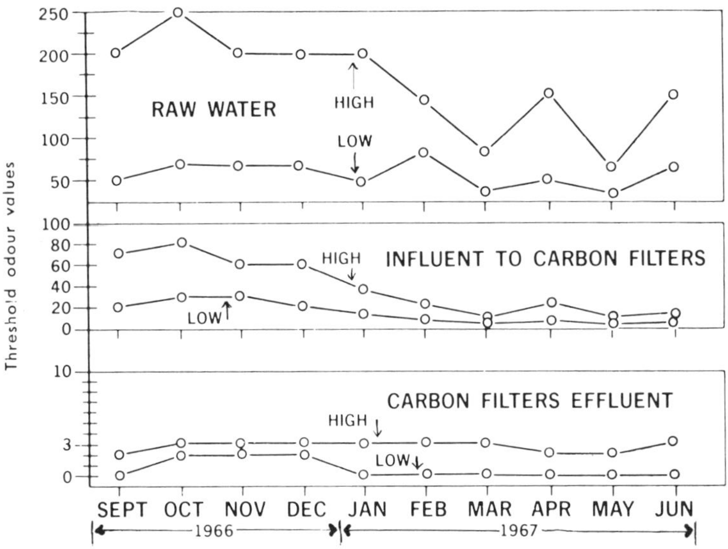

Figure 8–2 Threshold odor removal, Nitro, West Virginia. Reproduced from Hager and Fulker [2], courtesy of the Society for Water Treatment and Examination.

Several operational and design parameters were investigated for the Nitro, West Virginia, treatment plant [3]. The carbon beds contained 3,750 ft^3 (5 × 750 ft^2) of 8 × 30 mesh GAC (Pittsburgh Activated Carbon Co., Pittsburgh, PA). The design flow rate was 5 gpm/ft^2.

Early experiments with the carbon bed design were concerned with the effect of bed depth (or contact time) and carbon particle size [3]. Bed depth was increased by connecting several 5 ft columns in series. Kanawha River water was passed through these columns until breakthrough of odor occurred. A summary of these results is seen in Figure 8–3, where odor was detected in the effluent streams after 2, 8, 10, and 26 days of operation, respectively, for the 5, 10, 15, and 20 ft depths. Inasmuch as one 5 ft column had only 7.5 min contact time, the length of operation is fairly good. Turbidity was not a factor in these experiments. The finer-sized carbon (20 × 50 mesh) had a definite effect since odor did not appear until the nineteenth day and a TON (threshold odor number) of 4 was not reached until the twenty-eighth day. Flow rate was 5. gpm/ft^2 in these experiments (Figure 8–4). This is a surface area effect that may have been a factor in the selection of the 14 × 40 mesh carbon eventually installed at the Nitro plant.

Some fundamental design and operational information came from the early experiences at Nitro [4]. For example, the relationship between contact time within the carbon bed and the breakthrough of odor was established quickly. This is

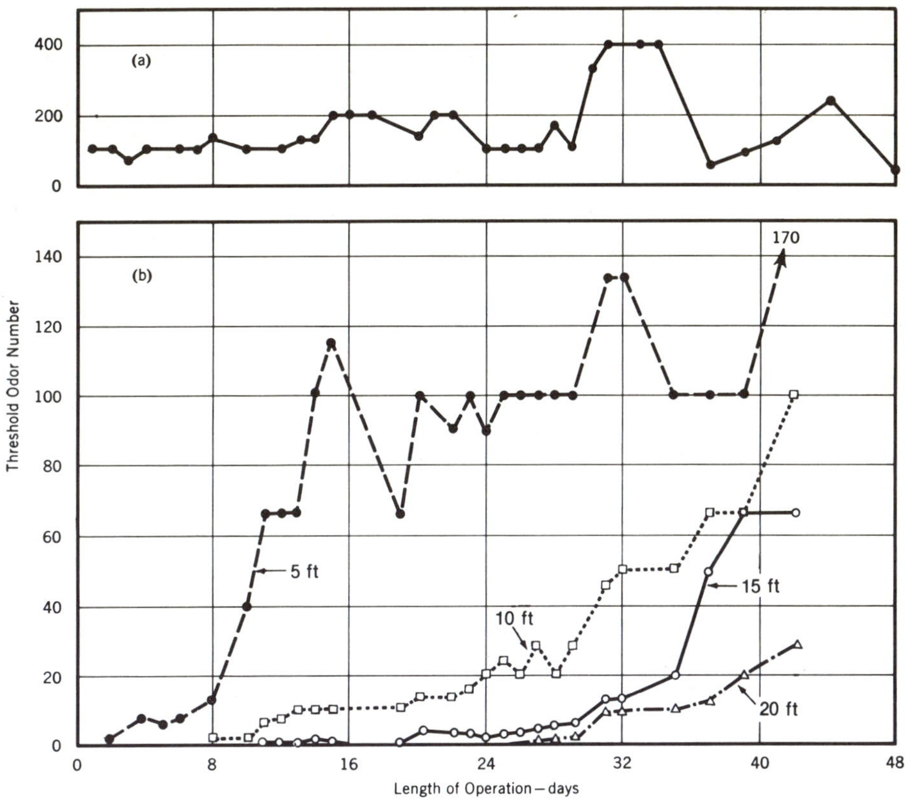

Figure 8–3 Effect of bed depth on passage of odor. The curve in (a) is for carbon bed influent. The curves in (b) are for effluents (at 10 gal/min/ft^2) at various depths in Experiment 1, as follows: 5 ft – – ; 10 ft ······· ; 15 ft —— ; and 20 ft ·—·— . Reproduced from Dostal et al. [3], courtesy of the American Water Works Association.

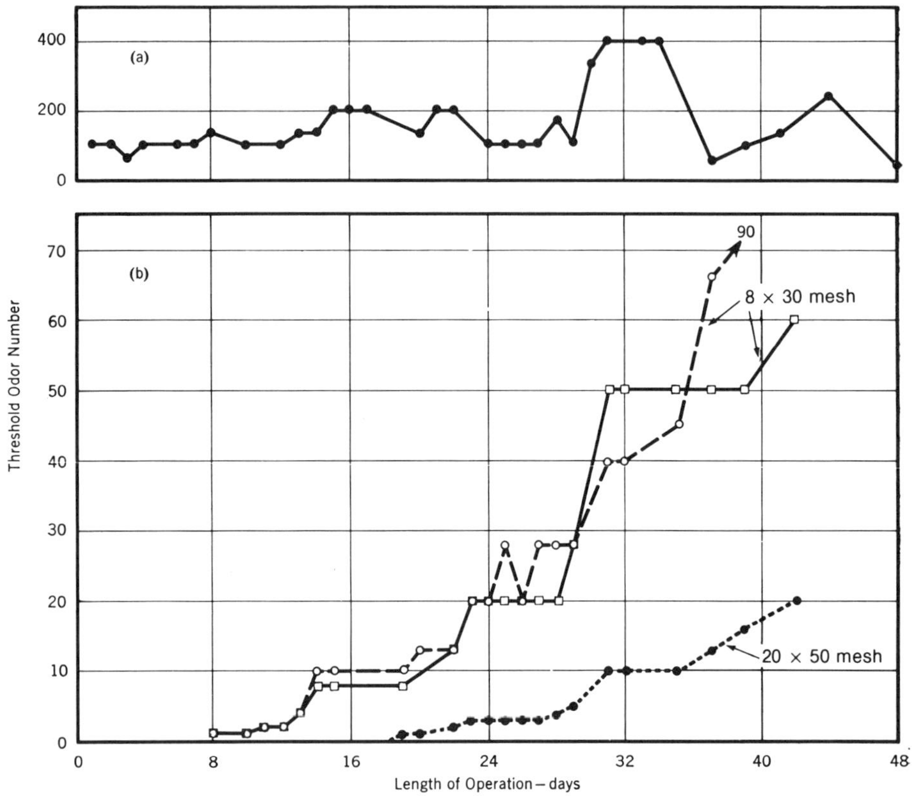

Figure 8–4 Effect of carbon size and activation on passage of odor. The curve in (a) is for carbon bed influent. The curves in (b) are for small carbon column effluent as follows: Experiment 2 (control)———; Experiment 3 (size) ·······; and Experiment 4 (activation) — — . Reproduced from Dostal et al. [3], courtesy of the American Works Association.

seen in Figure 8–5 where the semilog plot is for convenience only. These data came from the operation of the carbon bed as an adsorber and a filter. A period of 30 days at 10 gpm/ft^2 at a 20 foot depth can be considered excellent operation. Table 8–1 summarizes the design criteria for the GAC bed. These criteria were developed before the kinetics of adsorption had been researched extensively. However, the 2.5 to 5.0 gpm/ft^2 flow rates are still employed in the 1980s.

Omaha, Nebraska

The taste and odor qualities of Missouri River water historically have been extremely difficult treatment problems [5]. The causative agents are many and diverse and include silt, decaying vegetation, algae, actinomycetes, and wastewaters from packing plants, stockyards, and oil refineries. The taste and odor problem

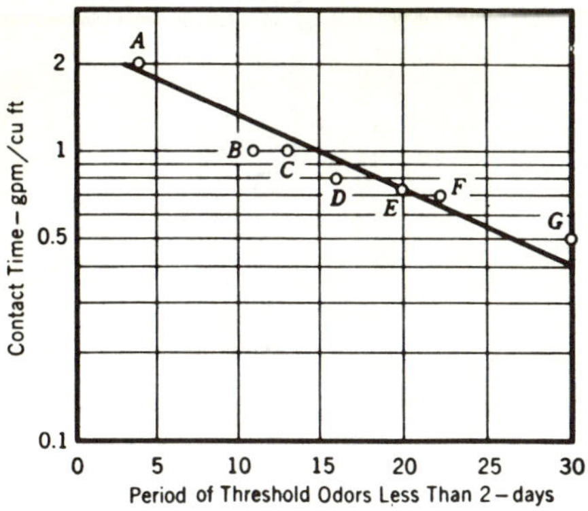

Figure 8-5 Relationship of contact time to reduction of threshold odor (Nitro, West Virginia). The influent threshold odor during the tests averaged 150. The data points represent the following: A, a 10 gal/min/ft^2 flow rate at a depth of 5 ft; B, 10 gal/min/ft^2 flow rate at 10 ft; C, 5 gal/min/ft^2 at 5 ft; D, 4 gal/min/ft^2 at 5 ft; E, 10 gal/min/ft^2 at 15 ft; F, 3.6 gal/min/ft^2 at 5 ft; and G, 10 gal/min/ft^2 at 20 ft. Reproduced from Hager and Flentje [4], courtesy of the American Water Works Association.

is especially apparent during heavy runoff in the early spring. In addition, there is usually severe turbidity due to silts and suspended sediments.

A schematic flow diagram of the Metropolitan Utilities Districts, Omaha, Nebraska, water treatment plant is seen in Figure 8-6 [5]. There are long contact times for the powder and carbon and alum treatments. Chlorine is added intermittently for taste and odor control. Applying it after the presedimentation basin was abandoned eventually in favor of applying it after the upflow sludge contact. This move reduced the chlorine demand and also avoided a chlorinous, or fishy, odor in the finished water. Potassium permanganate had been applied unsuccessfully for taste and odor control. Table 8-2 shows operational data for odor

Table 8-1 Summary of Design Criteria for GAC Filtration

System Factor	*Filtration-Adsorption*	*Adsorption Only*
Flow rate, gal/min/ft^2	$\leq$2.5	2.5–5.0
Bed depths, ft	2.5–5.0	5–10
Backwash, gal/min/ft^2	5–15	5–15
Minimum freeboard, %	30	30
Mesh size	8×30 or 14×40	8×30 or 14×40

Source: Reproduced from Hager and Flentje [4], courtesy of the American Water Works Association.

Table 8–2 Variability in Control of Threshold Odor Numbers, March 1963

Day of Month	*Odor*[a]	*Odor Numbers at Sampling Points*				*Tap Turbidity (ppm)*	*Carbon (ppm)*[b]	*$KMnO_4$ (ppm)*[c]	*Cl_2 (ppm)*[d]
		Raw Water	*Primary Influent*	*Primary Effluent*	*Laboratory Tap*				
8	M	5	5	2.5	2.5	0.10	6–4	1.5–1.0	
9	M	6		3.0	2.0	0.10	4		
10	E	8			2.0 M	0.10	4–8		
11	E and M	6 M	6	3.0	3.0	0.20	8–12	3–1–2	
					S and M				
12	B	22	6 B	5.0 M	4.0 M	0.30	12	2–3	
	O	24							
13	E	24	7 M	5.0	3.0	0.60	12	3–1–2	
14	E and M	13	6 E	12.0 OM	3.0	0.80	12	1–0	3
			and M						
15	E and M	21	8 OM	6.0 OM	8.0 M	0.75	14	2–4	3–5
16	M	21	6	10.0 OM	6.0 M	0.45	14	3–2–3	5–3
17	M	17	8	10.0	5.0	0.25	14	3–2	3–4

18	M	18	12	7.0	4.0	0.25	14	2–0	4–5
19	M	17 M	10	8.0	4.0	0.25	14–10–8		5–3
		21 M							
20	M	10	7	8	5	0.25	10–8		3–2.5
	M	12							
21	M	7	8	14 G	5 M	0.20	8–10		2.5
	M	12							
22	M	17	8	7	4 S and M	0.20	10		2.5–2
23	E	13	8 M	6	4	0.15	10		2.0
24	E	8	6 M	6	4	0.15	10–8		2.0
25	M	18	8	5	3.5	0.15	8–10		2.0
	M	17							
26	M	9	6	6	3.5	0.10	10–4		2.0
	M	12							
27	M	15	8	5	3.0	0.10	4		2.0

Source: Reproduced from Erdei [5], courtesy of the American Water Works Association.
[a]Odor description: B, barnyard; E, earthy; G, grassy; M, musty; O, offensive; S, sweet.
[b]Applied at presedimentation and mixing basin no. 2.
[c]Applied at clarification basin.
[d]Applied at primary effluent.

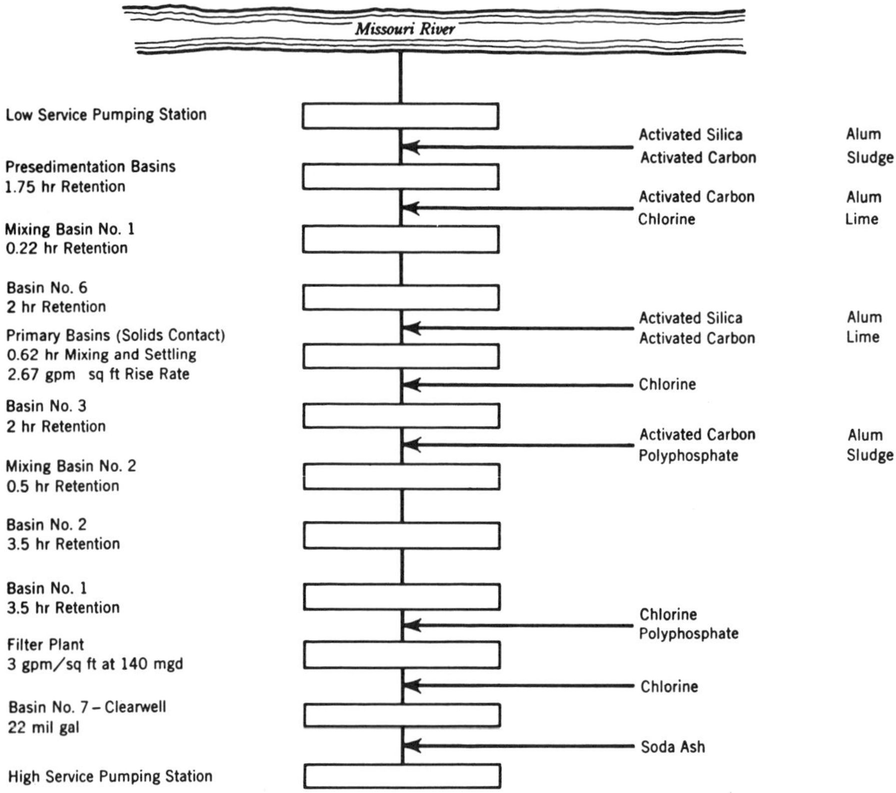

Figure 8–6 Schematic diagram of treatment at metropolitan utilities. All retention times are based on the maximum plant capacity of 140 mgd. Reproduced from Erdei [5], courtesy of the American Water Works Association.

control in the Missouri River water. That this is an extremely difficult treatment problem is seen in the odor values in the tap water. These data were obtained in March, when low water temperatures retard the rate of chemical reactions for odor removal. It is apparent that the carbon treatment alone did not remove the tastes and odors, but the Missouri River represents the concerted effort required to produce palatable water from highly contaminated raw sources.

Sulfide Odors

Occasionally a municipal water supply, usually from groundwater, is contaminated with hydrogen (poly) sulfides whose odor is rather familiar. This odor can be controlled rather easily with PAC and GAC. For example, the Galeta Water District, California, installed four pressurized GAC adsorbers that followed diatomaceous earth filters [6]. Cylindrical pressure tanks were filled with 8 feet of carbon that rested on stainless steel screens that were operated at 5 gpm/ft^2. These beds were operated for 2 years before reactivation. In contrast, one report showed that GAC was only partially effective in removing hydrogen (poly) sulfides [7].

Little Falls, New Jersey

Another pioneering effort in the development of GAC for taste and odor control, as well as removal of organic compounds in general, was conducted at the Passaic Valley Water Commission's plant at Little Falls, New Jersey [8]. Here, the Passaic River was (and still is) grossly contaminated with many diverse organics. During a plant reconstruction project, it was decided to replace the sand and gravel media in two 540 ft^2 gravity filters with two types of GACs for the experimental removal of tastes and odors. These filters were operated at 20 gpm/ft^2. The results from a 213 day test were not encouraging or conclusive. As seen in Figure 8-7, the organic compounds measured by the carbon chloroform extract (CCE) technique were less in the raw water than in the filtered water after 213 days. However, there was considerable removal of the CCE within the first 62 days of operation. There was no regeneration of the GAC during the 213 day experiment. In any event, the data gave an early indication of organics removal under plant conditions.

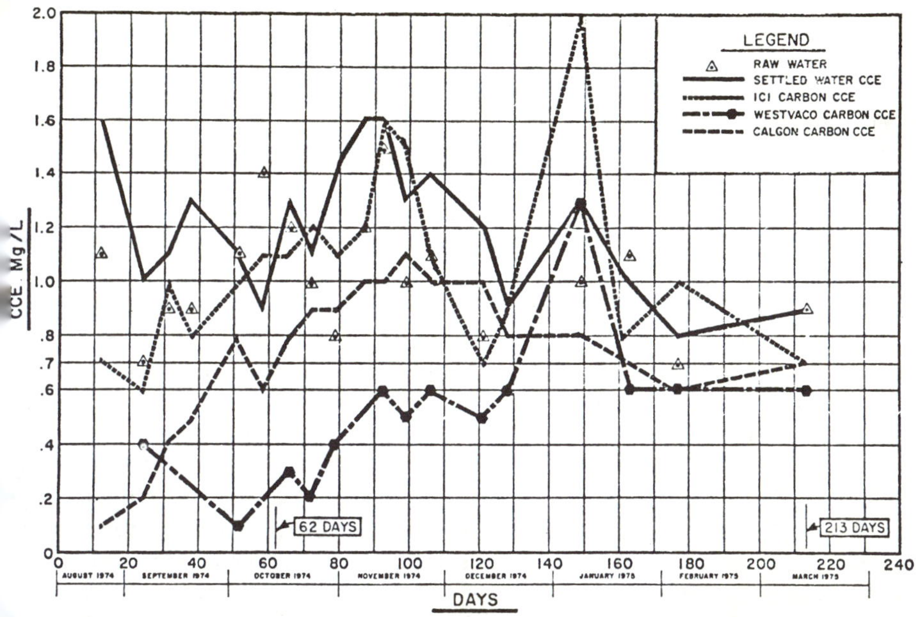

Figure 8-7 Passaic Valley Water Commission GAC sampling and analyses, 1974-1975 [8].

REMOVAL OF ORGANIC COMPOUNDS

Chloroform and Total Organic Carbon

There are several levels of investigations between the identification of an organic contaminant in a water supply and its subsequent removal at a treatment plant. Accordingly, at least three levels of investigation have been identified for the eventual design of full-scale systems for removal of trace quantities of organics:

a. small-scale laboratory research using synthetic mixtures;
b. medium-scale pilot plant studies using actual water which will be treated at the plant; and
c. full-scale testing of a modular unit. [9]

Many independent variables affect each of these levels. Small bench-scale studies acquire a lot of information in a short period of time. Adsorption isotherm and breakthrough curve investigations provide considerable information on the capacity and the breakthrough kinetics of an adsorbate. However, these data cannot be employed immediately to design a full-scale carbon adsorption unit. Additional tests are needed in systems with mixtures of organics at the concentrations found in the water. Medium-scale pilot experiments usually accomplish this point.

One of the most important items that evolve from pilot-scale studies is the hydraulic characteristic of the carbon system. This leads to a preliminary design of the eventual treatment facility. However, it is extremely desirable to construct and operate a smaller version of the full-scale plant before the final design is selected. A recommendation that a 1 to 10 mgd system should follow the pilot-scale study [9] is feasible only when there are sufficient financial resources.

To illustrate the hydraulics of the carbon system, a series of experiments was conducted by the Philadelphia Water Department [10]. In one study, carbon replaced sand in a rapid filter. This mode was compared to a carbon bed that followed the conventional filtration process. These means of GAC treatments were labeled the *filter/adsorber mode* and the *postcontactor mode,* respectively. There was a statistically significant difference in performance between the two types of carbon installations. The postcontactor vessel outperformed the filter/adsorber by 20 to 30% in removal of $CHCl_3$ and 11 to 16% in removal of TOC.

An important factor in the eventual design of a carbon bed is the EBCT (see also Chapter 4). Most pilot-scale studies incorporate this factor into their experiments. Such a study was conducted for EBCTs of 7.5, 1.5, 22.5, 30, and 60 minutes for breakthrough characteristics and mass loading curves for $CHCl_3$ and TOC [9]. Figure 8–8 shows a typical pilot-scale system for the Philadelphia study, and Table 8–3 gives the experimental conditions. The carbon was F-400 (Calgon) in two sizes: 8 × 30 mesh and 12 × 40 mesh. In Figure 8–8, columns 1 to 5 were employed for the effect of EBCT. Columns 6 and 7 were utilized for the influence of aqueous chlorine on the GAC process, and columns 8 and 9 were used to evaluate pretreatment.

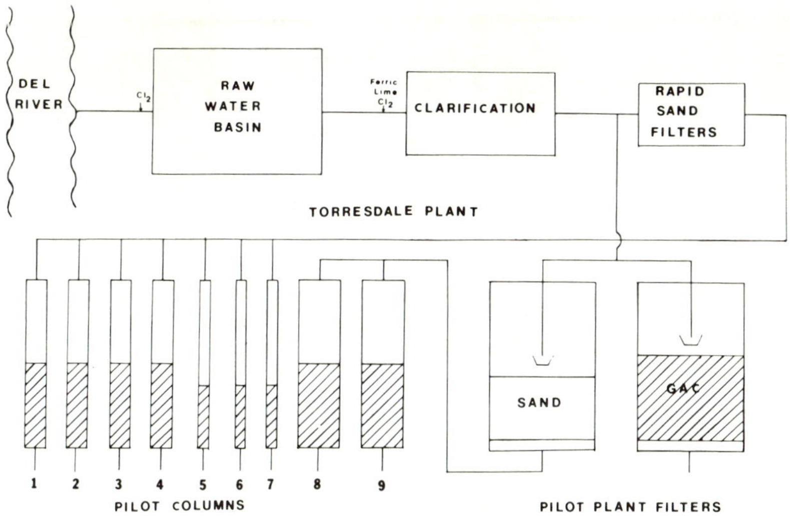

Figure 8–8 Pilot columns. Reproduced from Cairo et al. [9], courtesy of Ann Arbor Science Publishers, Inc.

Breakthrough curves are seen in Figure 8–9 for $CHCl_3$ for the five contact times of 7.5 to 60 minutes [9]. The influent concentrations of $CHCl_3$ ranged from 44 to 176 μg/l throughout the study period. It should be noted that this study did not use synthetic water; rather, Delaware River water, with its many organic contaminants, was used. Since different contact times were reached by varying the flow rate in each column, the columns with the shorter contact times and higher flow rates were exposed to a greater influent mass loading of $CHCl_3$ than the columns with longer contact times. This variation in influent mass loading of $CHCl_3$ should be considered when evaluating each column's performance.

GAC was apparently reasonably effective in removing the $CHCl_3$ at least initially into the breakthrough curves. The time to initial breakthrough corresponded closely to the EBCT. After breakthrough, the columns were saturated quickly and exhaustion was reached. It should be noted that the curves in Figure 8–9 describe the adsorption of only $CHCl_3$ in the presence of many organic compounds in the Delaware River water. This explains the erratic behavior of the curves on exhaustion. Also, there were several times when the effluent content of $CHCl_3$ exceeded the influent for the same time frame. This is the so-called chromatographic effect where competitive adsorption causes desorption of $CHCl_3$ in this case.

Figure 8–10 shows the $CHCl_3$ mass loading curves for the five EBCTs. It appears that there is a direct proportionality between contact time and the breakthrough characteristics and the mass loading removal of $CHCl_3$. Whatever is the mechanism of adsorption, it is apparently the same regardless of EBCT, flow rate, or linear velocity. The important design consideration inherent in this find-

Table 8–3 Pilot Column Experimental Conditions

Column Identification		*Dimension*			*Hydraulics*				*Remarks*	
Number	*Media*	*Size (in diameter)*	*Depth (ft)*	*Volume (ft^3)*	*EBCT (min)*	*Volumetric Loading (gal/min/ft^3)*	*Velocity (ft/min)*	*Surface Loading (gal/min/ft^2)*	*Start*	*Duration (wk)*
1	F-400	6	4	0.785	7.5	1.0	0.535	4.0	9/19/77	19
2	F-400	6	4	0.785	15.0	0.500	0.267	2.0	9/19/77	22
3	F-400	6	4	0.785	22.5	0.333	0.178	1.33	9/19/77	30
4	F-400	6	4	0.785	30.0	0.250	0.134	1.0	9/19/77	30
5	F-400	3	3	0.147	60.0	0.125	0.049	0.37	9/19/77	100
6	F-400	3	3	0.147	7.5	1.0	0.401	3.0	9/19/77	16
7	F-400	3	3	0.147	7.5	1.0	0.401	3.0	9/19/77	16
8	F-300	12	4	3.142	15.0	0.500	0.267	2.0	10/11/77	25
9	F-400	12	4	3.142	15.0	0.500	0.267	2.0	10/11/77	25

Source: Reproduced from Cairo et al. [9], courtesy of Ann Arbor Science Publishers, Inc.

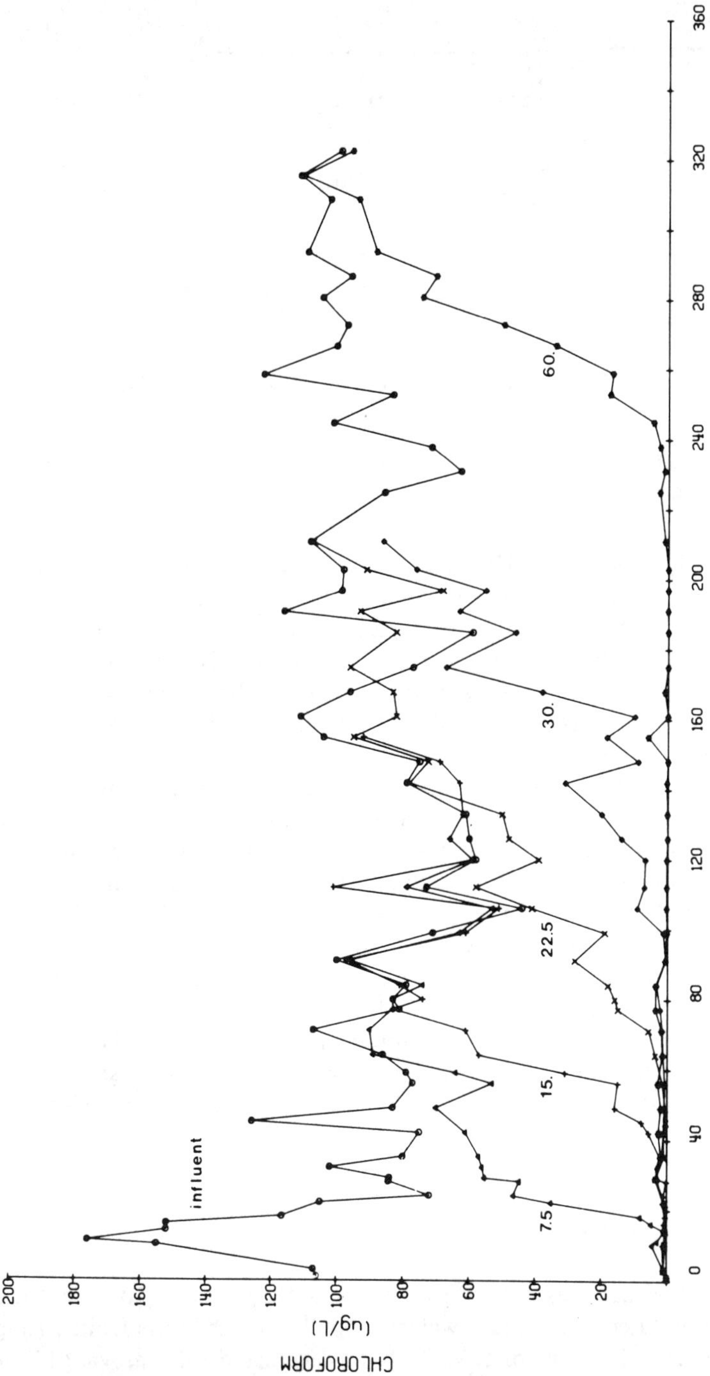

Figure 8–9 Chloroform breakthrough curves for EBCT from 7.5 to 60 min. Reproduced from Cairo et al. [9], courtesy of Ann Arbor Science Publishers, Inc.

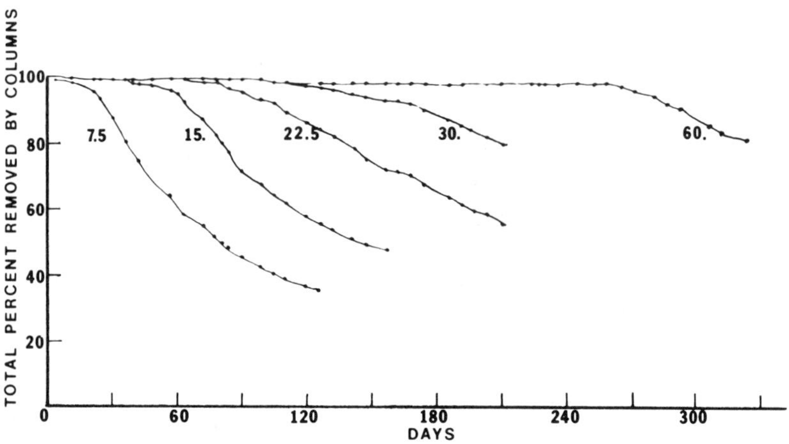

Figure 8–10 Chloroform mass loading curves. Reproduced from Cairo et al. [9], courtesy of Ann Arbor Science Publishers, Inc.

ing is that the daily removal of $CHCl_3$ per mass of carbon is the same, whereupon the amount of regeneration of carbon would be identical also.

Removal curves for the organic surrogate parameter, TOC, are seen in Figure 8–11 for the five EBCTs. These curves represent the adsorptive behavior of all organic compounds in the Delaware River, more or less. There were apparently two mechanisms of TOC removal operative in the columns [9]. In section 1, TOC removal was effected primarily by adsorption in the initial weeks from start-up of the column to the level portion of the breakthrough curves. After about 4 weeks, biological activity also began to remove TOC. In section 2, biological activity apparently was the predominant removal mechanism. The latter was followed by a determination of bacterial densities in the column's effluent. Each of the curves in Figure 8–11 levels off in section 2 where biological activity maintains the TOC removals. In fact, the 60 min EBCT column maintained a reasonably high removal, > 53%, through 2 years of operation [9]. Whether or not this would be a practical contact time for plant operation is an important consideration. It may be economically impractical to construct carbon beds of sufficient size for a 60 minute EBCT. In any event, these pilot-scale studies have demonstrated the effect of EBCT on $CHCl_3$ and TOC removal along with plant design implications.

Haloforms and Pesticides

The pilot-scale approach was utilized to investigate the possibility of converting the industrially polluted River Trent water to potable water in the United Kingdom [11]. Here, the efficiency of GAC beds was evaluated for removal of TOC and certain specific organics. Included in this study was a pilot plant investigation

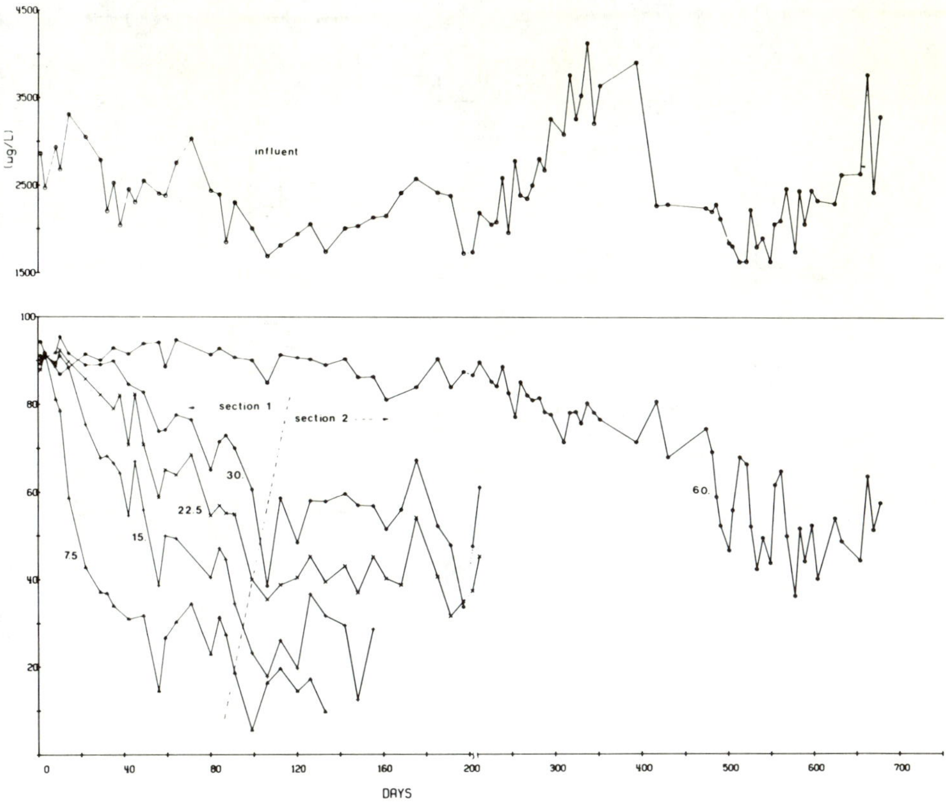

Figure 8–11 TOC removals for EBCT from 7.5 to 60 min. Reproduced from Cairo et al. [9], courtesy of Ann Arbor Science Publishers, Inc.

using several GACs for conventional treatment of River Thames water spiked with selected organic compounds.

The flow diagram for the River Trent water treatment is seen in Figure 8–12a. The GAC portion of this plant consisted of two 0.67 $m^2 \times 4$ m high bitumastic-lined steel tanks that were capable of series or parallel operation. The pilot plant for the River Thames water is seen in Figure 8–12b where a conventional flow diagram is employed. Here the GAC beds consisted of two columns of 0.3 m diameter by 2.5 m high, six of 0.15 m diameter $\times$ 4 m high, and one of 0.05 m diameter by 2.7 m high arranged in five parallel streams. This system was spiked with chloroform, bromodichloromethane, 1,2-dichlorobenzene, α BHC, γ BHC, and dieldrin. The final concentrations for these organics, in μg/l, were 200, 200, 20, 5, 5, and 5, respectively. Six types of carbons were investigated also: 207C (coconut), Anthrasorb CC 1230H (anthracite), Nuchar WV-G (bituminous coal), Haycarb (coconut), Filtrasorb 400 (coal), and ROW 0.8 Supra (peat). Mean particle diameters of these carbons ranged from 0.3 to 1.7 mm.

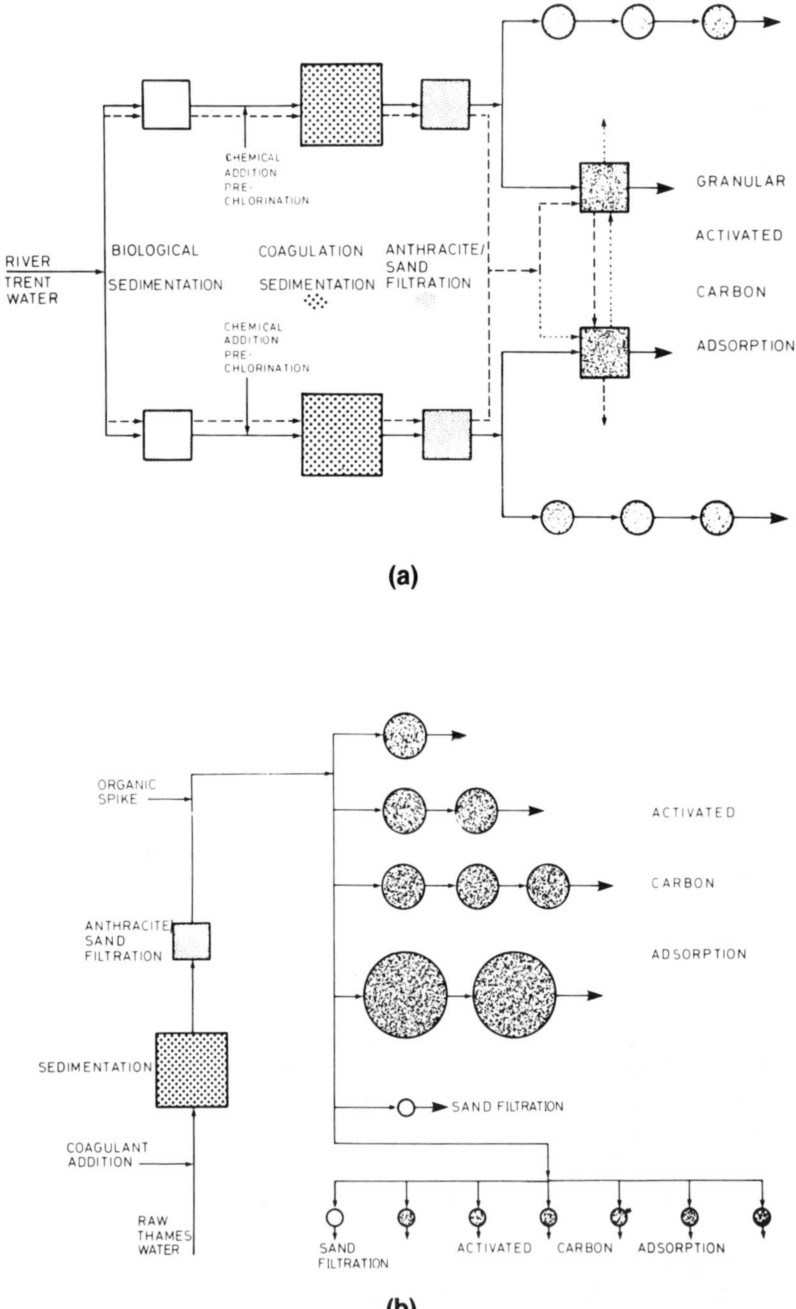

Figure 8–12 (a) Flow sheet of the GAC pilot plant treating River Trent water. (b) Flow sheet of the GAC pilot plant treating River Thames water. Reproduced from Hyde [11], courtesy of Ann Arbor Science Publishers, Inc.

Preliminary studies revealed that bromodichloromethane was removed better than chloroform and that dieldrin was adsorbed less than the other three chlorinated hydrocarbons [11]. Consequently, $CHCl_3$ and dieldrin were investigated in detail as being representative of their respective groups of organics.

For the six carbons, 207C was one of the best for removal of haloforms but one of the worst for pesticides (Figure 8–13) [11]. The reverse was observed for CC 1230H. Dieldrin yielded a greater percentage of removal on four of the carbons. Excessive biological growth on the carbon beds that was caused by continuous addition of methanol used for the organic spike was an operating problem. Filtrasorb 400 and Nuchar WV-G were the best for removal of organics (by TOC and ultraviolet adsorbance) in general and for the pesticides specifically. Haycarb and 207C were the most efficient for haloforms. These observations are

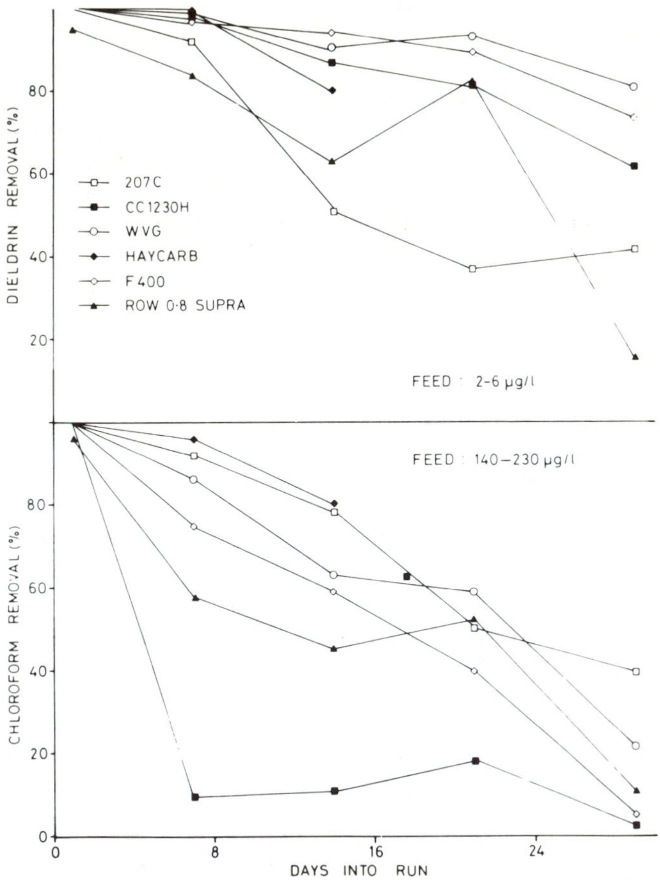

Figure 8–13 Comparison of chloroform and dieldrin removal by various carbons. Reproduced from Hyde [11], courtesy of Ann Arbor Science Publishers, Inc.

expected in view of the different sources of carbon with their various functional groups on their surfaces (see Chapter 5).

Design studies were conducted also for the pilot plants seen in Figures 8–12a and b [11]. These included the effect of contact time, the number of beds in series, and the downflow velocity using Filtrasorbs 300 and 400. Breakthrough curves for $CHCl_3$ were similar to those cited for the Philadelphia study [9] for the same contact time. In a single carbon bed operation, an increase in the contact time by increasing the bed depth and with a constant downflow velocity resulted in a decrease in the volume of water treated per unit mass of carbon. However, this treated water volume was increased by multibeds in series, which increases the overall contact time. Increasing the downflow velocity from 1.8 to 8.4 m/hr apparently decreased the capacity of carbon to adsorb the organics.

Chlorinated Benzenes, Total Organic Carbon, and Total Organic Halogens

Three full-scale GAC columns were employed for the removal of several organic compounds in an extensive 1 year study [12]. A parallel arrangement of the GAC system permitted direct comparison of fresh, once-regenerated, and exhausted Filtrasorb 300. This system is part of the Santa Clara (California) Valley Water District's Palo Alto Reclamation Facility for additional treatment of a chlorinated activated sludge effluent. This water is ultimately injected into an aquifer as part of a groundwater recharge system. The schematic flow diagram is seen in Figure 8–14 where the upflow GAC columns were operated at a superficial linear velocity of 4.4 gpm/ft^2 that provided an EBCT of 34 min. Removal of organic compounds was followed for individual species or collectively by TOC or TOX.

That there was an immediate breakthrough of 1.0 to 1.5 mg/l TOC on start-up of both the fresh and regenerated columns is seen in Figure 8–15 [12]. Apparently, 15 to 20% of the influent TOC was not amenable to adsorption by GAC. This is a common operating problem for this type of treatment facility. Similar observations were made with TOX. The cumulative removal of TOC by adsorption was 34 mg TOC/g for fresh C and 37 mg TOC/g for regenerated C. For TOX, the cumulative removals were 1.1 mg TOX/g for fresh C and 1.8 mg TOX/g for regenerated C. In addition, there was considerable removal by biodegradation.

Table 8–4 summarizes the removal behavior for individual compounds by GAC at the Palo Alto Reclamation Facility [12]. Chlorinated benzenes and naphthalenes were removed significantly throughout the 1 year study. Lesser removals were recorded for ethylbenzene, the xylenes, and heptaldehyde. Few differences were seen for the fresh and regenerated carbons whereas, in some cases, the exhausted carbons had considerable capacity to remove some of the compounds.

Another full-scale GAC system was investigated at the Cincinnati Water Works (CWW), using Ohio River Water. The water intake is located on this river

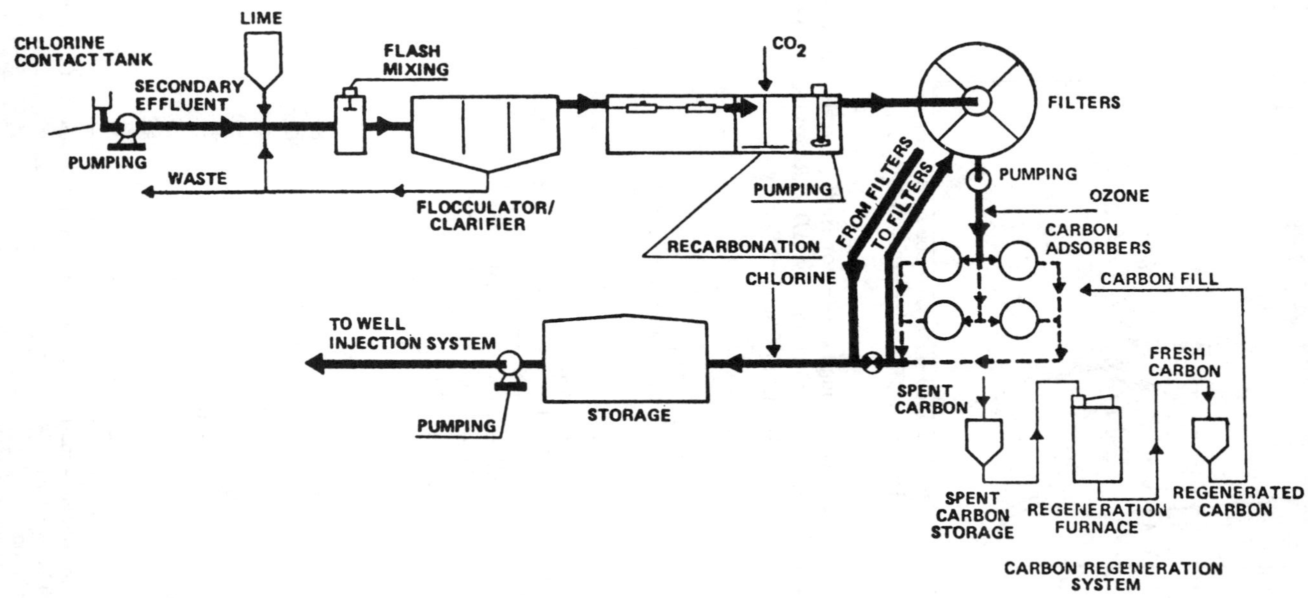

Figure 8–14 Schematic flow diagram of the Palo Alto Reclamation Facility. Reproduced from Summers and Roberts [12], courtesy of the American Chemical Society.

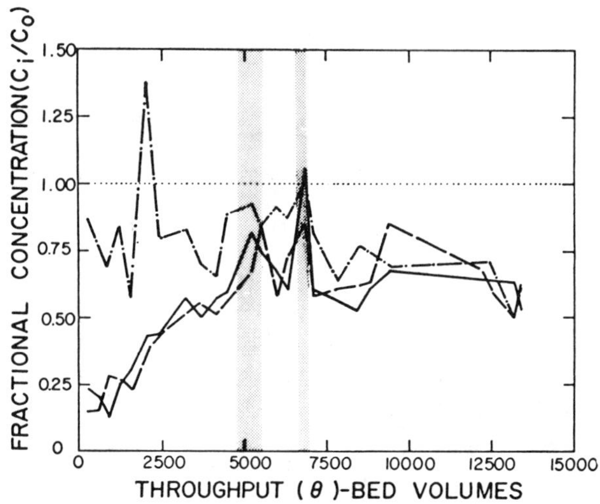

Figure 8-15 TOC fractional breakthrough curves for fresh (-), regenerated (--), and exhausted GAC (-•-). Conditions: □, high pH and low ΔDO. Reproduced from Summers and Roberts [12], courtesy of the American Chemical Society.

Table 8-4 Aromatic Compounds: Concentration and Removal

Compound	*Influent[a,b] Concentration (μg/l)*	*Average Percent Removal*		
		Fresh[a]	*Regenerated[a]*	*Exhausted[c]*
1,3-dichlorobenzene	1.47 ± 1.67	80	85	76
1,4-dichlorobenzene	2.50 ± 1.89	92	89	81
1,2-dichlorobenzene	2.73 ± 2.30	90	91	80
1,2,4-trichlorobenzene	8.33 ± 5.60	90	85	91
Naphthalene and 1,2,3-trichloro-benzene	3.22 ± 2.97	90	89	79
Chlorobenzene	0.56 ± 0.48	56	62	7
Ethylbenzene	0.16 ± 0.12	37	43	8
p-xylene	0.07 ± 0.04	37	37	−29
m-xylene	0.16 ± 0.11	20	11	−146
2-methylnaphthalene	0.10 ± 0.10	62	78	64
1-methylnaphthalene	0.12 ± 0.13	61	83	73
Heptaldehyde	0.17 ± 0.33	40	30	48

Source: Reproduced from Summers and Roberts [12], courtesy of the American Chemical Society.
[a]Average over 0–13,000 bed volumes.
[b]Mean ± standard deviation.
[c]Average over 13,000–17,500 bed volumes.

below six major tributaries, one of which is the infamous Kanawha River. The GAC systems (Figure 8–16) were arranged in two modes: the first was acting as a combined filter and adsorber and was labeled the sand replacement system. These GAC beds receive coagulated and settled water. The second mode used the carbon as an adsorber only since conventional sand filters were placed ahead of the C beds. This was labeled the postfiltration adsorption unit. The GAC was 12 × 40 mesh (brand name was not specified).

Figure 8–17 shows breakthrough curves for TOC removal at EBCTs of 4.5, 7.5, and 16.0 minutes for the CWW study [13]. The immediate breakthrough of TOC was noted here also. It is not certain whether this TOC is from short circuiting water or leachate from the carbon.

Small-scale pilot plant studies were continued on the GAC removals of several organic compounds from Delaware River water at Philadelphia [14]. Filtrasorb 400, 1 ft^3, was placed into two 7.5 cm × 1 m in columns through which the flow rate was 1 gpm/ft^3 and the contact time was 7.5 min. Table 8–5 is a summary of the breakthrough and adsorptive characteristics of the organics monitored in this study. These results would be typical of carbon beds receiving prechlorinated, coagulated, settled, and sand-filtered Delaware River water. Influent

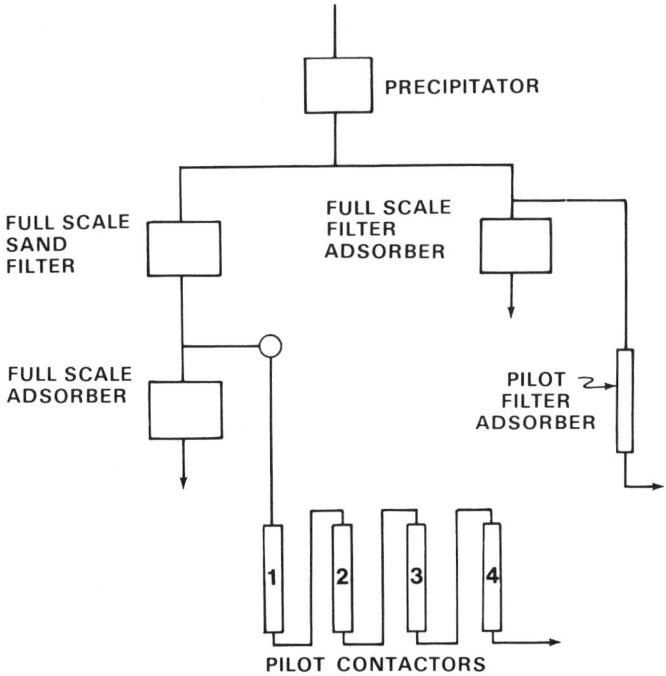

Figure 8–16 Schematic of carbon units for the Cincinnati Water Works. Reproduced from DeMarco et al. [13], courtesy of the American Chemical Society.

Table 8–5 Summary of Breakthrough and Adsorption Characteristics

Compound	*Influent Concentration*	*Column*	*Week of Maximum Breakthrough*	*Depth of Core Mass Centers (in)*	*Comments*
TOC	1.85 mg/l (week 13),	Cl_2	Immediate pseudo-steady state after 3 wk on both columns		Effluent ranges 1.0 to 1.5 mg/l (week 13)
	3.30 mg/l (week 2)	No Cl_2			1.2 to 1.7 mg/l (week 13)
$CHCl_3$ (instantaneous)	71 μg/l (week 14),	Cl_2	Both columns 50% exhaustion after 4 wk, 100% after 9 wk		Both effluents mimicked influent > 50 μg/l after 3 wk
	176 μg/l (week 2)	No Cl_2			
$CHCl_{3(1.5)}$	117 μg/l (week 12),	Cl_2	Instantaneous breakthroughs		
	307 μg/l (week 2)	No Cl_2	100 μg/l in fifth week		

$CHCl_1$ (precursor) as $CHCl_3$	23 μg/l (week 13),	Cl_2	>100%—6,7,8,11,12		
	155 μg/l (week 2)	No Cl_2	>100%—4,6,11,12,13		
p-DCB	<10–137 ng/l	Cl_2	Low levels—2.13	Diffuse—15,30	Inhibition top 7.5 cm (3 in)
		No Cl_2	None detected	Sharp—21,33	
Bisethane	<10–460 ng/l	Cl_2	7,11,14	Diffuse—15,27	Periodic effluent slugs
		No Cl_2	6,8,11,13	Sharp—21,33	Effluent follows influent
Diacetone-L-sorbase	<10–830 ng/l	Cl_2	11	Lower concentrations	Acid hydrolysis?
		No Cl_2	9,13	Higher concentrations	
Dibutyl phthalate	<5–480 ng/l	Cl_2	12	Diffuse —15,33	Inhibition top 15 cm (6 in)
		No Cl_2	3,6,11	Sharp—24,36	

Source: Reproduced from Yohe et al. [14], courtesy of the American Water Works Association.

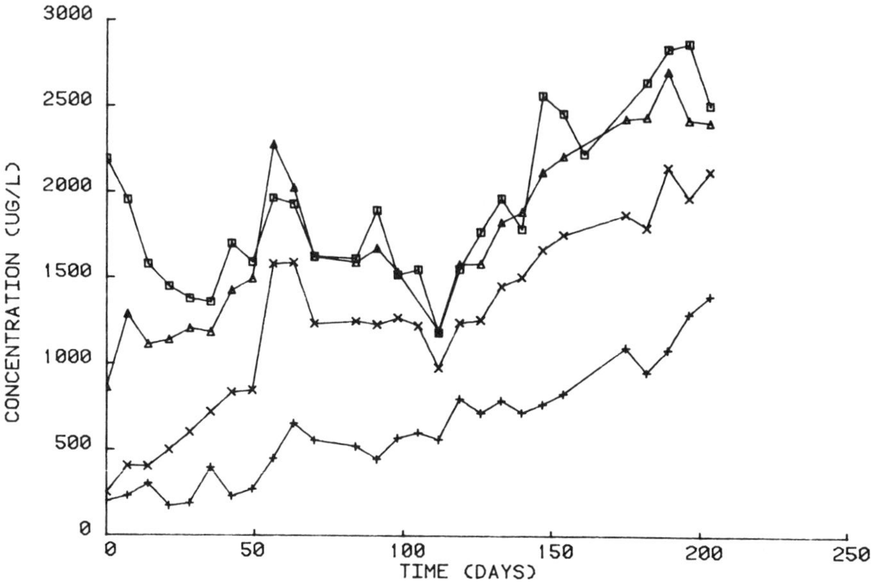

Figure 8–17 Effect of contact time on TOC. Key: □, sand filter effluent; Δ, contactor effluent (4.5 min EBCT); X, contactor effluent (7.5 min EBCT); and +, contactor effluent (16.0 min EBCT). Reproduced from DeMarco et al. [13], courtesy of the American Chemical Society.

concentrations of the organics are quite variable. Consequently, there was erratic behavior of the breakthrough curves. Frequently, the effluent concentrations exceeded the influent concentrations because of the chromatographic effect, where previously adsorbed compounds are displaced by incoming compounds. This was especially noticeable when the organic concentrations were 1 μg/l and less.

Organic Solvents

Laboratory and pilot-scale studies were conducted to determine the effects of aeration, adsorption, and boiling on water containing several organic solvents: TCE, PCE, 1,1,1-trichloroethane, cis-1,2-dichloroethylene, carbon tetrachloride, and 1,2-dichloroethane [15]. In summer and fall 1977, the U.S. EPA DWRD had constructed several pilot-scale adsorption columns at several sites in New England and New Jersey. One experimental column contained 75 cm of GAC (Filtrasorb 400), and a parallel column had 75 cm of a synthetic resin (Ambersorb XE-340) (see below). Figure 8–18 gives a composite of the adsorption isotherms for the six compounds (see also Figure 8–32). There was a spectrum of adsorptivities and efficiencies of removal of these compounds under competitive conditions. These were groundwaters with little or no effects from particulate matter.

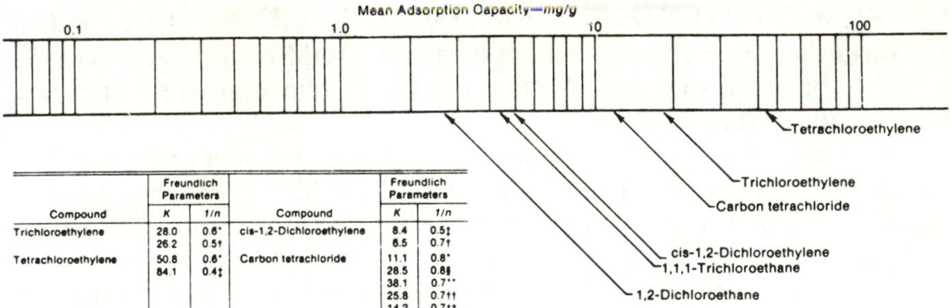

Compound	Freundlich Parameters		Compound	Freundlich Parameters	
	K	1/n		K	1/n
Trichloroethylene	28.0	0.6*	cis-1,2-Dichloroethylene	8.4	0.5‡
	26.2	0.5†		6.5	0.7†
Tetrachloroethylene	50.8	0.6*	Carbon tetrachloride	11.1	0.8*
	84.1	0.4‡		28.5	0.8§
				38.1	0.7**
				25.8	0.7††
				14.2	0.7‡‡
				14.8	0.4‡
1,1,1-Trichloroethane	2.5	0.3*	1,2-Dichloroethane	3.6	0.8*
	9.4	0.5‡		5.7	0.5‡

*Calgon Filtrasorb® 300[43]
†Calgon Filtrasorb® 300[44]
‡Witco Witcarb® 950[44]
§Norit[45]
**Calgon Filtrasorb® 400[44]
††Westvaco Nuchar® WV-G[45]
‡‡ICI Hydrodarco® 1030[45]

Figure 8–18 Comparison of isotherm adsorption capacities on activated carbon. Equilibrium concentration = 500 μg/l. Reproduced from Love and Eilers [15], courtesy of the American Water Works Association.

Trihalomethanes and Precursors

Early experience with the removal of THMs from water was reported from the Netherlands [16]. Forty percent of their water supply is taken from the Rhine and Meuse Rivers, the Yssel Lake, and the Haringvliet. In 1977, a new 5,000 m^3/hr plant was built at the Kralingen, Rotterdam Waterworks. Water is treated by coagulation and sedimentation in lamella separators, ozonation with 3 mg/l O_3, secondary iron dosing, double-layer filtration, carbon filtration, and safety chlorination with 0.8 mg/l Cl_2. Initially, this plant did not have carbon filtration, with the consequence that the total THM concentration averaged 103. μg/l in the raw water and 102. μg/l in the finished water. Table 8–6 shows the conditions for the carbon filter that was eventually installed, a treatment that was apparently

Table 8–6 Conditions for the Carbon Filters at Kralingen

Filter Condition	*Value*
Particle size, mm	0.8
Quantity of adsorbent per filter, m^3	116
Depth of adsorbent, m	4
Free board, m	3
Filter diameter, m	6
Contact time, min	12[a]

Source: Reproduced from Meijers et al. [16], courtesy of the American Water Works Association.
Note: Adsorbent used was Norit ROW 0.8 Supra and N.
[a]5 empty bed volumes/hr.

effective (Table 8-7). However, an interesting comment was that "the Rotterdam waterworks does not consider GAC treatment for THM removal economically feasible." Apparently they were able to control THM formation by manipulating the chlorination processes.

Other early studies of THM removal by carbon under plant conditions were reported by the American Water Works Service Co. [17]. This company operates more than 100 plants in 20 states. For example, the Davenport (Iowa) Water Co., East River station, is a 30 mgd clarification plant that employs flocculation, sedimentation, and filtration through GAC filters for treatment of Mississippi River water. Chlorine dioxide was used for disinfection. The removals of THMs were unexpectedly low and inefficient. For example, in May and July 1978, THMs were reduced by 23 and 31%, respectively, using carbon (Filtrasorb 400) that had been in service for 28 mo and had processed over 10^6 gal water per ft^3 of carbon (Table 8-8). These removals were confirmed by a pilot column packed with spent carbon (Table 8-9). This column and the plant's GAC filters were operated at 2 gpm/ft^2 with a 7.5 min contact time. Subsequently, the carbon was changed in July and August 1978, which effected a slight improvement in THM removal (Tables 8-8 and 8-10). However, the overall removal of THMs cannot be considered to be acceptable.

Three years of research were conducted at the 60 mgd Preston water treatment plant in Hialeah, Florida, where groundwater is obtained from wells 60 to 100 feet deep in limestone aquifers [18]. This groundwater had 10 mg/l TOC, and the THMFP ranged from 650. to 950. $\mu g/l$. After lime softening, about 18 mg/l Cl_2 was added for breakpoint formation. All these steps resulted in a THM content of 300 $\mu g/l$ at the consumer's tap. The pilot plant research involved four

Table 8-7 THM Concentrations at Kralingen with Carbon Filtration

Running Time of Carbon	$CHCl_3$[a]	$CHCl_2Br$	$CHClBr_2$	$CHBr_3$	*Total THMs*
1 mo after carbon filtration	1.2	0.8	0	0	2
1 mo after safety chlorination	3.2	3.7	5.4	5.7	18
2-5 mo after carbon filtration	17	11	4	0	32
2-5 mo after safety chlorination	18	12.5	13.5	8.5	52
7-10 mo after carbon filtration	5	3	1.5	0	9.5
7-10 mo (winter) after safety chlorination	6.5	6.5	9.5	6.5	29

Source: Reproduced from Meijers et al. [16], courtesy of the American Water Works Association.
[a]In $\mu g/l$.

ble 8–8 Davenport Water Co.: Initial Removal of Trihalomethanes by GAC Treatment

mpling ate	Sampling Location	Trihalomethanes (μg/l)					Percent Reduction
		$CHCl_3$	$CHBrCl_2$	$CHBr_2Cl$	$CHBr_3$	TTHM	
'8/78	Settled water effluent basin 1	93	4	<1	<1	97	
'8/78	Settled water effluent basin 2	97	4	<1	<1	101	
'8/78	Plant effluent	71	3	<1	2	76	23
'20/78	Settled water basins, 1 and 2	104	5	<1	<1	109	
'20/78	Filter 2[a]	73	2	<1	<1	75	31
20/78	Filter 9[b]	42	1	<1	<1	43	61
20/78	Plant effluent	42	2	<1	<1	44	60

urce: Reproduced from Blanck [17], courtesy of the American Water Works Association.
S-400, Calgon Corp., Pittsburgh, PA.
WS, Calgon Corp., Pittsburgh, PA.

ble 8–9 Davenport Water Co.: Pilot Column Test Using Exhausted Carbon

mpling ate	Sampling Location	Trihalomethanes (μg/l)					Percent Reduction
		$CHCl_3$	$CHBrCl_2$	$CHBr_2Cl$	$CHBr_2$	TTHM	
/10/78	Mississippi River water	12	<1	<1	<1	12	
/10/78	Settled water effluent basins 1 and 2	107	2	<1	<1	109	
/10/78	Pilot column effluent	81	1	<1	<1	82	25

urce: Reproduced from Blanck [17], courtesy of the American Water Works Association.
ote: FS-400, Calgon Corp., Pittsburgh, PA.

Table 8–10 Davenport Water Co.: Removal of Trihalomethanes by Renewed GAC Treatment

Sampling Date	Sampling Location	Trihalomethanes (μg/l)			Percent Reduction
		Quenched	Formation Potential	Terminal	
11/1978	Filter influent	152	87	239	44
	Filter effluent	120	15	135	
12/1978	Filter influent	93	56	149	30
	Filter effluent	97	8	105	
1/1979	Filter influent	71	26	97	29
	Filter effluent	62	7	69	

Source: Reproduced from Blanck [17], courtesy of the American Water Works Association.

types of GACs: Filtrasorb 400 (Cl) (12 × 40 mesh), Nuchar WV-G (C2) (12 × 40 mesh), Hydrodarco 1030 (C3) (10 × 30 mesh), and Witcarb Grade 950 (C4) (12 × 30 mesh). The first carbon was tested in beds 2.5, 5., 7.5, and 10 feet deep, with EBCTs of 6.2, 12.4, 18.6, and 24.8 minutes. The other three carbons were tested in a 2.5 foot deep bed with an EBCT of 6.2 minutes.

Typical breakthrough curves are seen in Figures 8–19a, b, and c, respectively, for $CHCl_3$, THMFP, and TOC for the four carbons in the Florida groundwater study [18]. These results are similar to those reached in other studies inasmuch as $CHCl_3$ is removed reasonably well by the carbons, but TOC and THMFP are not. On an equal weight basis, C4 was superior to the others with a 78% $CHCl_3$ removal (C1, 56%; C2, 57%; and C3, 36%). The reverse was observed for THMFP removal by the four carbons on an equal weight basis: C1, 54%; C2, 29%; C3, 16%; and C4, 8%. The effect of bed depth on removal of these organics is essentially an effect on bed life. For example, the bed lives of 2.5, 7.5, and 10 feet of C1 were about 10, 65, and 81 days, respectively, for removal of 100. μg/l THM.

Another important operational point came out of this study [18]. Rechlorination is necessary because of the high bacterial count in the GAC effluent.

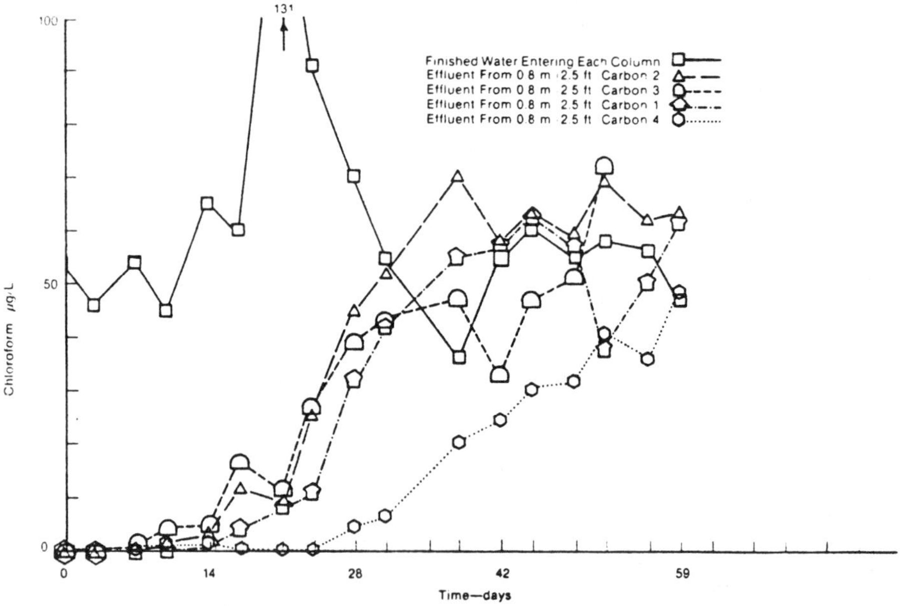

(a)

Figure 8–19 (a) Chloroform breakthrough curves through four types of GAC columns. (b) THMFP in finished water and in the effluent from four types of GACs. (c) TOC in finished water and removal by four types of GACs. Reproduced from Wood and DeMarco [18], courtesy of the American Water Works Association.

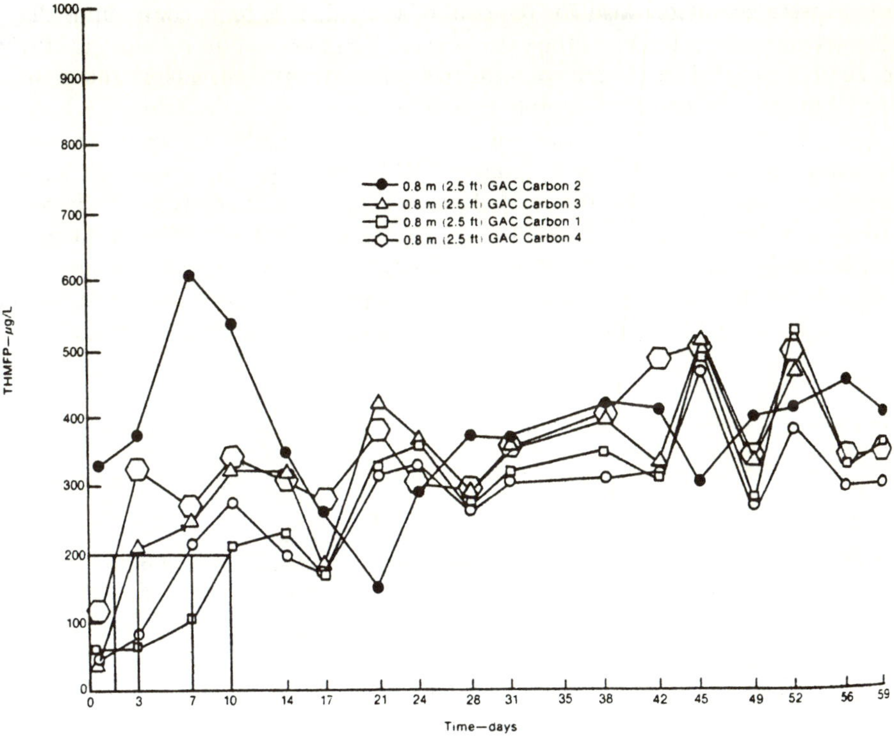

1000
900
800
700
600
500
400
300
200
100
0
THMFP—µg/L
0.8 m (2.5 ft) GAC Carbon 2
0.8 m (2.5 ft) GAC Carbon 3
0.8 m (2.5 ft) GAC Carbon 1
0.8 m (2.5 ft) GAC Carbon 4
0 3 7 10 14 17 21 24 28 31 35 38 42 45 49 52 56 59
Time—days

(b)

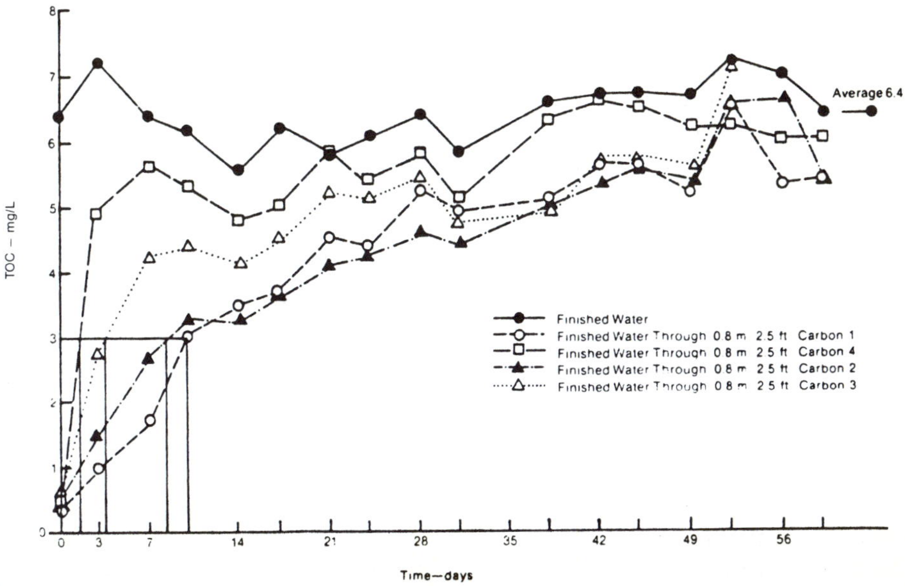

8
7
6
5
4
3
2
1
0
TOC - mg/L
Average 6.4
Finished Water
Finished Water Through 0.8 m 2.5 ft Carbon 1
Finished Water Through 0.8 m 2.5 ft Carbon 4
Finished Water Through 0.8 m 2.5 ft Carbon 2
Finished Water Through 0.8 m 2.5 ft Carbon 3
0 3 7 14 21 28 35 42 49 56
Time—days

(c)

Studies were conducted also for removal of individual organic compounds that were added to the influent of the carbon beds. Removal of 2-nitrobenzene, ethyl benzoate, naphthalene, 1-chloro-2-nitrobenzene, dimethylphthalate, and diisobutylphthalate ranged from complete to poor.

Five GACs, one anion exchange resin, and one carbonaceous resin were evaluated for their capability to remove THMs and TOC from Missouri River water in a pilot-scale operation [19]. Kansas City employs the typical Missouri River water treatment: coagulation-flocculation, sedimentation, lime softening, recarbonation, and filtration. $KMnO_4$ and Cl_2 are applied as pretreatment, and polymers supplement alum coagulation. In the experimental GAC units, bed depth was 3 feet mostly but 6 and 9 feet were studied also through a series mode. The

Table 8-11 Percentage of Total Influent Quantity of TTHM and TOC Removed by Various Adsorbents

Column Number	*Adsorbent*	*Bed Depth (ft)*	*TTHM (%)*		*TOC (%)*	
			Cumulative Removal	*Column Removal*	*Cumulative Removal*	*Column Removal*
1	Westvaco WV-G	3	80	80	51	51
2	Westvaco WV-G	6	98	18	66	15
3	Westvaco WV-G	9	99	1	75	9
4	Diamond Shamrock ES-561	3	29	29	7	7
5	Diamond Shamrock ES-561	3	14	15	20	13
6	Rohm & Haas XE-340	3		90		11
7	ICI HD-1030	3	83	83	37	37
8	ICI HD-1030	6	98	15	49	12
9	ICI HD-1030	9	99	1	54	5
10	Union Carbide LCK	3		82		30
11	Calgon Filtrasorb C	3		85		19
12	Norit ROW 0.8	3		68		45
13	Filter Sand	3		11		2
14	Westvaco WV-G (steamed)	3		92		56
15	Diamond Shamrock ES-561 (steamed)	3		22		7

Source: Reproduced from Badorek et al.[19], courtesy of Ann Arbor Science Publishers, Inc.

GACs were Westvaco Nuchar WV-G, ICI HD 1030, Union Carbide LCK, Filtrasorb C, and Norit Row 0.8. The anion exchange resin was ES-561 and the carbonaceous resin was Ambersorb XE-340. The columns were operated at the surface loading rate of 2 gpm/ft^2.

During the 3.5 month experimental period, the influent THM concentrations ranged from 5. to 211. μg/l, whereas the TOC was in the 2. to 4. mg/l bracket in the Missouri River water [19]. Table 8–11 gives the percentages of removal for five carbons and two resins. It appears that 99% removals of THMs can be achieved with a 9 foot column. However, the greatest removals occurred within the first column for the Westvaco WVG and ICI HD 1030 carbons. TOC removals were somewhat less in terms of percentages, although some were better than those reported in other studies. The carbonaceous resin, XE-340, effected a 90% removal of THMs whereas the anionic resin removed only 15%. This was an excellent comparative study under plant operating conditions.

Another comparative study was conducted on the performance of six commercially available GACs as final treatment in the Kralingen Plant of Rotterdam [20]. The major objective of this study was the removal by GAC of halogenated organic compounds, especially the nonvolatile TOXs resulting from pretreatment with Cl_2. Raw water is taken from the Meuse River and stored in open reservoirs for 3 months. Table 8–12 gives the characteristics of six carbons that were placed into 0.25 m diameter × 4.8 m columns. The carbon layer was 3 m, contact time

Table 8–12 Types of Carbons Tested

Number[a]	Type	*Effective Size (mm)*	*Uniformity*	*Iodine Adsorption (g/kg)* Given	Determined	*BET N_2 (m^2/g)*
1	European anthracite based	0.4–0.8	1.4	900	1,000	900
2	U.S. coal based	0.85	<2	950	950	900
3	European coal based	0.55–0.66	1.9	1,000	1,050	1,100
4	European petrol coke	—[b]	—[b]	1,100	1,050	1,100
5	U.S. lignite based	0.55–0.75	1.8	1,050	900	1,100
6	European peat based, extruded	Cylinders 0.8 mm diameter, variable length		1,100	1,050	1,000

Source: Reproduced from Rook [20], courtesy of the American Chemical Society.
[a]1 = Anthrasorb, type cc 1236, Thomas Ness, United Kingdom; 2 = GAC 30, Carborundum, United States; 3 = F-400, Chemviron, Belgium; 4 = Hydraffin LS Supra, Lurgi, Federal Republic of Germany; 5 = Nuchar WV-G. 42 × 1.68, Westvaco, United States; and 6 = ROW 0.8 Supra, Norit, The Netherlands.
[b]Only size range is given as 0.5–2.5 mm.

was 24 min, and the percolation rate was 7.5 m/hr. The results given in Tables 8–13 and 8–14 show that the coal- and lignite-based carbons are somewhat comparable in performance. Carbon 4 had the highest quantities of adsorption of the low molecular weight compounds ($CHCl_3$, AOCl, and AOBr) but was the poorest for the higher molecular weight materials (TOC, ultraviolet extinction, and THMFP). The microstructure of this carbon was revealed by scanning electron microscopy to have an abundance of small pores. Carbon 6 had a combination of very strong adsorption of small molecules with good adsorption of larger molecules.

A summary of the many bench-, pilot-, and full-scale studies for removal of THMs from water by GAC is seen in Table 8–15 for the United States [21]. The effect of EBCT is readily apparent although comparison is difficult because of the variation in the influent THM concentration. In any event, the longer the EBCT, the longer the time to exhaustion.

Another important operating variable is the individual adsorptivities of the

Table 8–13 Adsorption of Organohalogens in Milligrams per Kilogram Carbon after 10,000 Bed Volumes

Carbon Number	$CHCl_3$	*Carbon Number*	*AOCl*	*Carbon Number*	*AOBr*	*Carbon Number*	*AOI*
4	93	4	360	4	126	6	23
6	92	2	350	6	126	5	21.5
3	68	6	326	2	105	2	20.5
2	58	3	320	3	102	1	17.5
5	37	5	233	5	98	3	16
1	11	1	208	1	64	4	14

Source: Reproduced from Rook [20], courtesy of the American Chemical Society.

Table 8–14 Adsorption of Different Organics on Six Carbons (Ranked According to Effect)

Compound	*Low*	*Middle*	*High*
$CHCl_3$	1 5	2 3	6 4
AOCl[a]	1 5	3 6	2 4
AOBr[b]	1 5	3 2	6 4
TOC	4	5 2 3 1	6
UV extinction	4 6	2 3 1	5
THMFP	4	3 6 1 2	5

Source: Reproduced from Rook [20], courtesy of the American Chemical Society.
[a]Adsorbed organic chloride.
[b]Adsorbed organic bromide.

Table 8–15 Summary of Data on Adsorption of Total Trihalomethanes on Virgin GAC

Location	*Type of GAC*	*Type of System*[d]	*EBCT (min)*	*Cl/Br Ratio*[a,b]	*Influent TTHM Concentration*[b] *(μg/l)*	*Time to Exhaustion*[c] *(wk)*
Cincinnati, OH	WV-G 12 × 40	PC/PA	3.2	2.3	50	5
Cincinnati, OH	HD 10 × 30	PC/PA	3.2	1.9	71	4
Cincinnati, OH	WV-G 12 × 40	FS/SR	4.5	2.3	51	22
Cincinnati, OH	Filtrasorb® 400	PC/PA	5	2.3	121	6
Mt. Clemens, MI	HD 3000[e]	FS/SR	5.8	4.4	22	13
Mt. Clemens, MI	HD 3000[e]	FS/SR	5.8	4.6	25	12
Miami, FL	Filtrasorb® 400	PC/PA	6.2	1.8	156	12
Evansville, IN	HD 10 × 30	PC/PA	6.6	13.2	0.7	3
Huntington, WV	WVW 14 × 40	FS/SR	7.1	11.3	145	7
Cincinnati, OH	HD 10 × 30	PC/PA	7.5	2.9	31	8
Cincinnati, OH	WV-G	PC/PA	7.5	3.8	31	8
Cincinnati, OH	WV-G 20 × 50	FS/SR	7.5	4.1	51	22
Cincinnati, OH	WV-G 12 × 40	FS/SR	7.5	2.3	51	22
Philadelphia, PA	UNK[d]	PC/PA	7.5	INF[d]	80	9
Little Falls, NJ	HD 10 × 30	FS/PA	8	12	91	14
Little Falls, NJ	WVW	FS/PA	8	52	123	13
Little Falls, NJ	Filtrasorb® 400	FS/PA	8	12	91	14
Cincinnati, OH	HD 10 × 30	PC/PA	9	3.0	44	9
Newport, RI	Filtrasorb® 400	PC/PA	9	6.1	111	10
Cincinnati, OH	Filtrasorb® 200	PC/PA	9	6.3	54	9
Cincinnati, OH	WV-G 12 × 40	PC/SR	9.4	3.4	57	22
Evansville, IN	HD 10 × 30	PC/PA	9.6	19.9	0.7	3

(cont.)

Table 8–15 (*continued*)

Location	*Type of GAC*	*Type of System*[d]	*EBCT (min)*	*Cl/Br Ratio*[a,b]	*Influent TTHM Concentration*[b] *(μg/l)*	*Time to Exhaustion*[c] *(wk)*
Cincinnati, OH	Filtrasorb® 400	PC/PA	10	2.3	97	12
Cincinnati, OH	PICA-A[f]	PC/PA	10	0.72	48	12
Cincinnati, OH	PICA-B[f]	PC/PA	10	1.8	49	13
Beaver Falls, PA	Filtrasorb® C[g]	FS/SR	10.1	7.2	45	15
Jefferson Parish, LA	Filtrasorb® 400	PC/PA	10.9	30	3.5	8
Kansas City, MO	WV-G	PC/PA	11	11.7	14	12
Kansas City, MO	HD 10×30	PC/PA	11	4.6	18	12
Kansas City, MO	LCK[f]	PC/PA	11	5.0	29	12
Kansas City, MO	Norit ROW[f]	PC/PA	11	5.5	29	12
Kansas City, MO	Filtrasorb® C[g]	PC/PA	11	5.1	18	13
Jefferson Parish, LA	WV-G	PC/PA	11.2	54.5	5.0	7
Beaver Falls, PA	Filtrasorb® 400	FS/SR	11.3	6.4	67	10
Beaver Falls, PA	HD 8×16[e]	FS/SR	11.4	6.4	63	9
Cincinnati, OH	WV-G	PC/PA	11.8	4.7	8.7	16
Cincinnati, OH	HD 10×30	PC/PA	11.8	3.0	17	17
Miami, FL	Filtrasorb® 400	PC/PA	12.4	3.6	155	17
Jefferson Parish, LA	WV-G	FS/SR	13.6	5.7	6.6	18
Jefferson Parish, LA	WV-G	FS/SR	14	12.2	7.5	13
Cincinnati, OH	WV-G	PC/PA	16	13.6	17	17
Cincinnati, OH	HD 10×30	PC/PA	16	7.9	17	17
Jefferson Parish, LA	WV-G	PC/SR	17	10.3	5.0	16

Jefferson Parish, LA	WV-G	FS/PA	17.5	7.3	6.4	18
Jefferson Parish, LA	WV-G	FS/SR	18	INF	6.1	12
Jefferson Parish, LA	Filtrasorb® 400	FS/PA	19.3	9.0	3.3	14
Jefferson Parish, LA	WV-G	FS/PA	20	INF	6.1	15
Jefferson Parish, LA	Filtrasorb® 400	PC/PA	21.4	10.0	2.4	14
Jefferson Parish, LA	Filtrasorb® 400	PC/SR	21.6	33.0	3.8	15
Jefferson Parish, LA	Filtrasorb® 400	FS/SR	21.9	10.0	2.4	14
Jefferson Parish, LA	WV-G	PC/PA	22.1	74.8	4.3	16
Jefferson Parish, LA	Filtrasorb® 400	PC/PA	22.6	INF	5.5	14
Jefferson Parish, LA	WV-G	PC/PA	25.3	9.0	4.4	18
Jefferson Parish, LA	WV-G	FS/PA	26	20.8	4.0	>25
Jefferson Parish, LA	WV-G	PC/PA	32.6	88.8	3.0	26
Jefferson Parish, LA	Filtrasorb® 400	PC/PA	34.6	INF	3.6	18
Jefferson Parish, LA	WV-G	PC/PA	43.5	INF	1UNK	>26
Jefferson Parish, LA	Filtrasorb® 400	PC/PA	46.3	INF	UNK	>26

Source: Symons et al. [21].
[a]μg/l/μg/l.
[b]At time of exhaustion.
[c]Service time until effluent concentration nearly equals influent concentration.
[d]FS, full scale.
IH, in-house.
INF, infinite, chloroform only present.
NR, not reported.
PA, post-filter-adsorber.
PC, pilot column.
SR, sand replacement.
UNK, unknown.
[e]Manufactured by ICI America Inc., Atlas Chemicals Division, Wilmington, DE 19899.
[f]Foreign
[g]Not commercially available.

THMs. This is seen in Figure 8–20 where the more strongly adsorbed bromine-containing THMs have a smaller critical bed depth or a thinner mass transfer zone [21]. These bromine compounds could be removed over a longer time span and with a shallower bed than in the removal of $CHCl_3$.

Removal of precursors prior to chlorination is an obvious way to control THMs. Several bench- and pilot-scale studies have been reported for GAC treatment of various surface waters. Figures 8–21 and 8–22 show typical results from a pilot-scale treatment of Ohio River water [21]. For example, three major points are seen in Figure 8–21: (1) when the GAC is fresh, it adsorbed nearly all of the THMFP; (2) some precursors began to pass through the column almost immediately; and (3) bromine-containing THMs appear first after the precursors have passed through the column and react with the Cl_2. Apparently, the reaction between bromine and precursor is faster than Cl_2 and precursor. Data for the deeper lignite-based GAC are essentially the same as the coal-based GAC except two of the three columns showed good removals throughout the 30 week experimental period; that is, the effluent THMFP contents were less than the influent contents.

Table 8–16 gives a summary of GAC removal of THMFPs throughout the United States from bench-, pilot-, and full-scale treatments [21]. There is a scatter of performances, but the removals are greater with the longer EBCTs. It is difficult to understand the elevated pH values in the alkaline range. It is well known that carbon surfaces as well as some precursors with dissociated functional groups

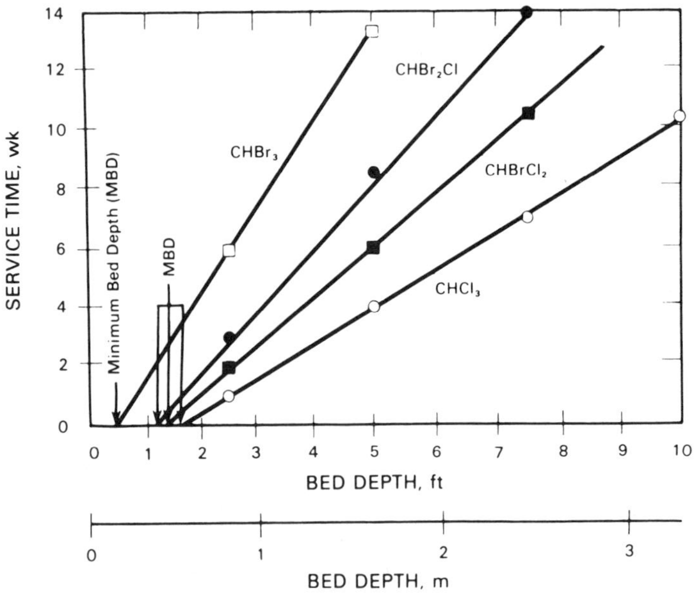

Figure 8–20 BDSTs to reach 2 μg/l in the effluent for the THMs being adsorbed by GAC [21].

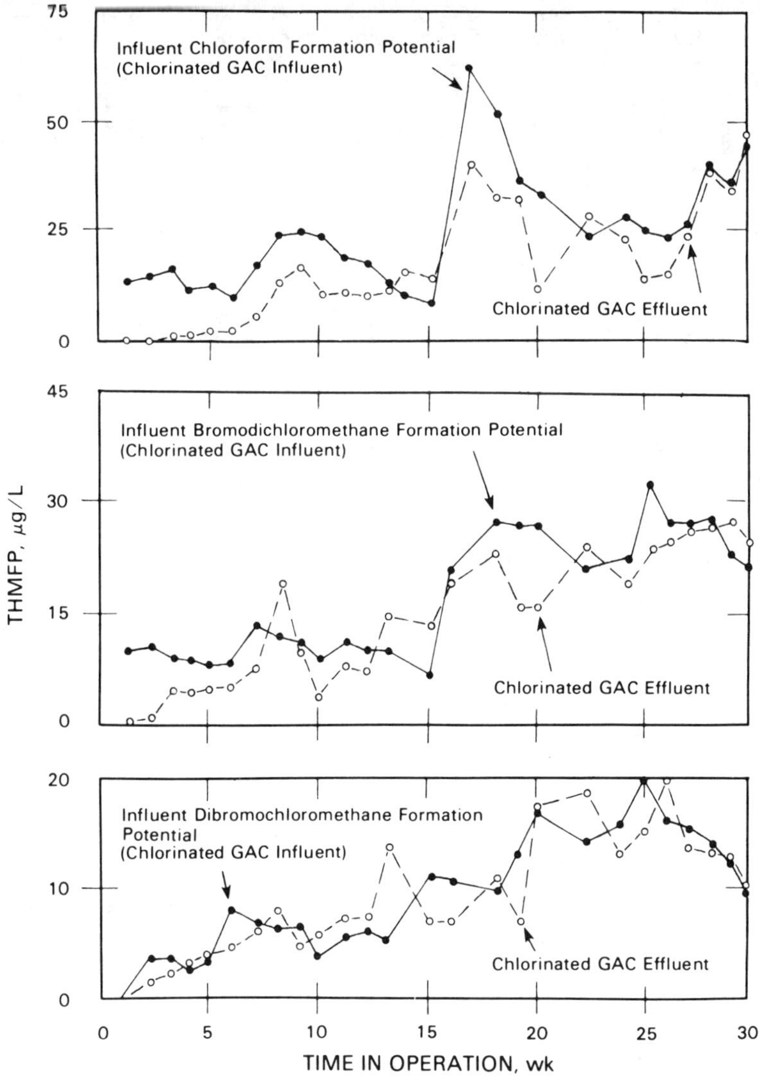

Figure 8–21 Removal of THM precursors from Ohio River water by coal-base GAC. Test period, March to October 1975; GAC type, Filtrasorb ®200; bed depth, 76 cm (30 in); hydraulic loading, 5 m/hr (2 gal/min/ft^2); EBCT, 9 min. THMFP conditions: pH 6.5; 20°C (68°F); storage time, 4 days [21].

would be negatively charged. The net effect would be to lower the efficiency of removal of THMFP compounds.

Some attempts have been made to remove THMs from water by PAC. For example, a 2 year investigation was conducted at the Kanawha Valley water treatment plant in Charleston, West Virginia, where an especially formulated PAC

Table 8–16 Summary of Performance Data for Removal of Trihalomethane Precursors by Virgin GAC Adsorption

Location	Type of GAC	Type of System[a]	EBCT (min)	Approximate Percent Initial THMFP Removal	Approximate Time to Steady-State Conditions[b] (wk)	Percent THMFP Removal at Steady-State Conditions	Influent THMFP at Steady-State Conditions (μg/l)	Sample Storage Conditions: Time (days)	Sample Storage Conditions: Temperature (°C)[h]	Sample Storage Conditions: pH
Cincinnati, OH	WV-G 12×30	PC/PF	3.2	97	4	6	281	7	29.4	9.2
Cincinnati, OH	HD 10×30	PC/PF	3.2	97	2	14	232	7	29.4	9.2
Evansville, IN	HD 10×30	PC/PA	3.7	84	6	16	58	3	17	8.0
Cincinnati, OH	WV-G 12×40	FS/SR	4.5	95	2	38	222	7	29.4	9.2
Mt. Clemens, MI	HD 3000[c]	FS/SR	5.8	81	4	32	51	5	25	7.3
Mt. Clemens, MI	HD 3000[c]	FS/SR	5.8	75	4	32	60	5	25	7.3
Evansville, IN	HD 10×30	PC/PA	6.6	87	6	26	58	3	17	8.0
Huntington, WV	WVW 14×40	FS/SR	7.1	93	6	11	120	7	20	8.3
Davenport, IA	Filtrasorb® 400	FS/SR	7.5	UNK[c]	>14	73	26	[e]	[e]	[e]
Cincinnati, OH	WV-G 12×40	FS/SR	7.5	97	2	50	222	7	29.4	9.2
Cincinnati, OH	WV-G 20×50	FS/SR	7.5	97	2	33	222	7	29.4	9.2
Cincinnati, OH	WV-G 12×40	PC/PF	7.5	98	4	41	281	7	29.4	9.2
Cincinnati, OH	HD 10×30	PC/PF	7.5	[e]	4	30	281	7	29.4	9.2
Cincinnati, OH	Filtrasorb® 200	PC/SR	9.0	97	16	0	27	4	20	6.5
Cincinnati, OH	Filtrasorb® 400	PC/SR	9.0	88	16	40	42	2	50	[e]
Cincinnati, OH	WV-G 12×40	PC/SR	9.4	95	6	37	244	7	29.4	9.2
Cincinnati, OH	WV-G 12×40	PC/SR	9.4	93	6	36	244	7	29.4	9.2
Evansville, IN	HD 10×30	PC/PA	9.6	89	6	31	58	3	17	8.0
Cincinnati, OH	Filtrasorb® 400	PC/PA	10.0	93	16	44	137	6	25	7.8
Beaver Falls, PA	Filtrasorb® C[g]	FS/SR	10.1	[e]	11	11	110	7	10–20	7.4
Jefferson Parish, LA	WV-G 12×40	PC/PA	11.0	68	18	0	251	5	30	10
Beaver Falls, PA	Filtrasorb® 400	FS/SR	11.3	[e]	11	19	110	7	10–20	7.4
Beaver Falls, PA	HD 8×16[c]	FS/SR	11.4	[e]	11	15	110	7	10–20	7.4

Cincinnati, OH	WV-G 12×40	PC/PF	11.8	98	8	40	230	7	29.4	9.2
Cincinnati, OH	HD 10×30	PC/PF	11.8	97	4	34	281	7	29.4	9.2
Jefferson Parish, LA	Filtrasorb® 400	PC/PA	12.0	77	20	17	273	5	30	10
Jefferson Parish, LA	WV-G 12×40	FS/SR	14.0	64	8	21	281	5	30	10
Jefferson Parish, LA	WV-G 12×40	FS/SR	14.0	74	18	11	319	5	30	10
Cincinnati, OH	WV-G 12×40	PC/PA	16.0	98	8	60	230	7	29.4	9.2
Cincinnati, OH	HD 10×30	PC/PA	16.0	97	5	35	259	7	29.4	9.2
Jefferson Parish, LA	WV-G 12×40	PC/SR	17.0	10	18	13	319	5	30	10
Jefferson Parish, LA	WV-G 12×40	FS/PA	18.0	65	21	14	192	5	30	10
Cincinnati, OH	HD 10×30	PC/SR	18.0	92	23	49	73	4	20	6.5
Jefferson Parish, LA	Filtrasorb® 400	FS/PA	19.0	53	21	25	365	5	30	10
Manchester, NH	WVW 8×30	FS/PA	21.7	78	8	52	138	7	28.5	8.0
Manchester, NH	WVW 8×30	FS/PA	21.7	82	8	47	133	3	28.5	8.0
Jefferson Parish, LA	WV-G 12×40	PC/PA	22.0	18	22	20	235	5	30	10
Jefferson Parish, LA	Filtrasorb® 400	PC/SR	22.0	82	19	44	343	5	30	10
Jefferson Parish, LA	Filtrasorb® 400	PC/PA	23.0	69	21	39	365	5	30	10
Jefferson Parish, LA	WV-G 12×40	PC/PA	23.0	55	21	20	192	5	30	10
Jefferson Parish, LA	Filtrasorb® 400	FS/SR	23.0	73	24	32	317	5	30	10
Jefferson Parish, LA	Filtrasorb® 400	PC/PA	23.0	81	21	34	365	5	30	10
Jefferson Parish, LA	WV-G 12×40	FS/PA	24.0	90	20	32	265	5	30	10
Jefferson Parish, LA	WV-G 12×40	PC/PA	33.0	58	24	28	149	5	30	10
Jefferson Parish, LA	Filtrasorb® 400	PC/PA	35.0	81	21	45	365	5	30	10
Jefferson Parish, LA	WV-G 12×40	PC/PA	44.0	56	11	61	137	5	30	10
Jefferson Parish, LA	Filtrasorb® 400	PC/PA	46.0	88	17	69	253	5	30	10

Source: Symons et al. [21].
[a]SR = sand replacement; FS = full scale; PA = post-filter-adsorber; PC = pilot column.
[b]Percentage of THMFP being removed is no longer declining.
[c]Manufactured by ICI America Inc., Atlas Chemical Division, Wilmington, DE 19899.
[d]Not reported.
[e]Unknown.
[f]In-house.
[g]Not commercially available.
[h]°F = °C × 1.8 + 32.

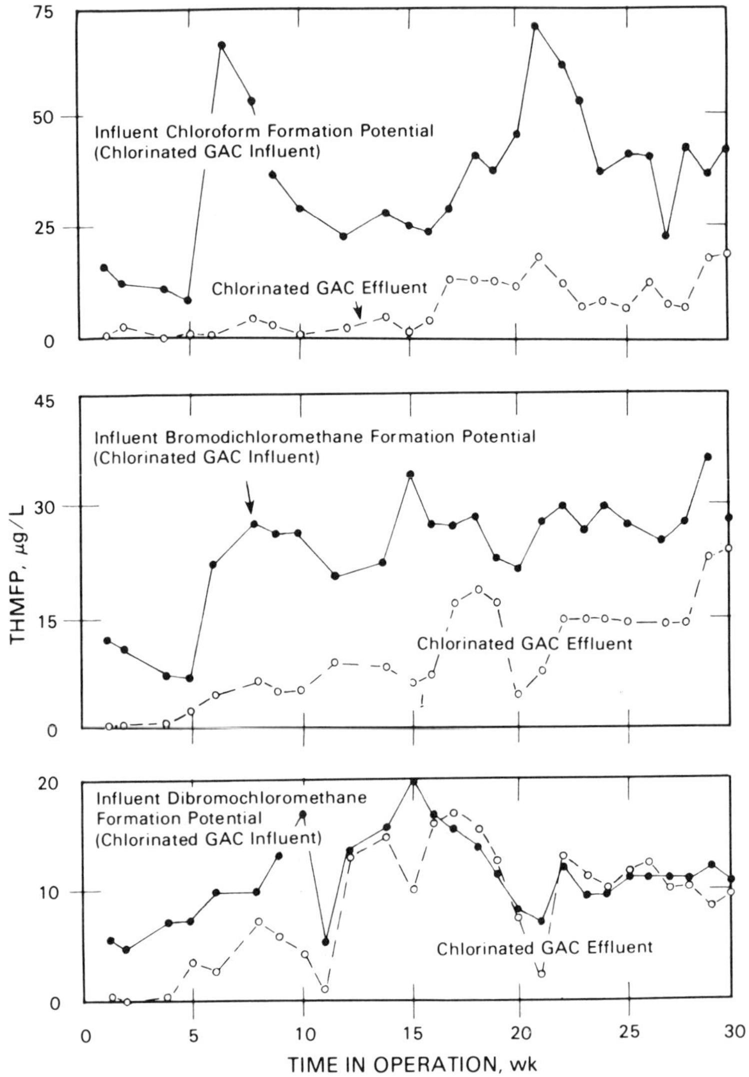

Figure 8–22 Removal of THMFP from Ohio River water by lignite-base GAC. Test period, May to December 1975; GAC type, HD 10 × 30; bed depth, 152 cm (62 in); hydraulic loading, 5 m/hr (2 gal/min/ft^2); EBCT, 18 min. THMFP conditions: pH 6.5; 20°C (68°F); storage time, 4 days [21].

(Westvaco Corp.) was employed [22]. This plant is rather conventional in its use of coagulation by a polyelectrolyte, sedimentation, and rapid sand filtration. Residence time in the entire plant is about 4 hours. The raw water is prechlorinated to 0.5 mg/l free Cl_2 residual leaving the flocculation basin. Low doses of PAC are added infrequently to control taste and odor problems. Past chlorination is practiced. The PAC was added as a wetted slurry.

Figures 8–23a, b, c, and d show the THM formation content reduction by PAC during several plant tests. At a dosage of 21.6 mg/l, the especially formulated carbon reduced the THM level from 37.4 to 23.8 μg/l (Figure 8–23a). There was also a 34.3% reduction of TOC in the same experiment. In the second experiment, there was an approximate 30 minute contact time between the carbon and water by virtue of a change of the point of chlorination (Figure 8–23b). This produced a greater removal of THMs at 45%. Figure 8–23c shows the effect of a lower quantity of PAC, 7.3 mg/l, whereas Figure 8–23d is a repeat of the first experiment. Here, the samples were analyzed at an independent laboratory. The THM data in Figure 8–23a apparently were high because of contamination problems in the Kanawha Valley's analytical equipment. These results confirmed, on a plant scale, that an especially formulated PAC can reduce THM precursor concentrations. In addition, TOC levels can be reduced. This study was confident that the 100 μg/l MCL for THM, at that time, could be reached [22]. PAC should

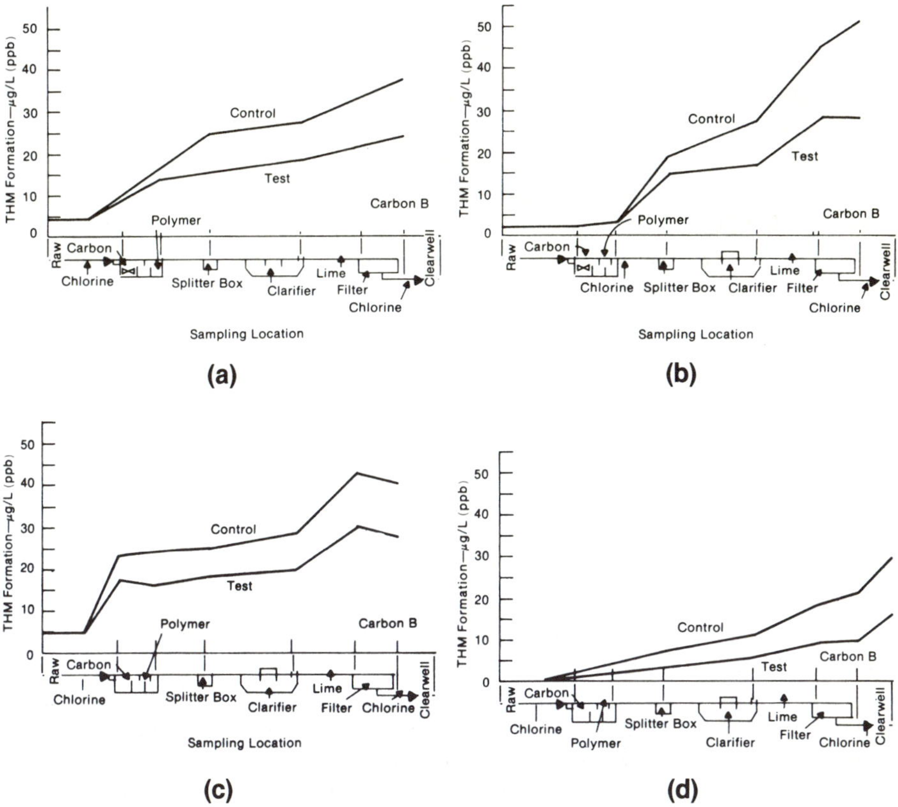

Figure 8–23 (a) THM formation reduction with 21.6 mg/l (ppm) PAC dosage, May 23, 1979. (b) THM formation reduction with 19.4 mg/l (ppm) PAC dosage, May 25, 1979. (c) THM formation reduction with 7.3 mg/l (ppm) PAC dosage, May 24, 1979. (d) Second sampling of THM formation reduction with 21.6 mg/l (ppm) PAC dosage, May 23, 1979. Reproduced from Anderson et al. [22], courtesy of the American Water Works Association.

Table 8–17 Removal of Organics at New Orleans, LA, with PAC

PAC Dosage	Removal Efficiency (%)								
	Volatiles[a] Study 1			*TTHM Study 2*			*HMW[b] Study 3*		
	1	*2[c]*	*3[c]*	*1*	*2[c]*	*3[c]*	*1*	*2[c]*	*3[c]*
5 mg/l	15	45/15	5/10	0	25/5	0/0	10	0/0	65/60
50 mg/l	100	100/100	0/100	30	40/40	50/30	5	0/35	35/50
500 mg/l	100	100/100	100/100	85	85/80	80/95	10	100/100	50/55
Finished water	0	35/15	0/0	20	20/25	50/40	20	5/45	70/65

Source: DiFilippo et al. [23].
[a]Only 1,2-dichloroethane was used because data for the other volatiles were erratic.
[b]Only atrazine was used because data for the phthalates were erratic.
[c]Data from reactors with and without a chlorine dose.

be applied at the raw water intake, the prechlorination point should be moved, and an alternative disinfectant to Cl_2 should be employed.

Four PACs, in a bench-scale study, were evaluated for their capability to remove a variety of organic compounds from raw water at New Orleans [23]. These carbons were Amoco PX-25, Westvaco Aqua Nuchar, Husky Industries Water Carb Plus, and Hydrodarco B (ICI). A contact time of 30 minutes was employed with PAC dosages of 5., 50., and 500. mg/l. Several experiments were conducted with raw water (study 1), coagulated/settled water (study 3) etc., with (study 2) and without Cl_2, etc. Table 8–17 is a summary of these experiments where the PAC is presumably the Hydrodarco B. The extremely high 500 mg/l dosage provided the greatest removals but may be beyond the economic capacity of most water treatment plants.

TREATMENT OF GROUNDWATERS

Groundwater represents greater than 95% of available fresh water in the United States, and this includes the Great Lakes. About 80% of all public water supplies rely on groundwater for potable water sources, and approximately 96% of all water employed for rural domestic purposes is obtained from groundwater. Usage of groundwater by public water utilities has almost tripled since 1950 from 3.5 bgd to 11 bgd in 1975 [24]. Concurrently, more and more contamination of these groundwaters by organic compounds has been reported (Table 8–18) [25].

Few options exist to avoid treatment of contaminated groundwaters. Frequently it is suggested to seek an alternative water supply. This approach may not be practical for several reasons: a pollution-free source may not be located nearby, and the cost of developing a new source may be prohibitive; a new groundwater supply may become contaminated eventually; and a neighboring municipal supply may not be capable of handling two systems. Consequently,

Table 8–18 Frequency of Occurrence of Organic Chemicals Found in Groundwater Supplies

Organic Chemicals	*Percent Frequency*[a]
Trihalomethanes	
Chloroform	70.3
Bromodichloromethane	69.2
Dibromochloromethane	64.5
Bromoform	36.3
Chlorinated hydrocarbon solvents	
Tetrachloroethylene	26.1
1,1,1-trichloroethane	22.2
1,1-dichloroethane	21.4
cis-1,2-dichloroethylene	21.4
Trichloroethylene	16.4
1,1-dichloroethylene	7.1
trans-1,2-dichloroethylene	7.1
Carbon tetrachloride	5.5
Other synthetic organics	
2,4-dichlorophenol	17.2
bis (2-chloroethyl) ether	8.7
Benzene	8.5
Pentachlorophenol	6.9
Toluene	<5.0
Xylene	<5.0
Methyl ethyl ketone	<5.0
Butyl acetate	<5.0

Source: Reproduced from Dyksen and Hess [25], courtesy of the American Water Works Association.
Note: Based on federal studies of finished waters.
[a]Represents the percent of time the compound was detected in the systems monitored.

treatment is recommended for the contaminated supply. Figure 8–24 shows the combined strategies for the treatment of groundwaters [25]. Inasmuch as most groundwaters do not have particulate matter, colored substances, and high bacterial counts, any treatment would be specific for organics. Essentially, this treatment consists of preaeration for removal of volatiles followed by adsorption by GAC and/or a macroreticular resin. A case study follows.

Large-Scale Systems

The groundwater supply of the township of Rockaway, New Jersey, was contaminated by several volatile organic chemicals (VOC) [26]. The township recently installed aeration and GAC adsorption facilities to remove the many VOCs. In

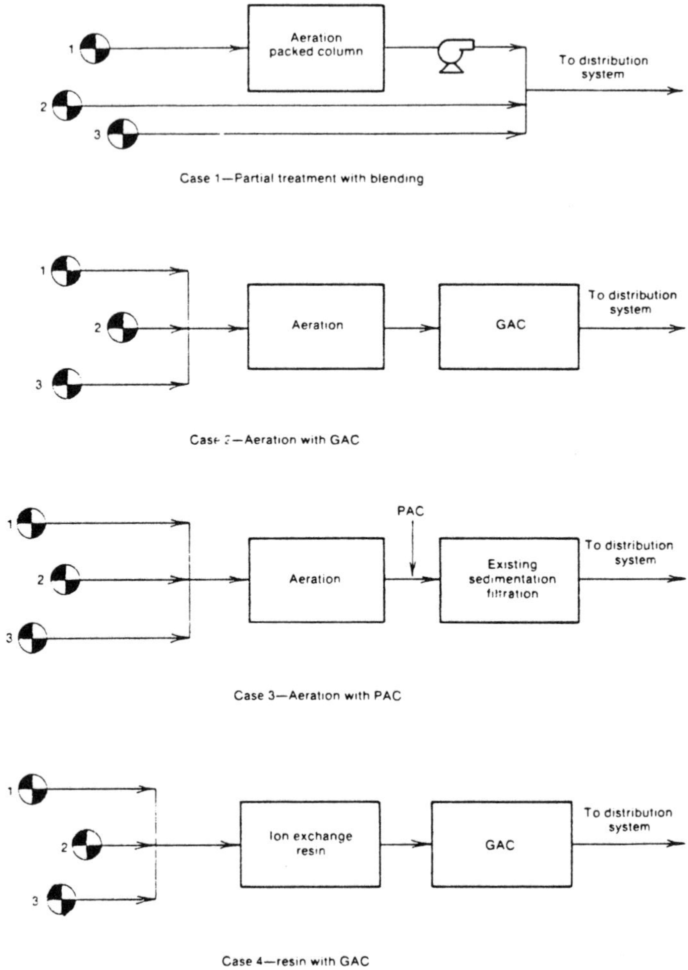

Figure 8-24 Combined strategies for treating groundwater supplies. Reproduced from Dyksen and Hess [25], courtesy of the American Water Works Association.

the absence of a federal regulation for VOCs, the New Jersey Department of Environmental Protection (N.J. DEP) uses the following guidelines: for a single VOC, the maximum allowable level is 50. μg/l, and 100. μg/l is the limit for total VOCs. Particular concern was expressed over the occurrence of TCE that ranged from 50. to 300. μg/l.

Figures 8-25a and b show the schematics and details of the Rockaway treatment system [26]. The essential aeration column dimensions and design criteria are 1,400 gpm (2 mgd), 9 ft diameter, 25 ft packing height, 35 ft overall height, 3 in polyethylene tellerettes, and a 200:1 air to water ratio. Air is blown upward through the countercurrent column with two 100 hp centrifugal blowers, each

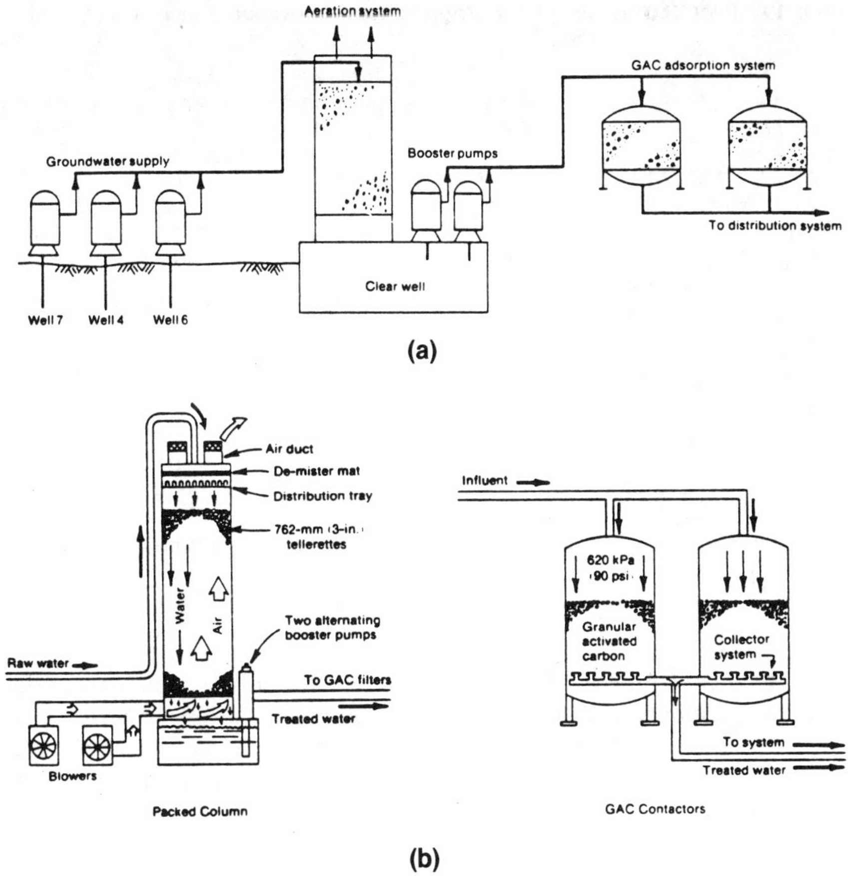

Figure 8–25 (a) Schematic of the Rockaway Township treatment system. (b) Schematic of the treatment processes. Reproduced from McKinnon and Dyksen [26], courtesy of the American Water Works Association.

rated at 27,000 ft^3/min at 16 in water gauge. The GAC system consists of two downflow pressure contactors that are 10 ft in diameter, 20 ft high, and contain 20,000 lb, each, of 8 × 30 mesh carbon (Calgon F-300). The carbon column's design parameters are rated capacity, 1,000 gpm; surface loading rate, 6.3 gpm/ft^2; and contact time, 12 min.

Pilot tests of the air-stripping process were conducted prior to installation of the final treatment system (Table 8–19) [26]. The percentage removals of individual compounds, in general, follow their Henry's Law constants; that is, the higher this constant, the more easily is a compound removed from water. All compounds were removed with acceptable efficiencies.

Operating data are seen in Figure 8–26 for removal of two organics, DIPE (diisopropyl ether) and MTBE (methyltertiary butyl ether), at the Rockaway

Table 8-19 Pilot Test Results of Air Stripping for Removal of Organic Compounds

Contaminant	*Air-to-Water Ratio*	*Influent Concentration (μg/l)*	*Effluent Concentration (μg/l)*	*Removal (percent)*
Diisopropyl ether (DIPE)	44:1	15.3	7.0	54
Methyltertiary butyl ether (MTBE)	44:1	14.0	7.8	44
Trichloroethylene (TCE)	44:1	217.9	39.6	82
Diisopropyl ether (DIPE)	75:1	14.4	5.5	62
Methyltertiary butyl ether (MTBE)	75:1	13.1	6.4	51
Trichloroethylene (TCE)	75:1	203.6	35.8	82
Diisopropyl ether (DIPE)	125:1	14.1	4.2	71
Methyltertiary butyl ether (MTBE)	125:1	13.1	4.8	63
Trichloroethylene (TCE)	125:1	203.6	27.0	87

Source: Reproduced from McKinnon and Dysken [26], courtesy of the American Water Works Association.

treatment system [26]. Initially, the GAC unit was operated without preaeration, which is a questionable strategy. These two compounds broke through the column rather quickly whereupon the carbon was replaced and not regenerated. Eventually, the aeration unit was installed and its effect on organics removal is rather obvious.

Small-Scale Systems

The need to treat groundwaters for domestic (home wells) situations has increased dramatically in the 1980s. Many units are commercially available for home use, but all are not efficient in the removal of organics from water. Table 8-20 is a summary of a study conducted by the Gulf South Research Institute and provides an excellent side-by-side comparison of 30 GAC units and 1 unit with an experimental resin [27]. The water for this study came from unspiked New Orleans taps in which the THM contents ranged from 110. to 239. μg/l. Efficiencies varied, but the units that removed the most organics had longer contact times and greater amounts of carbon than the others. There were concurrent studies for bacterial growth within some of the carbon units by the standard plate count (SPC). Influent SPC ranged from less than 10 to 14,000/ml whereas the effluent SPC ranged from less than 10 to 350,000/ml. Consequently, there is a bacteriological problem with these home treatment units, but this study did not address the health significance of these SPCs. Other studies also observed high SPCs in the effluents from home treatment units [28,29].

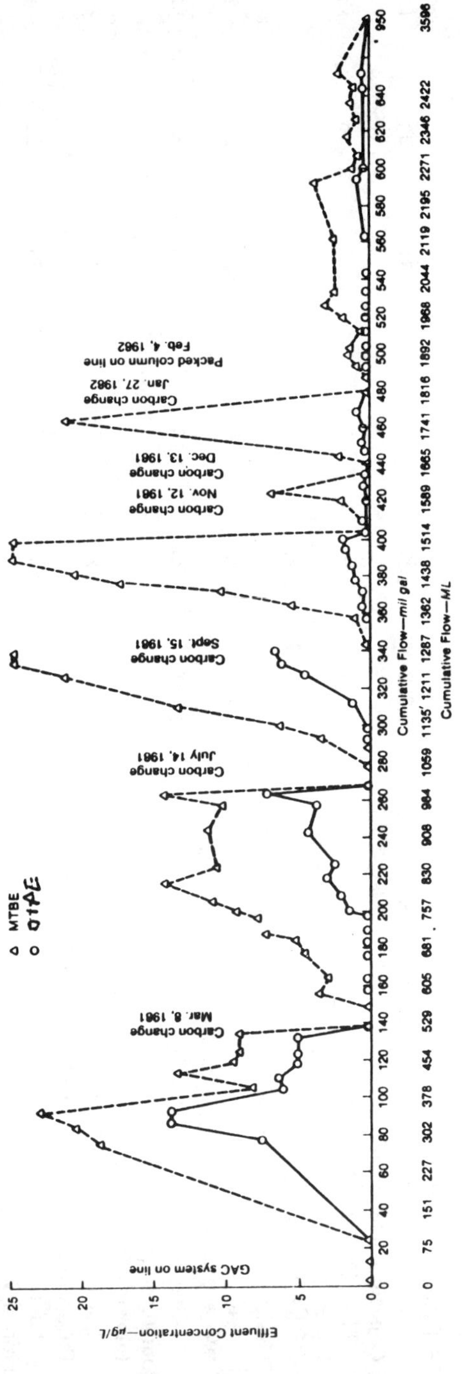

Figure 8–26 Breakthrough curves for DIPE and MTBE in Rockaway Township. Reproduced from McKinnon and Dyksen [26], courtesy of the American Water Works Association.

Table 8-20 THM and Nonpurgeable Total Organic Carbon (NPTOC) Reductions

Commercial Units Tested	Test Life[a] (l)	Test Life[a] (gal)	Estimated Average Contact Time (sec)	Weight of Carbon (g)	Average Reductions THM (percent)	Average Reductions NPTOC (percent)
1980 Data						
Faucet bypass						
Aquaguard, Model AGT-31, Cartridge T-3XL, Universal Water Systems, W. Chicago, Ill.	1,892	500	3.4	51	43	12
Concept Bacteriostatic Home Water Filter, New Tech Co., Canton, Mich.	151	40	1.6		16	18
Filter Fresh, Model FF-1, N. American Systems, Inc., Cleveland, Ohio	4,542	1,200	4.8	59	6	6
Hurley Town and Country, Hurley Chicago Co., Worth, Ill.	15,141	4,000	36.5	895	69	31
Water Washer, Countertop Model 1000, American Water Purification, Concord, Calif.	3,785	1,000	10.9	70	41	11
Pour-through						
Filbrook Pour-Thru Activated Carbon, PB Enterprises, North Pomfret, Vt.	1,135	300	43.8	97	74	30
Puritron Bacteriostatic Drinking Water Treatment Unit, Puritron of Florida, Miami, Fla.	3,785	1,000	14.1	30	21	6
Stationary						
AMF-Cuno Housing 1M, Cartridge AP-117, AMF Cuno Div., Meriden, Conn.	11,356	3,000	3.9	395	34	7
Filterite, Model 1 PC, Cartridge 1C-9, Filterite Corp., Timonium, Md.	11,356	3,000	1.2		18	8
Fulfo Water Filter, Model WC-10, The Carborundum Co., Lebanon, Ind.	11,356	3,000	2.6	208	15	11

Keystone, Model 3121 Housing with Model 310 Cartridge, Keystone Filter Div., Hatfield, Pa.	11,356	3,000	2.3		21	9
Line bypass						
Aquacell Bacteriostatic Water Treatment Unit, Aquatron Intl., Inc., New Orleans, La.	7,570	2,000	12.4	417	86	23
Aqualux Water Processor, Model CB-2, Aqualux Water Processing Co., Ft. Lauderdale, Fla.	7,570	2,000	35.2	1,150	98	23
Argenion Bacteriostatic Water Treatment Unit, Model 1, Comprehensive Marketing Concepts Co., Alexandria, Va.	7,570	2,000	8.1	146	23	
Continental Water Filter, Model 350, Continental Water Conditioning Co., New Orleans, La.	2,725	720	185	3,402	99	87
Culligan Super Guard, Model SG-2, Culligan Intl. Co., Northbrook, Ill.	15,141	4,000	39	1,708	89	28
Everpure, Model QC4-THM, Everpure Inc., Westmont, Ill.	3,785	1,000	43	765	99	55
Mariner Renaturalizer Water Units, Aqua Purification Systems, Inc., Ft. Lauderdale, Fla.	11,356	3,000	31.4	560	47	21
Polarisdynamic Water Unit, Polaris Water Purification, Inc., Pompano Beach, Fla.	9,463	2,500	26.6	527	61	18
Purogen Water Detoxifier, Aqua Manufacturing, Inc., Ft. Lauderdale, Fla.	9,463	2,500	11.6	859	38	6
Seagull IV (X-1), General Ecology, Inc., Lyonville, Pa.	3,785	1,000	14.8	300	81	41
Ultrapure Bacteriostatic, Water Purification Industries, Inc., Ft. Lauderdale, Fla.	11,356	3,000	28.0	1,103	40	20
Waterco, Model AS-5, Waterco Industries, Oakland Park, Fla.	11,356	3,000	5.0	495	25	
Other						
Wunderbar Portable Water Cleaner-Filter, The Clean Water Co., Seattle, Wash.	757	200	0.9		4	

(*cont.*)

Table 8–20 *(continued)*

Commercial Units Tested	*Test Life*[a] *(l)*	*Test Life*[a] *(gal)*	*Estimated Average Contact Time (sec)*	*Weight of Carbon (g)*	*Average Reductions THM (percent)*	*Average Reductions NPTOC (percent)*
1979 Data						
Faucet bypass						
Instapure, Model F1-C, Teledyne Water Pik, Ft. Collins, Colo.	757	200	1.6	27	24	11
Faucet (no bypass)						
Mini Aqua Filter, Mini Aqua Filter Ltd., New York, N.Y.	757	200	0.9	16	6	2
Pour-through						
H_2OK Portable Drinking Water Treatment Unit, Better Living Labs, Inc., Memphis, Tenn.	7,570	2,000	15	94	19	10
Stationary						
Sears Taste and Odor Filter, Sears Roebuck Co., Chicago, Ill.	12,946	3,420	4.5	398	46	12
Line bypass						
Aqualux Water Processor, Model HB, Aqualux Water Processing Co., Ft. Lauderdale, Fla.	7,570	2,000	23	575	45	28
System 1 Water Processor, Model SY1-34, Water Purification Systems, Inc., Plantation, Fla.	9,463	2,500	46	1,120	43	20
1980 Data						
Experimental unique resin						
Rohm and Haas Ambersorb XE-340, Rohm and Haas, Philadelphia, Pa.	13,248	3,500	50.6	2,353	93	6

Source: Reproduced from Bell et al. [27], courtesy of the American Water Works Association.

[a]Test life was set according to manufacturers' claims unless claims were unknown or practical aspects dictated a different length of test.

PRETREATMENT FOR REMOVAL OF ORGANIC COMPOUNDS

Treatment prior to adsorption is especially appropriate because the quantity of organic matter presented to the carbon is reduced considerably. Also, characteristics of the organic compounds may be rearranged whereby their adsorptive properties and their biological degradability are altered [30]. Several techniques are available: chemical coagulation, aeration for volatilization, and oxidation by O_3, ClO_2, and Cl_2. Most of the research on pretreatment has been on a full-scale water treatment plant. Consequently, the results are applicable immediately. An example is seen in Figure 8–27 where four adsorption isotherms are given for different pretreatments [31]. Ozonated Rhine River bank filtrate was coagulated with $FeCl_3$ and $Al_2(SO_4)_3$ prior to adsorption. The isotherm from the $FeCl_3$ coagulated water shows very little effect, whereas the $Al_2(SO_4)_3$ system gave a higher isotherm. This suggests a better adsorption of the remaining organic compounds after coagulation. Another example is seen in Figure 8–28 where five adsorption isotherms are given for Rhine River bank filtrate [32]. The combination of ozonation and clarification resulted in better adsorption of organics than either process alone.

Chemical Coagulation

A laboratory study was conducted for the chemical coagulation of groundwater spiked with humic and fulvic acids prior to adsorption on pulverized activated carbon (Calgon F-400) [33]. This was a batch-type, bench-scale investigation where the GAC was powdered "to reduce the time needed to reach equilibrium to between 3 and 5 days." The groundwater was moderately hard and contained 3.5 to 4 mg/l TOC. Adsorption isotherms are seen in Figures 8–29a, b, and c for several waters spiked with fulvic acid and with the naturally occurring TOC. Chemical coagulation with alum markedly increased the quantities of adsorbed organic matter. Exceptions occurred when there was little or no removal of TOC during coagulation.

There were several explanations of the observation that the adsorptive capacity was increased after large quantities of alum were added to the spiked groundwaters:

a. Removal of a large fraction of the TOC during coagulation resulted in lower initial TOC concentrations for the adsorption studies. If the organic substances were somewhat heterogeneous in nature, lower initial concentrations of TOC would result in apparent increases in adsorptive capacity if coagulation did not preferentially remove weakly adsorbing molecules.
b. Coagulation may have preferentially removed weakly adsorbing molecules, rendering the remaining fraction of TOC relatively more adsorbable.

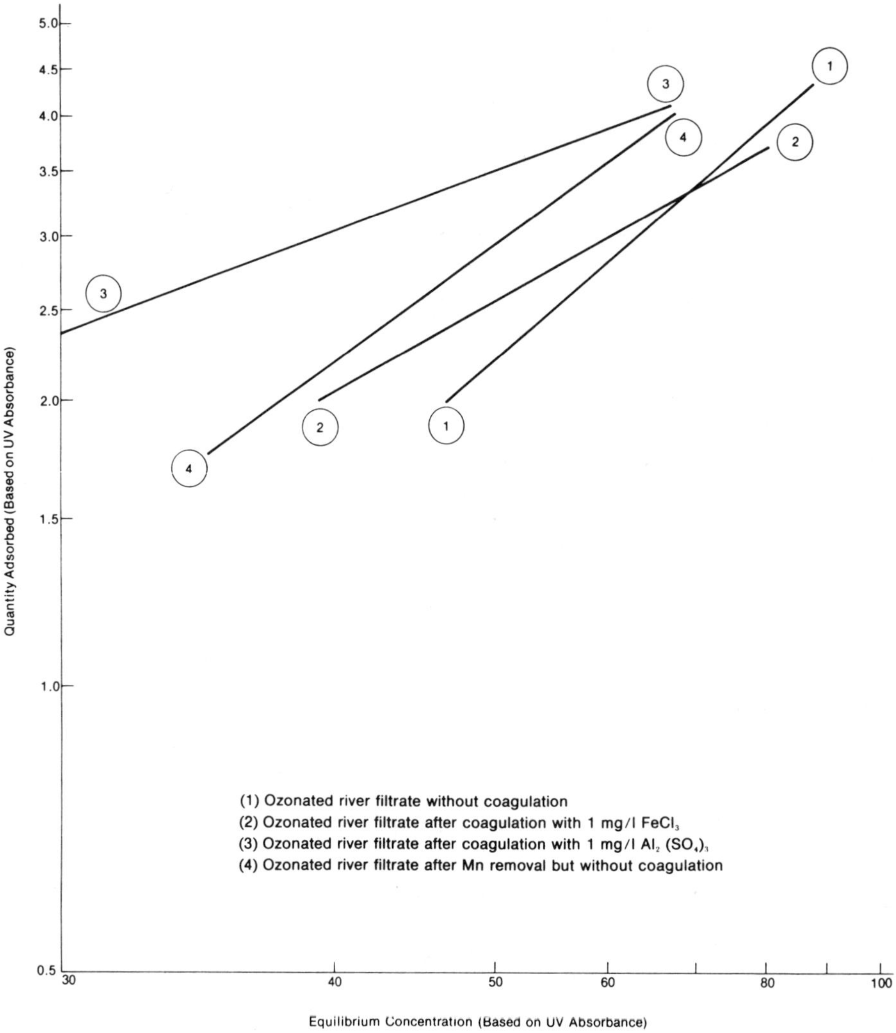

Figure 8-27 Adsorption isotherms after coagulation under various conditions. Reproduced from McCreary and Snoeyink [30], courtesy of the American Water Works Association.

c. Changes in either ionic strength or the concentrations of various inorganic ions may have altered the adsorbability of the organic matter.

d. Coagulation or precipitation of the organic matter may have continued during the five-day adsorption studies, with the organic matter being removed by the membrane filters along with the pulverized carbon particles.

e. Interactions between organic molecules and aluminum species may have occurred, either in solution or on the carbon surface, resulting in increased adsorption. [33]

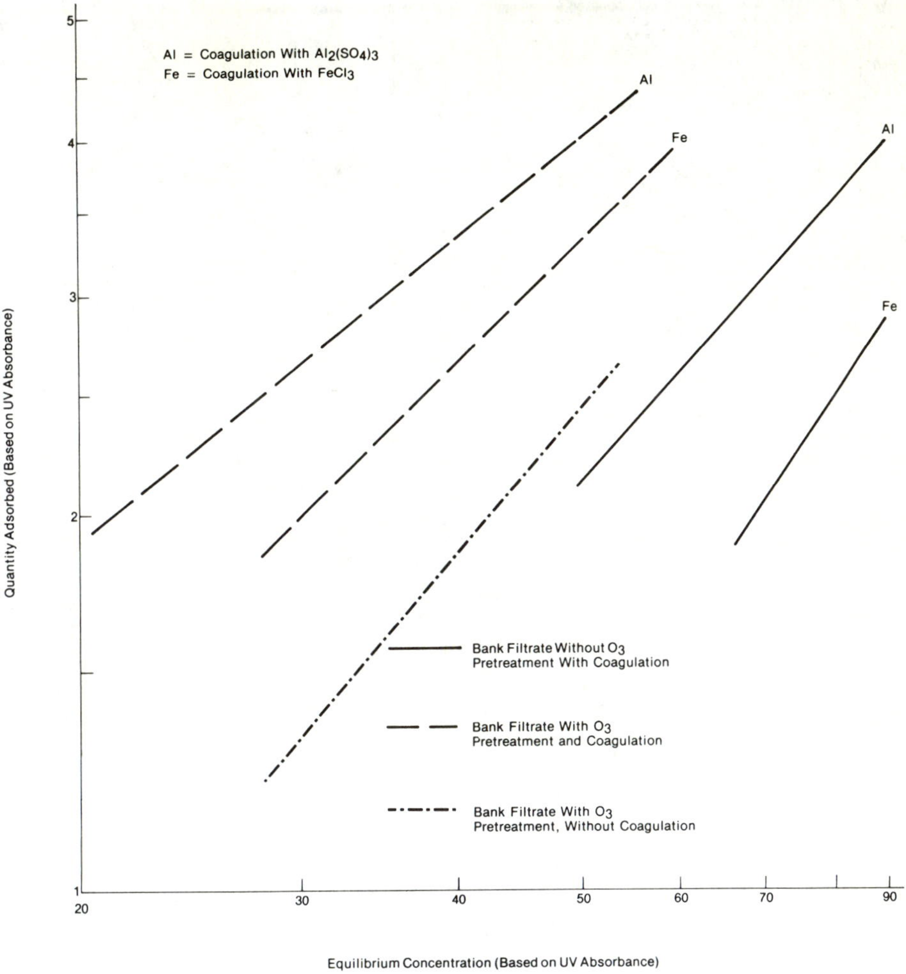

Figure 8–28 Adsorption isotherms after various pretreatments (Düisberg-Wittlaer). Reproduced from McCreary and Snoeyink [30], courtesy of the American Water Works Association.

A more rational explanation would be that chemical coagulation removes the larger molecular weight fractions of the humic and fulvic acids, leaving lower weight fractions dissolved in water. The latter would be more easily adsorbed in the absence of larger molecules that compete for the carbon's surface. In any event, there does seem to be greater adsorption after chemical coagulation.

Another laboratory bench-scale study evaluated pretreatment for eventual PAC removal of color, TOC, and THMFP from a surface water in the United States [34]. Pretreatment consisted of the traditional three-stage clarification process of coagulation, sedimentation, and mixed-media filtration. Alum and a

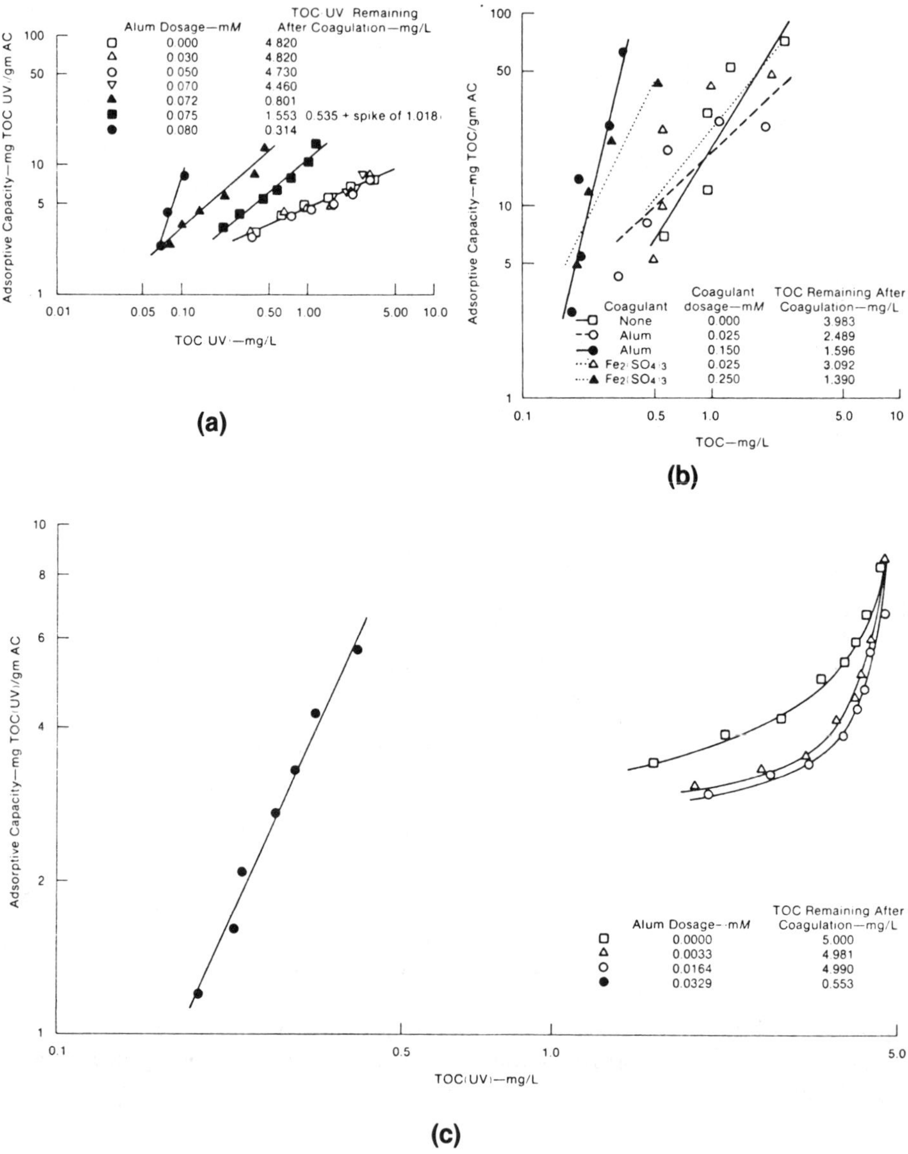

Figure 8-29 (a) Adsorption isotherms for untreated, pretreated, and spiked fulvic acid solutions. Coagulation at pH 6.0 with (Na_2SO_4) held constant at 0.6 mM; adsorption at pH 8.3 with 1.0 mM $NaHCO_3$ added. (b) Adsorption isotherms for untreated and pretreated groundwater; coagulation at pH 6.0; adsorption at pH 6.0 with 10.0 mM NaH_2PO_4 added. (c) Adsorption isotherms for untreated and pretreated podzol-soil fulvic acid solutions. Coagulation at pH 5.0 with (Na_2SO_4) held constant at 0.2 mM; 1 hour flocculation and 24 hours sedimentation prior to filtration; adsorption at pH 8.3 with 1.0 mM $NaHCO_3$ added. Reproduced from Randtke and Jepsen [33], courtesy of the American Water Works Association.

cationic polymer (Betz 1190) were used as primary coagulants. Sodium aluminate and an anionic polymer (Nalco 8182) were used as coagulant aids. Adsorption isotherm tests were performed on raw and pretreated waters using a PAC (Nuchar SA) (contact time > 12 hours).

Results from these tests indicated that large quantities of PAC were required to reduce significantly the various organics (Table 8–21a). Color in the pretreated water was reduced to the 5 to 10 color units, which is below the U.S. EPA's secondary MCL of 15. THMFP concentrations were reduced by more than 90%. Conversely, more than 1,000 mg/l of PAC were needed to produce color removal to the MCL. This kind of carbon dosage is unusually high and may have been needed because of the type that was used.

Pilot plant tests were conducted on the same surface water using the same pretreatment with subsequent processing of the effluent with either four GAC units (Westvaco C) in series or by ozone [34]. Each carbon column was 4 inches by 5 ft deep, and the surface loading was 2 gpm/ft^2. A summary of the ozonation tests is seen in Table 8–21b where color is removed satisfactorily to below the MCL. However, considerable quantities of TOC and THMFP were in the ozonator's effluent. This is unacceptable and would lead to THM formation in a distribution system if chlorine were applied for disinfection. In contrast, the GAC columns performed much better. Throughout the 50 day test, the effluent had < 5 color units, < 1. turbidity unit, average TOC of 2.3 mg/l, and an average THMFP of 156. μg/l.

Alum coagulation and lime softening exhibited variable effectiveness for removing the TOC associated with humic substances in Huron River water in a

Table 8–21a Results of Bench-Scale Carbon Adsorption Tests

		Equilibrium Concentration		
Type of Water Sample	*Carbon Dosage (mg/l)*	*Color (Co-Pt units)*	*TOC (mg/l)*	*THMFP (μg/l)*
Raw	0	350	48	1,295
	50	300	36	195[a]
	1,000	150	10	1,250
	5,000	20	5	–[b]
	10,000	20	3	265
Pretreated	0	80	15	–[b]
	50	80	11	–[b]
	100	40	12	–[b]
	500	20	8	450
	1,000	15	3	214
	5,000	4–10	2	–[b]
	10,000	5–10	2	114

Source: Reproduced from Argaman et al. [34], courtesy of the American Water Works Association.
[a]Analytical error suspected.
[b]Not determined.

Table 8–21b Summary of Initial Pilot-Plant Ozonation Tests

Run	*Contact Time (min)*	*Ozone Consumed (mg/l)*	*pH*		*Color (Co-Pt Units)*		*TOC (mg/l)*		*THMFP (μg/l)*	
			Influent	*Effluent*	*Influent*	*Effluent*	*Influent*	*Effluent*	*Influent*	*Effluent*
1	8.7	4.7	6.1		25	7	11	10	2,820	2,570
2	9.6	8.2	6.5		30	11	11	10	2,850	3,150
3	9.0	17	7.0		75	10	16	13	4,850	3,066
4	5.0	4.7	5.9		25	13	13	10	2,400	2,760
5	5.0	5.0	8.0	6.8	25	5	7	7		2,068
6	5.0	5.0	9.0	9.8	15/35[a]	10	11	8		2,852
7	5.0	5.0	10.0	10.2	25/65[a]	17	12	10		3,450

Source: Reproduced from Argaman et al. [34], courtesy of the American Water Works Association.
[a]Initial color/color after pH adjustment.

laboratory study [59]. For example, 50 mg/l of alum was effective in removing 80% of the TOC from the humic and fulvic acid solutions. However, the river water TOC was less amenable to removal by alum, 50 mg/l, where only a 30% reduction was achieved. Lime softening, 150 to 160 mg/l, of the Huron River water produced only a 50% reduction of humic and fulvic acid TOC. Pretreatment with alum or lime also affected the subsequent adsorption of TOC by GAC (F-400, mesh size not given). Freundlich isotherm parameters for solutions before and after lime and alum treatments are given in Table 8-22 [60]. These results indicate that lime and alum pretreatments yield humic and fulvic acid solutions with residual TOC components that adsorb differently than the TOC components of the same origin that were not pretreated. The effectiveness of pretreatment for removing TOC and for altering the adsorption characteristics of residual humics appears to depend on the source and type of humic material, at least for the Huron River.

Ozonation

A short-term pilot plant study was conducted for removal of natural organics in Lake Casitas in Ventura County, California [35]. The concern was THM for-

Table 8-22 Freundlich Isotherm Parameters for Treated and Untreated Humic Solutions

Solution	C_o *(TOC) (mg/l)*	K_F[a] *(mg/g)*	n
Huron River water (HRW)	5.4	15.8	1.3
HRW after lime (100 mg CaO/l)	4.2	21.7	1.3
HRW after alum (150 mg alum/l)	3.2	61.0	1.3
Humic acid in DDW	5.4	0.8	2.0
Humic acid in river water IBM (HA-IBM)	3.3	34.0	0.8
HA-IBM after lime (100 mg CaO/l)	3.1	37.0	0.7
HA-IBM after alum (40 mg alum/l)	3.4	27.5	5.0
Fulvic acid in river water IBM (FA-IBM)	3.1	43.5	1.1
FA-IBM after lime (70 mg CaO/l)	3.3	16.0	1.7
FA-IBM after alum (20 mg alum/l)	3.5	43.0	1.4
HRW after prechloramination and $KMnO_4$ addition[b]	6.1	14.1	1.4
HRW after lime and filtration[b]	3.9	12.2	2.0

Source: Reproduced from Weber and Jodellah [60], courtesy of the American Water Works Association.

Note: $q_e = K_F C_e^n$, where q_e is the adsorbed phase concentration (mg/g). C_e is the residual solution phase concentration (mg/l) at equilibrium, and K_F and n are characteristic constants.

[a]Calculated at an equilibrium solution concentration of 1.0 mg/l.

[b]Samples taken after different treatment stages at the Ann Arbor, Mich., water plant.

mation. Briefly, the pilot plant consisted of pretreatment with O_3 followed by GAC adsorption and filtration. The filtration unit was 88 inches deep and the surface loading rate was 6 gpm/ft^2. Figures 8–30a and b show that pretreatment with O_3 was not very effective. TOC was reduced from a maximum of 8 to 9 mg/l to the 3 to 4 mg/l range. However, when the raw water TOC was in the 2 to 4 mg/l range, O_3 had little effect for some unexplained reason. Similar results were observed for THMFP. The GAC column was reasonably effective throughout the brief study.

Laboratory and pilot-scale studies were conducted to examine the combined unit processes of oxidation by ozonation and adsorption by GAC using Delaware River water [36]. Also, the effect of chlorine and chlorine by-products on the O_3 and GAC treatments was studied. Figure 8–31 shows the schematic of the various experimental treatment systems at the Torresdale plant (Philadelphia). Table 8–23 contains the organic criteria established for the GAC regeneration. The GAC units remained in service for between 360 and 500 days. Neukrug et al. concluded:

> Enhanced TOC removal before treatment does not appear to increase the adsorptive capacity of the GAC for the trace organics of health concern. The cost of preozonation is not sufficiently offset by the lowered GAC operating costs associated with less frequent GAC regenerations, when the removal of the halogenated, volatile organic compounds is the controlling criteria [*sic*]. For TOC reductions, preozonation may be cost effective, depending on the exact criteria chosen.
>
> Chlorination appears to produce compounds resistant to further oxidation by ozone and should thus be applied only after ozonation.
>
> Chlorination and ozonation each affect the adsorptive capacity of the specific organic compounds differently.
>
> GAC effluents were effectively disinfected with chlorine. No bacterial regrowth was found after 5 days.
>
> There were no operational problems associated with maintaining a conventional treatment system on line for over 2 years without any type of predisinfection. [36]

In a pilot study that continued for 80 weeks, unozonated and ozonated waters were treated by GAC columns for their comparison of the removal of THMFP from the waters of Cross Lake, Louisiana [37]. This water was taken directly from the lake and went through a pretreatment of chemical coagulation and filtration prior to the ozonation and GAC units. The latter was F-400 in columns with an EBCT of 12 min at a flow rate of 2. gpm. The O_3 dosages ranged from 2 to 6.3 mg/l. Table 8–24 is a summary of the Cross Lake study where organic removal was measured by TOC and THMFP. The GAC columns alone removed 472 g THMFP whereas the O_3-GAC system removed 404 g. However, when the effect of direct oxidation was included, the latter value was increased to 543 g of THMFP. Statistical analysis indicated that the O_3-GAC system gave 10% better removal than GAC alone at the 95% confidence limit. TOC removal was similar.

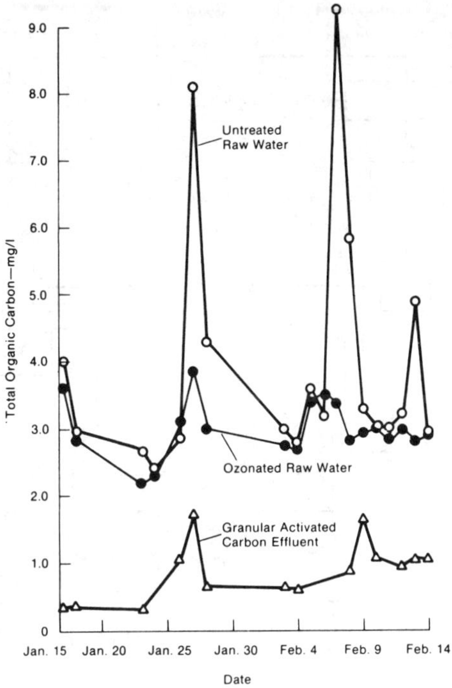

(a)

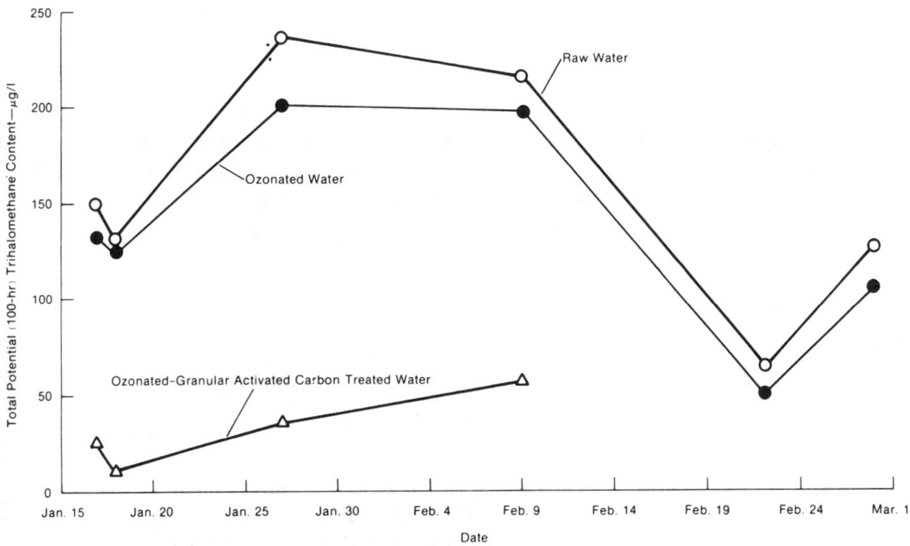

(b)

Figure 8–30 (a) TOC removal by ozone and GAC, winter 1977. (b) TPTHM removal by ozone and GAC, winter 1977. Reproduced from Barnett and Trussell [35], courtesy of the American Water Works Association.

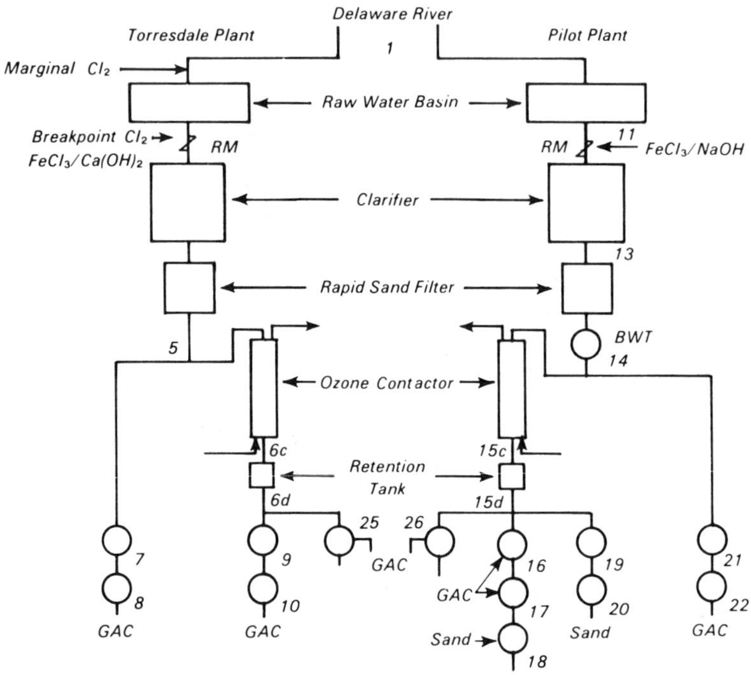

Figure 8–31 Schematic of the five advanced water treatment systems evaluated [36].

Aeration

Laboratory and pilot-scale investigations were conducted to determine the effects of aeration and adsorption on removal of six organic solvents from various ground- and surface waters [15]. These solvents were TCE, PCE, 1,1,1-trichloroethane, cis-1,2-dichloroethylene, carbon tetrachloride, and 1,2-dichloroethane.

Table 8–23 Organic Criteria for GAC Regeneration

Parameter	*Criteria Investigated*
DOC	Effluent levels of 0.5, 1.0, 1.5, and 2.0 mg/l
DOC, THMFP	50% cumulative breakthrough
Chloroform	Effluent levels of 1, 10, and 100 μg/l
1,2-dichloropropane	Effluent levels of 1 and 5 μg/l
All volatile organic compounds with average concentration ≥0.1 μg/l	Average time to initial breakthrough, 100% breakthrough

Source: Neukrug et al. [36].

Table 8–24 Total Cumulative Performance Data for GAC Columns

		Total Cumulative Parameter				Percent of Total Removed in GAC		Amount Removed by Ozone[a]		Percent of Total Removed by Ozone-GAC	
		Applied		Removed							
GAC Column	Ozone-Treated Water	TOC (kg)	THMFP (g)	TOC (kg)	THMFP (g)	TOC	THMFP	TOC (kg)	THMFP (g)	TOC	THMFP
5	No	24.17	969	6.23	258	25.8	26.6			25.8	26.6
6	No	17.94	711	4.70	214	26.1	30.1			26.1	30.1
5 + 6	No	24.17	969	10.93	472	45.2	48.7			45.2	48.7
3	Yes	22.04	830	6.10	204	27.7	24.6	2.13	139	36.5	50.8
4	Yes	15.94	626	4.38	200	27.5	31.9	2.13	139	36.3	50.6
3 + 4	Yes	22.04	830	10.48	404	47.5	48.7	2.13	139	52.2	56.0

Source: Reproduced from Glaze and Wallace [37], courtesy of the American Water Works Association.
Note: Data are for entire project period of 81 weeks.
[a]Amount removed in ozone contactor and detention basin.

In these studies, aeration was not necessarily used as a pretreatment, but the data have such a fundamental nature that it could be applied for pretreatment. Table 8–25 summarizes some diffused aeration of spiked Cincinnati tap water and an actual groundwater contaminated with some of these compounds. It appears that at least an 8:1 air to water ratio is needed to effect an appreciable reduction in the content of the solvents. Data from the groundwater study are given in Table 8–26 where the removals by aeration were in direct proportion to the Henry's Law constant for each compound. Figure 8–32 shows some typical operational data from a combined treatment of preaeration followed by GAC for the New Jersey groundwater (the prime in Figure 8–32 represents a total content for the five organics). It would appear that, after 30 weeks of service, the GAC column's

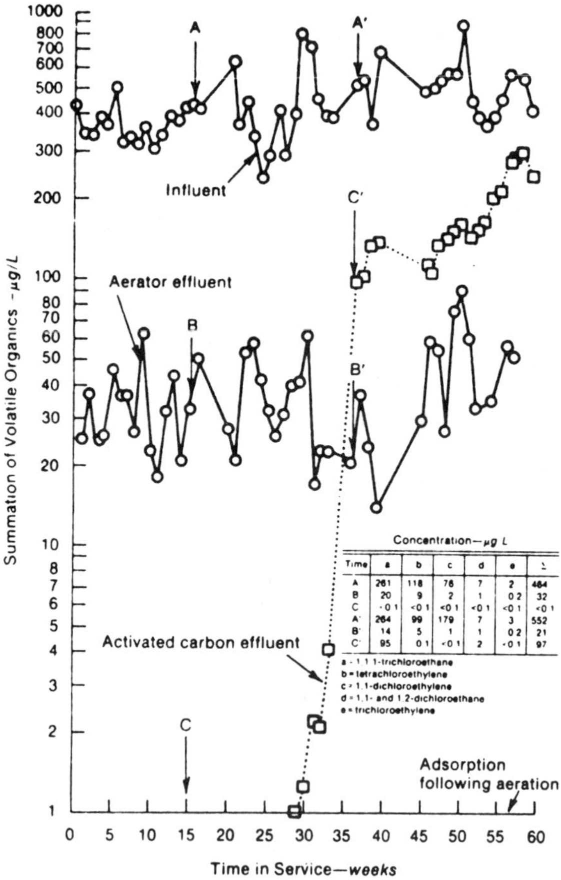

Time	a	b	c	d	e	Σ
A	261	118	76	7	2	464
B	20	9	2	1	0.2	32
C	<0.1	<0.1	<0.1	<0.1	<0.1	<0.1
A'	264	99	179	7	3	552
B'	14	5	1	1	0.2	21
C'	95	0.1	<0.1	2	<0.1	97

Figure 8–32 Removal of volatile organic compounds by aeration and adsorption on GAC. Pilot-scale study with New Jersey groundwater. Aeration: 4:1 (vol:vol); adsorption on GAC; EBCT = 18 min. Reproduced from Love and Eilers [15], courtesy of the American Water Works Association.

Table 8–25 Removal of Volatile Organic Compounds from Drinking Water by Using Diffused-Air Aeration (U.S. EPA-DWRD Pilot-Scale Studies)

Location of Study	*Compound*	*Average Influent Concentration (μg/l)*	*Average Effluent Concentration (μg/l) at Various Air-to-Water Ratios*[a]							*Remarks*
			1:1	*2:1*	*3:1*	*4:1*	*8:1*	*16:1*	*20:1*	
Cincinnati, Ohio	Tetrachloroethylene	1,025	698	416	304	156	16			The tap water was spiked with these solvents.
		636	161	177	46	34	8	<1	<1	
		338	139	103	47	34	4	1	2	
		114	32	17	7	4	<1	<1	<1	
		107	32	17	7	4	<1	<1	<1	
		17	3	2	1	1	<1	<1	<1	
	Trichloroethylene	1,064	796	614	508	319	53			
		397	223	273	102	82	22	<1	<1	
		241	136	110	61	53	8	2	3	
		110	40	28	18	9	3	<1	<1	
		73	22	14	8	6	1	<1	<1	
New Jersey	Tetrachloroethylene	94				9				The groundwater was contaminated by industrial discharge.
	Trichloroethylene	3				0.4				The pilot-scale column was run continuously for one year.
	1,1,1-trichloroethane	237				23				
	cis-1,2-dichloroethylene	0.5				<0.1				
	1,2-dichloroethane	1.4				0.8				

Source: Reproduced from Love and Eilers [15], courtesy of the American Water Works Association.
Note: All studies used a 4 cm (1.5 in) diameter glass column, a 10 min contact time, and a water depth of 0.8 m.
[a]Blank spaces indicate no tests were conducted.

Table 8–26 Effects of Aeration on a Solvent-Contaminated Groundwater

Contaminant	*Average Concentration (μg/l)* *Before Aeration*	*After Aeration*[a]	*Percent Removal*	*Henry's Law Constant*[b]
1,1-dichloroethylene	122	4	97	6.3
1,1,1-trichloroethane	237	23	90	1.2
Tetrachloroethylene	94	9	90	1.1
Trichloroethylene	3	0.4	87	0.5
cis-1,2-dichloroethylene	0.5	<0.1	>80	0.31
1,1-dichloroethane	6	1	83	0.24
1,2-dichloroethane	1.4	0.8	42	0.05

Source: Reproduced from Love and Eilers [15], courtesy of the American Water Association.
Note: U.S. EPA-DWRD study in New Jersey.
[a]Diffused-air aeration with a 10 min contact time, an air-to-water ratio (vol:vol) of 4:1, and a water depth of 0.8 m.
[b] $\frac{atm \cdot m^3}{mole}$ (presumably 25°C).

effluent total content was higher than the aerator's effluent (presumably the GAC's influent). Most of this total was 1,1,1-trichloroethane. Perhaps the chromatographic effect of displacing previously adsorbed molecules was responsible for this observation.

Several laboratory and pilot-scale tests have been conducted on aeration as a removal process or a pretreatment for THMs [21]. Typical results are seen in Table 8–27. Here, again, the air to water ratio is the most important consideration, with bed depth (contact time) being a secondary factor.

Chlorination

Pretreatment of raw waters with chlorine is not recommended because of the way chlorine reacts with various organic compounds to form THMs. However, there has been some research on the effect of Cl_2 on the GAC treatment of Delaware River water [38]. Two parallel GAC pilot columns were filled with F-400 with an EBCT of 7.5 min at a surface loading rate of 3. gpm/ft^2 for 113 days. One of the column's water was dechlorinated with sodium sulfite. That there is a slight reduction in the organic content of the chlorinated system is seen in Figures 8–33a and b and 8–34a and b. TOC, THM precursors, and instantaneous THM levels were slightly less in the chlorinated column. However, there was a relatively quick breakthrough of three categories of organics where precursors reached their influent content after 40 days of operation. These data suggest very little, if any, benefit from pretreatment with chlorine.

Table 8-27 Removal of THMs at North Miami Beach, Florida

	Run Number				
Variable	*1*	*2*	*3*	*4*	*5*
Packed bed depth					
Meters	1	2.1	2.5	2.1	2.1
Feet	3.3	7	8.3	7	7
Calculated air-to-water ratio (V/V)	10:1	10:1	10:1	32:1	40:1
$CHCl_3$					
Initial concentration (μg/l)	29	26	23	27	27
Final concentration (μg/l)	15	9.4	7.1	6.2	5.9
Percent removal	48	64	69	77	78
TTHM					
Initial concentration (μg/l)	40	35	29	35	33
Final concentration (μg/l)	22	12	9.3	7.0	5.9
Percent removal	45	66	68	80	82

Source: Symons et al. [21].
Note: Countercurrent flow, forced draft; tower aeration; water temperature = 27°C (81°F).

DESIGN AND ECONOMIC CONSIDERATIONS

In the late 1970s, the U.S. EPA attempted to mandate the use of GAC for communities with a population of 75,000 and greater. This move was quite controversial and serious challenges and questions were raised regarding the EPA's cost estimates for the use of GAC. Consequently, the EPA initiated a carefully designed investigation to determine water supply unit process cost curves "on a consistent and understandable basis" [39]. This study focused on treatments for the removal of contaminants from raw water that are included in the National Interim Primary Drinking Water Regulations. Also, cost of construction and maintenance curves were developed for these treatments. This resulted in curves for 99 different unit processes that, in turn, were divided into two categories: (1) large systems (1 to 200 mgd) and (2) small systems (2,500 gpd to 1 mgd) [39]. Only those costs associated with GAC are considered here. Also the reader should take into consideration the effect of inflation on costs since the original publication (1978).

To demonstrate application of construction and operation and maintenance cost curves, examples were developed for four types and capacities of GAC as a postadsorber [40]. These were: (1) 2 mgd plant with pressure carbon contactors and an off-site carbon reactivation at a regional facility; (2) 20 mgd plant with

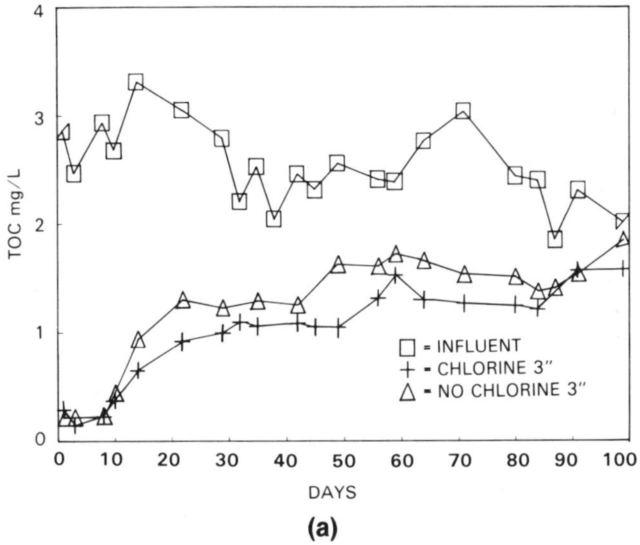

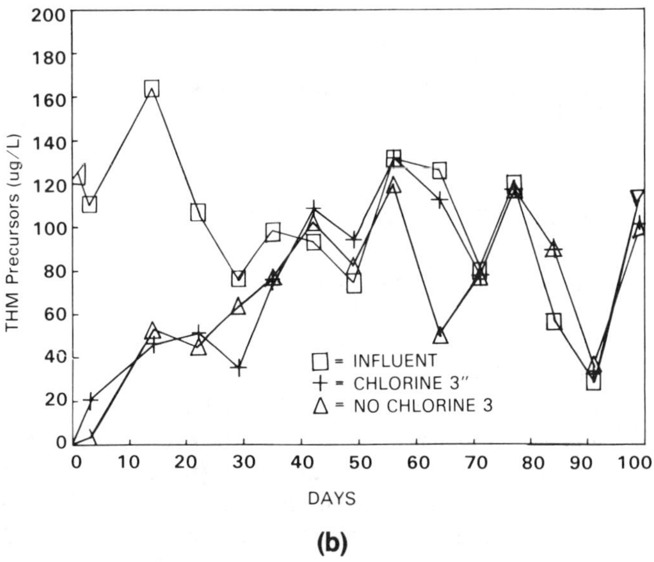

Figure 8–33 (a) TOC breakthrough from chlorinated and dechlorinated systems. (b) THM precursor breakthrough from chlorinated and dechlorinated systems. Reproduced from Yohe et al. [38], courtesy of Ann Arbor Science Publishers, Inc.

pressure carbon contactors and on-site infrared carbon reactivation; (3) 75 mgd plant with gravity, steel carbon contactors, and an on-site multiple hearth furnace carbon regeneration; and (4) 110 mgd plant with gravity, steel carbon contactors, and an on-site multiple hearth furnace carbon regeneration. Table 8–28 gives the design parameters (input variables) that were included to calculate

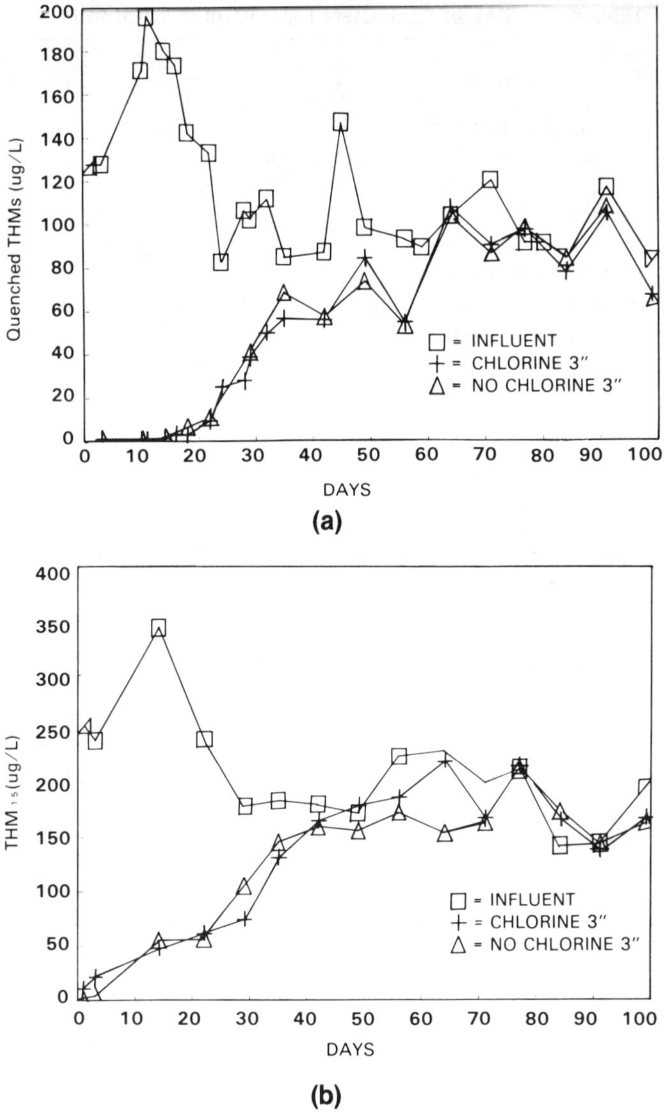

Figure 8–34 Instantaneous THM breakthrough from chlorinated and dechlorinated systems. (b) $THM_{1.5}$ breakthrough from chlorinated and dechlorinated systems. Reproduced from Yohe et al. [38], courtesy of Ann Arbor Science Publishers, Inc.

costs. Table 8–29 gives the assumed design configurations for various sized systems. An example of design criteria, capital cost calculation, and operation and maintenance costs for the 75 mgd plant is seen in Table 8–30. Construction costs are in October 1978 dollars. Annual costs for the operation of four GAC treatments are given in Tables 8–31a, b, c, and d. These data, in turn, were employed

Table 8–28 Design Parameters for Postfilter Adsorption

Design Parameter	*Level*
Activated carbon cost	$0.60/lb
Activated carbon loss per reactivation cycle	6%
Fuel cost	$1.75/mil Btu
Electric power cost	$0.03/kwh
Construction cost index	265.4
Producer price index	199.7
Direct hourly wage index	$10/hr
Amortization rate	8%
Amortization period	20 yr
Loss in adsorptive capacity	0%
Design capacity	70%
Empty bed contact time	18 min
Reactivation frequency	every 2.4 mo

Source: Reproduced from Gumerman et al. [40], courtesy of the American Water Works Association.

Table 8–29 Assumptions for Separate Postfilter Adsorption

	Design Capacity (mgd)				
Item	*1*	*5*	*10*	*100*	*150*
Number of contactors	3	6	12	40	60
Diameter of contactors (ft)	8	12	12	20	20
Depth of contactors (ft)	13	13	13	14	14
Volume of GAC per contactor (ft^2)	653	1,469	1,469	4,396	4,396
Minimum empty bed contact time (min)	18	18	18	18	18

Source: Reproduced from Gumerman et al. [40], courtesy of the American Water Works Association.

to draw cost curves (not shown here) that were related to specific design parameters such as volume of process, square feet of surface area, or feed rate. Also cost curves were formulated to allow for updating and adaptation to various regions in the United States through the use of economic indexes. Again, it should be noted that these costs are for GAC employed only as a postsand filter-adsorber without regard to influent concentrations of contaminants.

Another aspect of the costs of GAC treatment is the technique employed for organics removal [41]. Thus, two types of GAC systems were considered for the calculation of costs. One system used carbon in separate contactors after sand filtration (postfilter adsorber), and the other used GAC as a replacement for the

Table 8–30 Cost Calculation and Operation and Maintenance Requirements for a 75 Mgd Gravity, Steel GAC Plant

System and Design Criteria	*Design Parameter*	*Construction Cost (dollars)*	*Operating Parameter*	*Energy (Whr/yr)*	*Natural Gas (scf/yr)*	*Maintenance Material (dollars/yr)*	*Labor (hr/yr)*
Gravity, steel contactor—18 min EBCT, 30 ft diameter, 5 gpm/sq ft	12,533 cu ft/ contactor	3,029,620	125,330 cu ft	2,596,330	0	16,460	5,971
Initial carbon charge—26 lb/cu ft	3,258,690 lb	1,907,800		0	0	0	0
Backwash pumping—12 gpm/sq ft	7,520 gpm	101,710		0	0	0	0
Multiple hearth carbon reactivation furnace	800 sq ft	2,151,230	800 sq ft	909,700	109,446,800	11,630	10,956
Makeup carbon—6 percent/ regeneration, 5 times/yr		0	977,610 lb/yr	0	0	588,130	0
Subtotal		7,190,360		3,506,030	109,446,800	616,220	16,927
Sitework, interface piping, and roads at 5 percent		359,520					
Subsurface considerations		0					
Standby power		0					
Total construction cost		7,549,880					
General contractor's overhead and profit		754,990					
Subtotal		8,304,870					
Engineering at 10 percent		830,490					
Subtotal		9,135,360					
Land—15 acres at $2,000/ acre		30,000					
Legal, fiscal, and administrative		65,100					
Interest during construction— 8 percent		742,550					
Total capital cost		9,973,010					

Source: Reproduced from Gumerman et al. [40], courtesy of the American Water Works Association.

Table 8–31a Annual Cost for Mgd Pressure GAC Plant

Item	Total Cost/ Year (dollars)
Amortized capital at 8 percent, 20 years	69,430
Labor—1622 hr at $10/hr (total labor costs including fringe benefits)	16,220
Electricity—172,010 kwhr at $0.03 kwhr	5,160
Natural gas—3,973,960 scf at $0.00175/scf	6,950
Diesel fuel—1680 gal at $0.45/gal	760
Maintenance material	19,400
Total annual cost[a]	117,920

Source: Reproduced from Gumerman et al. [40], courtesy of the American Water Works Association.
[a]Dollars per 1,000 gal treated = $117,920/1,400 (365) = $0.23/1,000 gal.

Table 8–31b Annual Cost for 20 Mgd Pressure GAC Plant

Item	Total Cost/ Year (dollars)
Amortized capital at 8 percent, 20 years	400,060
Labor—4,828 hr at $10/hr (total labor costs including fringe benefits)	48,280
Electricity—1,816,760 kwhr at $0.03/kwhr	54,500
Maintenance material	178,340
Total annual cost[a]	681,180

Source: Reproduced from Gumerman et al. [40], courtesy of the American Water Works Association.
[a]Dollars per 1,000 gal treated = $681,180/14,000 (365) = $0.13/1,000 gal.

Table 8–31c Annual Cost for 75 Mgd Gravity, Steel GAC Plant

Item	Total Cost/ Year (dollars)
Amortized capital at 8 percent, 20 years	1,015,750
Labor—16,927 hr at $10/hr (total labor costs including fringe benefits)	169,270
Electricity—3,506,030 kwhr at $0.03/kwhr	105,180
Natural gas—109,446,800 scf at $0.00175/scf	191,530
Maintenance material	616,220
Total annual cost[a]	2,097,950

Source: Reproduced from Gumerman et al. [40], courtesy of the American Water Works Association.
[a]Dollars per 1,000 gal treated = $2,097,950/52,500 (365) = $0.11/

Table 8–31d Annual Cost for 110 Mgd Gravity, Steel GAC Plant

Item	Total Cost/ Year (dollars)
Amortized capital at 8 percent, 20 years	1,384,110
Labor—22,186 hr at $10/hr (total labor costs including fringe benefits)	221,860
Electricity—4,832,120 kwhr at $0.03/kwhr	144,960
Natural gas—158,149,050 scf at $0.00175/scf	276,760
Maintenance material	892,070
Total annual cost[a]	2,919,760

Source: Reproduced from Gumerman et al. [40], courtesy of the American Water Works Association.
[a]Dollars per 1,000 gal treated = $2,919,760/77,000 (365) = $0.10/

media in existing filter beds (sand replacement). Both systems were considered with on-site regeneration. Design parameters assumed for the postfilter adsorber are seen in Table 8–32. For the sand replacement filter, design parameters were the same except that a 10% loss of carbon per regeneration cycle and an EBCT of 9 minutes were assumed. Table 8–33 gives the unit costs for both types of systems, and Tables 8–34a and b show the amortized capital and operating and maintenance costs for both systems. These data indicate that postfilters are more capital intensive than sand replacement systems. Also, investments in sand replacement systems are more vulnerable to the effects of inflation than are the postfilter adsorbers because of fixed costs normally associated with filtration systems.

An important economic point was made in consideration of the costs of using alternative disinfectants to avoid THM formation [41].The following design parameters were assumed to be constant for three disinfectants: electric power, $0.03/kwh; construction cost index, 265.4; producer's price index, 199.7; direct hourly wage rate, $10; amortization period, 20 yr; capacity factor, 70%; contact time, 20 min. Units costs for Cl_2, O_3, and ClO_2 are summarized in Table 8–35, assuming equivalent disinfection capability, and chlorine appears to be the least expensive disinfectant based on 1979 dollars.

An extensive study of design and economic considerations of GAC systems for removal of organics was conducted at the Torresdale Water Treatment Plant

Table 8–32 Design Parameters for GAC

	Level	
Design Parameter	*Sand Replacement*	*Post-Filter Adsorption*
Activated carbon cost	$0.60/lb	$0.60/lb
Activated carbon loss per reactivation cycle	10 percent	6 percent
Fuel cost	$1.75/mil BTU	$1.75/mil BTU
Construction cost index	265.4	265.4
Producer's price index	199.7	199.7
Direct hourly wage rate	$10/hr	$10/hr
Amortization rate	8 percent	8 percent
Amortization period	20 years	20 years
Volume per filter	856 cu ft	
Loss in adsorptive capacity	0	0
Design capacity	70 percent	70 percent
Empty bed contact time	9 min	18 min
Reactivation frequency	Every 1.2 months	Every 2.4 months

Source: Reproduced from Clark and Dorsey [41], courtesy of the American Water Works Association.

Table 8–33 Amortized Capital and O&M Costs for GAC Systems

	Amortized Capital Cost				*O&M Cost*				*Total Cost*			
	Sand Replacement		*Filter Adsorber*		*Sand Replacement*		*Filter Adsorber*		*Sand Replacement*		*Filter Adsorber*	
Capacity (mgd)	*(¢/kl)*	*(¢/1,000 gal)*	*(¢/kl)*	*(¢/1,000 gal)*	*(¢/kl)*	*(¢/1,000 gal)*	*(¢/kl)*	*(¢/1,000 gal)*	*(¢/kl)*	*(¢/1,000 gal)*	*(¢/kl)*	*(¢/1,000 gal)*
1	6.4	24.8	7.6	29.4	5.1	19.5	5.8	22.1	11.5	44.3	13.4	51.5
5	2.3	9.0	3.3	12.7	3.4	13.0	3.3	12.6	5.7	22.0	6.6	25.3
10	1.6	6.2	2.6	9.8	3.0	11.7	2.8	10.8	4.6	17.9	5.4	20.6
100	0.5	2.0	1.3	4.9	2.3	8.7	1.9	7.4	2.8	10.7	3.2	12.3
150	0.4	1.7	1.2	4.6	2.2	8.4	1.8	7.1	2.6	10.1	3.0	11.7

Source: Reproduced from Clark and Dorsey [41], courtesy of the American Water Works Association.

Table 8–34a Amortized Capital and O&M Costs for GAC Systems Using Sand Replacement

	Reactivation				*Activated Carbon Fill*			
	Capital Cost		*O&M Cost*		*Capital Cost*		*O&M Cost*	
Capacity (mgd)	*(¢/kl)*	*(¢/1,000 gal)*	*(¢/kl)*	*(¢/1,000 gal)*	*(¢/kl)*	*(¢/1,000 gal)*	*(¢/kl)*	*(¢/1,000 gal)*
1	6.2	24.0	7.2	27.6	0.2	0.8	0	0
5	2.2	8.3	3.4	13.0	0.2	0.7	0	0
10	1.4	5.4	3.0	11.7	0.2	0.7	0	0
100	0.4	1.4	2.7	8.7	0.2	0.6	0	0
150	0.3	1.1	2.2	8.4	0.2	0.6	0	0

Source: Reproduced from Clark and Dorsey [41], courtesy of the American Water Works Association.

Table 8–34b Amortized Capital and O&M Costs for GAC Systems Using Filter Adsorbers

	Reactivation				*Contactors*				*Activated Carbon Fill*			
	Capital Cost		*O&M Cost*		*Capital Cost*		*O&M Cost*		*Capital Cost*		*O&M Cost*	
Capacity (mgd)	*(¢/kl)*	*(¢/1,000 gal)*	*(¢/kl)*	*(¢/1,000 gal)*	*(¢/kl)*	*(¢/1,000 gal)*	*(¢/kl)*	*(¢/1,000 gal)*	*(¢/kl)*	*(¢/1,000 gal)*	*(¢/kl)*	*(¢/1,000 gal)*
1	6.2	24.0	4.4	17.1	1.0	3.7	1.3	5.1	0.5	1.8	0	0
5	2.2	8.4	2.6	10.0	0.8	2.9	0.8	2.9	0.4	1.5	0	0
10	1.4	5.5	2.4	9.3	0.8	2.9	0.4	1.5	0.4	1.4	0	0
100	0.4	1.4	1.7	6.6	0.6	2.4	0.2	0.8	0.3	1.2	0	0
150	0.3	1.1	1.6	6.3	0.6	2.4	0.2	0.8	0.3	1.2	0	0

Source: Reproduced from Clark and Dorsey [41], courtesy of the American Water Works Association.

Table 8-35 Summary of Unit Costs of Disinfectants

Design Capacity		*Cost Item*					
		Chlorination (2 mg/l)		*Ozone (1 mg/l)*		*Chlorine-Dioxide (1 mg/l)*	
(Ml/day)	*(mgd)*	*(¢/kl)*	*(¢/1,000 gal)*	*(¢/kl)*	*(¢/1,000 gal)*	*(¢/kl)*	*(¢/1,000 gal)*
3.8	1	0.87	3.34	1.63	6.26	1.31	5.02
19	5	0.26	0.99	0.59	2.28	0.37	1.41
38	10	0.18	0.70	0.45	1.74	0.24	0.94
380	100	0.10	1.40	0.24	0.93	0.13	0.51
570	150	0.10	0.38	0.21	0.82	0.13	0.49

Source: Reproduced from Clark and Dorsey [41], courtesy of the American Water Works Association.

in Philadelphia [42]. This study was based on a pilot-scale operation wherein prechlorinated, clarified, and filtered water from the Delaware estuary was applied to several parallel columns that were operated at EBCTs of 7.5, 15., 22.5, 30., and 60. minutes. The following EPA design criteria were employed:

a. The concentration of any of the low-molecular-weight halogenated organics [excluding the THMs] that are determinable by a purge and trap, gas chromatographic, halogen detector type analysis shall not exceed 0.5 μg/l.
b. The initial removal of TOC must exceed 50%.
c. The increase in TOC concentration shall not exceed 0.5 mg/1 over the TOC concentration in the GAC column effluent at the start of the test.
d. In addition, other maximum contaminant levels such as THM and pesticides must be met. [42]

F-400 was the GAC employed in all columns. The 0.5 mg/l increase in TOC is the strongest criterion and becomes the limiting design factor. An example of operating results for this TOC standard is the bed life versus contact time: 3, 6, 9, 15, and 44 weeks for EBCTs of 7.5, 15, 22.5, 30, and 60 minutes, respectively. Another evaluated design factor was the mode of GAC operation; that is, one column was used as a sand replacement or as a filter-adsorber. Using an EBCT of 15 minutes, both modes produced the same bed life of 6 weeks for the TOC standard.

The major design factors for an economic analysis of GAC postcontactor treatment at Torresdale are given in Table 8-36 [42]. The rated capacity of the plant, 282 mdg, was selected as the design flow. There are few water treatment plants in the United States and elsewhere that would benefit from this GAC economic analysis; however, a normalization of the data on a million gallon per day basis may be useful for smaller treatment plants.

Table 8-36 GAC Design Parameters

Capital items	
Design flow	282 mgd
Pumping station	480 mgd
Regeneration facility load factor	70% at 75 $lb/ft^2/day$
Density of carbon	26 lb/ft^3
Price of carbon	$0.61/lb
Number of furnaces	3
Carbon buffer stock	30 day supply of daily losses
Spent and regenerated carbon storage	10 day of storage from largest furnace
O&M costs	
Carbon losses	7%
Fuel requirements	6,500 BTU/lb carbon
Fuel costs	$2.43/$10^6$ BTU
Steam requirements	1 lb steam/lb carbon
Steam costs	$4.00/1,000 lb
Electrical costs	$.04/kwh
Wage overhead and fringe benefits	90%
Financial items	
Interest rate	7%
Amortization period	23 yr
Bond issue costs	33% of capital
Revenue producing water	70% of water treated

Source: Reproduced from Quarino et al. [42], courtesy of Ann Arbor Science Publishers, Inc.

Capital costs associated with the GAC units with the five EBCTs are given in Table 8-37 [42]. The initial quantity of carbon, the pumping station, the contactors, and the regeneration facility are the major capital costs. They account for 90% of the capital at 7.5 minute contact time and 98% at 60 minutes. The decrease in cost of the regeneration facility as the contact time increases results from the proportionately higher bed life of the GAC at the higher EBCTs. Here, again, the reader should be reminded that the dollars in Table 8-37 are from 1978 to 1980. Corrections for inflation should be made if a similar GAC unit were constructed in the 1980s.

The first-year operation and maintenance costs for the various GAC units are seen in Table 8-38 [42]. The most significant item is the makeup carbon to compensate for a 7% loss. Gas for carbon regeneration and adsorber electric power are the next most costly items. Total operation and maintenance costs

Table 8-37 Philadelphia Water Department GAC Treatment Capital Costs for Torresdale

	Contact Time				
Item	*7.5 min*	*15.0 min*	*22.5 min*	*30.0 min*	*60.0 min*
Initial carbon fill	3,454,627	6,643,513	9,832,400	13,287,026	26,308,312
Pumping station	10,500,569	10,500,569	10,500,569	10,500,569	10,500,569
Contactors (gravity feed)	8,190,444	15,750,853	23,311,263	31,501,706	62,373,379
Regeneration facility	15,271,223	14,803,771	14,647,344	12,422,051	9,185,387
Carbon unloading facility	10,501	10,501	10,501	10,501	10,501
Carbon handling system	105,006	105,006	105,006	105,006	105,006
Carbon storage (buffer)	567,871	546,030	538,749	436,824	294,856
Carbon storage (spent)	1,287,689	1,238,162	1,221,653	990,530	668,608
Carbon storage (regenerated)	1,287,689	1,238,162	1,221,653	990,530	668,608
Backwash facility	525,028	525,028	525,028	525,028	525,028
Initial materials	104,721	104,721	104,721	104,721	104,721
Gas chromatograph	41,888	41,888	41,888	41,888	41,888
Subtotal	41,347,255	51,508,205	62,060,775	70,916,380	110,786,862
Contingencies 15%	5,683,894	6,729,704	7,834,257	8,644,403	12,671,783
Subtotal	47,031,150	58,237,908	69,895,032	79,560,783	123,458,645
Engineering 10%	4,703,115	5,823,791	6,989,503	7,956,078	12,345,864
Total capital bond issue	51,734,264	64,061,699	76,884,534	87,516,861	135,804,509
Total capital × 1.3333	68,979,019	85,415,599	102,512,712	116,689,148	181,072,679
Annual costs $/yr					
Amortized capital 23 yr 7%	6,119,400	7,577,553	9,094,305	10,351,952	16,063,668

Source: Reproduced from Quarino et al. [42], courtesy of Ann Arbor Science Publishers, Inc.
Note: Design flow is 282 mgd.

range from $8.8 million at a 7.5 minute EBCT to $5.6 million at 60 minutes. Again, these data reflect the increase in the ratio of bed life to EBCTs.

The Torresdale design and economic analysis also considered the effects of inflation, carbon losses, and sand replacement. For the latter, capital and operation and maintenance costs were less than when carbon is employed as a post-contactor only. Presumably, both were operating at the same removal of organics efficiency.

Thus, the design and mode of operation of GAC make up a rather complicated economic picture. The sand replacement mode was determined to be less

Table 8-38 Philadelphia Water Department GAC Treatment Operation and Maintenance Costs for the First Year of Operation at Torresdale

	Contact Time				
Item	*7.5 min*	*15.0 min*	*22.5 min*	*30.0 min*	*60.0 min*
Adsorber labor	312,550	312,550	312,550	312,550	312,550
Adsorber power	960,000	960,000	960,000	960,000	960,000
Adsorber materials	53,147	53,147	53,147	53,147	53,147
Regeneration labor	549,100	549,100	549,100	549,100	549,100
Regeneration power	266,479	256,230	252,814	204,984	138,364
Regeneration gas	1,484,670	1,427,567	1,408,533	1,142,054	770,886
Regeneration steam	375,985	361,524	356,703	289,219	195,223
Regeneration materials	63,329	62,370	62,045	57,183	49,074
Regeneration outside maintenance	610,849	592,151	585,894	496,882	367,415
Makeup carbon	4,013,637	3,859,267	3,807,810	3,087,413	2,084,004
Laboratory labor	141,550	141,550	141,550	141,550	141,550
Total annual O&M costs	8,831,296	8,575,456	8,490,146	7,294,083	5,621,314
Annual amortized capital costs	6,119,400	7,577,553	9,094,305	10,351,952	16,063,668
Total annual costs	14,950,696	16,153,009	17,584,451	17,646,035	21,684,982
Capital + O&M costs					

Source: Reproduced from Quarino et al. [42], courtesy of Ann Arbor Science Publishers, Inc.
Note: Design flow is 282 mgd.

expensive to construct and to operate than a postadsorber unit at an EBCT of 13.8 min [42]. A 7.5 minute EBCT for GAC as an adsorber only was the least difficult to construct and to operate but had the lowest bed life. Additional pilot-scale studies reduced the operation and maintenance costs of a 30 minute postcontactor to those of a 7.5 minute EBCT considering a medium rate of inflation over 23 years [42]. At a high rate of inflation, a 60 minute unit was calculated to be the most economical over 23 years. However, a 60 minute unit may not be a good investment inasmuch as the EPA criterion for the determination of bed life may change. And so it is with the economics of GAC treatment.

MICROBIOLOGY OF GRANULAR ACTIVATED CARBON

The GAC used in the treatment of potable water has been found to be an ideal habitat for bacterial growth [43]. GAC has a large surface area and pore structure and an affinity for adsorption of organic nutrients. Total coliforms apparently do not proliferate in GAC and have not been recovered in their effluents [43]. Rather, total bacterial counts in GAC beds and their effluents are significantly higher than those from sand filter effluents [44]. Accordingly, this microbial flora in the GAC represents a potential health risk of unknown proportions. Priorities for evaluating this microbial health risk were proposed by Cairo et al. [43]:

> Group 1 includes pathogenic organisms and opportunistic pathogens capable of infesting an altered or already debilitated host. Group 1A are those organisms that can inhibit or suppress the growth of existing coliform bacteria and therefore mask the survival of waterborne pathogens. Group 2 are endotoxins which are lipopolysaccharide-protein complexes produced in the cell walls of gram-negative aerobic bacteria. Group 3 bacteria are of concern because these could develop into a flora of chlorine resistant microbes that would present possible regrowth taste and odor, or chlorine demand problems in the distribution system. Group 4 includes harmless microbial indicators of the treatment process, and group 5 includes organisms that could be beneficial to the breakdown of organic compounds.

Bacteria in the GAC bed may have either a beneficial or a detrimental effect; that is, some may be involved in organics removal through biodegradation processes. The detrimental effects represent the health risk.

Four parameters were employed to determine the microbiological quality of a GAC system for treatment of Delaware River water [43]. They were membrane filter total coliforms, STP, biochemical identification of bacteria, and scanning electron microscopy. No total coliforms were found in the influent water to the GAC beds, and total bacterial levels ranged from 0 to 300/ml. In the GAC bed effluents, no total coliforms were found throughout the study period. However, total bacteria contents rose sharply in the first month to densities of 10^3 to 10^4/ml whereupon they decreased and remained at levels of about 10^1/ml (Figure 8–35a). The total bacterial count in a sand filter's effluent ranged from 0 to 9/ml in the same study period. Lower counts in the latter unit were due to prechlorination of the influent water that maintained a chlorine residual throughout the sand. Chlorine was removed in the upper layers of the carbon—hence, the higher counts.

Studies were conducted also to determine bacterial counts on the carbon particles [43]. Figure 8–35b shows the change with time in numbers of bacteria on the carbon particles and in the effluent of the same column. The upper curve in Figure 8–35b shows that the bacterial population follows a sigmoidal growth; that is, bacteria increase exponentially after a short lag phase until their environment limits their growth, after which there is a negative growth acceleration phase. At this point, the bacterial population stabilizes relative to the environmental limits. Figure 8–35c shows the effect of EBCT on bacterial levels in the GAC effluent. Normally, longer EBCT had higher bacterial counts in the effluents even after the volumetric flow rate was taken into consideration. Experiments were conducted also to demonstrate the effect of chlorination on the influents to the columns. As expected, dechlorinated systems had higher bacterial counts than chlorinated waters. Figure 8–36 shows a typical bacterial population

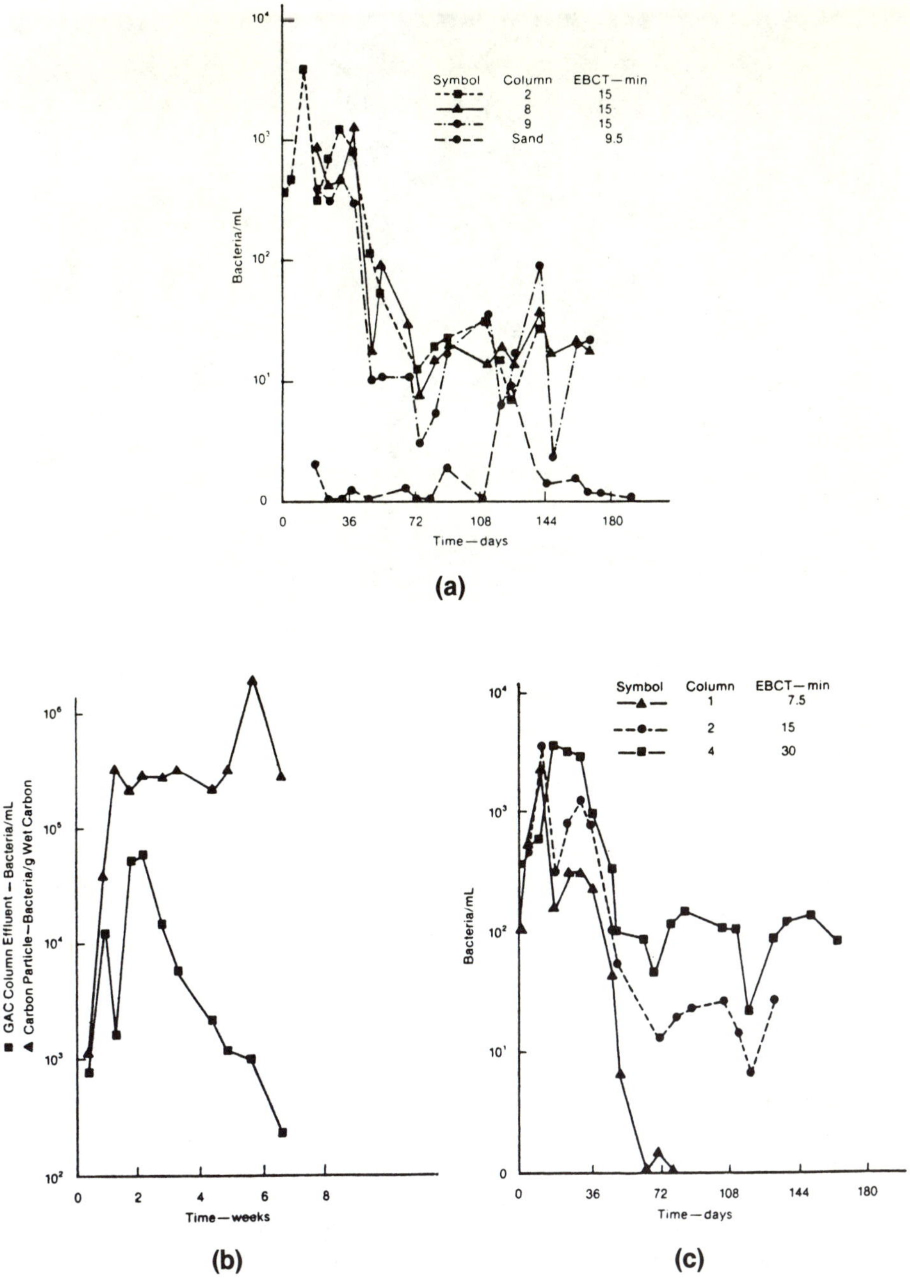

Figure 8–35 (a) Bacteria in the column effluents. (b) Bacteria on the GAC column and in the effluents from a GAC column. (c) Effect of contact time on bacterial levels in GAC effluent. Reproduced from Cairo et al. [43], courtesy of the American Water Works Association.

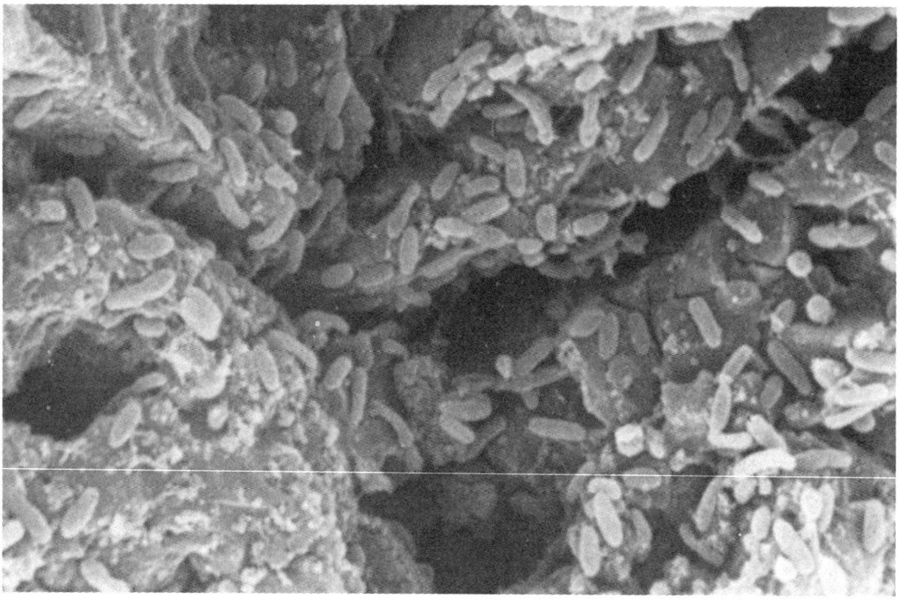

Figure 8–36 Electron micrograph: 4,000x magnification of carbon particle showing bacterial growth. Reproduced from Cairo et al. [43], courtesy of the American Water Works Association.

on GAC particles. They were found scattered over the carbon's surface with larger contents in the macropore areas. The predominant bacteria that were isolated and identified belong to the genus *Pseudomonas,* with several species of *Bacillus* present also in large numbers.

Microbial populations were characterized in effluent waters from two deep-bed GAC filters [45]. Ohio River water was the source: one column was pre-ozonated whereas the second was not disinfected. Results are given in Tables 8–39a and b of the microbiological assays of the two waters. Total bacterial population of the influent water varied from 1.4×10^2 to 1.86×10^4 colony-forming units (CFU)/ml with a mean of 4.26×10^3 for all samples. About 10% of these bacteria were recovered in the effluent where counts ranged from 1.10×10^2 to 1.30×10^3 CFU/ml. Enteric bacteria varied from 50 to 1,000 CFU/100 ml in the influent waters. There was an average recovery of 25% in the effluent's enteric bacteria. Coliform bacteria were detected only in the first week.

Table 8–39b gives some data for the untreated Ohio River water after passage through the GAC filter [45]. Bacterial populations ranged from none detected (ND) to 1.45×10^2 CFU/ml in the effluent. It is interesting to note that untreated bacterial counts were slightly lower than in the ozonated water. Here again, bacterial counts were lower in the effluent than in the influent.

One objective of the Ohio River water study was to speculate about possible

Table 8-39a Microbiological Analysis of Ozonated Water Filtered through GAC Filters

Sample Week	*Sample*	*Bacterial Population (CFU/ml)*	*Fungi (CFU/ml)*	*Enteric Bacteria (CFU/100 ml)*	*Pseudomonads (CFU/100 ml)*
1	Influent	2.65×10^3	ND	1,000	12
	Effluent	1.30×10^3	ND	100	11
2	Influent	1.39×10^3	ND	75	ND
	Effluent	2.20×10^2	ND	40	ND
3	Influent	1.86×10^4	4.0×10^1	500	ND
	Effluent	1.60×10^2	3.5×10^1	45	ND
4	Influent	1.78×10^3	5.0×10^1	736	ND
	Effluent	4.30×10^2	3.5×10^1	136	ND
5	Influent	1.03×10^3	7.5×10^1	50	ND
	Effluent	4.00×10^2	1.0×10^1	10	ND
6	Influent	1.40×10^2	8.0×10^1	450	ND
	Effluent	1.10×10^2	7.0×10^1	96	ND

Source: Reproduced from Brewer and Carmichael [45], courtesy of the American Water Works Association.
Note: CFU = colony-forming unit; ND = none detected.

Table 8-39b Microbiological Analysis of Untreated Water Filtered through GAC Filters

Sample Week	*Sample*	*Bacterial Population (CFU/ml)*	*Fungi (CFU/ml)*	*Enteric Bacteria (CFU/100 ml)*	*Pseudomonads (CFU/100 ml)*
1	Influent	7.5×10^3	ND	950	331
	Effluent	ND	ND	10	ND
2	Influent	3.1×10^3		250	4.5
	Effluent	1.45×10^2	ND	15	2.0
3	Influent	7.7×10^1	1.1×10^1	166	10.7
	Effluent	2.2×10^1	ND	30	ND
4	Influent	1.5×10^1	1.6×10^1	90.8	155
	Effluent	1.2×10^1	ND	20.5	ND
5	Influent	8.0×10^1	6.0×10^1	717	3.0
	Effluent	7.5×10^1	7.0×10^1	33.5	ND
6	Influent	8.8×10^1	2.9×10^1	296	12.0
	Effluent	7.0	1.1×10^1	69.3	ND

Source: Reproduced from Brewer and Carmichael [45], courtesy of the American Water Works Association.
Note: CFU = colony-forming unit; ND = none detected.

production of bacterial endotoxins in GAC filters [45]—that is, if certain bacteria colonize the GAC. An estimate of potential levels of total endotoxins (bound and free) in the water can be calculated from a regression equation [46]:

$$\log_{10} TE = 0.802 \log_{10} \text{enterics} + 1.625, \tag{8.1}$$

where $r = .91$ at a 1% significance level and TE = total endotoxin. This equation was applied to the treated and untreated waters of the Ohio River [45]. Endotoxin contents in the GAC effluent averaged 0.011 mg/l and 0.006 mg/l for the ozonated and untreated samples, respectively. These quantities should be considered as approximations at best. Also, the role of bacterial endotoxins as a toxic substance to humans through direct ingestion is uncertain [47].

Microbial population dynamics were determined on a GAC employed for treatment of surface impounded groundwater [48]. A pilot-scale study was designed for GAC removal of TCE from this water over a 6 month period. Three microbial areas of investigation were pursued: (1) pathogen retention and growth, (2) types and numbers of microorganisms, and (3) microbial sloughing from GAC columns into the effluents. Four GAC columns with an EBCT of 5 minutes were loaded at 4.4 to 4.5 gpm/ft^2. Table 8-40 shows the predominant bacteria that were isolated from the influent source water, GAC column effluents, and GAC cores. Sixty-four isolates were identified. None of the isolates was in the *Enterobacteriaceae* family. Also, no pathogenic enteric bacteria—that is, *Salmonella spp.* or *Shigella spp.*—were isolated from 2 liter influent or column effluent samples in 10 tests. *Pseudomonas aeruginosa* (a threat to compromised hosts) was sought but was rarely present in either the influent or column effluents.

With respect to types of microorganisms, most bacterial isolates were in one of two general groups: gram-negative rods, which are unable to ferment glucose, and gram-positive rods, which form spores [48]. In the first group are the genera *Pseudomonas, Acinetobacter, Alcaligenes, Achromobacter, Group IIk, Group Ve,* and *Group IIf.* Most of these are commonly found in soil and water.

Table 8-40 Predominant Isolates by Location

Influent	*Effluent*	*GAC Bed*
Pseudomonas acidovorans	*Pseudomonas cepacia*	*Group Ve* Biotype 1
Group IIk Biotype 1	*P. putida*	*Group Ve* Biotype 2
Group Ve Biotype 1	*Group IIk* Biotype 1	*Group IIk* Biotype 1
Acinetobacter calcoaceticus var. LWOFFI	*Group Ve* Biotype 2	*Alcaligenes denitrificans*
Alcaligenes denitrificans	*Group IIf*	*Achromobacter xylosidans* Biotype IIIa
Bacillus spp.	*Acinetobacter calcoaceticus* var. LWOFFI	*Bacillus cereus*
Chromobacterium spp.	*Micrococcus* spp.	*Bacillus* spp.
Actinomycete	*Actinomycete* fungus	*Micrococcus* spp. fungus

Source: Reproduced from Donlan and Yohe [48], courtesy of the American Chemical Society.

In the second group is the genus *Bacillus* that is isolated from numerous sources and is common in soil and water. The remainder of the bacterial isolates listed in Table 8-40 are commonly found in water.

Total plate counts (TPCs) were measured also for the GAC columns [48]. Mean TPCs of GAC column sections were about 6.8×10^5 microbes/g carbon, whereas the TPC of the influent water was 1.3×10^3/ml. This is a greater than 2 log increase in number, which indicates a sizeable increase in the microbial population within the column. Figure 8-37 shows the TPC with respect to column depth. Counts were always highest in the top 15 centimeters of the column. Also, there was considerable sloughing of microorganisms from the carbon columns. This is seen in Figure 8-38 for influent and effluent TPC over a 100 day period. In the first 20 days, effluent counts were significantly higher than influent counts, a finding that is contrary to another study cited earlier [45]. However, after 20 days, effluent levels were the same as or below the influent.

There is concern about microbial health hazards from the use of activated

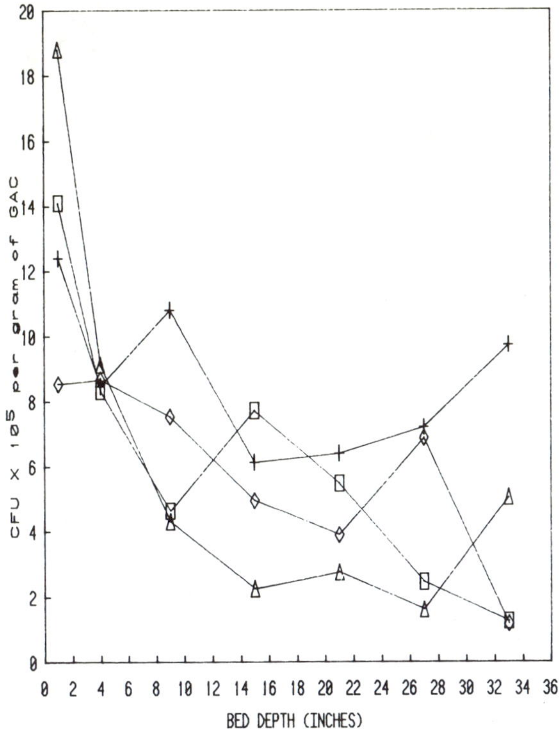

Figure 8-37 Plate counts versus bed depth. Key: Δ, Column 1 (WV-G, 176 days); ◇ Column 2 (WV-G, 176 days); +, Column 3 (F-400, 300 days); and □, Column 4 (HD 10 × 30, 142 days). Reproduced from Donlan and Yohe [48], courtesy of the American Chemical Society.

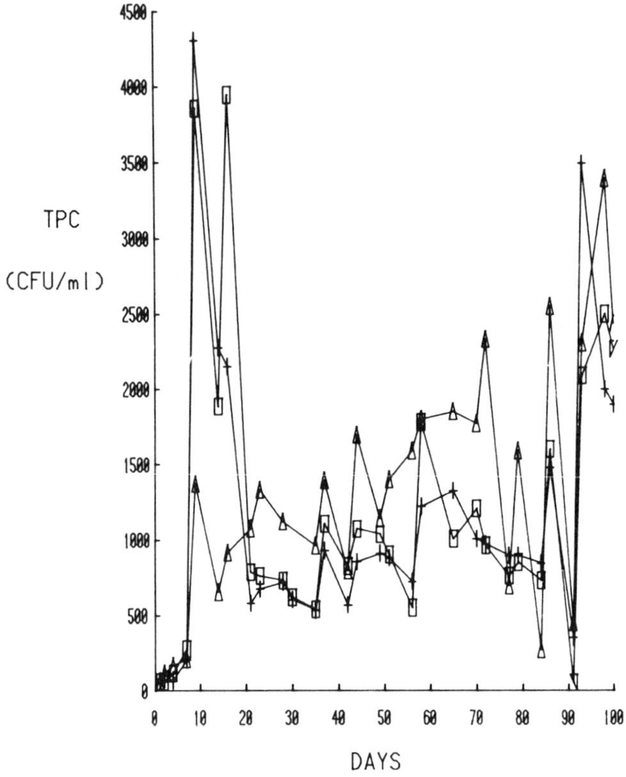

Figure 8-38 Total plate counts for 100 days. Key: Δ, influent; +, Column 1; and □, Column 2. Reproduced from Donlan and Yohe [48], courtesy of the American Chemical Society.

carbon filters in domestic water systems. A study was undertaken to investigate the effect of carbon filters on the microbial content of water [57]. Results indicated that the microbial content of filtered and unfiltered water increased to about the same extent on overnight standing and, in both cases, was reduced by flushing the next day. In addition, the use of activated carbon for the filtration of contaminated well water over a period of 11 weeks had no effect on the total coliform count. Under use conditions, activated carbon filters were found to have no significant effect on the number of bacteria present in the water.

REGENERATION PROCESSES

One of the advantages of GAC is its ability to be regenerated by one or more processes. A carbon system usually is composed of two distinct operations: (1) the contact or adsorption process and (2) a carbon regeneration system [13]. Figure 8-39 shows a typical equipment arrangement of the total system, and Figure

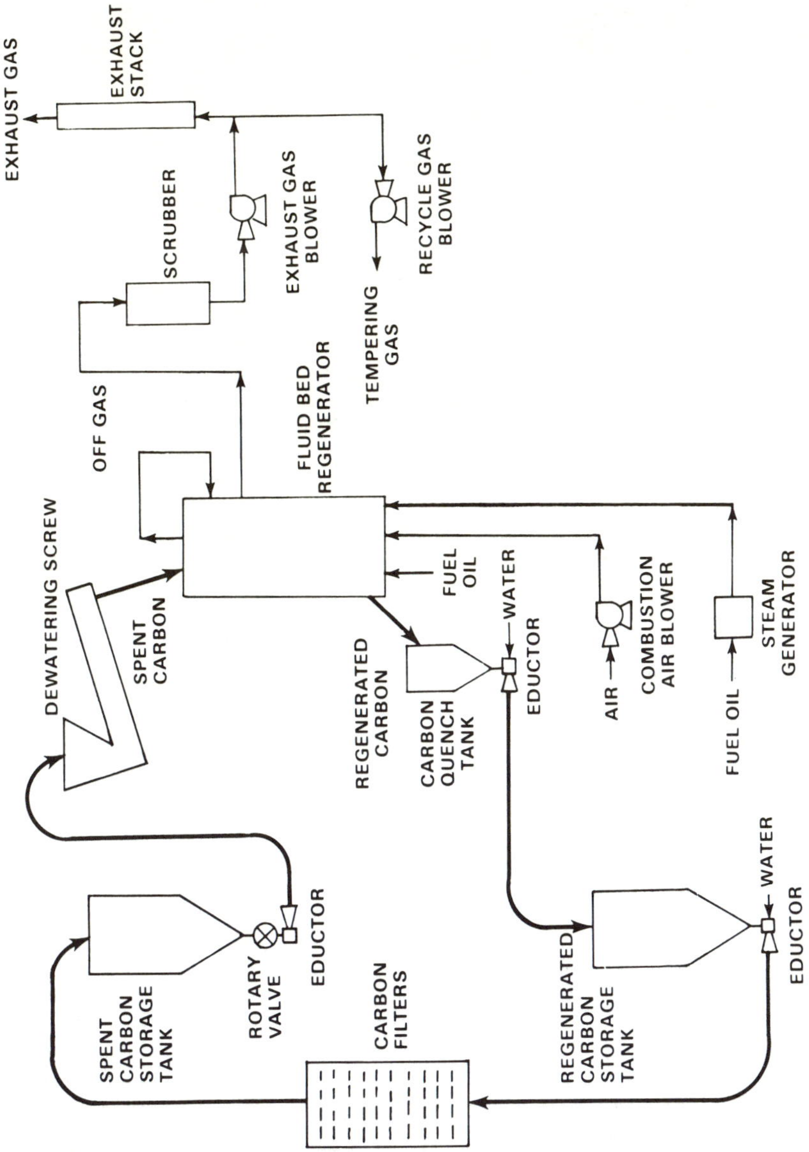

Figure 8–39 Carbon reactivation system. Reproduced from DeMarco et al. [13], courtesy of the American Chemical Society.

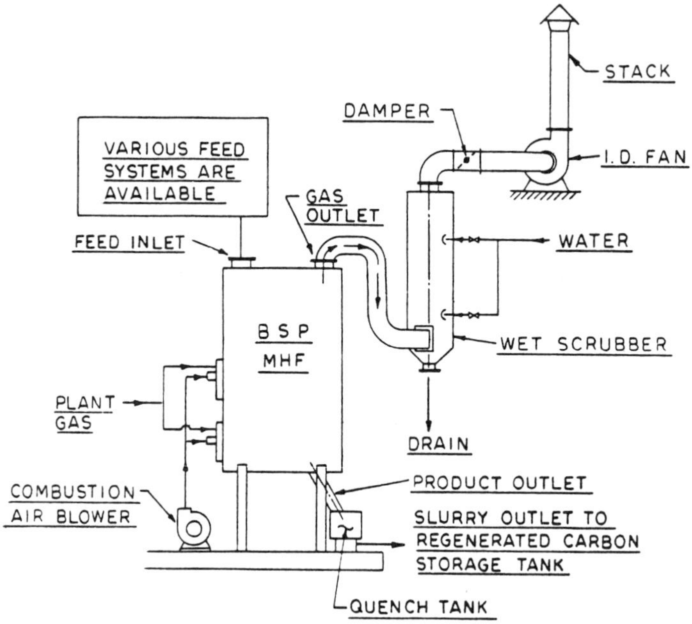

Figure 8–40 Carbon regeneration furnace. Reproduced from Lombana and Halahy [49], courtesy of Ann Arbor Science Publishers, Inc.

8–40 shows a carbon regeneration multiple hearth furnace. The latter consists of the following equipment groups [49]:

Conveying of the slurry;

Receiving, dewatering, and feed tanks;

Feeding;

Regeneration furnace [multiple hearth furnace (MHF)]

Quench tank and conveying;

Off-gas equipment;

Controls and instrumentation.

The carbon slurry must be received and dewatered before it is fed into the furnace. There are several ways to accomplish this. Usually, a carbon slurry is withdrawn from a column and conveyed to a storage tank and then to a dewatering tank prior to heating. The dewatering and storage are frequently combined into one tank (Figure 8–41). A surge tank is placed below the dewatering tank and prior to the furnace. Feeding equipment is usually one of three types: belt, screw feeder, and rotary valve type.

A typical multiple hearth furnace is seen in Figure 8–42. A furnace consists of a steel-sheet-lined refractory inside the unit. This refractory can be castable

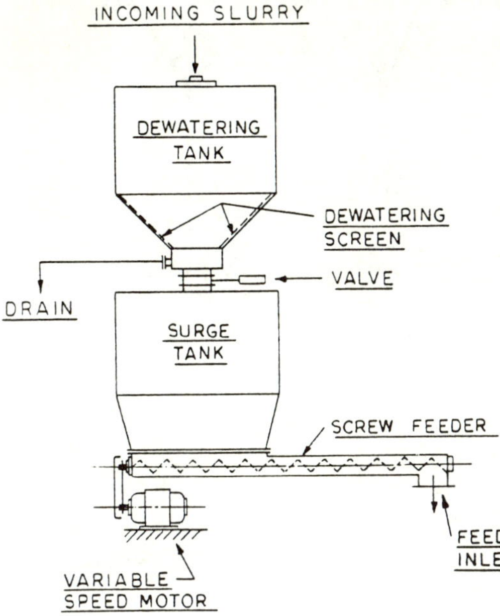

Figure 8-41 Dewatering and surge tanks. Reproduced from Lombana and Halahy [49], courtesy of Ann Arbor Science Publishers, Inc.

for small 30 inch units or can be brick as used in large units. A description of the furnace's interior follows [49]:

> The interior space of the furnace is divided by horizontal brick arches into separate compartments called hearths. Alternate hearths have holes at the periphery or at the center for the carbon to drop through from one hearth to the next. Through the center of the furnace goes a rotating shaft driven at the bottom by a speed reducer with variable-speed drive. It is scaled at the top and bottom by special sand seals to prevent air or gas leakage. The shaft is hollow and has sockets where arms, called rabble arms, are fitted. An inner tube in each arm and in the shaft provide the means for air cooling of both to prevent damage by the intense heat. This cooling air is blown in through a special connection at the bottom of the shaft. The arms are, in turn, fitted with rabble teeth, placed at an angle, and which, when the shaft is rotated, impart a motion to the carbon, moving it "in" in some hearths and "out" in the others. The hearths that move the material out and which have the peripheral holes, called drop holes, are fitted with lute caps. A lute cap is just a disc attached to the shaft which prevents the material dropping from the hearth above. Burners are attached at certain locations in the shell to heat the furnace to the required temperature. These burners can be of the nozzle mixing or premixed type and are normally set so excess air can also be introduced through them for the oxidation of the organic impurities in the carbon. The burner system includes the usual complement of accessories such as spark-ignited pilots, solenoid valves, mix-

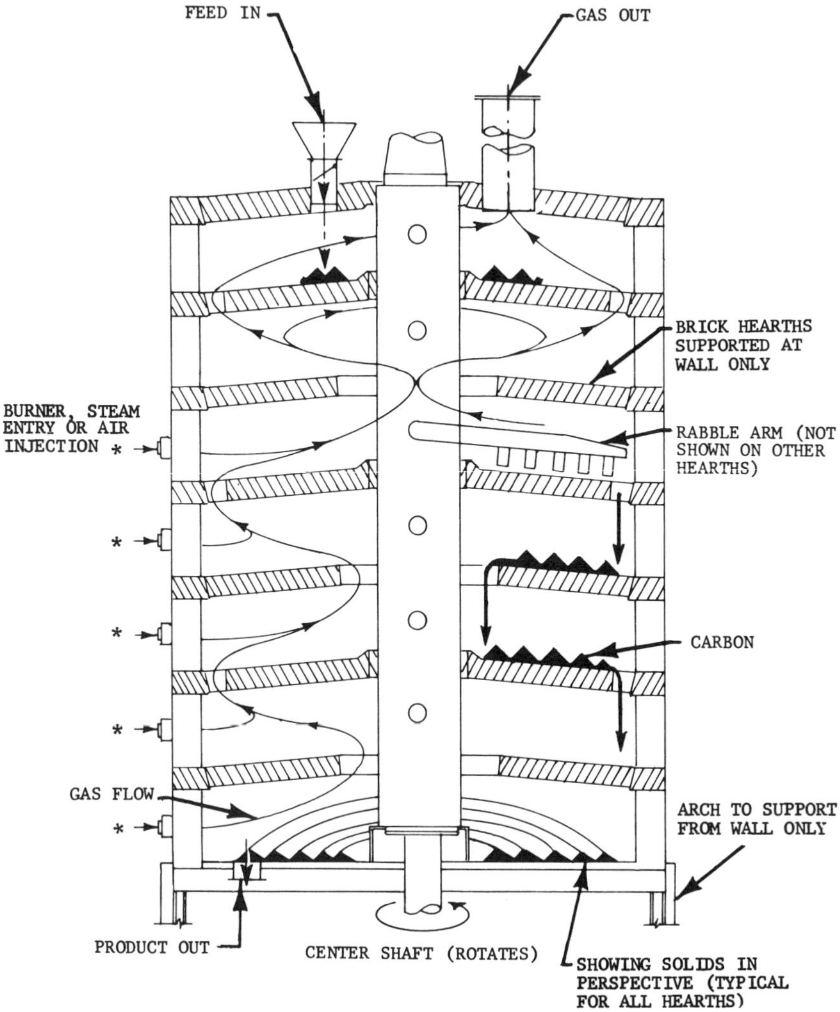

Figure 8-42 Furnace. Reproduced from von Dreusche [50], courtesy of Ann Arbor Science Publishers, Inc.

ers, regulators, safety shutoff valves, combustion air blowers, etc. Fuel used can either be No. 2 diesel oil or gas, either natural, propane or butane. The shell is also provided with nozzles for injecting steam and air in certain hearths to aid in the regeneration of the carbon.

Temperatures in the furnace are usually 1,600 to 1,800°F. Exit temperatures are in the 600 to 700°F range.

After the carbon is discharged from the furnace, it falls into a quench tank that is filled with water up to a level set by a controller or float valve. In this step, the red hot carbon should be isolated from the air. From the quench tank, the carbon is returned to a feed tank and then into the carbon bed.

In multiple hearth furnace regeneration of spent carbon, the loss of carbon capacity due to oxidation and attrition is a significant factor in the overall economics of GAC water treatment. Also, furnaces frequently break down, which requires standby capacity. It is well known from the carbon adsorption method era that organic solvents such as chloroform and ethyl alcohol solubilize or desorb compounds from carbon. There are, however, several disadvantages to the employment of organic solvents on a large scale for carbon regeneration. They are expensive and must be recovered for reuse. Also, there may be health hazards to humans. However, there are other solvents for carbon regeneration.

The desired solvent characteristics for efficient and economic regeneration of GAC are high solubility for the adsorbates, favorable mass transfer properties for rapid desorption, and high volatility for subsequent separation of solutes [51]. Since liquid solvents have high solubility but slow diffusion and gases have rapid mass transfer but low solubilities, supercritical fluids represent a good compromise. The density of supercritical fluids is about one-third of normal liquids, which provides for good solubility. But the density is low enough for high diffusivity and rapid mass transfer.

Carbon dioxide has been found to be an excellent candidate for a supercritical fluid because its critical temperature is near ambient (31°C) and its critical pressure is 72.9 atm [51]. Supercritical CO_2 is a good solvent for varied substances such as octadecanol, stearic acid, polyvinyl chloride, and naphthalene. Favorable mass transfer characteristics have been demonstrated in a study of I_2 dissolution in CO_2 [52].

The degree to which supercritical fluid CO_2 regenerates activated carbon was determined by exposing a column to repetitive adsorption and desorption cycles [51]. Figure 8–43 shows the experimental system. Adsorption loading was determined by integrating the breakthrough curves. The experimental column contained 0.37 g of F-300 screened to 100–120 mesh, while another column contained 7 g of 20 mesh F-300. Figure 8–44 shows the rapid and sharp breakthrough curves for phenol on GAC and for four adsorption-desorption cycles. It should be noted that cycles 2, 3, and 4 had slightly less adsorptive capacities than cycle 1. The rate of desorption of phenol from the small-sized particles of F-300 is seen in Figure 8–45. The phenol concentration in the regenerate fluid peaks quickly after the start of regeneration, decreases rapidly until 80 to 90% of the solute has been removed, then decreases slowly until regeneration is complete. Desorption of phenol is very rapid, with 50% removal within 20 minutes and 90% within 1 hour. Two other carbons (Anthrasorb CC1230EH and Amoco GX-31) and a carbonized synthetic resin (Ambersorb XE-348) were studied also and yielded results similar to F-300. It appears that supercritical CO_2 regeneration of activated carbon is feasible provided this solvent is recyclable. The authors of this study were confident that this can be accomplished [51].

Some users, however, prefer solvent regeneration of activated carbon [53]. The arguments against thermal regeneration cite the 5 to 10% attrition loss of

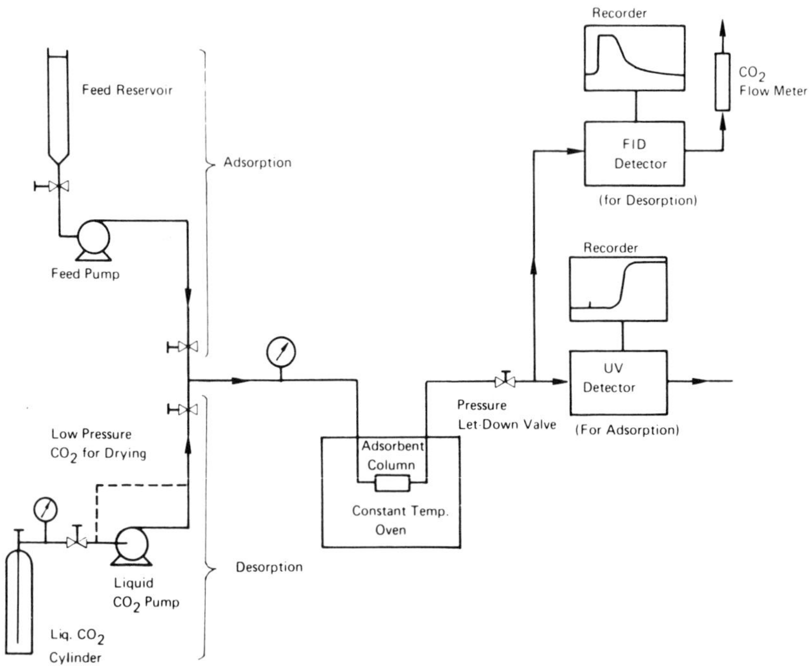

Figure 8–43 Schematic of adsorption/desorption apparatus. Reproduced from Modell et al. [51], courtesy of Ann Arbor Science Publishers, Inc.

carbon during each cycle as well as significant fuel costs. Arguments for solvent regeneration are as follows [53]:

> (a) It can be done relatively rapidly in situ: thus it minimizes downtime for the adsorbers and completely eliminates unloading, transporting and repacking of the carbon. (b) No carbon attrition occurs: thus the cost of replacing attrited carbon is avoided. (c) No degradation of the carbon surface or pore structure due to repetitive pyrolysis and reoxidation can occur. Only a modest loss of adsorption capacity over the first few cycles (but not thereafter) occurs due to irreversible solvent adsorption. Hence, the carbon can last almost indefinitely. (d) Recovery of valuable adsorbates is possible: or, if they are not valuable, they can be easily disposed of by combusting them along with residual solvent (after concentration by distillation). (e) With properly chosen solvents, solvent removal from the columns and subsequent recovery by distillation can be done easily.

Consequently, 19 solvents were evaluated in batch tests involving desorption of a representative organic adsorbate (phenol) from PAC. Three of the better solvents (acetone, dimethylformamide, methanol) were tested further in fixed beds of GAC.

Batch desorption tests were conducted on Norit A-PAC saturated with phenol [53]. Subsequently, the solvent-carbon mixture was stirred in a covered

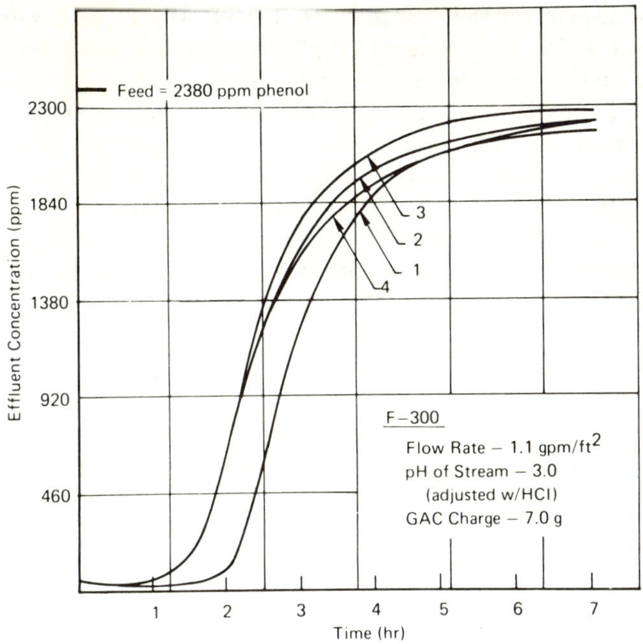

Figure 8-44 Adsorption of phenol on F-300 (numbers on curves refer to adsorption cycle). Reproduced from Modell et al. [51], courtesy of Ann Arbor Science Publishers, Inc.

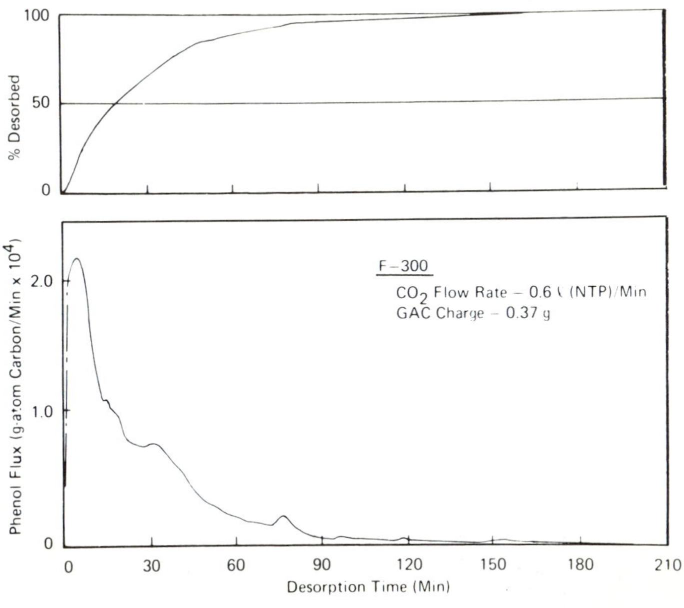

Figure 8-45 Desorption of phenol from GAC with supercritical CO_2. Reproduced from Modell et al. [51], courtesy of Ann Arbor Science Publishers, Inc.

beaker by a magnetic mixer for 30 minutes. Table 8-41 gives results of the extent to which each of 19 solvents desorbed phenol from the PAC. The percentage removals apparently represent a once only extraction.

The next consideration in solvent regeneration of activated carbon is its solubility in water and adsorptive characteristics on carbon [53]. Solvents with a limited solubility in water clearly could cause severe problems in flushing the carbon bed with water. Also, solvents with low water solubilities have a tendency to adsorb strongly onto carbon. From the list in Table 8-41, those solvents with less than complete miscibility with water are methanol, ethanol, propanol, acetone, and dimethylformamide (DMF). Of these, methanol, acetone, and DMF were chosen for further evaluation.

Fixed bed experiments were conducted with CAL-type GAC, 20 × 28 mesh, onto which 0.24 g phenol per g of C was adsorbed [53]. Subsequently, 300 ml of the three solvents were individually applied to the column at a rate of about 20 ml/min. Figure 8-46 shows the desorption curves. Percentages of desorption were DMF, 93%; acetone, 90%; and methanol, 82%.

Table 8-41 Batch Desportion Results (0.1 g Phenol Loaded onto 0.5 g Carbon)

Solvent (50 ml)	*% of Adsorbed Phenol Desorbed*
Dimethylformamide	95.0
95% ethanol	88.3
Acetone	85.7
Toluene	82.5
Benzene	76.5
Ethyl acetate	76.3
Dioxane	74.7
n-butanol	74.2
2-propanol	73.7
Isoamyl alcohol	71.7
Methanol	70.6
Chloroform	69.6
Tetrachloroethane	66.9
Dichloroethane	62.3
Carbon disulfide	57.3
Formaldehyde	55.0
Carbon tetrachloride	41.0
Distilled water	18.3
Acetic acid	4.7

Source: Reproduced from Cooney et al. [53], courtesy of Pergamon Press.

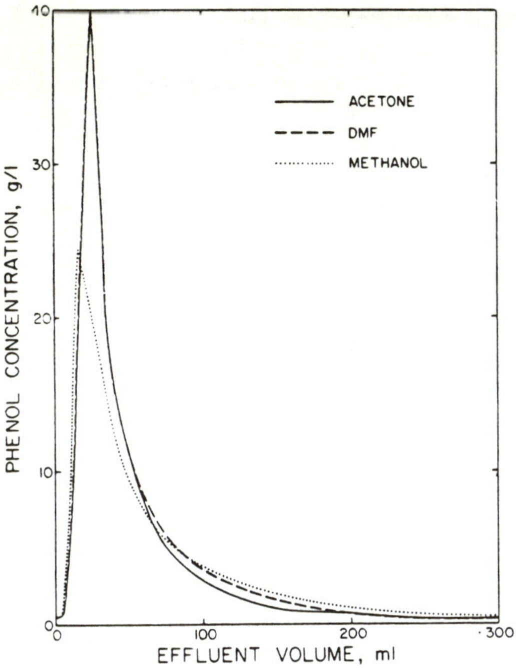

Figure 8–46 Desorption of phenol by various solvents at room temperature and 20 ml/min flow rate. Reproduced from Cooney et al. [53], courtesy of Pergamon Press.

Important questions are how much solvent remains on the carbon after water rinsing, and does the solvent interfere with adsorption of the phenol on the next cycle? For methanol, a water wash of 300 ml was effective in restoring almost all the phenol's adsorptive capacity. For acetone and DMF, a hot water wash or steam stripping should be more effective than water.

A detailed study was conducted on chemical regeneration of exhausted activated carbon for a wide range of adsorbates and regenerants [54]. F-400, 0.85 to 1 mm, was used throughout the study for adsorption of phenol, aniline, benzyl alcohol, benzaldehyde, and nitrobenzene. A 72 hour contact time was employed for the adsorptive phase, and this was followed by a 2 hour desorption period by the regenerant. Capacity of the regenerated carbon was evaluated also. A regeneration efficiency (RE%) was calculated from the ratio of $(A_r/A_o) \times 100$, where A_r is the adsorptive capacity of the regenerated carbon and A_o is the original capacity of the virgin carbon.

Tables 8–42 to 8–48 give the regeneration efficiencies for the many solvents employed in this study [54]. The effect of oxidizing agents on RE is seen in Table 8–42 where these values are low and, in some cases, negative. A negative RE indicates that the carbon did not adsorb any organics for the A_r value; rather, additional desorption occurred that yielded a higher equilibrium solute concen-

Table 8-42 Oxidizing Agents and Their REs

	RE (%) for Carbon Exhausted with				
Regenerant	*Aniline*	*Phenol*	*Benzyl Alcohol*	*Benzaldehyde*	*Nitrobenzene*
10% $KMnO_4$	37.0	41.3	−18.2	28.9	−7.6
10% $K_2Cr_2O_7$	0	38.0	33.6	36.2	−6.9
0.35% NaOCl	68.0	56.0	−4.5	22.4	−7.4

Source: Reproduced from Martin and Ng, [54], courtesy of Pergamon Press.

tration at the end of the phase than at the beginning. It is not surprising that treatment with oxidizing agents gave low results. This process increases the number of surface oxides on the carbon's surface that, in turn, lowers the adsorptive capacity (see also Chapter 5). Table 8-43 shows the effect of 10% HCl and 6M NaOH on the RE values. For the most part, the RE values were low due to simple $[H^+]$ effects on the acidic functional groups on the carbon's surface. Also, there may be some competitive adsorptive effects from H^+ and $\overline{O}H$ ions.

Eight carboxylic acids were employed for carbon regeneration [54]. REs in excess of 90% were achieved in many instances, especially with phenol (Table 8-44). However, the data suggest that as the molecular weight of the acid increases, the RE values decrease. This is an obvious situation where the larger molecules of the acid compete with the adsorbates for adsorption.

Four amine solvents were examined for their ability to regenerate carbon [54]. The results are seen in Table 8-45 where RE values ranged from 42.7% (aniline) to 95.9% (benzaldehyde). The molecular weight effect is evident especially where the larger molecules of benzaldehyde and nitrobenzene overcome the molecular weights of the amines.

Three chloromethanes were investigated for their capacity to regenerate the F-400 [54]. Again, the efficacy of these solvents as regenerants appears to be affected by their molecular weight. RE values decrease when the molecular weight of the solvents increases for the five adsorbates (Table 8-46). Again, the molecular weight of the adsorbate in relation to that of the regenerant affects the RE value.

Table 8-47 gives the RE values for acetone and benzene, and Table 8-48 contains these values for the alcohols [54]. As expected, acetone was a better

Table 8-43 HCl and NaOH and Their REs

	RE (%) for Carbon Exhuasted with				
Regenerant	*Aniline*	*Phenol*	*Benzyl Alcohol*	*Benzaldehyde*	*Nitrobenzene*
10% HCl	72.2	52.8	25.7	13.0	15.2
6M NaOH	29.0	44.0	8.0	22.5	14.7

Source: Reproduced from Martin and Ng, [54], courtesy of Pergamon Press.

ble 8–44 Carboxylic Acids and Their REs

generant	Molecular Weight of Acid	RE (%) for Carbon Exhausted with				
		Aniline	Phenol	Benzyl Alcohol	Benzaldehyde	Nitrobenzene
rmic acid	46.03	84.0	98.1	98.2	90.5	91.8
etic acid	60.05	72.0	99.1	72.7	91.8	87.0
opionic acid	74.08	79.0	99.1	60.9	82.3	83.6
utyric acid	88.11	78.0	97.2	62.7	93.9	81.0
-butyric acid	88.11	67.0	74.1	50.0	59.9	77.5
valeric acid	102.13	46.0	74.1	34.5	74.8	84.8
-valeric acid	102.13	73.0	75.0	53.6	85.0	84.5
exanoic acid	116.16	34.0	70.4	16.4	70.1	69.1

urce: Reprodued from Martin and Ng, [54], courtesy of Pergamon Press.

ble 8–45 Amines and Their REs

generant	Molecular Weight of Amine	RE (%) for Carbon Exhausted with				
		Aniline	Phenol	Benzyl Alcohol	Benzaldehyde	Nitrobenzene
ropylamine	59.11	79.6	71.3	89.4	95.9	93.6
hanolamine	61.08	62.1	64.8	71.7	62.6	71.7
outylamine	73.14	57.3	70.4	80.5	81.6	93.6
iethylamine	101.19	42.7	54.6	65.5	95.9	90.6

urce: Reproduced from Martin and Ng, [54], courtesy of Pergamon Press.

ble 8–46 Chloromethanes and Their REs

generant	Molecular Weight of Chloromethane	RE (%) for Carbon Exhausted with				
		Aniline	Phenol	Benzyl Alcohol	Benzaldehyde	Nitrobenzene
chloromethane	84.93	84.0	72.5	60.4	91.2	94.3
loroform	119.38	49.0	49.5	43.2	83.7	89.5
rbon tetrachloride	153.82	21.0	16.5	1.4	29.3	47.8

urce: Reproduced from Martin and Ng, [54], courtesy of Pergamon Press.

regenerant than benzene. All six alcohols performed well as regenerants and gave the highest results of all the solvents in this study. There was no discernible correlation with alcohol molecular weight and RE. This observation may be due to low molecular weights of the alcohols in relation to that of the five adsorbates; that is, RE values were expected to be high for the alcohols. Also, these solvents are miscible with water and are reasonably soluble. Thus, they would be removed easily by a water rinse following regeneration.

It would appear from this study that, as expected, solvents vary in their capacity to regenerate carbon and adsorbates vary in their desorptive characteristics [54]. Consequently, there is no universal solvent for carbon regeneration. Also, solvent recovery would have to be considered.

A mechanically stirred graphite microfurnace was developed for thermal regeneration of spent PAC or PAC-activated sludges [55]. Three PACs, Amoco PX-21, Aqua Nuchar-SA, and Hydrodarco-C, were evaluated for their comparative virgin and regenerated adsorptive performances. With spent PAC alone, 100% and greater recovery of wastewater adsorptive capacity was achieved always from a 20 minute, 825°C regeneration with a 40% N_2, 50% H_2O, and 10% CO_2 purge gas. In some cases, adsorptive capacity recoveries exceeded 200%. These procedures were applied to wastewater treatment.

Table 8-47 Acetone and Benzene and Their REs

		RE (%) for Carbon Exhausted with				
Regenerant	*Molecular Weight of Regenerant*	*Aniline*	*Phenol*	*Benzyl Alcohol*	*Benzaldehyde*	*Nitrobenzene*
Acetone	58.08	26.4	93.6	81.8	63.9	85.7
Benzene	78.11	12.0	60.2	−4.5	55.6	74.0

Source: Reproduced from Martin and Ng, [54], courtesy of Pergamon Press.

Table 8-48 Alcohols and Their REs

		RE (%) for Carbon Exhausted with				
Regenerant	*Molecular Weight of Alcohol*	*Aniline*	*Phenol*	*Benzyl Alcohol*	*Benzaldehyde*	*Nitrobenzene*
Methanol	32.04	90.7	97.2	74.3	73.5	85.8
Ethanol	46.07	77.8	92.6	69.9	78.8	85.0
1-propanol	60.10	85.8	90.7	66.4	78.1	85.4
2-propanol	60.10	68.5	100.0	66.4	75.1	81.1
1-butanol	74.12	75.0	94.4	74.3	76.2	92.7
2-butanol	74.12	78.7	91.7	66.4	74.8	85.4

Source: Reproduced from Martin and Ng, [54], courtesy of Pergamon Press.

Laboratory and pilot column studies were conducted to evaluate the effect of regeneration on the adsorptive mechanism and capacity of activated carbon [58]. It was suggested that at least 10 regeneration cycles should be evaluated. In the laboratory study, a regeneration temperature of 542°C (1,000°F) did not restore the fundamental kinetic characteristics of virgin carbon (F-300 or F-400) for adsorption of methyl ethyl ketone (MEK) or *n*-butanol. At a higher temperature, 850°C (1,550°F), regeneration conditions produced breakthrough curves (not shown) that were virtually identical to the virgin carbon for these two compounds. However, a third temperature, 962°C (1,750°F), produced a quicker regeneration, but this resulted in a lower overall adsorption capacity.

The pilot column study used the same three regeneration temperatures cited earlier, with $CHCl_3$ as the model compound [58]. The most significant results were obtained from the breakthrough curves for $CHCl_3$ on F-400 operating in a postcontactor mode. It can be seen in Figures 8–47a, b, and c that the 962°C regenerated carbon produced an effluent concentration of $CHCl_3$ twice as large as the virgin, 542°C, and 850°C regenerated carbons. Pilot column experiments were conducted also with TOC as a surrogate parameter. Here, the breakthrough curves (not shown) for all virgin and regenerated F-400 GACs were virtually identical, whereas the 962°C regeneration was slightly better than that for the other carbons for the F-300 columns.

A plausible explanation for these observations is that the conventional manner of regeneration, 962°C, may be too harsh a condition for potable water carbon. This temperature was developed mainly for carbon used in treatment of wastewaters with high concentrations of organics. Also, this temperature may fracture the micropore structure and weaken the macropore structure. It would appear from this study that the 850°C regeneration may be the optimum temperature for potable water carbon.

COAGULATION

To remove PAC quickly from water, it is necessary to add a chemical coagulant. This step frequently occurs as a routine practice where raw waters are coagulated for turbidity removal. However, it may be necessary to remove only PAC whereupon an optimum coagulant dosage is desired. The latter was achieved through laboratory experiments where alum was added to various suspensions of carbon ranging from 2. to 500. mg/l at pH 7.0 and 2 meq/l alkalinity [56]. The essential results of this study are given in Table 8–49 where the alum doses are well within those employed at water treatment plants.

BACKWASHING GAC FILTERS

There is a wide variety of backwashing procedures as practiced in the United States [59]. This step in the treatment of drinking water may be the most important and most costly. It is certainly as important to restore the filtration capacity as well as the adsorptive capacity.

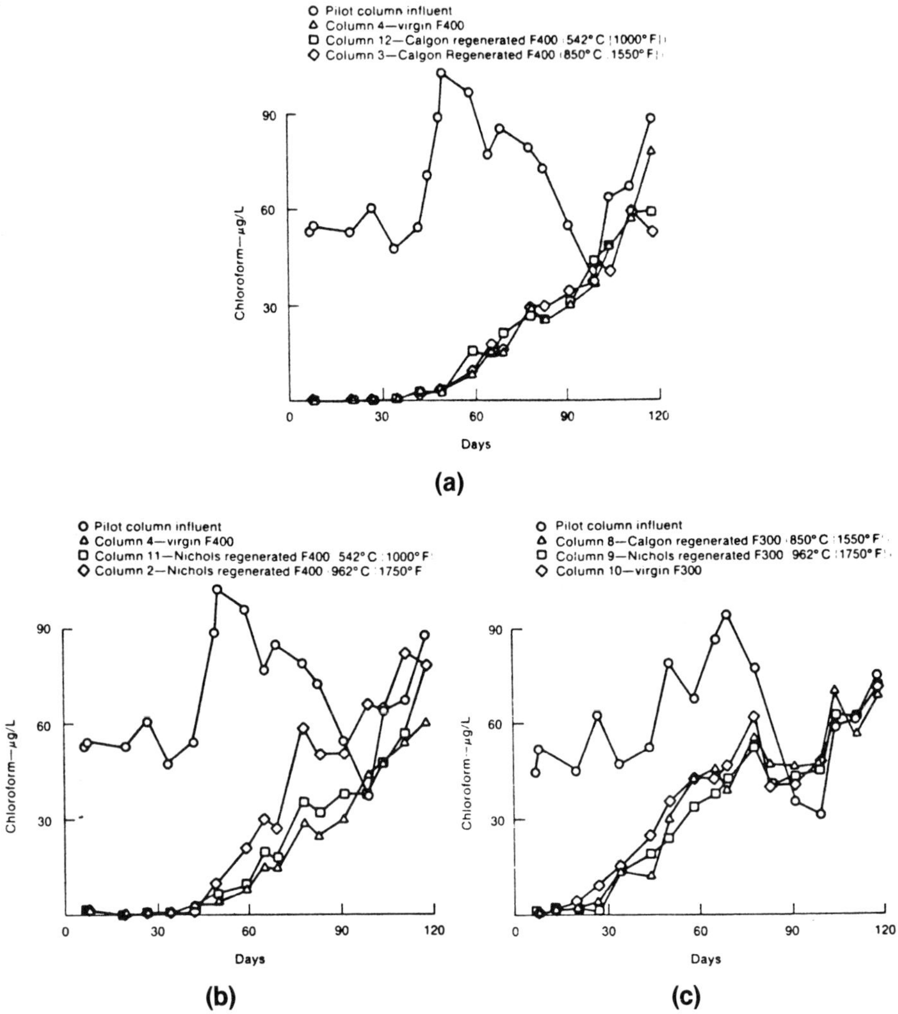

Figure 8-47 (a) Chloroform breakthrough curves for virgin and Calgon regenerated F-400 carbons. (b) Chloroform breakthrough curves for virgin and Nichols regenerated F-400 carbons. (c) Chloroform breakthrough curves for virgin and Calgon and Nichols regenerated F-300 carbons. Reproduced from Cairo et al. [58], courtesy of the American Water Works Association.

The many categories of backwashing are:

1. High-rate backwash—for the purpose of this report, providing full-bed fluidization and substantial bed expansion (20–50 percent) is normally preceded and followed by a lower rate backwash. This backwash system can be used for single or multimedia filters.

Table 8-49 Coagulation of Carbon Suspensions

Suspension	Concentration of Carbon (mg/l)	Initial Residual Turbidity (units)	*Alum Dose to Achieve* 50% Turbidity Reduction	Residual Turbidity of One	Minimum Residual Turbidity (units)	Minimum Residual Turbidity (units)
I-1	2	0.36	7.2	—	25	0.04
I-2	5	1.3	7.0	5.8	30	0.06
I-3	25	5.6	6.8	8.0	25	0.14
I-4	50	5.5	3.9	7.8	30	0.27
I-5	100	5.9	3.9	7.6	25	0.22
I-6	150	6.0	2.8	9.0	25	0.21
I-7	300	6.2	2.9	10.2	35	0.45
I-8	500	5.9	5.2	12.2	25	0.75

Source: Reproduced from Letterman et al. [56], courtesy of the American Water Works Association.
Note: pH, 7.0; alkalinity, 2 me/l.

2. Low-rate backwash with little or no bed fluidization or bed expansion—auxiliary scour is essential to low-rate backwash. This backwash system can be used for single medium filters only.
3. Water backwash with surface-wash auxiliary.
4. Water backwash with air auxiliary.
   a. Air scour followed by low-rate water backwash—for use with single-medium filters only.
   b. Air scour followed by high-rate water backwash—for use with single and multimedia filters.
   c. Simultaneous air scour and low-rate water backwash, followed by low-rate water backwash alone—for use with single-medium filters only.
   d. Simultaneous air scour and low-rate water backwash, followed by high-rate water backwash alone—for use with single and multimedia filters. [59]

The complexity of the backwash problem is compounded by the types of media being used and special needs related to the various media. For example, the use of air-scour auxiliary during backwash causes mixing of the media. On the other hand, the use of high-rate water backwash causes some stratification of the media in which the fine-sized grains of a sand filter collect at the upper layers. In the case of dual-media (coal over silica sand) or triple- (mixed-) media filters (coal over silica sand over garnet sand), the finer fractions of each component collect in the upper layers of that component. Thus, if air-scour auxiliary is used with a sand filter, and if filtrate quality is not sacrificed, it may be desirable not to restratify since the coarser surface media of the unstratified filter produce longer filter runs. On the other hand, if air-scour auxiliary is used with dual or triple media, full fluidization and restratification are essential to place the media in their desired positions within the filter bed. [59]

Effectiveness of the above backwashing system must be judged by consideration of such factors as:

1. Media size—coarse media will react differently to high or low upflow velocity or to surface wash or air scour than fine media.
2. Media shape—rounded grains are generally thought to be easier to clean than angular grains or flat grains.
3. Media density—denser material needs higher velocities to suspend it in the upflow, increasing the hydrodynamic forces at the water-filter-grain interfaces.
4. Water quality—colored, turbid, or softened water and primary or secondary sewage behave differently regarding ability to form mudballs or to attach the dirt firmly to the grains.
5. Coagulant used—the amount and type of coagulants used (metallic coagulant or polyelectrolyte) change the "adhesiveness" of the film formed around the grains. Weak or strong flocs will also behave differently with regard to ease of backwashing. [59]

Current U.S. practice still employs sand filters, but dual- or triple- (mixed) media are being utilized on a wider basis. The latter use high filtration rates. Backwash has traditionally been at velocities sufficient to expand the bed by 20–30%. This used with sand and dual-media filters.

On high quality surface waters, water-fluidization backwash alone is usually effective. An example is where color removal is the primary treatment objective. However, this washing technique is not effective on highly turbid waters. Here, a surface-wash auxiliary is employed. On the other hand, air scour is being used on a wider basis. In the U.S., a typical air rate is 55–91 m/hr whereas the washwater rate needed to expand and restratify the bed is typically in the 36–55 m/hr.

Mixed-media beds have created special backwashing problems. Here it is essential that all media be fluidized during backwash to ensure that all layers are stratified into their proper place. This is especially necessary where air scour is employed during part of the backwash. The upflow velocity must be high enough to raise all the course grains located at the bottom of each layer. In the U.S., a high-rate backwash is needed for dual- and triple-media filters. Also, temperature affects the rate of backwash. For each degree Celsius change in water temperature, a 2% change in the backwash velocity is required to maintain the same degree of expansion.

REFERENCES

1. Hager, D.G. *Environ. Sci. Technol. 1,* 287 (1967).
2. Hager, D.G., and R.D. Fulker. *J. Soc. Water Treat. Exam. 17,* 41 (1968).
3. Dostal, K.A., et al. *J. Amer. Water Works Assoc. 57,* 663 (1965).
4. Hager, D.G., and M.E. Flentje. *J. Amer. Water Works Assoc. 57,* 1440 (1965).
5. Erdei, J. *J. Amer. Water Works Assoc. 55,* 1506 (1963).
6. Lawrence, C.H. *Water Wastes Eng. 5,* 46 (1968).
7. Monsitz, J.T., and L.D. Ainesworth. *Public Works 101,* 113 (1970).

8. Inhoffer, W.R. "Use of Granular Activated Carbon at Passaic Valley Water Commission." Paper presented at the American Water Works Association Water Technology Conference, Atlanta, Ga. (December 9, 1975).
9. Cairo, P.R., et al. "The Application of Bench-Scale and Pilot-Scale Studies for Control of Chemical Contaminants in Drinking Water," Chapter 1 in *Activated Carbon Adsorption of Organics from the Aqueous Phase*, Vol. 2. M.J. McGuire and I.H. Suffet, eds. Ann Arbor Science Publishers, Ann Arbor, Mich. (1980), p. 3.
10. Cairo, P.R., et al. "Development of Criteria for the Design of Full Scale Carbon Adsorption Systems." *Proc. Amer. Water Works Assoc. Water Quality Technol. Conf.* (1978).
11. Hyde, R.A. "Removal of Haloforms and Pesticides by Granular Activated Carbon," Chapter 6 in *Activated Carbon Adsorption of Organics from the Aqueous Phase,* Vol. 2. M.J. McGuire and I.H. Suffet, eds. Ann Arbor Science Publishers, Ann Arbor, Mich. (1980), p. 137.
12. Summers, R.S., and P.V. Roberts. "Dynamic Behavior of Organics in Full-Scale Granular Activated-Carbon Columns," Chapter 22 in *Treatment of Water by Granular Activated Carbon.* M.J. McGuire and I.H. Suffet, eds. Advances in Chemistry Series 202. American Chemical Society, Washington, D.C. (1983), p. 503.
13. DeMarco, J., et al. "Experiences in Operating a Full-Scale Granular Activated Carbon System with On-Site Reactivation," Chapter 23 in *Treatment of Water by Granular Activated Carbon.* M.J. McGuire and I.H. Suffet, eds. Advances in Chemistry Series 202. American Chemical Society, Washington, D.C. (1983), p. 525.
14. Yohe, T.L., et al. *J. Amer. Water Works Assoc. 73,* 402 (1981).
15. Love, O.T. Jr., and R.G. Eilers. *J. Amer. Water Works Assoc. 74,* 413 (1982).
16. Meijers, A.P., et al. *J. Amer. Water Works Assoc. 71,* 628 (1979).
17. Blanck, C.A. *J. Amer. Water Works Assoc. 71,* 525 (1979).
18. Wood, P.R., and J. DeMarco. *J. Amer. Water Works Assoc. 71,* 674 (1979).
19. Badorek, D.L., et al. "Removal of Trihalomethanes and Total Organic Carbon from Drinking Water by Adsorption onto Several Granular Activated Carbons," Chapter 3 in *Activated Carbon Adsorption of Organics from the Aqueous Phase,* Vol. 2. M.J. McGuire and I.H. Suffet, eds. Ann Arbor Science Publishers, Ann Arbor, Mich. (1980), p. 71.
20. Rook, J.J. "Comparison of the Removal of Halogenated and Other Organic Compounds by Six Types of Carbon in Pilot Filters," Chapter 20 in *Treatment of Water by Granular Activated Carbon.* M.J. McGuire and I.H. Suffet, eds. Advances in Chemistry Series 202. American Chemical Society, Washington, D.C. (1983), p. 455.
21. Symons, J.M., et al. "Treatment Techniques for Controlling Trihalomethanes in Drinking Water." EPA-600/2-81-156. U.S. EPA, Cincinnati, Ohio (1981).
22. Anderson, M.C., et al. *J. Amer. Water Works Assoc. 73,* 432 (1981).
23. DiFilippo, J.D., et al. "Evaluation of Powdered Activated Carbon for Removal of Trace Organics at New Orleans, LA." EPA-600/52-81-027. U.S. EPA, Cincinnati, Ohio (1981).
24. Murray, C.R., and E.B. Reeves. "Estimated Use of Water in the US in 1975." Circular 765. U.S. Department of the Interior, Geological Survey, Washington, D.C. (1977).
25. Dyksen, J.E., and A.F. Hess III. *J. Amer. Water Works Assoc. 74,* 394 (1982).
26. McKinnon, R.J., and J.E. Dyksen. *J. Amer. Water Works Assoc. 76,* 42 (1984).
27. Bell, F.A. Jr., et al. *J. Amer. Water Works Assoc. 76,* 126 (1984).
28. Taylor, R.H., et al. *J. Amer. Water Works Assoc. 71,* 577 (1979).

29. Wallis, C., et al. *Water Res. 8,* 111 (1974).
30. McCreary, J.J., and V.L. Snoeyink. *J. Amer. Water Works Assoc. 69,* 437 (1977).
31. Scheidtmann, H., et al. *Gas Wasserfach Wasser/Abwasser, 114,* 267 (1973).
32. Sontheimer, H. and D. Maier, *Gas Wasserfach, Wasser/Abwasser 113,* 187 (1972).
33. Randtke, S.J., and C.P. Jepsen. *J. Amer. Water Works Assoc. 73,* 411 (1981).
34. Argaman, Y., et al. *J. Amer. Water Works Assoc. 76,* 172 (1984).
35. Barnett, R.H., and A.R. Trussell. *J. Amer. Water Works Assoc. 70,* 660 (1978).
36. Neukrug, H.M., et al. "Removing Organics from Philadelphia Drinking Water by Combined Ozonation and Adsorption." EPA-600/52-83-048. Municipal Environmental Research Laboratory, Cincinnati, Ohio (1983), p. 1.
37. Glaze, W.H., and J.L. Wallace. *J. Amer. Water Works Assoc. 76,* 68 (1984).
38. Yohe, T.L., et al. "Monitoring and Analyses of Aqueous Chlorine Effects on Granular Activated Carbon Pilot Contactors," Chapter 2 in *Activated Carbon Adsorption of Organics from the Aqueous Phase,* Vol. 2. M.J. McGuire and I.H. Suffet, eds. Ann Arbor Science Publishers, Ann Arbor, Mich. (1980), p. 27.
39. Gumerman, R.C., et al. "Estimating Costs for Water Treatment as a Function of Size and Treatment Efficiency." EPA-600/12-78-182. U.S. EPA, Cincinnati, Ohio (1978), p. 1.
40. Gumerman, R.C., et al. *J. Amer. Water Works Assoc. 71,* 690 (1979).
41. Clark, R.M., and P. Dorsey. *J. Amer. Water Works Assoc. 72,* 450 (1980).
42. Quarino, C.F., et al. "Design and Economic Considerations of Granular Activated Carbon Systems for Removal of Trihalomethanes and Synthetic Organic Contaminants," Chapter 22 in *Activated Carbon Adsorption of Organics from the Aqueous Phase,* Vol. 2. M.J. McGuire and I.H. Suffet, eds. Ann Arbor Science Publishers, Ann Arbor, Mich. (1980), p. 507.
43. Cairo, P.R., et al. *J. Amer. Water Works Assoc. 71,* 660 (1979).
44. McElhaney, J.B., and W.R. McKeon. *Proc. Amer. Water Works Assoc. 6th Water Quality Technol. Conf.* (1978), p. 35.
45. Brewer, W.S., and W.W. Carmichael. *J. Amer. Water Works Assoc. 71,* 738 (1979).
46. Evans, T.M., et al. *Appl. Envir. Microbiol. 35,* 376 (1978).
47. Weinbaun, G., et al. "Bacterial Endotoxins." In *Microbial Toxins.* Academic Press, New York (1971).
48. Donlan, M., and T.L. Yohe. "Microbial Population Dynamics on Granular Activated Carbon Used for Treating Surface Impounded Groundwater," Chapter 15 in *Treatment of Water by Granular Activated Carbon.* M.J. McGuire and I.H. Suffet, eds. Advances in Chemistry Series 202. American Chemical Society, Washington, D.C. (1983), p. 337.
49. Lombana, L.A., and D. Halahy. "Carbon Regeneration Systems," Chapter 25 in *Carbon Adsorption Handbook.* P.N. Cheremisinoff and F. Ellerbusch, eds. Ann Arbor Science Publishers, Ann Arbor, Mich. (1978), p. 905.
50. von Dreusche, C. Jr. "Process Aspects of Regeneration in a Multiple-Hearth Furnace," Chapter 26 in *Carbon Adsorption Handbook.* P.N. Cheremisinoff and F. Ellerbusch, eds. Ann Arbor Science Publishers, Ann Arbor, Mich. (1978), p. 923.
51. Modell, M., et al. "Regeneration of Activated Carbon with Supercritical Carbon Dioxide," Chapter 21 in *Activated Carbon Adsorption of Organics from the Aqueous Phase,* Vol. 1. I.H. Suffet and M.J. McGuire, eds. Ann Arbor Science Publishers, Ann Arbor, Mich. (1980), p. 447.
52. Rance, R.W., and E.L. Cussler. *Amer. Inst. Chem. Eng. J. 20,* 353 (1974).
53. Cooney, D.O., et al. *Water Res. 17,* 403 (1983).

54. Martin, R.J., and W.J. Ng. *Water Res. 18,* 59 (1984).
55. Clifford, D., et al. *Water Res. 17,* 1125 (1983).
56. Letterman, R.D., et al. *J. Amer. Water Works Assoc. 62,* 652 (1970).
57. Fiore, J.V., and R.A. Babineau. *Appl. Environ. Microbiol. 34,* 541 (1977).
58. Cairo, P.R., et al. *J. Amer. Water Works Assoc. 74,* 94 (1982).
59. Weber, W.J. Jr., and A.M. Jodellah. *J. Amer. Water Works Assoc. 77,* 132 (1985).

9

Biological Activated Carbon Treatment of Drinking Water

The term *biological activated carbon* (BAC) has been applied to drinking water treatment as a process consisting of the sequential combination of oxygenation and/or chemical oxidation; sand, anthracite, or multimedia filtration; optional oxygenation; and GAC adsorption [1]. A research group in Germany was perhaps the first to examine the beneficial effect of microbial activity on the performance of a GAC bed [2]. However, another research group in the United States had discussed the beneficial effects of microbial activity on GAC treatment of wastewater [3]. Considerable experience has been achieved in Europe with BAC in drinking water treatment, whereas domestic and industrial wastewater treatments have been the major applications in the United States.

A typical flow diagram of a BAC system is seen in Figure 9–1 [4]. The usual process pretreats the water with ozone (O_3) ahead of sand filtration. Chemical coagulation, pH adjustment, and other factors may precede the O_3 in drinking water treatment. The schematic in Figure 9–1 shows an aeration tank between the filtration and GAC bed. In this option, dissolved oxygen is added to maximize the aerobic microbial activity of the GAC. This may be more important in wastewater treatment than in drinking water treatment. It must be understood that the effluents from BAC are disinfected chemically to eliminate any microorganisms leaving the bed (see also Chapter 8).

Benefits claimed for the BAC process are biological nitrification (conversion of NH_4^+-N to NO_3^-, which is not important in drinking water treatment), biological oxidation of dissolved organic compounds, and partial biological regeneration of the GAC [5]. It is essential to have chemical oxidation as a pretreatment step to partially oxidize the dissolved organic compounds, lower the molecular weights of organics not readily adsorbed by the GAC, and prevent excessive growth of algae and phytoplankton in the filter media [5]. Ozone is the usual oxidant and some interesting claims are made for its efficiency (see following section).

Two mechanisms have been proposed for which biological activity is employed in BAC beds [5]. Microorganisms are present in the larger activated carbon macropores near the external surface. These macropores represent

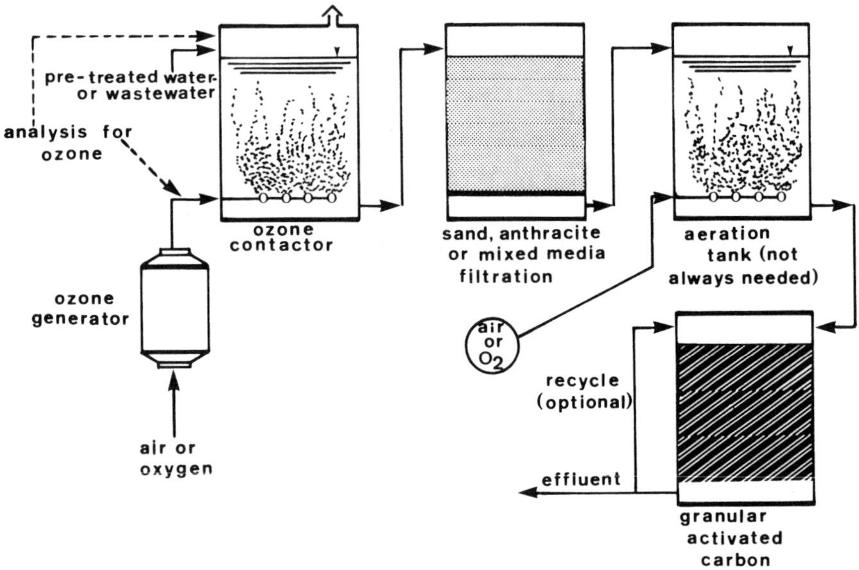

Figure 9–1 Biological activated carbon system. Reproduced from Rice and Robson [4], courtesy of Ann Arbor Science Publishers, Inc.

approximately 1% of the total available surface area of most GACs. Dissolved organic compounds, therefore, are adsorbed on the external surface, in smaller macropores, and in micropores where there are no bacteria present because of their relatively large size. These two mechanisms are simply (1) biological degradation of the organics and (2) adsorption of the organics in other parts of the carbon.

PRETREATMENT: CHEMICAL OXIDATION

The primary intent of chemical oxidation is to convert as much of the dissolved and/or colloidal organic matter as possible with as little oxidant as possible. This conversion presumably makes the compounds more available for biodegradation. Complete oxidation of the organics is neither intended nor practiced because of the relatively slow reaction kinetics relative to the hydraulic design of the pretreatment unit. Another aspect of this chemical oxidation is conversion of the initial organic molecule "to other compounds normally identified only in the latter stages of the biological oxidation process, effectively skipping several enzymatic oxidation stages" [4]. Apparently there is experimental evidence to support this statement [4].

In consideration of chemical oxidation as the initial step of the BAC process, several points should be remembered:

a. The relative ease of chemical oxidation of the starting material,
b. The relative ease of chemical oxidation of intermediate oxidation products,
c. The stability of the "end products" to further chemical oxidation,
d. The cost-effectiveness of additional oxidant necessary after the initial oxidation reaction has slowed,
e. The cost-effectiveness of increasing detention time in the presence of excess oxidant after the initial oxidation reaction has slowed. [4]

The third point requires additional explanation. It has been shown that when phenol (C_6H_5OH) is ozonized, many intermediate organic oxidation products are formed [6]. This includes catechol, which is an intermediate in the β-ketoadipate pathway of bacterial oxidation, and oxalic acid ($C_2O_4H_2$), which is a stable end product toward further chemical oxidation within the hydraulic contact time. Presumably this example of oxalic acid can be oxidized or adsorbed within the BAC.

Chemical oxidants that have been employed prior to BAC are O_3, Cl_2, ClO_2, $KMnO_4$, and H_2O_2 [4]. Each has its advantages and disadvantages. For example, $KMnO_4$ produces colloidal $MnO_{2(s)}$, which adds to sludge disposal and filtration problems. Chlorine and, in some cases, ClO_2 yield chlorinated organics that may be less biodegradable than nonchlorinated compounds. Conversely, O_3, H_2O_2, Cl_2, and ClO_2 produce only small quantities of insoluble by-products known as *microflocculants* [4]. These by-products would be undesirable additions to GAC beds.

Ozone

There are conflicting opinions regarding the effectiveness of preozonation as a prerequisite to enhance microbial activity on GAC. For example, a small increase of biodegradable organics by 0.1 to 0.2 mg/l was reported from preozonation [7]. There was an apparent substantial improvement in final water quality in pilot- and full-scale tests at the Mülheim plant (discussed later) [8]. Here, an O_3 quantity of 2. mg/l improved BAC performance as measured by removal of ultraviolet-absorbing substances and organic carbon. Thus, there are two questions here: one is concerned with O_3 producing a more biodegradable organic carbon, and the second is concerned with the BAC's performance. For drinking water treatment, data are sparse.

There is evidence that ozonation decreases the adsorption of organics. Figures 9–2a and b show this effect at dosages as low as 0.45 mg/l [9,10]. Ozone tends to yield more polar organics such as carboxylic acids and aldehydes that are more weakly adsorbed on abiotic carbon. The adsorption isotherms in Figures 9–2a and b were measured for ultraviolet-absorbing substances and not for organic carbon; that is, the organic carbon may not have been altered appreciably despite a decrease in the ultraviolet absorbance.

A short-term pilot plant operation was conducted for 2 months on the preozonation and GAC treatment of water from Lake Casitas, California (a drink-

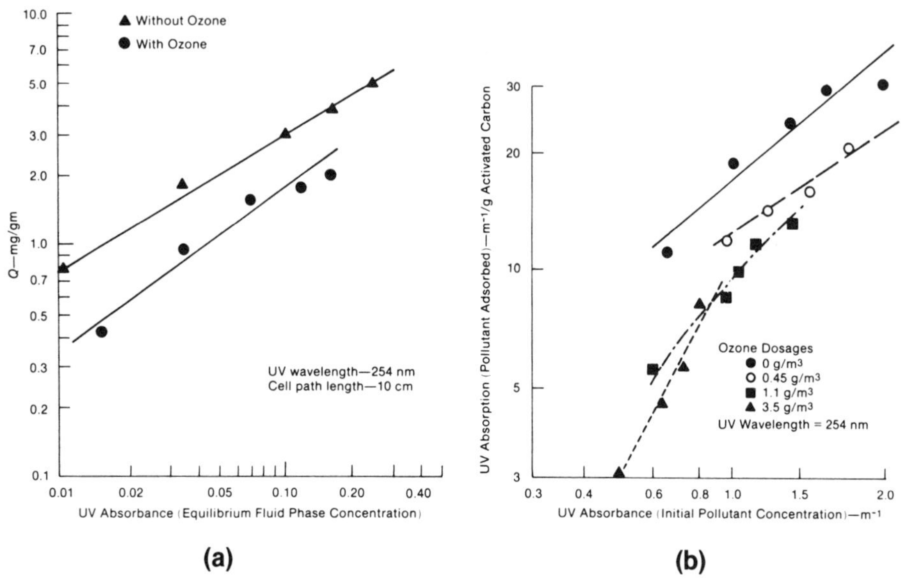

Figure 9–2 (a) Effect of ozonation of equilibrium adsorption of organics as measured by UV absorbance. Reproduced from Benedek [9], courtesy of the American Water Works Association. (b) Effect of ozonation on equilibrium adsorption of organics as measured by UV absorbance. Reproduced from Kühn et al. [10] in [9], courtesy of the American Water Works Association.

ing water reservoir) [11]. The organic problem was a precursor to THMs. Water entering the pilot plant was treated with 2 mg/l O_3 (3 minutes contact time), which lowered the TOC content from a maximum of 8 to 9 mg/l to a 3 to 4 mg/l range. However, when the raw water's TOC contents were in the 2 to 4 mg/l range, O_3 had very little effect. Following ozonation, the water was passed through a GAC filter (may not have been BAC in a strict sense) containing 2.23 m of 0.68 effective size particles. Figure 8–30a shows TOC levels in the appropriate waters for 4 weeks of operation. Figure 8–30a also gives the total potential THM contents for the specific waters. This short-term study demonstrated only a slight removal of organics (Figure 8–30b). However, the O_3 dosage was low, 2 mg/l, and the contact time was short, 3 minutes.

Another, longer-term pilot plant study was conducted for removal of THM precursors from raw drinking water (Cross Lake, Louisiana) [12]. This pilot plant consisted of a Sinflex strainer, a 10-GPM-rated Neptune-Microfloc "Waterboy" with four chemical feed options, tube settler, and multimedia filter. After this filter, an O_3 contactor was placed ahead of several GAC columns. At 6.0 GPM, retention time in the contact basin was approximately 3.3 minutes. One flow configuration had a 4.0 GPM stream through an O_3 detention basin of 30 minutes. For this study, Figure 9–3 shows that oxidation with low doses of O_3 was not an effective method for reducing THM precursors. Only a 5 to 15% reduction of the THMFP was recorded with settled/filtered water as the test medium.

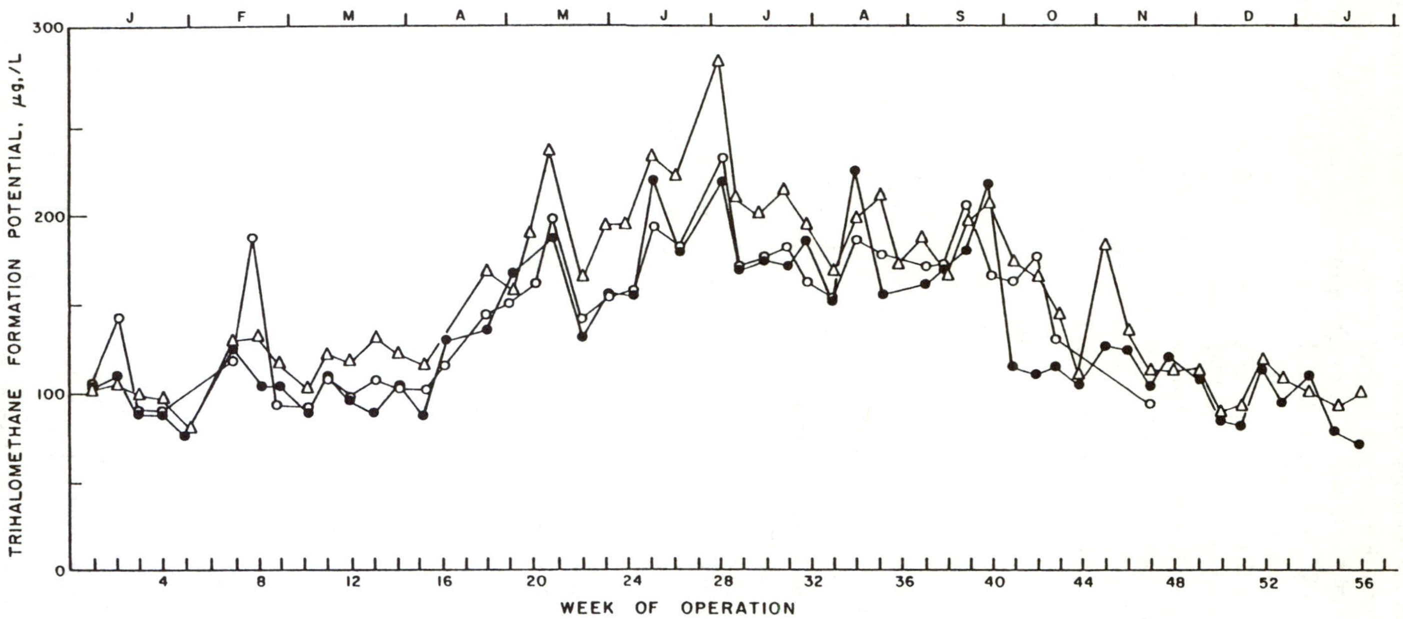

Figure 9–3 Effect of ozonation on THMFP of settled/filtered water. Ozone dose: 6.3 mg/1 for weeks 1 to 6; 2.0 mg/1 for weeks 7 to 19; and 2.5 mg/1 thereafter. Key: Δ settled/filtered water, ○ ozonated water with 3 min contact time, and ● ozonated water with 40 min contact time. Reproduced from Glaze et al. [12], courtesy of the American Chemical Society.

A 3 year pilot-scale investigation of advanced water treatment systems was conducted on Delaware River water in Philadelphia (see also Chapter 8) [13]. Following conventional treatment, effluents from the rapid sand filters were applied to the O_3 and GAC contactors. The physical and operating characteristics of the O_3 systems are seen in Table 9–1. According to this study, "the ozonation systems achieved their desired purposes of oxidation and disinfection." O_3 application was inefficient, as seen by the 25% transfer efficiency [13,p.158]. This problem was countered by a long O_3-H_2O contact time of 47 minutes. Changes in concentrations of organics through the O_3 contactors are given in Tables 9–2a and b. Oxidation and volatilization were responsible for removal of the organics. The latter mechanism apparently removed the lower molecular weight $CHCl_3$ and 1,2-dichloropropane but had little effect on TOC. The changes in TOC contents seen in Tables 9–2a and b reflect those organics that were oxidized to $CO_{2(g)}$ whereupon any products of partial oxidation would have been measured as TOC.

Bench-scale and pilot plant experiments were conducted for removal of organic matter and color in the Intracoastal Waterway (ICWW) waters near Myrtle Beach, South Carolina [14]. Results of these bench-scale ozonation tests are seen in Table 9–3 (no information was given on contact time). These data show that O_3 reduced the color of the raw and conventionally pretreated water. However, there was very little removal of organics, as seen by the TOC and THMFP data. Apparently, O_3 modified the color characteristics of the organics.

An extremely definitive laboratory study was conducted for the effect of ozone on the biological degradation and activated carbon adsorption of natural and synthetic organics in water [15]. Ozonation and biodegradation studies were conducted on water from the Grand River (southern Ontario), Spencer Creek (southern Ontario), and dilute aqueous nitroaniline solution. The Grand River water was pretreated with alum coagulation, sedimentation, and sand filtration, and the Spencer Creek water was sand filtered only prior to ozonation. Subse-

Table 9–1 Operating Parameters of Ozonation System

Parameter	*Value*
Conact time	
Contactor (min)	22
Retention tank (min)	25
Flow rate	
Gas (scfh)	60–80
Water [l/sec (gpm)]	0.13 (2.0)
Gas-to-water ratio	4:1 to 5:1
Ozone concentration	
Applied dosage (mg/l, air)	1.2
Transfer efficiency (percent)	25
Residual in water (mg/l water)	0.3

Source: Reproduced from Neukrug et al. [13], courtesy of the American Water Works Association.

Table 9–2a Changes in Concentrations of Organics as a Result of Ozonation in the Chlorinated Torresdale Plant

	Concentration in Rapid Sand Filter Effluent[a]		*Concentration in Ozone Effluent*[b]		*Reduction in Concentration as a Result of Ozonation*
Parameter	*Average (μg/l)*	*Range (μg/l)*	*Average (μg/l)*	*Range (μg/l)*	*(percent)*
TOC	2,559	1,782–3,131	2,452	1,504–2,967	4
Chloroform	73	20–163	46	16–110	37
Dichloropropane	6.1	0.1–24	4.8	0.1–22	21

Note: Samples taken 4/80–4/81.
[a]The rapid sand filter effluent serves as the influent to the ozone contactor and the nonozonated GAC contactor.
[b]The ozone effluent serves as the influent to the ozone-GAC contactor.

Table 9–2b Changes in Concentrations of Organics as a Result of Ozonation in the Nonchlorinated Pilot Plant

	Concentration in Rapid Sand Filter Effluent[a]		*Concentration in Ozone Effluent*[b]		*Reduction in Concentration as a Result of Ozonation*
Parameter	*Average (μg/l)*	*Range (μg/l)*	*Average (μg/l)*	*Range (μg/l)*	*(percent)*
TOC	1,826	1,015–2,442	1,622	930–2,459	11
Chloroform	4.3	0.3–43	2.8	0.2–26	35
Dichloropropane	4.5	0.2–19	3.0	0.1–16	33

Source: Reproduced from Neukrug et al. [13], courtesy of the American Water Works Association.
Note: Samples taken 4/80–4/81.
[a]The rapid sand filter effluent serves as the influent to the ozone contactor and the nonozonated GAC contactor.
[b]The ozone effluent serves as the influent to the ozone-GAC contactor.

quently, ozone was supplied to the contactor at a constant rate of 0.6 mg/l/min for a given contact time. TOC, COD, and ultraviolet analyses were used to follow the course of biodegradation of the soluble organic matter. Grand River water after pretreatment had TOC values ranging from 3.15 to 6.15 mg/l and COD values of 7.3 to 12.7 mg/l. Spencer Creek water's TOC and COD values were slightly higher, 8.6 and 19.0 mg/l, respectively. A COD/TOC ratio was employed to indicate the extent to which organics were oxidized.

The kinetics of organic matter removal by ozone followed a pseudo-first-order equation:

$$\frac{d[\mathrm{TOC}]}{dt_c} = -k_{\mathrm{TOC}}[\mathrm{TOC}]; \tag{9.1}$$

Table 9-3 Results of Bench-Scale Ozonation Tests

Type of Water Sample	*Ozone Dosage* Delivered (mg/l)	*Ozone Dosage* Consumed (mg/l)	*Color (Co-Pt units)*	*TOC (mg/l)*	*THMFP (μg/l)*
Raw	0	0	350	43	1,295
	113	46	200	38	–[a]
	178	46	100	35	–[a]
	329	46	50	32	–[a]
	510	–[a]	30	31	–[a]
	736	–[a]	30	29	1,100
Pretreated	0	0	30	12	–[a]
	57	22	5	11	–[a]
	122	22	5	9	–[a]
	275	22	5	9	–[a]
	543	–[a]	5	9	1,000

Source: Reproduced from Argaman et al. [14], courtesy of the American Water Works Association.
[a]Not determined.

that is, a semilog plot of [TOC] versus time yielded more or less linear curves whose slopes were the reaction rate constant values, k_{TOC} (presumably to the log 10 base), that are seen in Table 9-4. Contact times ranged from 35 to 70 minutes. Evidence for changes in the organic composition, presumably by oxidation, is seen in Figures 9-4 and 9-5 [15]. There is a decrease in TOC with an increase in

Table 9-4 Pseudo-First-Order Reaction Rate Constants (k_{TOC}) for TOC Removal

Water Source and Compounds	*Run Number*	k_{TOC} *(min⁻¹)*
Grand River Water	1	0.0041
	2	0.0073
	3	—
	4	0.0108
	5	0.0155
	6	0.0221
	7A	0.0233
	7B	0.0048
Spencer Creek Water	8	0.0099
	9	0.0102
p-nitroaniline	10	0.0047
o-nitroaniline	11	0.0041

Source: Reproduced from Stephenson et al. [15], courtesy of Lewis Publishers.

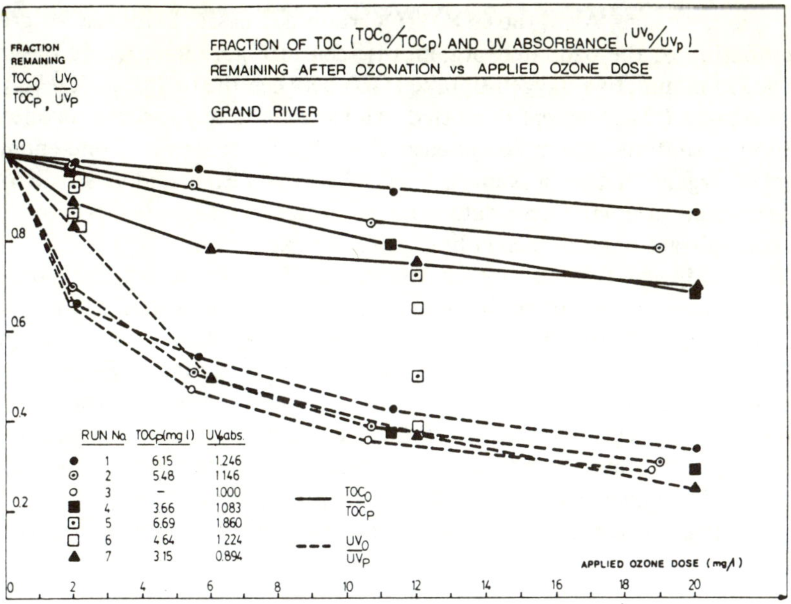

Figure 9–4 Fraction of TOC (TOC_o/TOC_p) and UV absorbance (UV_o/UV_p) remaining after ozonation versus applied ozone dose in the Grand River. Reproduced from Stephenson et al. [15], courtesy of Lewis Publishers.

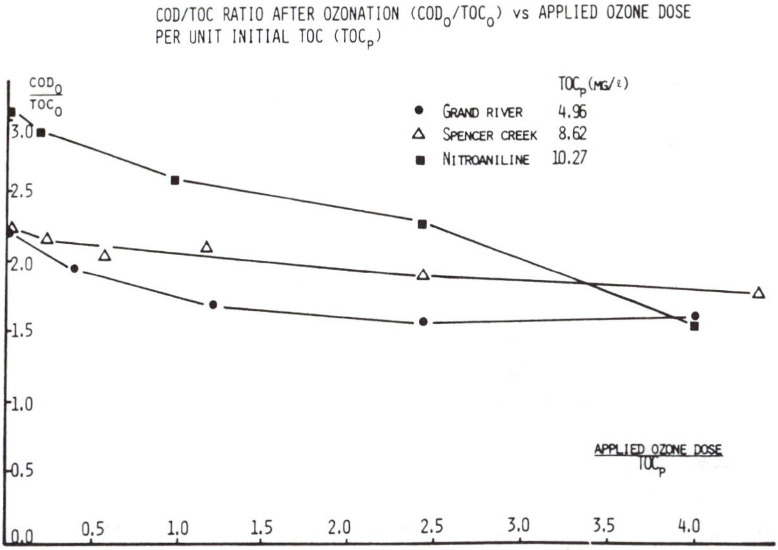

Figure 9–5 COD/TOC ratio after ozonation (COD_o/TOC_p) versus applied ozone dose per unit initial *TOC* (TOC_p). Reproduced from Stephenson et al. [15], courtesy of Lewis Publishers.

applied ozone dosage. Also, the COD/TOC ratio decreases. Evidence for greater biodegradation of the ozonated organics is seen in Figure 9-6, where the mean difference in cumulative oxygen uptake (*OU*) between the ozonated (OU_{OB}) and the unozonated (OU_B) waters is plotted against time (days) for four ozone dosages. Biodegradation appears to increase with [O_3]. Evidence for ozone-enhanced removal of organic matter was given by TOC removal expressed as a fraction of the unozonated sample. These data are summarized in Table 9-5 where "large TOC removals are possible depending upon the ozone dose" [15].

This study was followed by determination of carbon adsorption isotherms and breakthrough curves on pretreated and ozonated waters from the Grand River (southern Ontario) [16]. Filtrasorb 400, 200 mesh, was utilized for the isotherm studies that occurred over 5 days of agitation at 20°C. Granular F-400 was employed in the column studies. Figures 9-7a, b, c, and d are isotherms from the ozonated and biodegraded waters. It was necessary to present composite isotherms with 95% confidence limits because the data were difficult to interpret on individual samples. Figure 9-7a shows the pretreated and ozonated, 2 mg/l, isotherm where the scatter of data produced wide confidence limits. Figure 9-7b

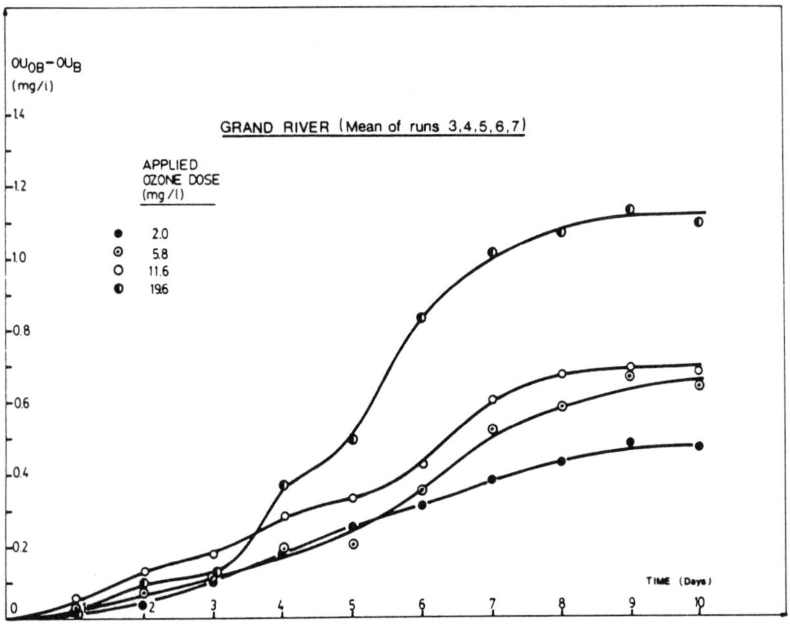

Figure 9-6 Mean difference in cumulative oxygen uptake (OU) between the ozonated (OU_{OB}) and the unozonated (OU_B) samples for various applied ozone doses versus incubation time in the Grand River. Reproduced from Stephenson et al. [15], courtesy of Lewis Publishers.

Table 9-5 Overall TOC Removal for the Combined Ozonation-Biodegradation Process

Water Source and Compound	*Range of Applied Ozone Dose (mg/l)*	*Range of TOC Removal (%)*
Grand River Water	1.9–20.1	10–66
Spencer Creek Water	2.1–40.0	9–54
o-nitroaniline	10.0–41.2	22–74

Source: Reproduced from Stephenson et al. [15], courtesy of Lewis Publishers.

shows the effect of biodegradation on the adsorption isotherm where the confidence limits are narrowed quite substantially. Also, adsorptivity is enhanced where more TOC is adsorbed per unit of equilibrium TOC in solution. Figure 9–7c suggests that ozonation increases adsorptivity initially (2 mg/l O_3), and as the O_3 dosage is increased to 12 mg/l, adsorption decreases. Figure 9–7d shows three isotherms, two of which included ozonation. These isotherms indicate that biodegradation "evens out the differences observed immediately after ozonation" [16]. Additional statistical analysis of the isothermal data strongly indicated that the effect of ozone on adsorptivity was insignificant. Essentially the same observations were reported for the column studies using TOC as a surrogate parameter. Several experiments were conducted on the effects of ozone on carbon adsorption of PAH. In this situation, ozone was very efficient at contents as low as 2. mg/l.

It appears from the studies discussed here that O_3 is somewhat limited in its ability to remove organics prior to BAC. The effectiveness of O_3 on GAC performance is discussed later.

Chlorination

Various investigators have suggested that prechlorination decreased the number of biodegradable organic compounds [7,8,10]. The best example was provided by pilot- and full-scale tests of GAC at Mülheim, Germany [8]. Parallel operation of preozonated GAC beds, for 3 months with and without breakpoint chlorination, demonstrated clearly that GAC in the latter process was more effective in removal of ultraviolet-absorbing compounds. Opposite results were obtained at the Rotterdam Waterworks where it was estimated that 99% of applied chlorine was used for oxidation that produced more, rather than fewer, biodegradable compounds [7].

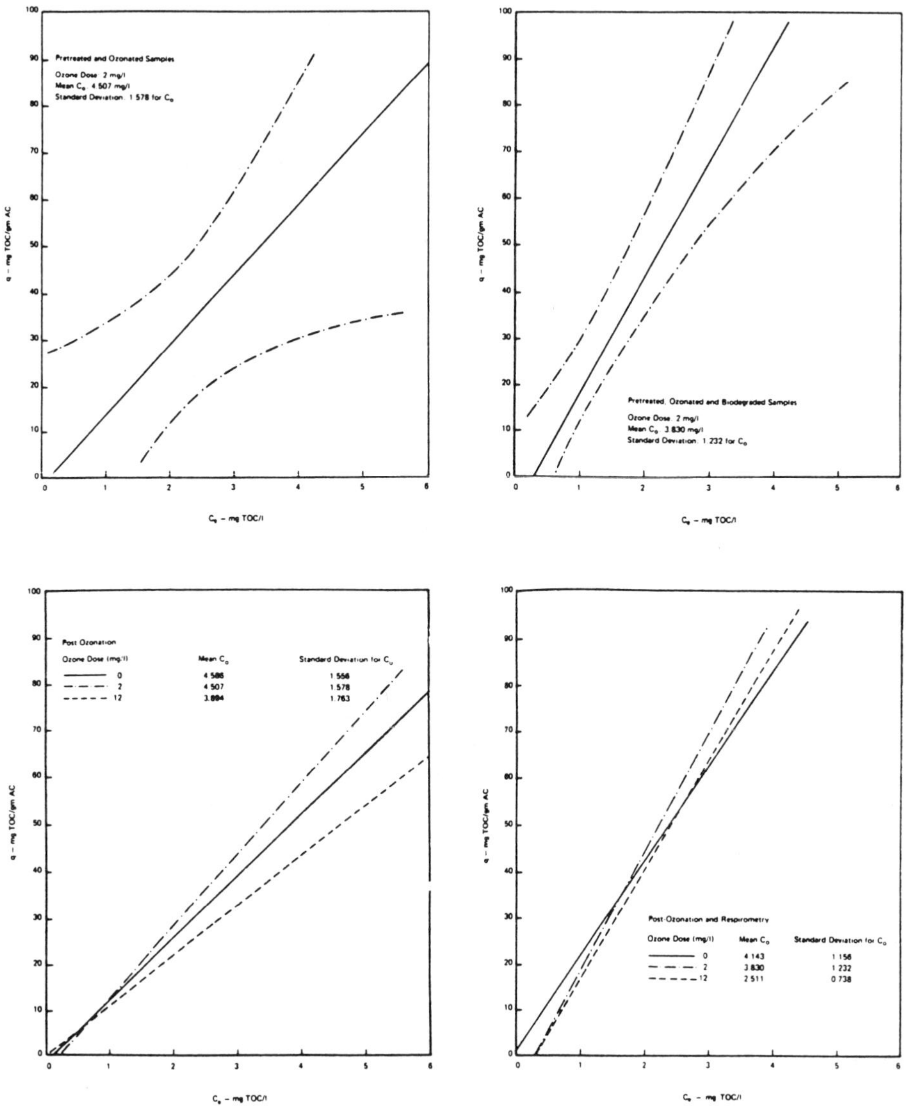

Figure 9–7 Composite activated carbon adsorption isotherms (least squares fitted), showing 95% confidence limits. Reproduced from Benedek, et al. [16] courtesy of Lewis Publishers.

There is an absence of data on the effectiveness of prechlorination for organics removal or molecular modification prior to GAC adsorption. However, some data are given in the next section for the effects of prechlorination on GAC performance.

PERFORMANCE

In an evaluation of BAC removal of organic compounds from water, two factors must be considered: adsorption and biodegradation. In some cases it is difficult to separate these two removal mechanisms.

A pilot plant study was conducted for the removal by GAC of various organic compounds from Delaware River water at Philadelphia [17]. In this case, the GAC was not operated strictly with biodegradation as one of the removal mechanisms, but high levels of bacteria were found in all carbon columns. Figure 8–33a shows the TOC breakthrough from chlorinated and dechlorinated systems for 100 days of operation. TOC removals were 22% higher in the chlorinated columns and took slightly longer to break through. However, there was no effect of chlorine on removals of THMs (85% chloroform and 14% bromodichloromethane) and THM precursors. Here again, there is a beneficial effect of preoxidation on the collective organics parameter, TOC. However, this effect was not observed with individual compounds.

GAC columns used in the Cross Lake, Louisiana, study were 30 cm in diameter and contained 1.2 m of GAC (F-400) [12]. At 2.0 GPM, the EBCT was 12 minutes. Figure 9–8 shows typical GAC performance curves for removal of THMFP in a cumulative plot. There is a slight increase in the GACs performance. In another plot of the data, Figure 9–9 shows a slightly lesser effect of preozonation on THMFP removal. Figures 9–10a and b show the breakthrough curves for Figures 9–8 and 9–9 [18]. Table 8–24 gives the total cumulative data for the unozonated and preozonated GAC columns. It was concluded that preozonation did not have much of an effect on the removal of THM precursors with the 2. to 2.5 mg/l dosages of O_3. During an 80 week operation of the pilot plant, the GAC columns outperformed the O_3 GAC beds in terms of cumulative TOC or THMFP removal. However, the percentage reductions were almost identical for the two systems.

Opposite results to the Cross Lake, Louisiana, study were reported from the Delaware River experiments [13]. Here, preozonation increased the bed life of GAC for TOC removal significantly, as seen in the breakthrough curves in Figures 9–11a and b. The patterns of TOC removal in these curves may be described by three phases. In the first phase, 85 to 95% of the TOC is removed by the carbon bed. Here, adsorptive mechanisms "are so strong during this phase that those organics not removed may be considered unadsorbable and nonbiodegradable" [13]. The second breakthrough phase is characterized by diminished adsorptive forces and a gradual increase in TOC effluent contents. Steady-state TOC removal constitutes the third breakthrough phase where biodegradability and slow adsorption kinetics may be operative when the carbon surface is nearly saturated. It is argued that O_3 tended to increase the TOC's biodegradability. Conversely, preozonation did not significantly increase GAC bed life for removal of volatile halogenated organics. In fact, breakthrough of these organics occurred earlier for the O_3 GAC systems than for GAC-only systems. For removing tastes and odors, O_3, either alone or with GAC, had very little effect.

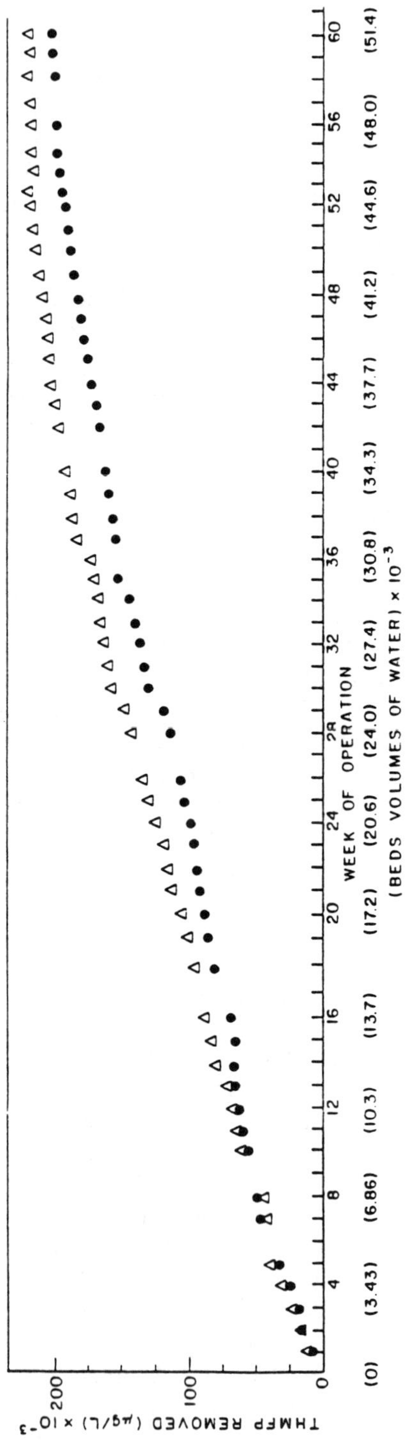

Figure 9–8 Cumulative THMFP removed in unozonated, column 3 (●), and ozonated, column 5 (Δ), GAC. Reproduced from Glaze et al. [12], courtesy of the American Chemical Society.

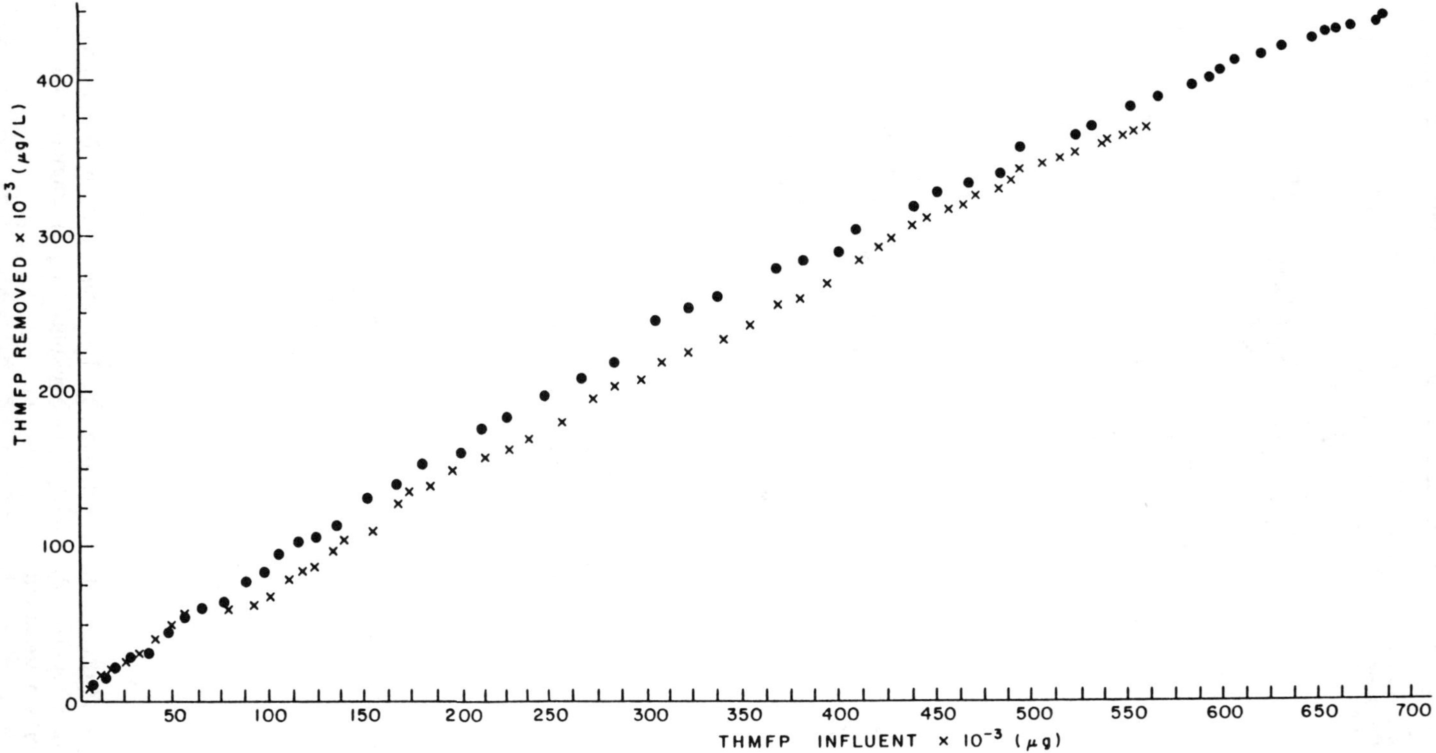

Figure 9–9 Cumulative THMFP removed versus THMFP influent to GAC column pairs. Key: X, GAC 3 + 4, ozonated settled/filtered water; ●, GAC 5 + 6, settled/filtered water. Reproduced from Glaze et al. [12], courtesy of the American Chemical Society.

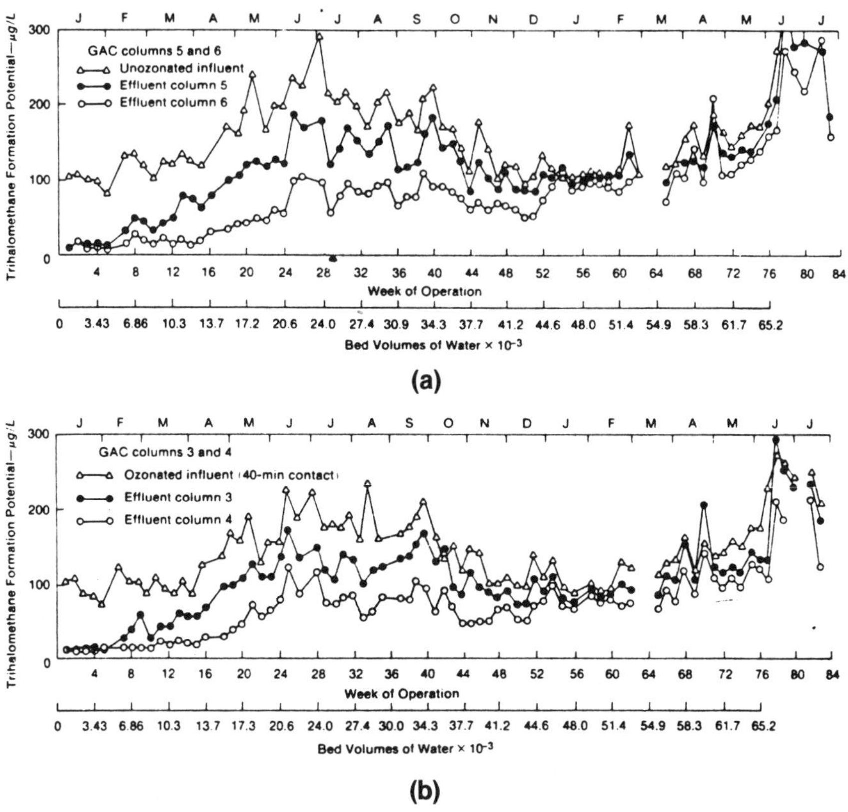

Figure 9–10 (a) THMFP values for settled-filtered water and effluents of GAC column set. (b) THMFP values for ozone-treated water and effluents of GAC column set. Reproduced from Glaze and Wallace [18], courtesy of the American Water Works Association.

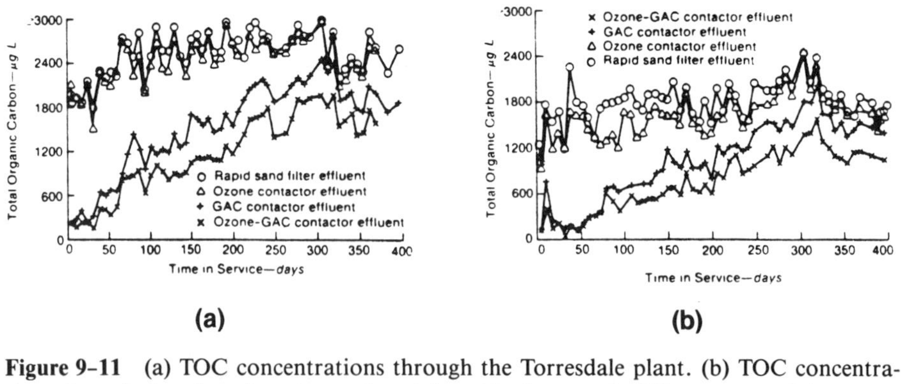

Figure 9–11 (a) TOC concentrations through the Torresdale plant. (b) TOC concentrations through the pilot plant. Reproduced from Neukrug et al. [13], courtesy of the American Water Works Association.

OPERATING EXPERIENCES AND CASE STUDIES

The Mülheim Process

One of the first applications of biological treatment of potable water was with the Ruhr River at Mülheim, Federal Republic of Germany [8]. This river was heavily polluted in the 1970s with NH_4^+-N, 4 to 6 mg/l, and breakpoint chlorination was utilized for its removal. In turn, organochloro compounds were formed. Conventional treatment was ineffective in removing these compounds. Consequently, the process schematic is seen in Figure 9–12, and the process parameters are given in Table 9–6. It is interesting to note that the O_3 dosage, an average of 1. mg/l, depended on the raw water's turbidity and especially on the algae count. Table 9–7 gives the operational data of the new plant (April to July 1977) compared to the conventional process (1975, 1976). Lower dissolved organic carbon (DOC) and ultraviolet-absorbing substances were achieved by the new O_3 BAC process.

Additional data about the performance of the Mülheim process is seen in Figure 9–13 [4]. Various types of GACs were tested in a pilot plant operation, and individual performances are noted in the figure. Approximately 75% of the total organic matter was removed by biological oxidation, and the remainder was removed by adsorption. How these percentages were calculated was not mentioned. There was a decline in performance in the third half-year of operation

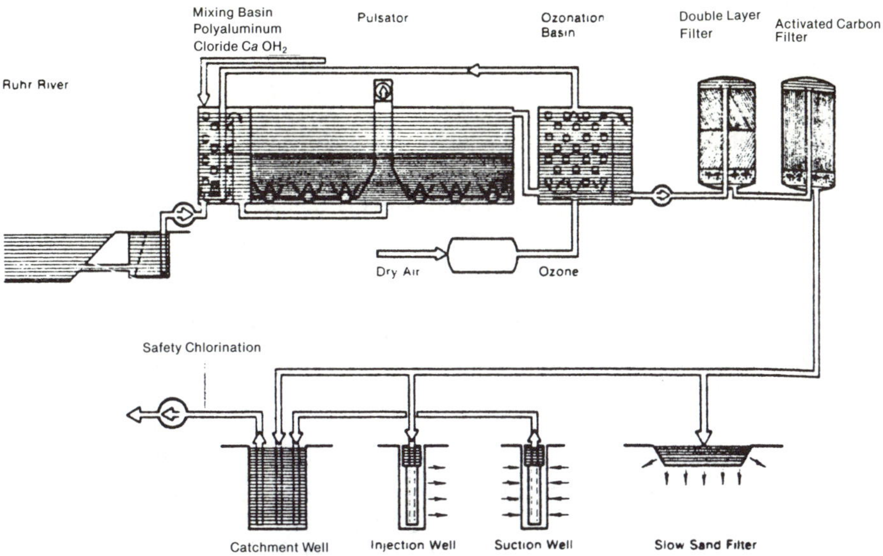

Figure 9–12 Process scheme for surface water treatment at the Dohne Water Utility in Mülheim after the introduction of combined chemical and biological oxidation. Reproduced from Sontheimer et al. [8], courtesy of the American Water Works Association.

Table 9-6 Process Parameters at the Dohne Water Utility (Mülheim) before and after Change of Treatment

Classic Treatment	*Treatment Step*	*New Treatment*
10–50 mg/l Cl_2	Preoxidation	1 mg/l O_3
3–4 mg/l Al^{3+}	Dosing	4–6 mg/l Al^{3+}
0.1 kW/m^3	Power input	2.5 kW/m^3
Retention time ≅0.5 min		Retention time ≅0.5 min
5–15 mg/l $Ca(OH)_2$	Flocculation	5–15 mg/l $Ca(OH)_2$
Retention time ≅1.5 hr	Sedimentation	Retention time ≅1.5 hr
	Ozonation	2 mg/l O_3 Retention time ≅5 min
v = 10.7 m/hr[a]	Filtration with preflocculation	v = 9 m/hr 0.2 mg/l Al^{3+} 0.1 mg/l polyelectrolyte
v = 22 m/hr h = 2 m[b]	Activated carbon filter	v = 18/hr h = 2 m (4 m)
Retention time 12–50 hr	Ground passage	Retention time = 12–50 hr
0.4–0.8 mg/l Cl_2	Safety chlorination	0.2–0.3 mg/l Cl_2

Source: Reproduced from Sontheimer et al. [8], courtesy of the American Water Works Association.
[a]v = filter velocity.
[b]h = bed height.

that was explained by an improvement in the quality of Ruhr River water: the influent DOC had decreased from 2.5 to 1.8 mg/l.

A unique feature of the Mülheim process is that after passage through the GAC columns, the water goes into 15 injection wells for storage underground where it remains for 12 to 15 hours (Figure 9-12). Table 9-8 compares the mean DOC values from the old process (1975) with mean values during May to August 1977. Spent carbon from the old process was used during this time. Ammonia levels were reduced from an average of 1.07 mg/l in the influent to zero after ground passage [4].

In operating a BAC process, it is necessary to consider the bacterial quality of the water eluting from the GAC column. Table 9-9 gives the bacterial counts at various points in the Mülheim process [4]. These data show that effluents from the old GAC columns had considerable bacterial activity that the new columns did not. It was noted in future operation of the Mülheim process that ground passage of the treated water will be eliminated [4]. Instead, it will serve as a storage reservoir.

Table 9-7 Mean Dissolved Organic Carbon and Ultraviolet Extinction Values for the Different Treatment Steps

	1975			*1976*			*April to July 1977*		
Treatment Step	*Dissolved Organic Carbon (DOC) (mg/l)*	*Ultraviolet (UV) (254 nm) (m^{-1})*	*UV/ DOC*	*DOC (mg/l)*	*UV (254 nm) (m^{-1})*	*UV/ DOC*	*DOC (mg/l)*	*UV (254 nm) (m^{-1})*	*UV/ DOC*
Raw water (Ruhr)	3.9	6.8	1.7	5.0	9.1	1.8	3.6	6.1	1.7
After flocculation and sedimentation	3.2	4.5	1.4	4.0	5.5	1.4	2.9	3.2	1.1
After filtration	3.2	4.4	1.4	3.8	5.4	1.4	2.6	1.8	0.7
After activated carbon	3.0	4.0	1.3	3.7	5.3	1.4	2.3[a]	1.6[a]	0.7[a]
After ground passage	1.8	3.1	1.7	2.1	4.0	1.9	0.9	1.6	1.6

Source: Reproduced from Sontheimer et al. [8], courtesy of the American Water Works Association.
[a]Filters filled with fully loaded activated carbon, used in the previous treatment process.

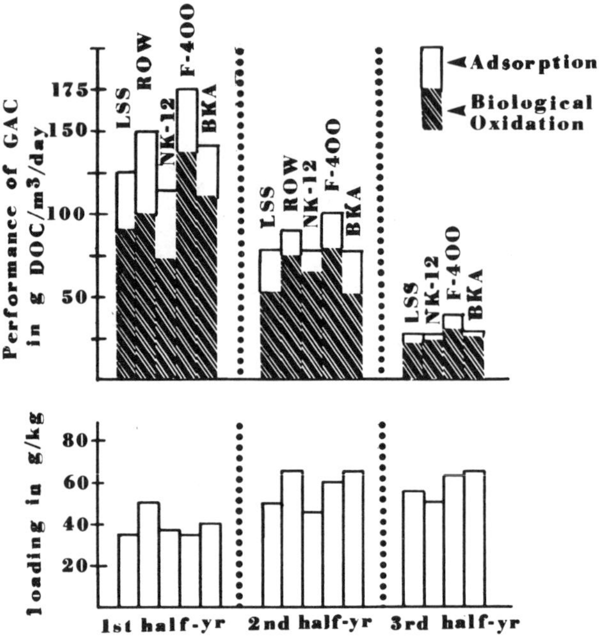

Figure 9–13 Performance and loading of BAC adsorbers. Reproduced from Rice and Robson [4], courtesy of Ann Arbor Science Publishers, Inc.

Mery-sur-Oise and Choisy-le-Roi, France

An interesting application of O_3 and BAC was reported for three water treatment plants in suburban Paris, France [19]. Figure 9–14 shows the flow diagram of one of these plants in which O_3 is added in three steps to maximize the biological removal of NH_4^+-N and organics in the preoxidized reservoirs, the biologically operating sand filters, and the biologically operating GAC adsorbers. One of the O_3 applications is to ensure elimination of viruses with a residual of 0.4 mg/l for 10 min. NH_4^+-N is removed biologically by nitrification, partly in the storage tank and partly in the flocculators and sand filters. Ozonation ahead of the latter two units had a dual role. The water is saturated with O_2, which enhances nitrification in the sand filters and assists in eliminating organic materials during the coagulation-sedimentation-filtration processes. No data were given for these statements other than a low quantity of O_3 (0.5 to 0.8 mg/l) combined with a coagulant removed about 50% of the organic materials. Without preozonation, a 35 to 40% reduction was achieved. Another qualitative comment was that preozonation ahead of the GAC stabilized organics removal at 30%. Without preozonation, organics removal in the GAC became nil with time. The TOC contents of the raw water ranged from 5. to 7. mg/l, and NH_4^+-N was 0.3 to 0.8 mg/l.

Table 9-8 Comparison of Old and New Plant Processes at the Dohne Plant for Removal of DOC and NH_4^+

	Old Process 1975	*New Process May–August 1977*	
Treatment Step	*DOC (mg/l)*	*DOC (mg/l)*	*NH_4^+ (mg/l)*
Ruhr River water	3.9	4.0	1.07
After flocculation	3.2	3.1	0.82
After O_3 + sand filtration	—	2.8	0.22
After GAC contacting	3.0	—	—
After ground passage	1.8	0.9	0

Source: Reproduced from Rice and Robson [4], courtesy of Ann Arbor Science Publishers, Inc.

Table 9-9 Bacterial Count Data at Dohne Plant with BAC Treatment Process

	With Old GAC				*April 1978 (new GAC)*	
	(Total Counts/ml)		*(E. coli/100 ml)*			
Sampling Point	*(M_g)*	*(σ_g)*	*(M_g)*	*(σ_g)*	*(Counts/ ml)*	*E. coli/ 100 ml)*
Raw water	14,490	2.0	1,620	1.7	20,000	—
After flocculation + sedimentation	2,340	4.2	6.7	3.2	20,000	—
After sand filtration	6,010	4.9	$<<1$	—	90	0
After GAC	3,700	4.0	$<<1$	—	57	0
After ground passage	27	2.3	$<<1$	—	22	2.2[a]

Source: Reproduced from Rice and Robson [4], courtesy of Ann Arbor Science Publishers, Inc.
Note: M_g = geometric mean; σ_g = geometric standard deviation.
[a]About 20% of the water present after ground passage comes from infiltration from the Ruhr River. This infiltrated water is the source of the *E. coli* found.

The Düsseldorf Process

In the lower Rhine River region of the Federal Republic of Germany, the raw water contains many organic contaminants in which there are many chlorinated compounds [4]. There are three water treatment plants at Düsseldorf whose general schematic is seen in Figure 9-15. In the initial step, raw water is permitted some biological treatment from river bank filtration. It takes about 3 weeks for the water to pass through the river sand banks, whereupon it is pumped from wells (10 to 30 m deep) located 50 to 100 m from shoreline and then into the

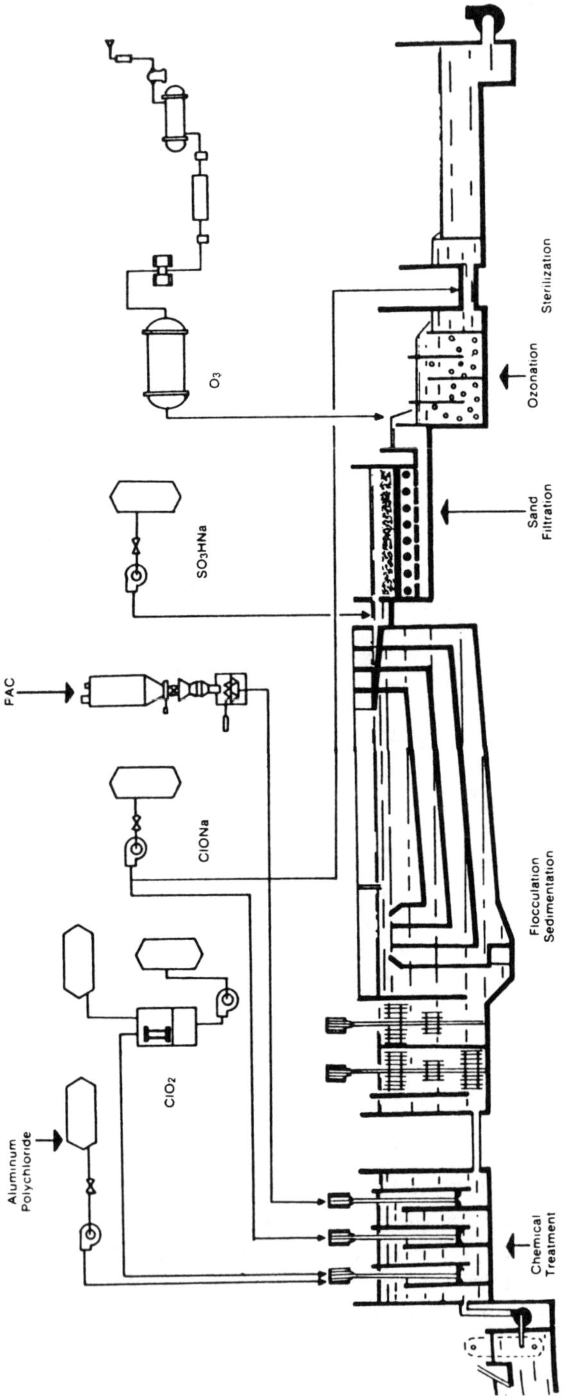

Figure 9–14 Improvements at the Mery-sur-Oise and Choisy-le-Roi Treatment Plants. Reproduced from Schulhof [19], courtesy of the American Water Works Association.

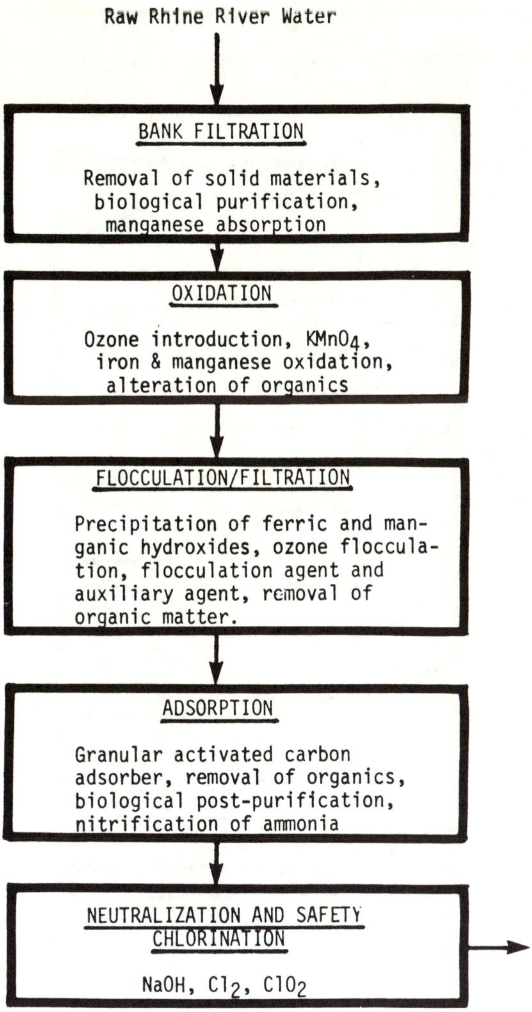

Figure 9–15 General procedural plan for the treatment of drinking water in the bank filtration works on the lower Rhine (Düsseldorf area). Reproduced from Rice and Robson [4], courtesy of Ann Arbor Science Publishers, Inc.

treatment plant. Changes in the Rhine River water quality are seen in Figure 9–16 as it passes through the various treatment steps. About 60 to 70% of the organics are removed in the river sand bank.

The second step in the Düsseldorf process is ozonation with 1. to 3. mg/l at the contact chamber that then proceeds to a holding tank for an additional

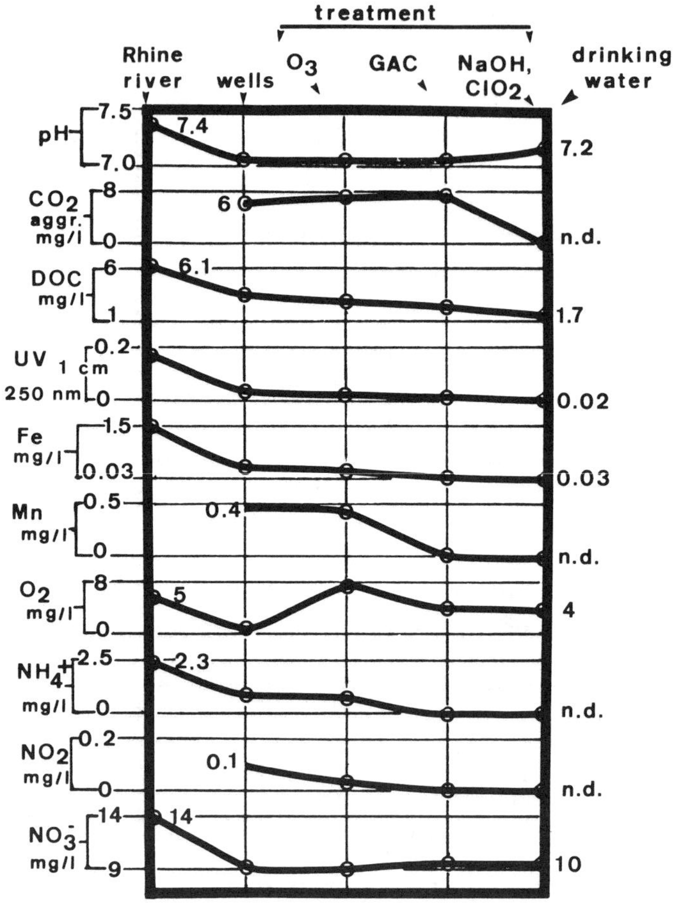

Figure 9–16 Changes in Rhine River water quality parameters during passage through the Düsseldorf water treatment process. Reproduced from Rice and Robson [4], courtesy of Ann Arbor Science Publishers, Inc.

30 minutes. Oxidation occurs here for Fe, Mn, and some dissolved organics (see Figure 9–16). Following the holding tank, water passes through a two-stage pressure filter. The first stage is a 1.5 m deep bed of nonactivated carbon (0.5 to 2.5 mm grain size). Here, NH_4^+-N is converted to $\overline{NO}_3$ and some particulate matter is removed by filtration. In the second layer of the pressure filters, a mixture of the GACs, L S Supra and F-300, is employed (mean diameter 1.4 to 1.6 mm) at a depth of 2.5 m. Here, adsorption of organics occurs as well as some additional nitrification. The hydraulic loading rate on the filter-adsorber is 4.8 gal/min/ft^2. A mixture of Cl_2 and ClO_2, 0.3 mg/l, is added for disinfection. The percentage removals across the Düsseldorf plant are given in Figure 9–17.

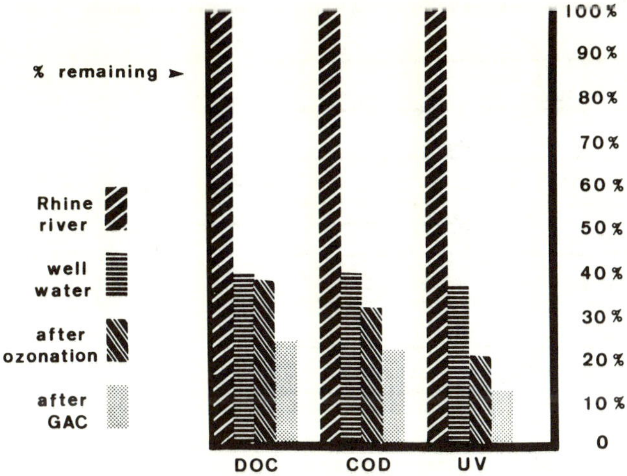

Figure 9–17 Changes in organics parameters in Düsseldorf water treatment plants. Reproduced from Rice and Robson [4], courtesy of Ann Arbor Science Publishers, Inc.

Rotterdam, The Netherlands

Raw water for the Kralingen water treatment plant at Rotterdam comes from the River Maas and goes through two storage (Biesbosch) reservoirs prior to treatment [4]. The new Kralingen plant became operational in March 1977, and its schematic, Figure 9–18, shows a high degree of treatment. This undoubtedly resulted from the early research of J.J. Rook on the THM problem [20]. Water flows sequentially through two storage reservoirs. Each is 20 m deep for a combined retention time of 3 months. The second reservoir was fitted with aeration devices, primarily to control algae, to prevent stratification, and to promote biological decomposition of organic matter and conversion of NH_4^+-N to $N\overline{O}_3$. Table 9–10 shows the raw and treated water quality for the Kralingen plant where the NH_4^+-N/$N\overline{O}_3$ data are confusing. It appears that the nitrate content (mg/l as $N\overline{O}_3$ presumably) is unaffected by treatment. There is some reduction in the NH_4^+-N contents.

After reservoir storage, the water is coagulated with $FeSO_4$ for a low turbidity water. Clarified water then flows to a 5 minute O_3 contacting chamber, 3. mg/l dosage, and subsequently to an 11 minute retention tank. Following ozonation, the water flows to multimedia filters (20 cm anthracite on sand supported on gravel). Water then is pumped into pressurized GAC columns with Norit carbon, which has an EBCT of 10 minutes. Removal of organics is seen in Table 9–10 through the COD and TOC parameters. Biological activity in the GAC units slowly increased during the first year of operation. However, biological activity was operative in the multimedia filters as well as the GAC presumably for degradation of organics. No evidence was given to support this contention.

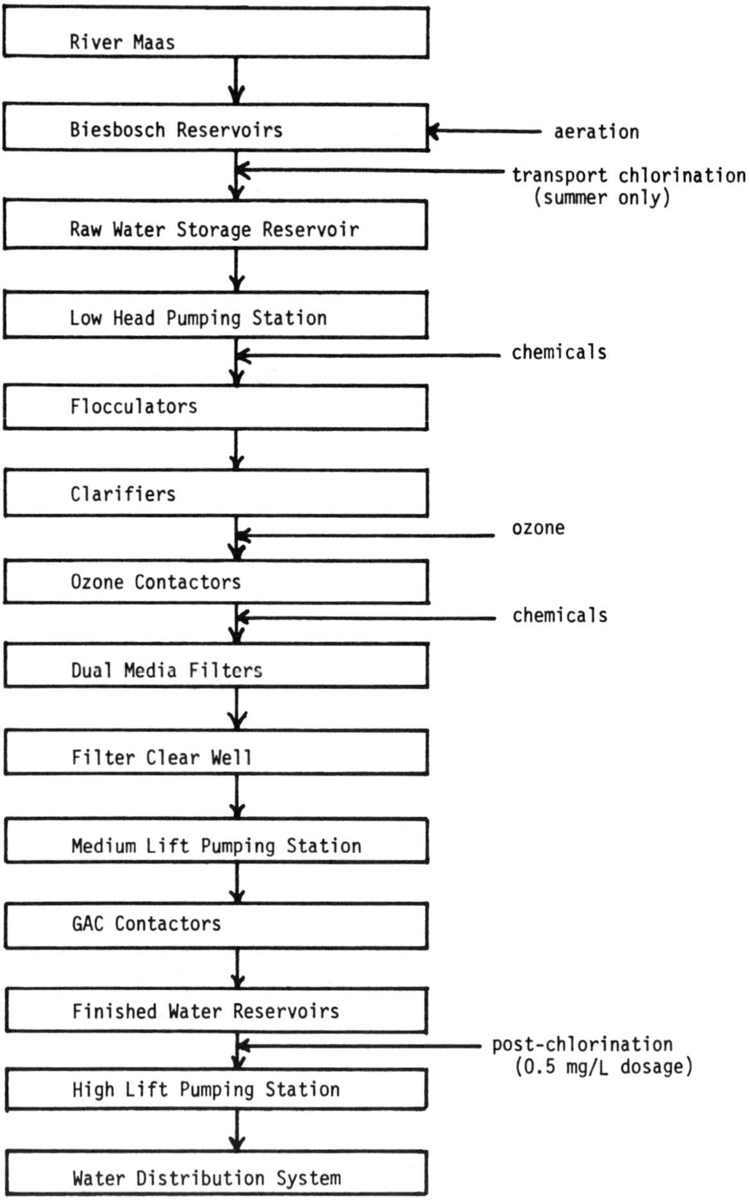

Figure 9–18 Process flow diagram of the Kralingen Water Treatment Plant, Rotterdam, The Netherlands. Reproduced from Rice and Robson [4], courtesy of Ann Arbor Science Publishers, Inc.

Table 9–10 Analysis of Raw and Finished Waters at Kralingen, January 1978

	Raw Water[a]			*Finished Water*		
Parameter	*Minimum*	*Average*	*Maximum*	*Minimum*	*Average*	*Maximum*
Color (mg Pt/l)	9	12	14	<1	<1	1
Temperature (°C)	5.5	6.0	7.5	5.5	6.0	7.5
Turbidity (JTU)	0.60	1.6	3.9	0.05	0.10	0.25
$KMnO_4$ demand (mg/l)	10	11	12	3	4	5
Nitrite (mg/l)	0.019	0.050	0.070	—	0.000	—
Nitrate (mg/l)	18.5	18.8	19.0	19.5	19.9	20.0
Sulfate (mg/l)	78	83	90	113	116	119
NH_4^+ saline (mg/l)	0.20	0.26	0.33	0.04	0.06	0.11
NH_4^+ albuminoid (mg/l)	0.17	0.20	0.22	0.09	0.11	0.13
Fe (mg/l)	0.02	0.03	0.05	0.01	0.02	0.04
Mn (μg/l)	6	7	9	2	3	3
DO (mg/l)	11.2	12.2	12.6	11.2	12.4	13.0
Phenol (μg/l)	1	1	2	<1	<1	<1
Detergents (μg/l)	45	50	55	10	11	15
COD (mg/l)	8	8	9	2	4	5
TOC (mg/l)	3.5	3.9	4.1	2.2	2.8	3.7
Br (μg/l)	120	125	135	60	65	80
Hexachlorobenzene (μg/l)	—	<0.01	—	<0.01	<0.01	<0.01
HcH, μg/l	<0.01	<0.01	<0.01	—	—	—
HcH, μg/l	<0.01	<0.01	<0.01	0.01	0.01	0.01
HcH, μg/l	0.01	0.01	0.01	0.01	0.01	0.01
Cholinesterase inhibitors, in parathion equivalents (μg/l)	0.04	0.07	0.08	0.02	0.03	0.04

Source: Reproduced from Rice and Robson [4], courtesy of Ann Arbor Science Publishers, Inc.
[a]Raw water = water influent to the plant.

DESIGN AND OPERATIONAL CONSIDERATIONS

Preliminary Design

If a BAC treatment is considered for drinking water for a situation where no plant exists or for a modification, then laboratory and pilot plant studies should be conducted prior to any design plans. The techniques for predesign studies are described in great detail in Chapter 8 [4,21]. For BAC, preliminary design considerations establish criteria for the final design of the facilities. The heart of the BAC process is the GAC bed. Consequently, any final design for BAC will be similar to GAC.

First, a determination related primarily to the organics content of the raw water quality should be made for pretreatment. A surrogate organic C parameter is sufficient for this determination. Such parameters include COD, TOC, DOC, and ultraviolet absorbance. Whether or not pretreatment is beneficial can be ascertained through laboratory studies. As seen in the foregoing discussion, chemical oxidation is usually the choice. It is almost unanimous that chlorination should not be used. The reasons for this are well documented.

Second, the hydraulic profile of the system will affect many final design decisions and modes of operation for the lifetime of the plant [4]; that is, the treatment process may include a single or several pumping steps to transport water from raw supply to the distribution system. This should not normally be a problem, but extensive pumping will add to the cost of treatment.

Third, major design parameters for GAC adsorbers are the EBCT and the system's configuration—that is, parallel, series, upflow, downflow, or pulsed bed [4]. Here again, laboratory column studies provide most of the data for the EBCT. However, these studies should be confirmed by comparison to similar situations or by pilot plant experiments. Table 9–11 shows some GAC design criteria from experiences in the Federal Republic of Germany [22]. It appears that BAC has

Table 9–11 Design Criteria for GAC Filters in Germany

Treatment Method	*Filtration Velocity (m/hr)*	*Bed Height (m)*	*Empty Bed Retention Time (min)*	*Throughput Ratio before Regeneration (m^3/m^3)*
Dechlorination	25–35	2	2–4	>1,000,000
Taste and odor removal	20–30	2–3	6–10	100,000
Organics removal	10–15	2–3	8–15	25,000
Biological activated carbon	8–12	2–4	15–25	100,000

Source: Reproduced from Sontheimer [22], courtesy of the American Water Works Association.

the lowest liquid loading rate, greatest depths, longest EBCT, and an intermediate bed volume to regeneration. Once the adsorber configuration is selected, then design judgment is needed in sizing the bed and regeneration system to accommodate unexpected surges in flow or high contents in the influent water, future changes in quantity and quality of influent, possible increases in effluent water quality requirements, and possible decrease in GAC performance after repeated regeneration cycles [4].

Detailed Design

After these preliminary steps, some designers recommend additional extensive literature reviews and facility visits to avoid repetition of previous design errors [4,23]. Tables 9–12 and 9–13 list one person's opinion of the insights gained from such surveys [23]. Some construction details are that steel and reinforced concrete are viable construction materials. Guidelines for downflow adsorbers are as follows [24]:

a. Deep bed, factory-fabricated steel pressure vessels with a maximum 12 ft. diameter may be used for GAC bed volumes ranging from 2,000 to 50,000 ft^3.
b. Shallow bed, reinforced concrete, gravity filter-type units may be used for GAC bed volumes ranging from 1,000 to 200,000 ft^3. Shallow beds are appropriate when long service cycles between GAC regenerations can be expected based on pilot plant testing.
c. Deep bed, site-fabricated steel, 20 to 30 ft. diameter, open gravity tanks may be used for carbon volumes ranging from 6,000 to 200,000 ft^3 or larger.

Some concerted attention should be given to the GAC regeneration system also [4]. Key portions of an on-site system are GAC transfer system, spent GAC storage tank, dewatering screw, furnaces, quench tank, and regenerated GAC storage tank (see also Chapter 8). The essential design consideration here is that the piping materials should be able to accommodate corrosion and erosion of the liquid and GAC. Also, provisions should be made for deposition and plugging of the slurry transfer lines.

An important consideration in the design and operation of GAC/BAC beds is the backwashing step. Either an air scour or a surface wash is essential. Air scour is the initial step during the backwash sequence of all European drinking water treatment plants using BAC. This scour breaks up the agglomerated GAC and prevents mudball formation.

ECONOMIC CONSIDERATIONS

Results from a pilot plant study, Torresdale, in Philadelphia were employed as a basis for an economic study that compared the costs for GAC alone, O_3 alone, and O_3 followed by GAC [13]. Two types of GACs were used: a GAC filter-

Table 9–12 Design Principles for GAC Water Treatment Filters

Pretreatment Needs
Capability for vigorous preparation to strip out organics.
Capability to optimize for soluble organic removal—for example, pH control to 6 to 6.5 if alum is used.
Capability to add ozone prior to flocculation and sedimentation, conventional and GAC filters.
Capability to add PAC prior to flocculation and sedimentation system and a GAC topping to the conventional filter.
GAC Filter
Following conventional filtration.
Gravity, downflow mode with two in series.
Use of ready inventory and spent carbon bins so that all columns can be recharged quickly with regenerated GAC as quickly as transfer can be accomplished.
Off-steam washing of carbon fines if at all possible.
Air scour with hydraulic backwash.
Corrosion protection, rigorous inspection, protection against defects.
Capability to isolate system and vessels therein, and add super dose of chlorine or some other disinfection/sterilizing compound, capability to promote either on-stream or off-stream biological activity with direct application of oxygenated water to the carbon filter.
Downstream stilling well to capture lost carbon with capability to collect and return inventory lost in the event of a bottom failure.
Ease of access to GAC bed and cleanout of chambers and conduits downstream of carbon filter.
Design plant for service with variable depths of carbon; start operation with minimum depth and run this way for first six to twelve months of operation.
Surge control for abnormal backwash situation with recycle fed gradually to upstream flocculation basin.
Regeneration Facilities
Elsewhere, by carbon manufacturer if at all possible.
If not, rigorous performance guarantee warranty by manufacturer of system for GAC losses.
Other
Flexible, clear vinyl tubing if applicable.
PVC, ABS, fiberglass, stainless steel, and so forth where applicable and possibly for linings and conduits.
Ceramic lined centrifugal pumps for low concentration carbon slurries; eductor pumps for high carbon slurries.
Check anybody's recommendation for similar application at existing water and wastewater GAC facilities.

Source: Reproduced from Mulbarger [23], courtesy of the American Water Works Association.

Table 9-13 Engineering Principles for GAC Applications

There is no substitute for pretreatment.
Do everything possible to minimize activated carbon handling, suspension, transfer, and regeneration.
Activated carbons are poor media for filtration but excellent media for adsorption and biological growth.
Avoid pressure conditions in reactors, if possible, and long conduits and low velocities, radical bends and low spots without cleanouts.
Design for upset, minimum and maximum conditions on a unit process and plant basis; establish and design for backup conditions that will allow the plant to achieve its primary goals with the GAC filter/adsorber out of service.
Wet activated carbon can establish conditions for galvanic corrosion of steel and can deoxygenate the atmosphere of closed vessels; anaerobic conditions can accelerate both (and can be considered inevitable in some way in any attached growth aerobic biological system); dry activated carbon fines or dust can be potentially explosive under proper conditions; workers coming in contact with carbon fines and dust should have available protective clothing, eye protection, and breathing devices; activated carbon is abrasive and subject to attrition and false bridging. Anticipate the inspection needs of any carbon vessel, and provide adequate space for access of workers with self-contained breathing apparatus and positive means of ventilation.
Pay particular attention to regeneration facilities, exhaust gas ducts, and air pollution control system for proper operation and corrosive conditions; physically separate the furnace from all other operations; provide reliability of afterburner for odor control and organic destruction.
Use instrumentation as much as possible for control, monitoring, and verification of operating and static conditions; always provide the capability for manual overrides.
Plan on extensive operator training if regeneration is to be incorporated.
If the carbon is or may be in contact with anything, paint its container black.

Source: Reproduced from Mulbarger [23], courtesy of the American Water Works Association.

adsorber and a GAC postfiltration adsorber. GAC and O_3 – GAC system costs were based on the bed life and use rates given in Tables 9-14 and 9-15. All cost estimates were based on preliminary system designs and on the best available cost information using December 1982 dollars.

Components that affect the overall economics of a GAC facility are: capital—carbon contactors, regeneration facility, pumping station, handling systems, storage; and GAC; and operating and maintenance—personnel, power, makeup GAC, and other materials [13]. Plant capacities were based on a maximum daily flow of 282 mgd except the pumping station and related piping that were sized for a peak hourly flow of 350 mgd. Operating and maintenance costs were based on an average daily flow of 230 mgd.

Table 9-14 GAC Bed Life and Use Rate for TOC Criteria

	Torresdale Plant						Pilot Plant					
	GAC			Ozone-GAC			GAC			Ozone-GAC		
		Use Rate			Use Rate			Use Rate			Use Rate	
Criteria	*Bed Life (day)*	*(g/ m³)*	*(lb/mil gal)*	*Bed Life (day)*	*(g/ m³)*	*(lb/mil gal)*	*Bed Life (day)*	*(g/ m³)*	*(lb/mil gal)*	*Bed Life (day)*	*(g/ m³)*	*(lb/mil gal)*
≤0.5 mg TOC/l	38	110	916	57	73	611	76	55	458	84	50	414
≤1.0 mg TOC/l	69	61	505	148	28	235	146	29	238	225	19	155

Source: Reproduced from Neukrug et al. [13], courtesy of the American Water Works Association.

Table 9–15 GAC Bed Life and Use Rate for Volatile Halogenated Organics

Criteria	Torresdale Plant						Pilot Plant					
	GAC			Ozone-GAC			GAC			Ozone-GAC		
		Use Rate			Use Rate			Use Rate			Use Rate	
	Bed Life (day)	(g/m^3)	(lb/mil gal)	Bed Life (day)	(g/m^3)	(lb/mil gal)	Bed Life (day)	(g/m^3)	(lb/mil gal)	Bed Life (day)	(g/m^3)	(lb/mil gal)
Chloroform												
≤1 μg/l	35	119	995	32	131	1,088						
≤10 μg/l	51	82	683	44	95	791						
≤100 μg/l	78	54	446	80[a]	52	435						
DCP												
≤1 μg/l	109	38	319	113	37	308	141	30	247	148	28	235
≤5 μg/l	203	21	172	272	15	128	272	15	128	273	15	128

Source: Reproduced from Neukrug et al. [13], courtesy of the American Water Works Association.
[a]Highest effluent level recorded on day 80 when chloroform concentration was 97.6 μg/l.

Table 9–16 Design Criteria for GAC Systems

Criteria	*Postfiltration Contactor*		*Filter-Adsorber*	
	Rated	*Average*	*Rated*	*Average*
Flow [Ml/d (M gal/day)]	1,067 (282)	871 (230)	1,067 (282)	871 (230)
EBCT (min)	15	18.5	15	18.5
Surface loading rate (mm/sec) (gal/min/ft^2)	3.3 (4.9)	2.7 (4.0)	1.4 (2.0)	1.1 (1.6)
Surface area [m^2 (ft^2)]	74 (800)		97 (1,044)	
Bed depth [cm (ft)]	305 (10)		122 (4)	
Number of contactor pairs	25		47	
Pumping station capacity [Ml/d (M gal/day)]	1,325 (350)		–[a]	

Source: Reproduced from Neukrug et al. [13], courtesy of the American Water Works Association.
[a]Not required.

Table 9–16 shows the design criteria used for the GAC postfiltration contactors and filter-adsorbers [13]. Twenty-five contactor pairs were required to achieve the 15 minute EBCT at the design flow of 282 mgd. The 94 rapid sand filters at Torresdale were modified to GAC filter-adsorbers that also had a 15 minute EBCT at 282 mgd.

Capital costs for the regeneration furnace and related facilities were based on the multiple hearth design [13]. Furnace capacity was based on a hearth-loading rate of 75 lb carbon/ft^2/day. Excess furnace capacity was provided to allow for downtime.

Of the O&M costs, power is the major factor (Table 9–17). The principal user of energy is the regeneration facility. Large quantities of energy are needed also for the filter and adsorber system. The postfiltration contactor system consumes more energy than the filter-adsorber system since it requires a pumping station, backwash facility, and continued operation of existing rapid sand filters [13].

Table 9–17 Power Requirements of a GAC Facility

Use	*Type of Energy*	*Cost*
Process energy	Electricity	\$0.056/kwh
Regeneration heat	Gas	\$5.74/10^6 BTU[a]
Regeneration steam	Steam	\$30.87/1,000 kg[b] (\$14.00/1,000 lb)

Source: Reproduced from Neukrug et al. [13], courtesy of the American Water Works Association.
[a]9,444 BTU required to regenerate 1 kg carbon (4,250 BTU/lb carbon).
[b]1 kg (1 lb) steam required to regenerate 1 kg (1 lb) carbon.

Makeup carbon is a major contributor to operations and maintenance costs [25]. Carbon losses vary from 5 to 15%, but as handling techniques improve, these losses should decrease to 7%.

Table 9–18 is a summary of costs for two GAC systems that are based on a regeneration cycle of once every 6 months. The major difference in the capital costs of the two systems is due to the need for a new pumping station and backwash facility for the postfiltration contactor system. Also, modification of the

Table 9–18 GAC System Costs Based on a Six Month Regeneration Frequency

Cost Component	*Cost for Postfiltration Contactor (thousand dollars)*	*Cost for Filter-Adsorber (thousand dollars)*
Capital		
Initial GAC fill	7,000	6,870
Pumping station	11,068	0
Contactors	20,898	15,489
Regeneration facility	6,644	6,561
Unloading-handling	161	161
Storage	881	864
Backwash facility	731	0
Initial materials	144	144
Subtotal	47,527	30,089
Engineering and contingencies	11,440	4,284
Total capital	58,967	34,373
Bond costs	20,280	11,821
Total bond issue	79,247	46,194
Annual amortized capital	9,789	5,706
Operations and maintenance		
Contactor		
Labor	228	103
Power	1,097	694
Materials	65	65
Regeneration		
Labor	952	952
Power	850	834
Materials	51	50
Outside maintenance	266	262
Makeup carbon	982	964
Laboratory labor	102	102
Total	4,593	4,026
Total first-year cost[a]	14,382	9,732

Source: Reproduced from Neukrug et al. [13], courtesy of the American Water Works Association.

[a]Total first-year costs are equivalent to \$171/3.78 ml (1 mil gal) treated water or \$1.97/28 m^3 (1,000 ft^3) billed for the postfiltration contactor system and \$116/3.78 ml (1 mil gal) treated water or \$1.33/28 m^3 (1,000 ft^3) billed for the filter-adsorber system.

rapid sand filters is relatively low cost. Operating and maintenance costs are given also in Table 9-18. Regeneration costs are slightly higher for the postfiltration contactors mainly because of the need for a greater quantity of GAC. Data were given also for costs based on frequency of carbon regeneration [13].

Table 9-19 Ozonation System Costs

Cost Component	*Cost (thousand dollars)*
Capital	
Ozone equipment	5,790
Ozone contactors	800
Equipment housing	135
Subtotal	6,725
Engineering and contingencies	1,782
Total capital	8,507
Bond costs	2,892
Total bond issue	11,399
Annual amortized capital	1,408
Operations and maintenance	
Energy	597
Labor	47
Materials	58
Total	702
Total first-year cost[a]	2,110

Source: Reproduced from Neukrug et al. [13], courtesy of American Water Works Association.

[a]Total first-year cost is equivalent to \$25.12/3.78 Ml (1 mil gal) treated water or \$0.29/28 m^3 (1,000 ft^3) billed.

Table 9-20 Cost Range of Ozonation Systems (from Air)

Ozone Generation Capacity (lb/day of Ozone)	*Approximate Capital Cost (\$/lb production capacity/day)*
100	850-1,150
200	600-800
300	500-700
400	450-650
500	400-600
1,000	300-500

Source: Reproduced from Rice and Robson [4], courtesy of Ann Arbor Science Publishers.

Note: 1979 costs.

Ozonation costs are independent of any GAC system. Table 9–19 gives the O_3 generating equipment and operating costs based on an average dosage of 2 mg/l (for Torresdale this is 4,700 lb O_3/day) and a maximum dosage of 4 mg/l (9,400 lb O_3/day) [13]. Twelve O_3 generators, 900 lb/day capacity, satisfied these requirements with a 10% safety margin. Costs for O_3 contacting were based on a contact time of 10 minutes followed by another 10 minutes in a holding tank. A 90% O_3 transfer efficiency was assumed.

These costs are associated with a large-volume facility. It is doubtful if costs are linear with the volume of treated water. Intuitively, costs for GAC, BAC, and O_3 should be proportionately higher for small volumes less than 100 mgd. Some of this non-linearity is seen in Table 9–20, which provides a 1979 cost range of ozonation systems [4].

REFERENCES

1. Rice, R.G., et al. "A Review of the Status of Preozonation of Granular Activated Carbon for Removal of Dissolved Organics and Ammonia from Water and Wastewater," Chapter 14 in *Carbon Adsorption Handbook.* P.N. Cheremisinoff and F. Ellerbusch, eds. Ann Arbor Science Publishers, Ann Arbor, Mich. (1978), p. 485.
2. Ererhardt, M., et al. Report No. 7. Engler-Bunte-Institute, Department of Water Chemistry, Karhsruhe University, Federal Republic of Germany (1979).
3. Weber, W.J. Jr., et al. *J. Water Pollut. Cont. Fed. 42,* 83 (1970).
4. Rice, R.G., and C.M. Robson. *Biological Activated Carbon.* Ann Arbor Science Publishers, Ann Arbor, Mich. (1982), p. 37.
5. Miller, G.W., et al. "Large-Scale Applications of Granular Activated Carbon with Ozone Pretreatment," Chapter 13 in *Activated Carbon Adsorption of Organics from the Aqueous Phase,* Vol. 2. M.J. McGuire and I.H. Suffet, eds. Ann Arbor Science Publishers, Ann Arbor, Mich. (1980), p. 323.
6. Gould, J.P., and W.J. Weber, Jr. *J. Water Pollut. Cont. Fed. 48,* 47 (1976).
7. *J. Amer. Water Works Assoc. 73,* 447 (1981).
8. Sontheimer, H., et al. *J. Amer. Water Works Assoc. 70,* 393 (1978).
9. Benedek, A. *J. Amer. Water Works Assoc. 73,* 447 (1981).
10. Kühn, W., et al. *J. Amer. Water Works Assoc. 70,* 326 (1978).
11. Barnett, R.H., and A.R. Trussell. *J. Amer. Water Works Assoc. 70,* 660 (1978).
12. Glaze, W.H., et al. "Pilot-Scale Evaluation of Ozone-Granular Activated Carbon Combinations for Trihalomethane Precursor Removal," Chapter 13 in *Treatment of Water by Granular Activated Carbon.* M.J. McGuire and I.H. Suffet, eds. Advances in Chemistry Series 202. American Chemical Society, Washington, D.C. (1983), p. 303.
13. Neukrug, H.M., et al. *J. Amer. Water Works Assoc. 76,* 158 (1984).
14. Argaman, Y. *J. Amer. Water Works Assoc. 76,* 172 (1984).
15. Stephenson, P., et al. *Ozone Sci. Eng. 1,* 263 (1979).
16. Benedek, A., et al. *Ozone Sci. Eng. 1,* 347 (1979).
17. Yohe, T.L., et al. "Monitoring and Analysis of Aqueous Chlorine Effects on Granular Activated Carbon Pilot Contactors," Chapter 2 in *Activated Carbon Adsorption of Organics from the Aqueous Phase.* Vol. 2. M.J. McGuire and I.H. Suffet, eds. Ann Arbor Science Publishers, Ann Arbor, Mich. (1980), p. 27.

18. Glaze, W.H., and J.L. Wallace. *J. Amer. Water Works Assoc. 76,* 68 (1984).
19. Schulhof, P. *J. Amer. Water Works Assoc. 72,* 428 (1980).
20. Rook, J.J. *Water Treat. Exam. 23,* 234 (1974).
21. Faust, S.D., and O.M. Aly. *Chemistry of Water Treatment.* Butterworths, Stoneham, Mass. (1983).
22. Sontheimer, H. *J. Amer. Water Works Assoc. 71,* 618 (1979).
23. Mulbarger, M.C. *Proc. Amer. Water Works Assoc. 1980 Conf., Part II,* 1135 (1981).
24. Culp, R.L., et al. EPA Report No. 600/2-81-177. U.S. EPA, MERL, Cincinnati, Ohio (1981).
25. Culp, R.L., and R.M. Clark. *J. Amer. Water Works Assoc. 75,* 398 (1983).

10

Adsorption by Macroreticular Resins

There has been a surge in the use of synthetic resins as sorbents for the removal of organic compounds from water. Advantages of their application are that they can be tailor made for specific situations and can be regenerated in situ [1]. Several commercial resins have a variety of surface properties. Spent resins can be regenerated with nonaqueous solvents or by aqueous solutions of acids, bases, or salts. The nonaqueous solvents can be regenerated, and the adsorbate can be recovered if desired. This chapter reviews the individual performance of these resins and compares their adsorptive properties to those of GAC.

PHYSICAL AND CHEMICAL PROPERTIES

Types

Most resin adsorbents are polystyrene-divinylbenzene copolymers or polymethacrylates cross linked with an appropriate nonaromatic material (Figure 10–1) [2]. These adsorbents, commonly termed *macroreticular resins,* are characterized by being highly porous and having a large surface area. The pore size distribution is controlled during manufacturing by varying the level of cross linking. Another important feature is the absence of any ionic functional groups in their structure. Therefore, they can serve as hydrophobic adsorbents capable of removing hydrophobic solutes from solution via van der Waals interactions.

Polymeric adsorbents have been used successfully in concentrating organic contaminants in raw and finished waters for analytical purposes. High recoveries of the sorbed organic by desorption in an organic solvent for subsequent analysis contributes to the high efficiency of the analytical procedures. They are used also for removing chlorinated pesticides and dyes from industrial wastes. Regeneration is accomplished with acetone, isopropanol, or methanol, and the spent regenerant can be reclaimed for reuse. Adsorption studies from solutions of relatively high concentrations of phenols, carboxylic acids, and alkylbenzenesulfonates indicate that the styrene-divinylbenzene and polymethacrylate polymer

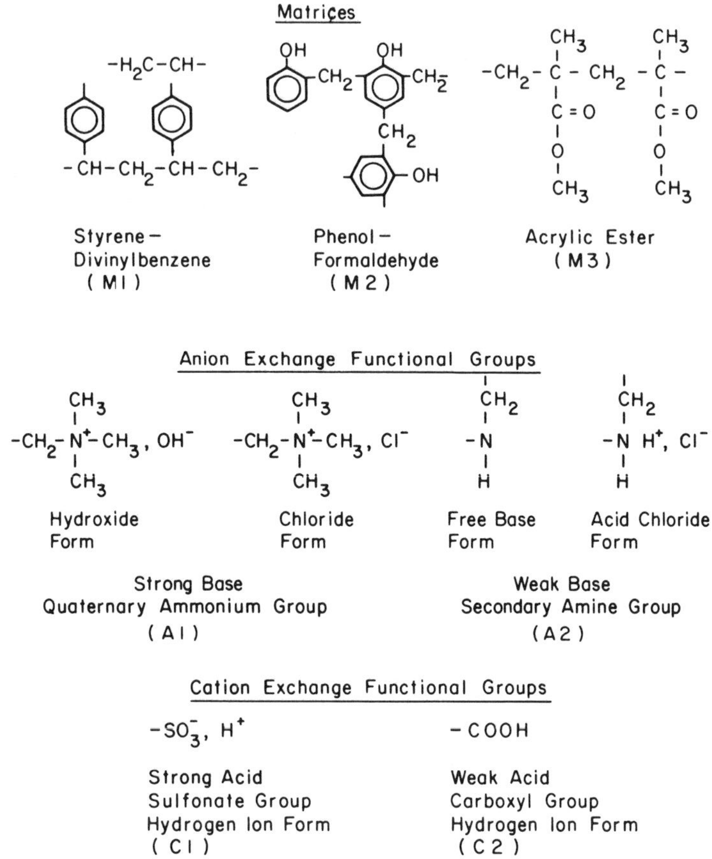

Figure 10–1 Matrixes and functional groups of resins commonly used for water purification. Reproduced from Kim et al. [2], courtesy of the Water Pollution Control Federation.

adsorbents can sorb these compounds but do so less effectively than activated carbon [1].

Physical Properties

Table 10–1 shows typical properties of some polymeric resins. It is noted that their porosities and surface areas are considerably less than those of activated carbon (see Chapter 5). Table 10–2 shows the physical properties of three synthetic carbonaceous adsorbents that are manufactured by partial pyrolysis of beads of macroreticular copolymers under such conditions that the polymer chars and retains the macroporous structure. These resins were developed in an effort to improve their adsorptive characteristics, especially in the micropore structure.

Table 10-1 Typical Properties of Amberlite Polymeric Adsorbents

Resin Type	*Chemical Nature*	*Helium Porosity Volume*		*BET Surface Area (m^2/g)*	*Average Pore Diameter (Å)*	*Skeletal Density (g/cm^3)*	*Nominal Mesh Sizes*
		%	*cm^3/g*				
Nonpolar[a]							
XAD-1	Polystyrene	37	0.69	100	200	1.06	20–50
XAD-2	Polystyrene	42	0.69	330	90	1.08	20–50
XAD-4	Polystyrene	51	0.99	750	50	1.09	20–50
Intermediate polarity[a]							
XAD-7	Acrylic ester	55	1.08	450	80	1.25	20–50
XAD-8	Acrylic ester	52	0.82	140	250	1.26	20–50
Synthetic carbonaceous adsorbent[b] XE-340	Polystyrene (partially pyrolyzed)			400	6–40 Å 18 vol.% 40–100 Å 13 vol. % 100–300 Å 69 vol. %	1.34	20–50

Source: "Drinking Water and Health," vol. 2, National Academy Press, Washington, D.C. (1980).
[a]From (3).
[b]From (4).

Table 10-2 Typical Physical Properties of Ambersorb™ XE-340, Ambersorb XE-347, and Ambersorb XE-348

Property	*Ambersorb XE-340*	*Amersorb XE-347*	*Ambersorb XE-348*
Appearance	Black, spherical nondusting	Black, spherical nondusting	Black, spherical nondusting
Total surface area (N_2, BET method) (m^2/gm)	400	350	500
Bulk density (lb/ft^3)	37	43	37
Bulk density (g/cm^3)	0.60	0.70	0.60
Particle density (g/cm^3) (Hg displacement)	0.92	1.05	0.91
Skeletal density (g/cm^3) (He displacement)	1.34	1.85	1.95
Pore volume (cm^3/g)	0.34	0.41	0.58
Particle size (U.S. Sieve Series)	20–50	20–50	20–50
Crush strength (kg/particle)	>3.0	>3.0	1.0
Ash content (%)	<0.5	<0.5	<0.5
Diameter range (vol. %)			
<6 Å	0	50	16
6–40 Å	18	0	21
40–100 Å	13	0	9
100–300 Å	69	50	51
>300 Å	0	0	3

Source: Rohm and Haas Co. [4].

REMOVAL OF INDIVIDUAL COMPOUNDS

Humic Substances

For approximately 75 years, ion exchange resins have been utilized widely to treat drinking water for various purposes including softening, dealkalization, deionization, and removal of color, metals, nitrates, ammonia, and cyanides. Both cation and anion exchangers are employed. The removal of color (humic substances) has been observed for many years and was called "fouling of resins." Recent research has been directed toward the use of polymeric resins as a deliberate treatment process for the removal of color, humic substances, and THM precursors.

An early study was concerned with the adsorption of commercially avail-

Table 10–3 Adsorption of Humic Acids by Anion Exchange Resins

Resin	*Humic Acid Removed (%)*	
	pH 2.2	*pH 7*
IRA-410[a]	97	70
IRA-910[a]	92	52
A-57[b]	79	22
A-30B[b]	71	44
ES-340[b]	82	59

Source: Davis [5].
[a]Produced by Rohm and Haas Co.
[b]Produced by Diamond Shamrock Co.

able humic acids at pH values of 2.2 and 7. by several anion exchange resins in batch equilibrium systems [5]. Data are summarized in Table 10–3. The strong base anion exchange resins, IRA-410 and IRA-910, removed considerably more than the intermediate strength resins, A-57, A-30B, ES-340, and a weak resin, Statabed 93 (R & H Co.). Cation exchange resins had little or no affinity for humic substances.

Two weak base resins, A20S Imacti (Akzo Chemicals) and 368 PR Duolite (Diamond Shamrock), were investigated in a pilot plant study for removal of THM precursors from Meuse River water [6]. In this 2 year study the second resin was found to be more effective than the first. A 70 to 85% reduction in $CHCl_3$-forming potential was achieved. However, the removal of THMFP was less, 60%.

There was a pilot column study of the removal of XE-340, GAC, and IRA-904 anion exchange resins of purgeable organics and THM precursors from raw groundwater after lime softening and in finished water at Hialeah, Florida [7]. Here, the THMFP in the influent to the columns varied from 600. to 868. μg/l in raw water, from 394. to 580. μg/l in lime-softened water, and from 274. to 451. μg/l in finished water. It is interesting to note that the lime-softening process removed an average of 30% of the THM precursors.

Table 10–4 is a summary of the pilot plant data for the Hialeah case [7]. A 75 cm deep bed of IRA-904 resin removed more THM precursors than the GAC (F-400) or the XE-340 from raw and lime-softened water. However, the GAC removed slightly more precursors from the finished water. In the raw water application, there was a considerable leakage of THM precursors, 100 μg/l, at the start of an IRA-904 bed. In any event, the THMFP removal, on a weight basis, is extremely low. This is seen in Figures 10–2 and 10–3 for the two resins [8]. Amberlite IRA-904 was initially quite effective for removal of THMFPs in the raw source, but it was unable to remove any precursor material from water that had been lime softened.

Table 10–4 Removal of THM Precursors by Pilot Columns of XE-340, IR-904, and GAC after Different Stages of Water Treatment at Hialeah, Florida

	Average THMFP Removal			
Pilot Columns	*Total THMFP Entering Column (g)*	*Percent at End of Run*	*Volume Basis (g/column at 49 day)*	*Weight Basis (g/100 g at 49 day)*
Raw water				
GAC	7.1	29 (119)[a]	0.99	0.56
XE-340	7.1	24 (119)	0.71	0.33
IR-904	2.6	46 (49)	1.21	0.44
IR-904[b]	2.6	55 (49)	1.43	0.26
Lime-softened water				
XE-340	5.0	4 (119)	0.09	0.04
IR-904	1.7	32 (49)	0.55	0.21
Finished water				
XE-340	3.8	0 (119)	0.00	0.00
IR-904	1.2	13 (49)	0.19	0.07
GAC	1.2	18 (49)	0.23	0.13

Source: Reproduced from Wood and DeMarco [7], courtesy of Ann Arbor Science Publishers, Inc.
Note: Based on breakthrough data for a bed depth of 75 cm.
[a]Number of days in parentheses.
[b]Bed depth of 150 cm.

Trihalomethanes

Some pilot plant column studies have been conducted for the removal of individual THMs by a synthetic resin (Ambersorb XE-340) [7]. The results are summarized in Table 10–5, where $CHCl_3$ breaks through very quickly but $CHBr_3$ was adsorbed rather well and took a considerably longer period for breakthrough. The EBCT was 6.2 minutes, which may be inadequate.

Batch adsorption experiments were conducted by using polyethylene (PE), neoprene (NEO), polyvinylchloride (PVC), and polyurethane (PU) foam as adsorbent materials for the removal of methylene chloride, chloroform, and carbon tetrachloride from water. An 8 hour agitation contact time was employed, presumably at room temperature. Results are summarized in Table 10–6, where the adsorption efficiencies for the three compounds on the plastics were considerably less than GAC (8 × 30 mesh, Nuchar). Some appreciable removals were achieved with CCl_4 on PVC and PU when the initial concentration was 10. mg/l. It is doubtful, however, that these plastics will find any pragmatic application for water treatment.

A pilot plant project was conducted for the Ambersorb XE-340 removal of TCE, PCE, cis-1,2-dichloroethylene, and 1,1,1-trichloroethane from the drink-

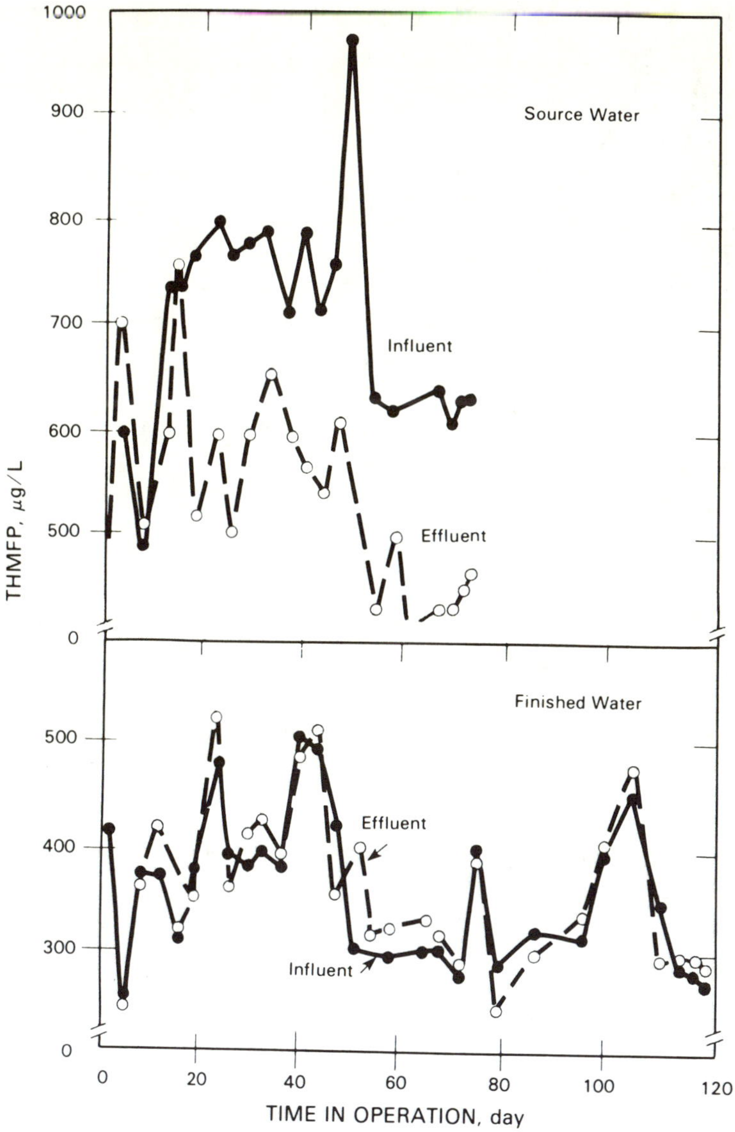

Figure 10–2 Removal of THM precursors by Ambersorb® XE-340; EBCT, 6.1 min. THMFP conditions: pH 9.0; 22°C (72°F); storage time, 6 days [8].

ing water supply (groundwater) for Glen Cove, New York [10]. Three columns were employed at a flow rate of 4 gal/min/ft^3. One column was operated downflow and the other two were in an upflow mode. The downflow column apparently produced the best results, with breakthrough times to reach 5 μg/l: 43 days for cis-1,2-dichloroethylene (from 38 to 79 μg/l), 110 days for TCE (from 3 to

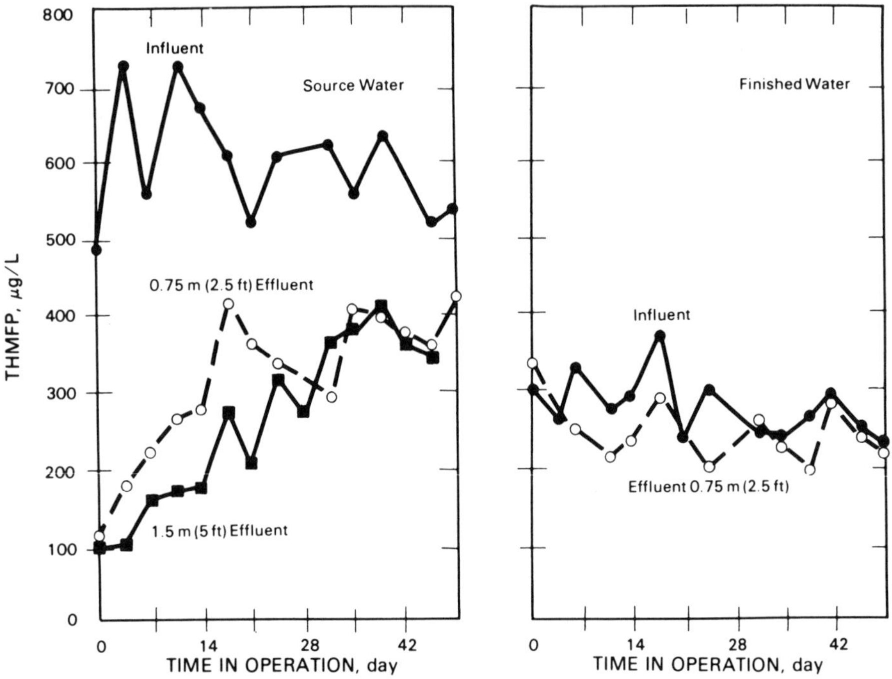

Figure 10–3 Removal of THM precursors by Amberlite® IRA-904; EBCT, 6.1 and 12.2 min. THMFP conditions: pH 9.0; 22°C (72°F); storage time, 6 days [8].

Table 10–5 Removal of THMs in Miami, Florida, by a Synthetic Resin

Compound	*Average Influent Concentration (μg/l)*	*Time to 2 μg/l Breakthrough (wk)*	*Time to Exhaustion[a] (wk)*
$CHCl_3$	80	3	156
	69	3	150
	64	0	150
$CHBrCl_2$	37	20	216
	43	20	210
	42	22	CE[b]
$CHBr_2Cl$	12	47	260
	25	45	260
	27	25	CE
$CHBr_3$	1.9	63	CE

Source: Reproduced from Wood and DeMarco [7], courtesy of Ann Arbor Science Publishers, Inc.
Note: Ambersorb® XE-340; EBCT = 6.2 min; bed depth = 0.8 m (2.5 ft).
[a]Sometimes predicted by extrapolation.
[b]Cannot extrapolate.

Table 10–6 Removal Efficiencies Determined from Batch Adsorption Experiments

Concentration (mg/l)	*Adsorbent Material*	*Methylene Chloride (percent)*	*Chloroform (percent)*	*Carbon Tetrachloride (percent)*
0.10	PE	0	5.52	2.56
	NEO	0	17.18	10.26
	PVC	5.66	25.77	29.74
	PU	3.77	28.83	29.74
	GAC	33.02	69.94	74.36
1.0	PE	0	4.96	27.59
	NEO	0	25.87	36.37
	PVC	14.76	45.58	44.98
	PU	5.95	46.78	40.76
	GAC	89.29	89.94	86.83
10.0	PE	0	9.68	38.68
	NEO	0	30.84	65.20
	PVC	18.01	50.02	78.38
	PU	8.69	53.53	73.87
	GAC	91.86	94.77	92.03

Source: Reproduced from Cook and Hartz [9], courtesy of the American Water Works Association.

7 μg/l), and 142 days for PCE (from 33 to 207 μg/l). Upflow steam was used for regeneration of the resin. Air stripping and GAC were investigated also in this study.

Phenols

The use of macroreticular resins for the removal of organics is not necessarily a recent development in water treatment. There were several publications in the 1950s for the removal of phenols by anion exchange resins [11,12,13]. Statements like the following were made: "Organic ion exchange resins have been shown to sorb certain organic solutes beyond their exchange capacities for such materials. The quaternary ammonium anion exchangers have a rather high nonexchange capacity for phenolic compounds." The resin, Dowex 1-X7.5, 20–50 mesh, Cl^- form, was employed to obtain adsorption isotherms for phenol and PNP [11]. Experimental conditions were 25°C and 20 hours of equilibration. Figures 10–4 and 10–5 show the isotherms for phenol and PNP, respectively. These are typical isotherms and are similar in shape to those obtained with GAC (see Chapter 1). The maximum quantities of adsorption in Figures 10–4 and 10–5 were 941. mg/g for phenol and 1,113. mg/g for PNP, respectively. These amounts compare favorably to PAC and GAC.

Later studies were conducted for adsorption of phenol and *p*-cresol and other organics on two Rohm and Haas resins, XAD-2 and XAD-7 [1]. Typical

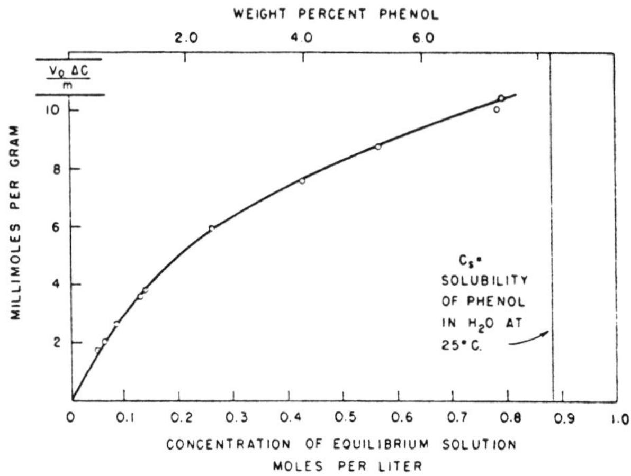

Figure 10–4 Phenol sorption from aqueous solution on Dowex 1, Cl^- at 25°C. Reproduced from Anderson and Hansen [11], courtesy of the American Chemical Society.

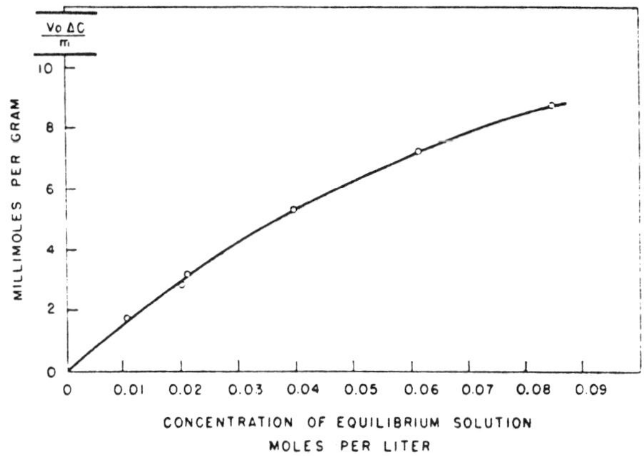

Figure 10–5 *p*-nitrophenol sorption from aqueous solution on Dowex 1, Cl^- at 25°C. Reproduced from Anderson and Hansen [11], courtesy of the American Chemical Society.

Freundlich adsorption isotherms for these compounds are seen in Figure 10–6. No experimental conditions other than temperature were given. Quantities of phenol and *p*-cresol adsorption are given in Table 10–7. Here, again, this was an early study and the XAD resins were in an early experimental stage.

Another study investigated the adsorption of phenols and chlorophenols by Amberlite XAD-4 (surface area of 780 m^2/g) in a column mode [14]. No experimental conditions other than flow rate were given. Figure 10–7 shows the breakthrough curves for four phenols. At a flow rate of 0.5 $gal/min/ft^3$, XAD-4

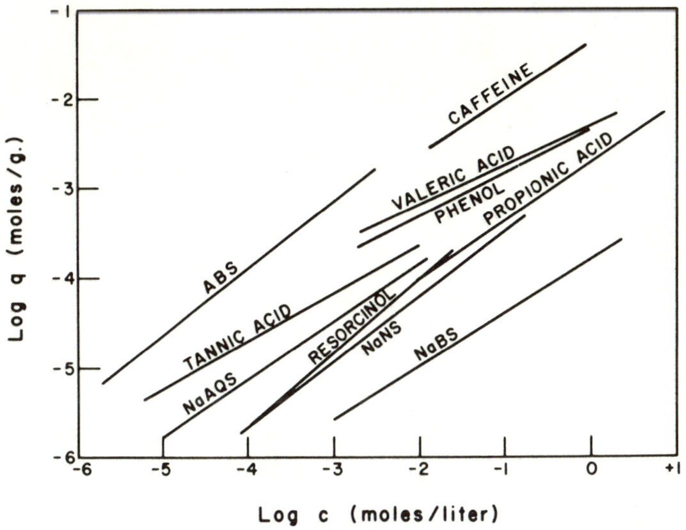

Figure 10–6 Freundlich isotherms for adsorption of organic compounds by Amberlite XAD-2 at 25°C. Reproduced from Gustafson and Paleos [1], courtesy of Marcel Dekker, Inc.

Table 10–7 Adsorption of Phenol and *p*-Cresol by Amberlite XAD-2 and Amberlite XAD-7 at 25°C

Resin	*Solute*	*Equilibrium Solute Concentration (mol/l)*	*Grams Solute Adsorbed per Gram Resin*
XAD-7	Phenol	0.0152	0.0764
XAD-2	Phenol	0.0306	0.0511
XAD-7	*p*-cresol	0.0146	0.1431
XAD-2	*p*-cresol	0.0265	0.0924

Source: Reproduced from Gustafson and Paleos [1], courtesy of Marcel Dekker, Inc.

adsorbs about 0.78 lb/ft^3 of phenol at 0% leakage, 2.4 lb/ft^3 for 3-chlorophenol, 5.1 lb/ft^3 for 2,4-DCP, and 12 lb/ft^3 for 2,4,6-TCP. The latter compare favorably with GAC.

Several polymeric resin adsorbents were examined for this capacity to adsorb phenol from water in a column operation at room temperature [15]. The results are summarized in Table 10–8. Those resins without a commercial name were manufactured in Japan. However, the Amberlite IRA-400 adsorbed the greatest amount of phenol.

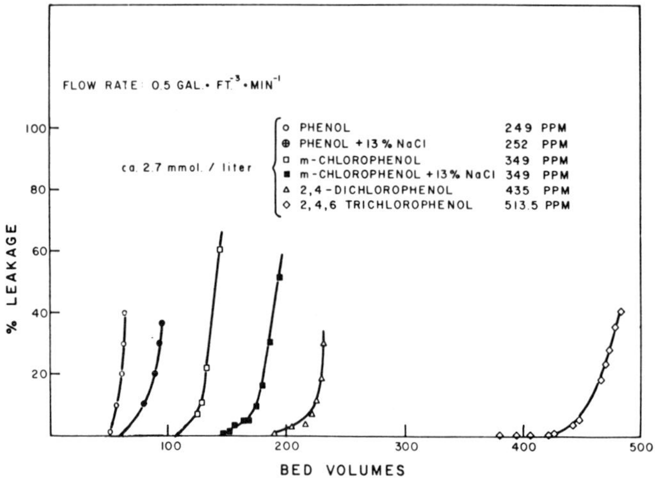

Figure 10–7 Adsorption of phenol and substituted chlorophenols on Amberlite XAD-4. Reproduced from Kennedy [14], courtesy of the American Chemical Society.

Table 10–8 Breakthrough Capacities of Polymeric Adsorbents for Phenol Adsorption

	Adsorption Capacity	
Polymeric Adsorbent	*meg/dry g*	*mequiv/ dry g*
Pulverized copolymer of 4-vinylpyridine with divinylbenzene containing 72 mol % 4-vinylpyridine	275	2.92
Pulverized copolymer of 2-vinylpyridine with divinylbenzene containing 71 mol % 2-vinylpyridine	240	2.60
Strong base anion exchange resin in the hydroxide form (Amberlite IRA-400)	405	4.30
Weak base anion exchange resin in the free base form (Amberlite IR-45)	272	2.89
Porous styrene-divinylbenzene resin with no ion exchange functional group (Amberlite XAD-4)[a]	150	1.59
Porous styrene-divinylbenzene resin with no ion exchange functional group (Amberlite XAD-2)[b]	89	0.95
Pulverized copolymer of styrene with divinylbenzene containing 57 mol % styrene	0	0.00

Source: Reproduced from Kawabata and Ohira [15], courtesy of the American Chemical Society.
Note: Determined by the continuous flow column method using 3,000–3,500 ppm of aqueous phenol solution. The flow rate was 3 bed vol/h.
[a]Claimed to have a surface area of 784 m^2/g.
[b]Claimed to have a surface area of 300 m^2/g.

Detergents

In an early resin adsorption study, the degrees of binding of two sulfonates by Amberlites XAD-1 and XAD-2 and EXP-500 (526 m^2/g) were measured under various concentrations and temperature conditions [1,16]. Equilibrium (25°C) and column studies were conducted for sodium anthraquinonesulfonate (AQS) and sodium dodecylbenzenesulfonate (ABS). Freundlich isotherms were obtained for AQS on the three resins, with the degree of adsorption increasing with surface area as expected: EXP-500 > XAD-2 > XAD-1 (Table 10–9). The Freundlich isotherms, however, for adsorption of ABS on the three resins were not linear, and no explanation was offered. Perhaps the ABS molecule was too large for the resin's pore structure. Column studies gave unexpected results for ABS adsorption by the three resins. As seen in Figure 10–8, ABS broke through the EXP-500 rather rapidly, whereas it took longer for the other two resins and a GAC (CAL 12 × 40 mesh). Poor kinetics of adsorption are obvious for the EXP-500.

Two weak base phenol formaldehyde anion exchange resins were found to have an excellent capacity for removing the anionic organic compounds: benzenesulfonate (BS), *p*-toluenesulfonate (PTS), and DBS from solutions at pH values of 2 and 4. The two resins were Duolites A-7 and S-37 (Diamond Shamrock). Equilibrium studies were conducted under agitation for 3 days at room temperature (19 to 23°C). Column studies were conducted also at a flow rate of 102.4 ml/min. Figure 10–9a shows the adsorption isotherms for BS, PTS, and DBS on Duolite A-7 at the pH values of 2 and 4 (unlikely for actual water treat-

Table 10–9 Efficiency of Utilization of Surface Areas of Amberlite XAD-1, Amberlite XAD-2, and EXP-500 in the Adsorption of AQS, ABS, and Caffeine

Solute	*Resin*	*Surface Area (m^2/g)*	*Moles Solute Adsorbed per Square Meter*
AQS[a]	XAD-1	100	7.30×10^{-8}
AQS	XAD-2	354	5.08×10^{-8}
AQS	EXP-500	526	7.81×10^{-8}
ABS[b]	XAD-1	100	1.59×10^{-6}
ABS	XAD-2	354	0.86×10^{-6}
ABS	EXP-500	526	1.00×10^{-6}
Caffeine	XAD-1	100	2.13×10^{-7}
Caffeine	XAD-2	354	2.24×10^{-7}
Caffeine	EXP-500	526	4.01×10^{-7}

Source: Reproduced from Gustafson and Paleos [1], courtesy of Marcel Dekker, Inc.
[a]AQS = sodium anthraquinonesulfonate
[b]ABS = sodium dodecylbenzenesulfonate

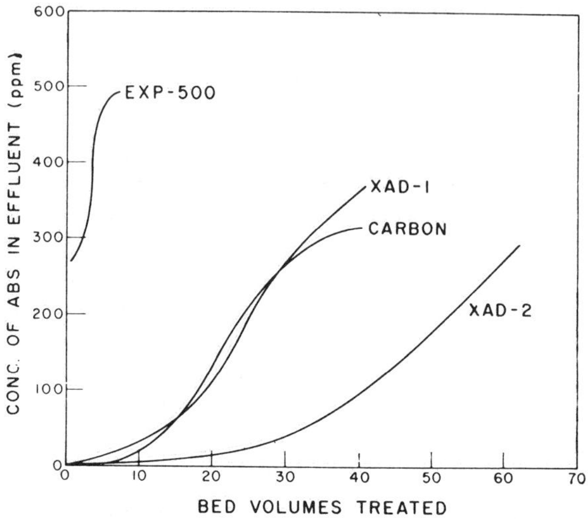

Figure 10–8 Adsorption of ABS by styrene-DVB copolymers and activated carbon in column operations at room temperature. Influent concentration: 680 ppm flow rate, 2.0 gal/ft³/min. Reproduced from Gustafson et al. [16], courtesy of the American Chemical Society.

ment plant operation). The isotherms are not linear over the broad concentration range, but this may be due to the high-low solute concentration effect that was observed with phenols on GAC (see Chapter 6). The lower adsorption capacities seen at pH 4 are explained by protolysis of the sulfonic group into its anionic form that are more difficult to adsorb on a weak base resin. Similar isotherms are seen in Figure 10–9b for BS and DBS on Duolite S-37. A comparison of isotherms for the Duolite resins, GAC, and XAD-2 is seen in Figure 10–9c. It appears that the greatest capacities lie with the Duolites.

COMPARATIVE STUDIES: RESINS AND GRANULAR ACTIVATED CARBON

Humic Substances

Many comparative studies have been conducted on the adsorption of various organic compounds by polymeric resins and GAC. Batch- and bench-scale column tests were used to compare the performance of five types of commercially available activated carbons and seven types of resins for the removal of humic and fulvic acids from water [18]. The latter were the soil and leaf fulvic acids and soil and commercial humic acids. The GACs were representative of those commercially available (brand names were not given). They were manufactured

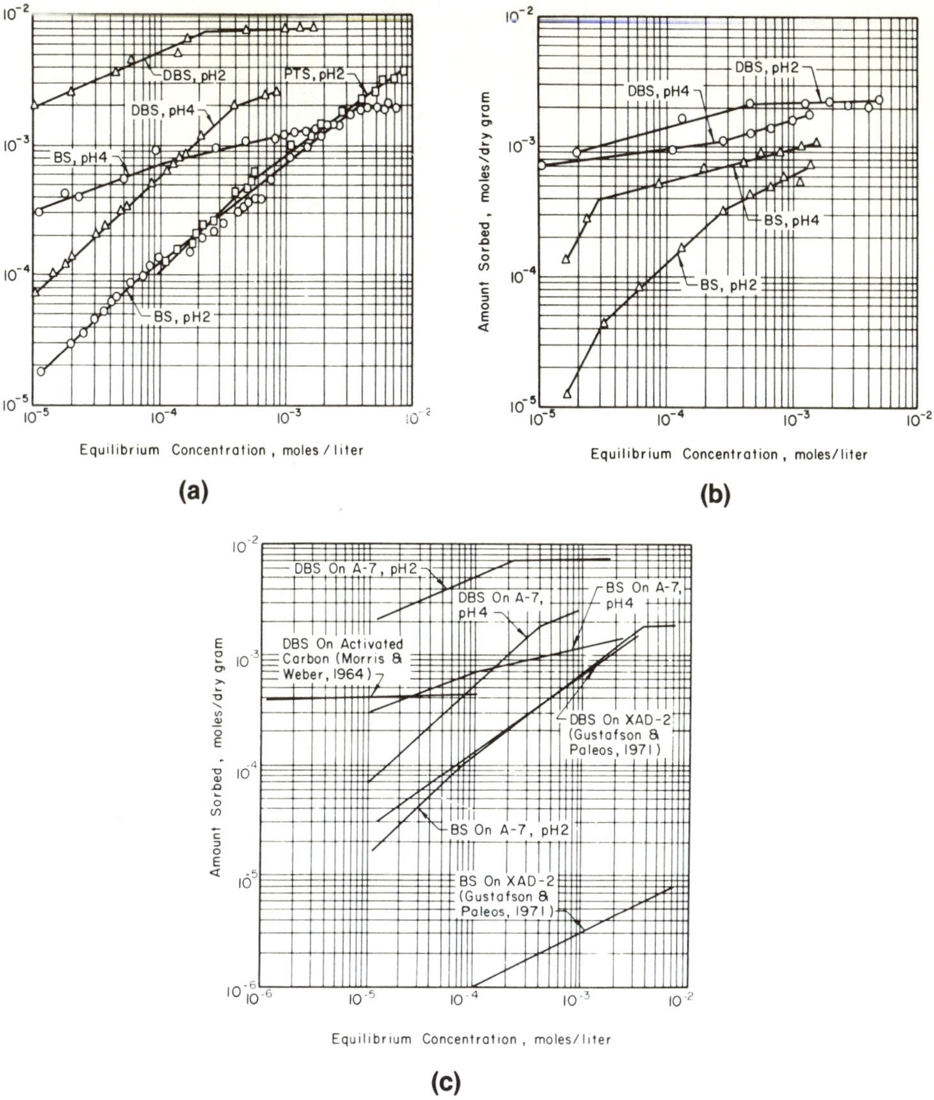

Figure 10–9 (a) BS, PTS, and DBS sorption isotherms on Duolite A-7. (b) BS and DBS sorption isotherms on Duolite S-37. (c) BS and DBS sorption isotherms on Duolite A-7, Amberlite XAD-2, and activated carbon. Reproduced from Hinrichs and Snoeyink [17], courtesy of Pergamon Press.

either from bituminous coal or lignite. Surface areas ranged from 650 to 1,100 m^2/g. Seven commercially available resins were evaluated. Their properties are given in Table 10–10. Of course, adsorption isotherms were developed from continuously agitated systems over a 7 to 8 day period at room temperature. Figure 10–10 shows the log-log isotherms for humic acid adsorption on three carbons

Table 10–10 Adsorbent Resin Properties

Resin	*Matrix*	*Functional Group*	*Exchange Capacity*[a] *(meq/g)*	*BET-N, Surface Area* *(m^2/g)*	*Commercial Name*	*Manufacturer*
R-1a	Phenol-formaldehyde	Amine	9.1		A-7	Diamond Shamrock, Redwood City, Calif.
R-1b	Phenol-formaldehyde	Amine	3.5		ES-561	Diamond Shamrock, Redwood City, Calif.
R-2	Styrene-divinylbenzene	None		300	XAD-2	Rohm and Haas, Philadelphia, Pa.
R-3	Acrylic	None		200	XAD-8	Rohm and Haas, Philadelphia, Pa.
R-4	Styrene-divinylbenzene	Quaternary ammonium	2.6	60	IRA-904	Rohm and Haas, Philadelphia, Pa.
R-5	Acrylic	Quaternary ammonium	4.4	0.1	IRA-458	Rohm and Haas, Philadelphia, Pa.
CR	Carbonized resin	None		400	XE-340	Rohm and Haas, Philadelphia, Pa.

Source: Reproduced from Boening et al. [18], courtesy of the American Water Works Association.
[a]Dry exchange media.

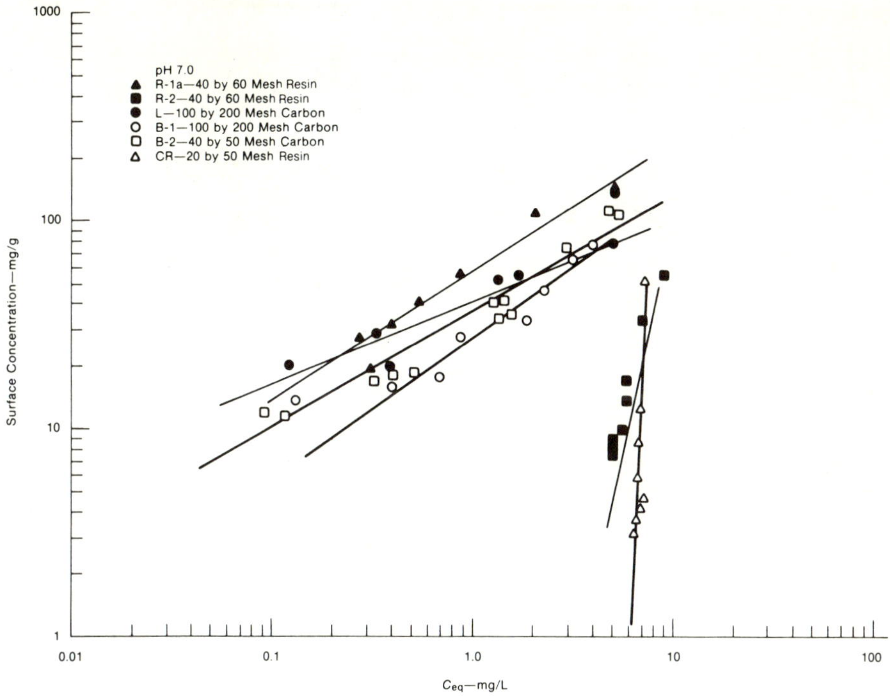

Figure 10–10 Commercial humic acid isotherms for Resins R-1a, R-2, and CR and for carbons B-1, B-2, and L. Reproduced from Boening et al. [18], courtesy of the American Water Works Association.

and three resins. It appears that at least one of the resins, R-1a, compares favorably with the three carbons.

Figure 10–11 shows the adsorption isotherms for humic acid on three resins as a function of pH [18]. The isotherm for resin R-1a at pH 7 is reproduced from Figure 10–10 for comparative purposes. The capacity of resin R-1b is lower than that of resin R-1a by an approximate factor of two at an equilibrium concentration of humic acid of about 1. mg/l. When the pH value is either decreased from 8.3 to 7.0 or 5.5 or increased from 8.3 to 9.5, the capacity of resin R-1b decreases. This observation was ascribed to association of the amine functional groups under acidic conditions and to their dissociation under alkaline conditions to negative sites. Here again, resin R-4 was clearly superior to the other three resins. Also, it does not show any great variation in adsorptive capacity due to $[H_3^+O]$, which is desirable as a water treatment mode. Resin R-4 is a styrene-DVB polymer with quaternary ammonium functional groups. It had one of the lowest surface areas of the experimental resins, suggesting external surface adsorption of the humic acid.

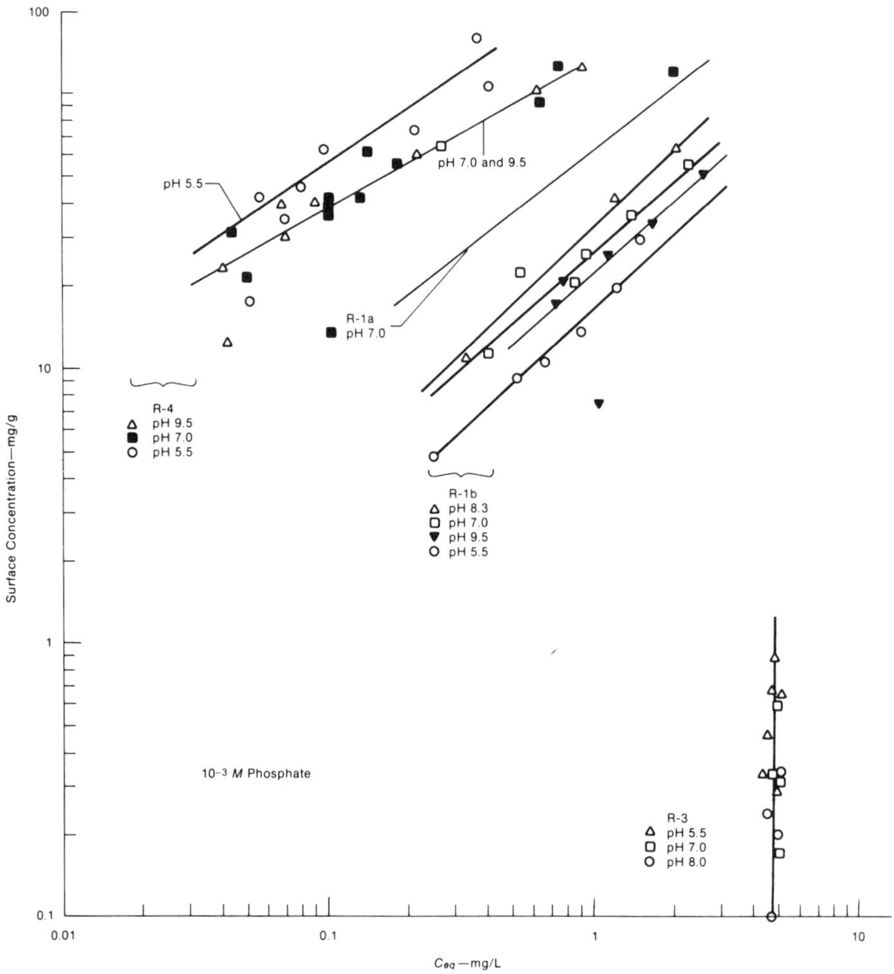

Figure 10-11 Commercial humic acid isotherms as a function of pH of Resins R-1a, R-1b, R-3, and R-4. Reproduced from Boening et al. [18], courtesy of the American Water Works Association.

Trihalomethanes

Some comparative studies have been reported for the removal of the low molecular weight THMs by GAC and polymer resins. Adsorption isotherms and bench-scale column studies were employed to compare the performance of five commercially available activated carbons and three resins for the removal of chloroform from water [19]. Four of the carbons were bituminous coal and one was lignite. The three resins were R-1a, R-2, and CR, whose properties are given in Table 10–10. Adsorption isotherms were developed from continuously agitated systems for at least 7 days, presumably at room temperature.

Figure 10–12 shows the $CHCl_3$ isotherms for one carbon and the carbonaceous resin CR in systems containing only $CHCl_3$ and others in which commercial humic acid was added. The capacities of the resin and carbons are very close. It is interesting to note that the relatively large molecule of humic acid had no significant effect on the $CHCl_3$ adsorptive capacity of the resin whereas there was a decrease by a factor of 1.5 on carbon L. The absence of a competitive effect between humic acid and $CHCl_3$ on the CR resin was consistent with previous observations that humic acids are not adsorbed by this resin [18]. However, the absence of strong competitive effects on the lignite carbon was surprising because of its affinity for humic acid. This was explained plausibly by invoking adsorption of the two molecules at separate sites on the carbon.

An extensive pilot plant study was conducted for the comparative removal of THMs and other volatile organics by one GAC (F-400) and two resins (XE-340 and IRA-904) from water at the Preston Water Treatment Plant at Hialeah (Greater Miami), Florida [7]. Several bench-scale columns were operated at a flow rate of 3 gal/min/ft^2 with an EBCT of 6.2 min for a 2.5 ft deep bed. Deeper beds were obtained through a series operation. The adsorption of several organics was reported from this study; however, only the data from two compounds are given here. Table 10–11 is a summary of the $CHCl_3$ adsorption data that were obtained from breakthrough curves under a variety of experimental conditions. It appears that $CHCl_3$ broke through the resin columns more quickly than the GAC columns but that the quantities adsorbed were about the same. Similar results were reported for bromodichloromethane in Table 10–12 although the amount adsorbed per 100 g of adsorbent at the end of the test was greater for XE-340 than for the GAC. It should be noted that these adsorption data were

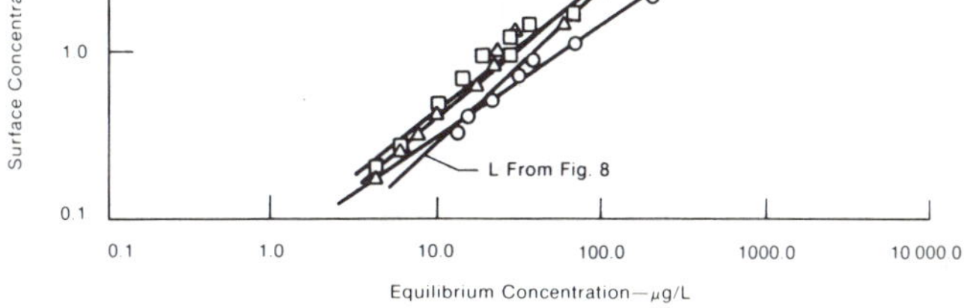

Figure 10–12 Chloroform adsorption isotherms on Resin CR and Carbon L. Reproduced from Chudyk et al. [19], courtesy of the American Water Works Association.

Table 10–11 Summarized Chloroform Adsorption Data

Experimental Design Number	*Bed Depth (ft)*	*Adsorbent*	*Average Influent (μg/l)*	*Column Breakthrough (days)*	*Column Saturation (days)*	*MTZ (in)*	*Test Duration (days)*	*Total Entering Each Column during Test (g)*	*Total Adsorbed by Each Column at End of Test (g)*	*Adsorbed by Each Column at Saturation (g)*	*% Adsorbed at End of Test*	*% Adsorbed at Saturation*	*Adsorption per 100 g Adsorbent at End of Test (g)*	*Adsorption per 100 g Adsorbent at Saturation (g)*	*(cc)*
Lime-softened water															
1	2.5	XE-340	3.4	Questionable			117	0.0355							
1R	2.5	XE-340	1.2	7?	152?	28.6	122	0.013	0.0081	0.0085	62	52	0.0037	0.00395	0.0027
Finished water															
1	2.5	XE-340	80.2	3	156	29	117	0.838	0.497	0.569	59	51	0.231	0.265	0.177
1R	2.5	XE-340	69.3	3	150	29	122	0.736	0.456	0.473	62	52	0.212	0.22	0.148
2	2.5	XE-340	64	0	150	30	63	0.36	0.247	0.429	69	50	0.115	0.2	0.134
3	2.5	GAC	57	7	22	21	53	0.27	0.074	0.074	27	66	0.042	0.042	0.028
3	2.5	IRA-904	57				53	Average increase of 1.75X							
4	2.5	GAC	67.3	8	23.5	20	122	0.733	0.094	0.094	13	67	0.534	0.0534	0.0358
4	5	GAC	67.3	29	49	24.5	122	0.733	0.235	0.235	32	80	0.067	0.067	0.0449
4	7.5	GAC	67.3	49	76	32	122	0.733	0.376	0.376	51	82	0.071	0.071	0.048
4	10	GAC	67.3	72	98	31.8	122	0.733	0.511	0.511	70	87	0.073	0.073	0.049

Source: Reproduced from Wood and DeMarco [20], courtesy of Ann Arbor Science Publishers, Inc.

Table 10–12 Summarized Bromodichloromethane Adsorption Data

Experimental Design Number	Bed Depth (ft)	Adsorbent	Average Influent (μg/l)	Column Breakthrough (days)	Column Saturation (days)	MTZ (in)	Test Duration (days)	Total Entering Each Column during Test (g)	Total Adsorbed by Each Column at End of Test (g)	Adsorbed by Each Column at Saturation (g)	% Adsorbed at End of Test	% Adsorbed at Saturation	Adsorption per 100 g Adsorbent at End of Test (g)	Adsorption per 100 g Adsorbent at Saturation (g)	Adsorption per 100 g Adsorbent at Saturation (cc)
Lime-softened water															
1	2.5	XE-340	2.4	69	None		117	0.012	0.01		83		0.005		
1R	2.5	XE-340	?		63		Data too erratic to calculate				83				
Finished water															
1	2.5	XE-340	37.1	20	216	27	117	0.388	0.301	0.391	78	55	0.14	0.182	0.0907
1R	2.5	XE-340	42.7	20	210	27	122	0.465	0.36	0.439	77	54	0.167	0.204	0.1017
2	2.5	XE-340	42.4	22	Can't extrapolate		63	0.239	0.225		94		0.105		
3	2.5	GAC	39	16	53	21	53	0.185	0.121	0.121	65	65	0.069	0.069	0.0344
3	2.5	IRA-904	39				53	Average increase of 1.13X							
4	2.5	GAC	47	14	56	22.5	122	0.512	0.147	0.147	29	63	0.084	0.084	0.0419
4	5	GAC	47	42	98	34.3	122	0.512	0.293	0.293	57	71	0.083	0.083	0.0414
4	7.5	GAC	47	73	139	42.7	122	0.512	0.444	0.444	87	76	0.084	0.084	0.0419
4	10	GAC	47	105	Can't extrapolate		122	0.512	0.51		99		0.072		

Source: Reproduced from Wood and DeMarco [20], courtesy of Ann Arbor Science Publishers, Inc.

obtained in a competitive situation where at least 16 low molecular weight organics were present. It was concluded in this study that XE-340 had approximately three times the adsorptive capacity of GAC for the individual halogenated organics.

This conclusion was supported by an independent adsorption isotherm study (continuous agitation for 48 hours) of $CHCl_3$ [20]. Table 10–13 is a summary of this study where the $CHCl_3$ capacity (mg/g) from an equilibrium concentration of 2 ppm for XE-340 was up to five times greater than that for five carbons. Apparently the XE-340's surface is more conducive to $CHCl_3$ adsorption than those of the other adsorbents.

Phenols

Very few comparative studies of phenol adsorption on resins and GAC have been reported. Adsorption isotherms were developed for PNP on Duolite A-7 and one

Table 10–13 Physical and Adsorption Properties of Various Adsorbents

Number	*Adsorbent*	*Type*	*Chloroform Capacity (mg/g) at 2 ppm*	*Surface Area[a] (m^2/g)*	*Volume Increase (%) in Liquid Chloroform (±3)*
1	Ambersorb XE-340	Carbonaceous	50	440	13
2	Ambersorb XE-347	Carbonaceous	21	400	3
3	Ambersorb XE-348	Carbonaceous	28	550	2
4	XE-225A	Very low cross-linked polymeric	6	50	520
5	Amberlite XAD-1	Low cross-linked polymeric	10	100	42
6	Amberlite XAD-2	Medium cross-linked polymeric	4	300	13
7	Amberlite XAD-4	High cross-linked polymeric	6	750	23
8	Filtrasorb 300	GAC	10	1,000[b]	—
9	Filtrasorb 400	GAC	12	1,100[b]	0
10	Nuchar WV-G	GAC	12	1,000[b]	2
11	BPL (12 × 30)	GAC	17	1,100[b]	1
12	Kuhrea	Spherical carbon	13	780	1

Source: Reproduced from Neely [21], courtesy of Ann Arbor Science Publishers, Inc.
[a]Measured on Micromeretics Model 2200 high-speed analyzer.
[b]Values quoted by manufacturers.

GAC (60 × 80 mesh) under continuous agitation for 3 to 4 days at 19.5 to 22°C [2]. Figure 10–13 shows the log-log plots for seven pH values for PNP on A-7. Isotherms for the GAC and XAD-7 from other studies were added for comparative purposes. It is seen that the adsorptive capacity of the GAC is greater than for either resin when the pH value is greater than PNP's *pKa* value (7.15). This was observed also for pH values less than the *pKa* value, but the A-7 isotherms are closer to the carbon isotherm (pH 4.1). No pH effect was reported for the XAD-7 resin because it is an acrylic base resin with no ion exchange functional groups. There were very strong pH effects on the A-7 resin at 3.6 and 9.3. Apparently there is a competitive effect between H_3^+O ions and PNP at the lower pH value, whereas there are repulsive forces between the phenolate anion and the resin matrix at the higher pH value.

Two activated carbons and eight synthetic resins were compared for their adsorptive capacities for phenol [21]. Adsorption isotherms were developed from continuously agitated systems for 8 days presumably at room temperature. It was necessary to develop an isothermal equation because neither the Langmuir nor Freundlich equations were sufficient for the broad range of equilibrium concentrations employed in the study:

$$C_e = ai \quad qe^{(a_2 q_e^{a_3} + a_4)} \qquad (10.1)$$

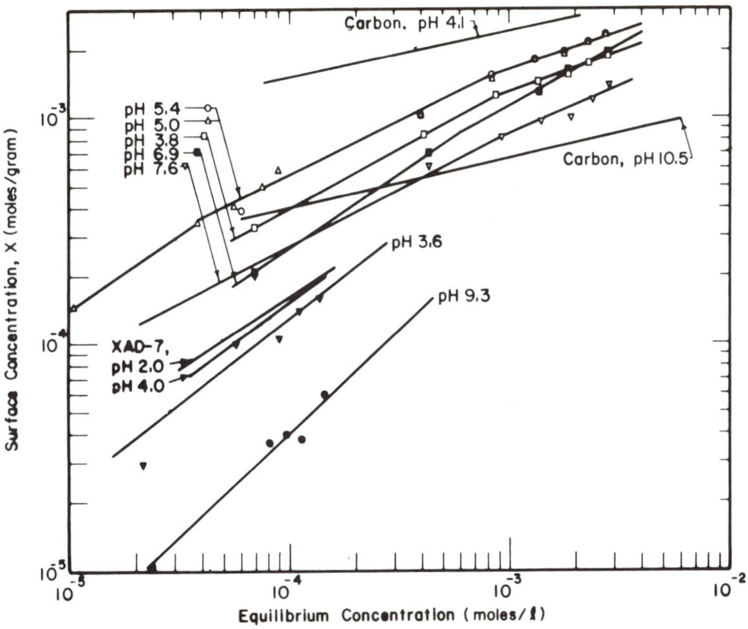

Figure 10–13 *p*-nitrophenol isotherms on A-7. Reproduced from Kim et al. [2], courtesy of the Water Pollution Control Federation.

where C_e = equilibrium concentration, mg/l; q_e = corresponding adsorption capacity, mg/g; and *ai* (i = 1,4) = coefficients determined by a multiple nonlinear curve-fitting technique predicated on the minimization of the sum of squares of residuals. This equation is essentially an extension of the Freundlich isothermal relationship (see Chapter 1). Figure 10–14 shows the isotherms for phenol on the various adsorbents. It appears that three of the resins, XE-347, XE-348, and XE-340, are similar in effectiveness to the two carbons. These finite bath studies also developed the kinetics of adsorption as a precursor to column experiments. The film transfer coefficients and effective surface diffusion coefficients for phenol and the various adsorbents are given in Table 10–14 (see also Chapter 6). As seen, these two coefficients are somewhat related or influenced by the shape and roughness factor in that the rough and irregular surfaces of the two carbons produce the highest diffusion and film transfer coefficients. These values were employed to model breakthrough curves from column studies (not given here).

Two phenols, two sulfonates, and CCl_4 were subjected to batch and column studies to determine their adsorption on eight resins and two GACs [22,23]. Adsorption isotherms were presented in terms of a two-coefficient Dubinin equation [24]:

$$W = W_o \exp\left(\frac{-B\epsilon^2}{B}\right), \tag{10.2}$$

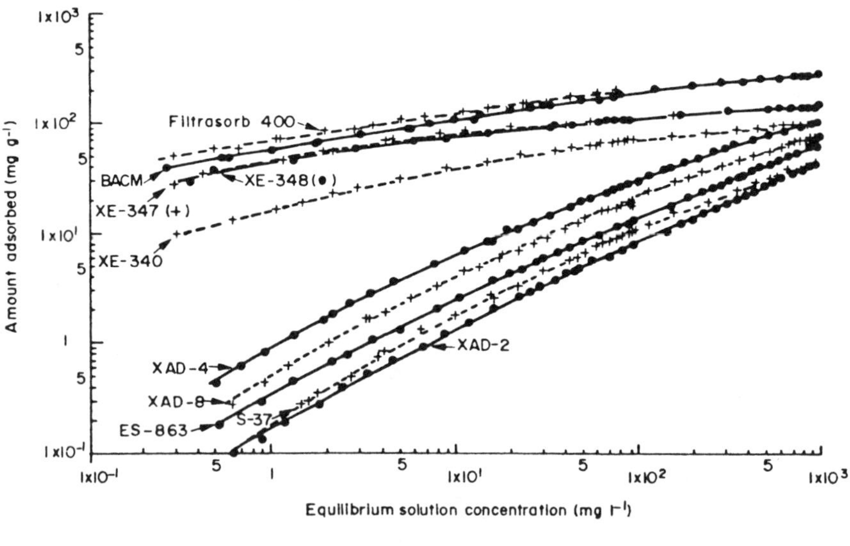

Figure 10–14 Isotherms for phenol on various adsorbents. The solid and dashed lines represent the mathematical description of the data (by equation). Reproduced from Van Vliet et al. [22], courtesy of the Pergamon Press.

Table 10–14 Finite Bath Rate Coefficients for Phenol

Adsorbent	$D_s \times 10^9$ ($cm^2\ sec^{-1}$)[a]	$k_{fb} \times 10^3$ ($cm^2\ sec^{-1}$)[b]	*Shape and Roughness Factor* (η, *dimensionless*)[c]
BACM carbon	2.18	7.04	1.00
Filtrasorb 400 carbon[d]	1.24	6.88	0.98
Ambersorb XE-348	0.696	4.88	0.69
Amberlite XAD-4	20.6	4.35	0.62
Ambersorb XE-347	0.572	1.32	0.19
Ambersorb XE-340	0.484	1.01	0.14

Source: Reproduced from Van Vliet et al. [22], courtesy of Pergamon Press.
Note: Stirrer speed = 1,200 rpm; adsorbent particle size fraction = 0.30–0.42 mm.
[a]D_s = effective intraparticle surface diffusion coefficient.
[b]k_{fb}= film transfer coefficient derived from CMB rate experiments.
[c]η: defined as the ratio of the k_{fb} for a particular adsorbent to k_{fb} for the BACM carbon.
[d]Based on the 100 mg/l initial concentration isotherm.

where $W = Q/10p$; Q = the solid-phase loading of adsorbate in mg adsorbate/g adsorbent; p = the bulk density of the adsorbate in g/cm³; W_o = the maximum solid-phase loading of adsorbate in cm³ adsorbate/100 g adsorbent as $C \rightarrow C_s$; $\epsilon = RT \ln (C_s/C)$ in kJ/mol; R is the universal gas constant in kJ/mol-°K; T = the absolute temperature in °K; C = the solution-phase concentration of adsorbate in mg/l; C_s = the solubility of the adsorbate in mg/l; and B is the net adsorbate-adsorbent affinity coefficient in $(kJ/mol)^2$. Tables 10–15 and 10–16 summarize the values for B and W_o and compare the maximum adsorptive capacities for the 5 compounds on the 10 adsorbents. The coefficient B is considered to reflect the net adsorbate-adsorbent affinity that is consistent with the definition of a gas-phase equivalent and to embrace the effects of distribution of pore volume of this affinity [24]. In any event, the comparative adsorption data are seen in Table 10–16. With few exceptions, the two GACs are the most effective adsorbents.

Miscellaneous Compounds

Aroclors 1242 and 1254 were studied in continuously agitated aqueous batch systems for 30 minutes for their adsorption on two GACs, two resins, two polyurethane foams, and PVC [25]. Figure 10–15 shows the adsorption data that were not linear on a log-log plot. The two carbons and XAD-2 had the greatest adsorption capacities. The atypical isotherms are due to the low solubility of the PCBs in water and perhaps to the analytical problems associated with these compounds in water.

Adsorption isotherms and bench-scale column studies were employed to compare the performance of several GACs and three resins for removal of 2-

Table 10–15 Affinity Coefficients (*B*) Values (kJ/mol)[2]

Adsorbent						
Identification	*Trade Name*	*Phenol*	*p-Chlorophenol*	*Carbon Tetrachloride*	*p-Toluene-sulfonate*	*Dodecyl-benzene-sulfonate*
A	F-400[a]	462.07	546.25	184.280	719.66	1,261.30
B	BACM[b]	414.22	538.66	324.960	680.93	874.06
C	XE-340[c]	355.16	477.60	178.830	248.19	568.14
D	XE-347[c]	549.94	787.84	214.640	599.57	739.04
E	XE-348[c]	523.47	832.51	301.270	426.50	855.95
F	XAD-2[c]	118.86	135.76	43.473	217.85	370.73
G	XAD-4[c]	139.04	178.99	36.178	185.57	454.06
H	XAD-8[c]	128.81	151.79	28.942	182.27	305.86
I	ES-863[d]	125.49	138.47	31.560	172.55	306.38
J	S-37[d]	118.02	118.09	37.503	217.22	514.74

Source: Reproduced from Weber and Van Vliet [24], courtesy of the American Water Works Association.
[a]Calgon Corp., Pittsburgh, PA
[b]Mitsubishi International Corp., New York, NY
[c]Rohm and Haas Co., Philadelphia, PA
[d]Diamond Shamrock Chemical Co., Dallas, TX

Table 10–16 Maximum Capacity (W_o) Values

Adsorbent						
Identification	*Trade Name*	*Phenol*	p-*Chlorophenol*	*Carbon Tetrachloride*	p-*Toluene-sulfonate*	*Dodecyl-benzene-sulfonate*
A	F-400[a]	33.780	36.25	41.940	7.3700	44.210
B	BACM[b]	34.130	36.03	42.930	7.5920	42.830
C	XE-340[c]	13.550	12.98	11.380	0.6163	3.875
D	XE-347[c]	16.850	16.74	1.740	0.9342	3.707
E	XE-348[c]	17.570	16.59	10.500	3.6520	12.880
F	XAD-2[c]	8.219	11.31	19.160	0.8246	26.110
G	XAD-4[c]	20.570	20.97	40.820	2.6820	76.330
H	XAD-8[c]	16.420	21.34	13.450	0.6809	30.440
I	ES-863[d]	12.030	13.89	26.730	1.6670	67.630
J	S-37[d]	10.460	14.59	1.791	9.5970	41.610

Source: Reproduced from Weber and Van Vliet [24], courtesy of the American Water Works Association.
[a]Calgon Corp., Pittsburgh, PA
[b]Mitsubishi International Corp., New York, NY
[c]Rohm and Haas Co., Philadelphia, PA
[d]Diamond Shamrock Co., Dallas, TX
Note: Expressed as cm^3/100g dry absorbent (as $C \rightarrow C_s$).

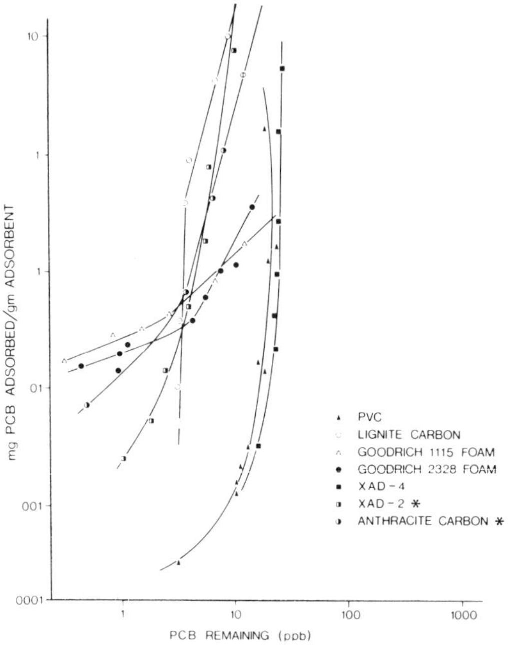

Figure 10–15 Adsorption of Aroclor 1254 on PVC lignite carbon, anthracite carbon, polyurethane foams, and Amberlite XAD-2 and XAD-4. Asterisk indicates adsorption of Aroclor 1242 rather than 1254. Reproduced from Lawrence and Tosine [26], courtesy of the American Chemical Society.

methylisoborneol (MIB) from water [19]. MIB is one of the compounds that causes an earthy or musty odor in raw water supplies. Some of the adsorption studies were conducted with humic substances in the experimental solutions to examine their effect on the adsorption process. Figure 10–16 gives the adsorption isotherms for three resins and one GAC in the 100 × 200 mesh size. It is apparent that none of the resins adsorbs as much MIB as the one carbon. Resin R-1a had essentially no adsorptive capacity for MIB. The presence of humic acid did not affect the adsorption process appreciably.

Several low molecular weight organic solvents were adsorbed by the resin, Ambersorb XE-340, in aqueous solutions via laboratory studies [26]. Table 10–17 is a summary of column experiments where several bed depths and EBCTs were utilized. "The removal of organic contaminants by means of the Ambersorb resin looks promising. Studies have shown that the capacity of this resin for most of these contaminants [Table 10–17] is typically two to three times that of GAC (on a weight basis) for control to breakthru." Concern was expressed about economical regeneration of the resin.

Table 10–18 is a summary of the relative effectiveness of aeration and adsorption for removal of the six organic solvents from water [26]. The resin treat-

Table 10–17 Adsorption of Trichloroethylene and Related Solvents by the Synthetic Resin: Summary

Compound	*Average Concentration (μg/l)*	*Bed Depth (m)*	*Bed Depth (ft)*	*Empty Bed Contact Time (min)*	*Loading to 0.1 μg/l in Effluent (except as noted) (m^3/m^3)*
Trichloroethylene	215	0.3	1	2	83,700[a]
	210	0.6	2	4	78,600[a]
	210	1.2	4	7.5	>53,300[a]
	177	0.8	2.5	9	>20,160
	4	0.8	2.5	8.5	>123,340
	3	0.2	0.8	5	>117,000
Tetrachloro-ethylene	41	0.3	1	2	>99,000[b]
	51	0.6	2	4	78,600[b]
	65	1.2	4	7.5	>53,300[b]
	70	0.3	1	2	106,000[b]
	94	0.8	2.5	5	112,900
	1,400	0.8	2.5	9	17,920
	3	0.8	2.5	8.5	>123,340
	2	0.8	2.5	9	>20,160
1,1,1-trichloro-ethane	5	1.2	4	7.5	39,300[b]
	33	0.8	2.5	9	56,000
	237	0.2	0.8	5	82,600
	23	0.2	0.8	5	>100,800
	1	0.8	2.5	9	>20,160
Carbon tetrachloride	19	0.8	2.5	5	7,560
	19	0.8	2.5	10	15,120
cis-1,2-dichloro-ethylene	40	0.3	1	2	37,200[b]
	38	0.6	2	4	39,500[b]
	40	1.2	4	7.5	19,700[b]
	40	0.3	1	2	36,400[b]
	25	0.8	2.5	6	14,400
	22	0.8	2.5	6	7,200
	16	0.8	2.5	6	11,500
	6	0.8	2.5	9	>20,160
	2	0.8	2.5	8.5	>59,000 but <123,340
1,2- dichloroethane	1	0.2	0.8	5	108,860

Source: Reproduced from Love and Eilers [27], courtesy of the American Water Works Association.
Note: Ambersorb® XE-340 is synthetic resin used.
[a]20 μg/l in effluent
[b]5 μg/l in effluent

Table 10–18 Relative Effectiveness of Treatment

	Aeration		*Adsorption*			
			GAC		*Synthetic Resin*	
Compound	*Predicted from Henry's Law Constant*	*Pilot-Scale Experience*	*Predicted from Isotherm Data*	*Pilot-Scale Experience*	*Pilot-Scale Experience*	*Boiling*
Trichloroethylene	3	2	2	1	1	5
Tetrachloro-ethylene	2	1	1	2	2	1
1,1,1-trichloro-ethane	5	5	6	5	5	4
Carbon tetrachloride	1	3	3	3	6	3
cis-1,2-dichloro-ethylene	4	4	5	6	4	2
1,2-dichloro-ethane	6	6	4	4	3	

Source: Reproduced from Love and Eilers [27], courtesy of the American Water Works Association.
Note: 1 = best performance; 6 = poorest performance.

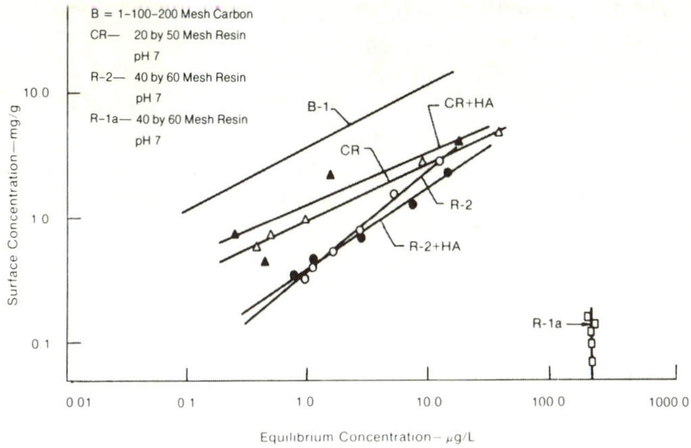

Figure 10–16 MIB adsorption isotherms on Resins R-1a, R-2, and CR. Reproduced from Chudyk et al. [19], courtesy of the American Water Works Association.

ment appears to be as effective as GAC and slightly better than aeration. Here, again, regeneration of the resin by steam may limit the widespread application of resins for water treatment.

REFERENCES

1. Gustafson, R.L., and J. Paleos. "Interactions Responsible for the Selective Adsorption of Organics on Organic Surfaces," Chapter 10 in *Organic Compounds in Aquatic Environments.* S.D. Faust and J.V. Hunter, eds. Marcel Dekker, New York (1971), p. 213.
2. Kim, B.R., et al. *J. Water Pollut. Cont. Fed. 48,* 120 (1976).
3. Kunin, R. *Soc. Plast. Eng. Technol. Pap. 22,* 248 (1976).
4. Rohm and Haas Co. "Ambersorb Carbonaceous Adsorbents." Bulletin 1E-231. Philadelphia (1977).
5. Davis, T.A. "Electro-regenerated Ion Exchange Deionization of Drinking Water." Report No. EPA/600/1-77-035. U.S. EPA, Cincinnati, Ohio (1977).
6. Rook, J.J., and S. Evans. *J. Amer. Water Works Assoc. 71,* 521 (1979).
7. Wood, P.R., and J. DeMarco. "Effectiveness of Various Adsorbents in Removing Organic Compounds from Water-Removing Purgeable Halogenated Organics," Chapter 4 in *Activated Carbon Adsorption of Organics from the Aqueous Phase,* Vol. 2. M.J. McGuire and I.H. Suffet, eds. Ann Arbor Science Publishers, Ann Arbor, Mich. (1980), p. 85.
8. Symons, J.M., et al. "Treatment Techniques for Controlling Trihalomethanes in Drinking Water." EPA-600/2-81-156. U.S. EPA, Cincinnati, Ohio (1981).
9. Cook, F.C. Jr., and K.E. Hartz. *J. Amer. Water Works Assoc. 75,* 423 (1983).
10. Feige, W.H., and D. Ruggiero. "Removal of Organics Contaminants from Drinking Water Supply at Glen Cove, N.Y.: Phase 11." EPA-600/52-82-027. U.S. EPA, Cincinnati, Ohio (1982).

11. Anderson, R.E., and R.D. Hansen. *Ind. Eng. Chem. 47,* 71 (1955).
12. Chasanov, M.G., et al. *Ind. Eng. Chem. 48,* 305 (1956).
13. Pollio, F.X., and R. Kunin. *Environ. Sci. Technol. 1,* 160 (1967).
14. Kennedy, D.C. *Ind. Eng. Chem. Prod. Res. Devel. 12,* 56 (1973).
15. Kawabata, N., and K. Ohira. *Environ. Sci. Technol. 13,* 1396 (1979).
16. Gustafson, R.L., et al. *Ind. Eng. Chem. Prod. Res. Devel. 7,* 107 (1968).
17. Hinrichs, R.L., and V.L. Snoeyink. *Water Res. 10,* 79 (1976).
18. Boening, P.H., et al. *J. Amer. Water Works Assoc. 72,* 54 (1980).
19. Chudyk, W.A., et al. *J. Amer. Water Works Assoc. 71,* 529 (1979).
20. Neely, J.W. "A Model for the Removal of Trihalomethanes from Water by Ambersorb XE-340," Chapter 17 in *Activated Carbon Adsorption of Organics from the Aqueous Phase,* Vol. 2. M.J. McGuire and I.H. Suffet, eds. Ann Arbor Science Publishers, Ann Arbor, Mich. (1980), p. 417.
21. Van Vliet, B.M., et al. *Water Res. 14,* 1719 (1980).
22. Weber, W.J. Jr., and B.M. Van Vliet. *J. Amer. Water Works Assoc. 73,* 420 (1981).
23. Weber, W.J. Jr., and B.M. Van Vliet. *J. Amer. Water Works Assoc. 73,* 426 (1981).
24. Dubinin, M.M. *Chem. Rev. 60,* 235 (1960).
25. Lawrence, J., and H.M. Tosine. *Environ. Sci. Technol. 10,* 381 (1976).
26. Love, O.T. Jr., and R.G. Eilers. *J. Amer. Water Works Assoc. 74,* 413 (1982).
27. Slejko, F.L., and G.F. Meigs. "Economic Analysis of Employing Ambersorb XE-340 Carbonaceous Adsorbent in Trace Organics Removal from Drinking Water," Chapter 19 in *Activated Carbon Adsorption of Organics from the Aqueous Phase,* Vol. 2. M.J. McGuire and I.H. Suffet, eds. Ann Arbor Science Publishers, Ann Arbor, Mich. (1980), p. 443.

Index

Acidity, surface, 182–185, 189–191
Activated carbon. *See also* Biological activated carbon process; Granulated activated carbon; Powdered activated carbon
  ash content of, 42, 170–172
  electrophoretic characteristics of, 188–191
  physical properties of, 14, 170–180
  predicting performance of, 22, 35–36
  preparation of, 167–170
  surface of, 14–15, 172, 176, 180–191
Adsorbate properties, 25–26, 80–83
Adsorber design, 138–143, 399
Adsorption
  competitive, 41, 42, 266–267, 274, 339
  heat of, 2, 14, 230. *See also* Kinetics
  models of, 25–62, 114–115
  principles of, 1–2, 25–33
  by resins, 471–502
Adsorption equilibria, 14–22
Adsorption isotherms. *See also* Freundlich adsorption isotherm; Langmuir adsorption isotherm
  effects on, 442–443
  equilibrium of, 14–22, 132–133
  Gibbs, 7–10, 55
  and phenol, 86–87, 108–109, 185, 492–495
  prediction of, 33–34, 56, 58
  uses of, 14–16, 33, 125
Adsorption rates
  effect of concentration on, 66, 74–79
  effect of molecular size on, 66–67, 80–84, 96
  effect of temperature on, 67, 75–76
  effect of transportation on, 65, 66–67, 84–85
  in fixed bed models, 100–101
  of mixed systems, 94–100
  of organics, 65–67
  of single compounds, 67–94
Adsorptive capacity. *See also* Polanyi adsorption potential theory
  factors affecting, 22, 25–26, 27, 169, 185, 379–381
  and mixture of solutes, 263, 274–276
  single-solute, 174–274
Aeration, 329, 373–374, 388–392
Affinity coefficients, 53
Affinity of solutes, 30, 32, 33
Air-stripping process, 372–373
Alkylbenzenesulfonates, 67–68, 76–77, 82–83, 94, 98–100, 254–256
Alum coagulation. *See* Coagulation
Amundson, N. R., 86
Anion exhange resin, 358–359
Aqueous solution regeneration, 471
Arbuckle, W. B., 58–61
Arrhenius equation, 75–76
Arsenic, 287–289
Ash content, 42, 170–172

Backwashing, 425–428, 461
Bacterial activity, 374, 405–412, 433–434, 450
Barium (Ba), 289
Basal planes, 185
Base titration procedure, 185
Batch desorption studies, 418–420
Batch kinetic studies, 67–99, 118–119, 317

Batch systems, 67–100
  design and performance of, 65, 66, 123–129
  multiple-stage, 124, 129
  and single compounds, 68–94
Baylis, J., 313–314
Bed depth, critical, 133, 146–148, 330–331. *See also* Mass transfer zone
Bed depth-service time (BDST) method, 145–148
  based on Bohart-Adams model, 145–147, 158–161
  and design of GAC systems, 158–163
  Hutchins' modifications of, 147–148, 154, 161–163
  and series or pulsed beds, 151–152
BET adsorption isotherm, 20–22
Biochemical oxygen demand (BOD), 193
Biocomponent systems, 269
Biodegradation, 438–439, 442
Biological activated carbon (BAC) process, 433–470
  case studies of, 449–459
  design factors in, 460–461
  performance of, 445–448
  pretreatment step in, 434–444
Black Pearls, 185
Bohart-Adams method, 145–147, 158–161
Boltzmann equation, 13, 38
Boundary conditions, 114
BP, 251
Branched pore model, 118
Breakthrough, TOC, 208–210
Breakthrough curves, 130–134
  for $CHCl_3$, 339
  to determine capacity of carbon, 134, 143
  to determine critical bed depth, 133–134
  for humic acid, 219
  for MIB, 248
  for PNP/BA systems, 281–283
Breakthrough point, 133–134
Burnoff, 167–168, 172

Cadmium, 289–294
Carbamates, 241, 244
Carbon. *See also* Activated carbon; Dissolved organic carbon; Total organic carbon removal
  activation of, 168–170, 180–182. *See also* Activated carbon
  capacity of, 143
  functional surface groups of, 180–191
  makeup costs for, 467–468
  properties of, 167–192
  regeneration of, 148, 414, 417, 418
Carbon black, 180
Carbon columns. *See* Column studies
Carbon dioxide (CO_2), 180, 182, 417
Carbon dosage rate, 124–129, 138, 143–145, 148, 152
Carbonization, 168
Carbon slurry, 123–124, 414, 461
Carbon tetrachloride (CCl_4), 96–97, 210, 211
Characteristic curves, 36, 38–39
Chemical oxygen demand (COD), 193
Chemisorption, 1–2
Chloramines, 320–322
Chlorinated hydrocarbons, 241
Chlorination
  as disinfectant, 399
  and odor removal, 224–226
  of organic compounds, 207, 386
  as a pretreatment, 392–393, 443–444
Chlorine reactions
  background on, 313–314
  kinetics of, 314–320
  particle size and, 317–320
  and THM removal, 124, 222, 354, 392
Chloroform removal, 338–342
Chlorophenols, 266–267, 480–481
Chromatographic effect, 339, 352, 392
Chromium, 294–298
Cincinnati Water Works, 346–349
Coagulation
  effect of, on adsorption, 221–222, 241, 379–381
  and PAC removal, 241, 244–245, 381–383, 425
  as pretreatment, 222, 379–385
  and TOC removal, 381, 383–384
Color removal, 217–219, 381–383, 438, 474
Column adsorbents, 129–134
Column studies
  in competitive situations, 276–283
  design approaches and, 142–155
  of fulvic and humic acids, 219–224

of organic solvents, 259–261
of pesticides, 246–247
of petrochemicals, 239–240
of phenolic compounds, 232–237
Competitive systems. *See* Mixed system adsorption
Concentration profiles. *See* Breakthrough curves
Constant pattern approximation, 133
Contactors
fixed bed, 134–135, 138
pulsed bed, 135–138
Contact period. *See* Empty bed contact time
Copper, 298–299
Correlation curves, 39, 42
Correspondence, theory of, 53–58
Countercurrent treatment, 124, 129

DDT, 241
Deashing, 172
Debye-Huckel equation, 48
Dechlorination. *See* Chlorine reactions
Detergents, 68–70, 73, 254, 483–484. *See also* Alkylbenzenesulfonates
Diffusion
and fixed bed adsorption, 106–120
in mixed bed systems, 94–99
processes controlled by, 85–94
as step in adsorption, 65–67, 75–76, 80
surface, 66–67, 108, 114
Dipole interactions, 3
Disinfection, 399, 433
Dispersion, 3–4, 32, 132
Displacement effects, 274
Dissolved organic carbon (DOC), 274, 434
Divided feed system. *See* Countercurrent treatment
Downflow beds. *See* Fixed beds; Gravity beds
Downflow design studies, 346, 461
Drinking water treatment, 325, 433–470
Dual particle diffusion (DPD) model, 114–118
Dusseldorf, Germany, 453–456

Edeskuty, F. J., 86
Electron spin resonance (ESR) spectroscopy, 180
Electrophoretic characteristics, 188–191
Empty bed contact time (EBCT)
carbon dosages and, 124–129
and design of carbon bed, 338, 342
and $CHCl_3$ removal, 338–342
and GAC filters, 208, 338–339
vs. other methods, 161, 163
in pilot-scale studies, 143–145, 156–158, 338–342, 346, 349
Endotoxins, 410
Entropy, 2
Environmental Protection Agency (EPA), 210, 393–405
Equilibrium studies
of commercial carbons, 207
of fulvic and humic acids, 217–219
and the mass transfer zone, 132–133
in mixed systems, 262–276
of organic pesticides, 241–246
of other organic compounds, 254–259
of petrochemicals, 238–239
of phenolic compounds, 226–232
Expanded films, 11–12
External diffusion, 65–67, 85–94, 100, 106–107, 281

Feed concentration, 149–150
Fick's First Law, 88–89
Film diffusion, 65, 66, 67, 100, 106–107. *See also* External diffusion
Films, surface, 10–14, 16–18
Filter/adsorber mode, 338
Fixed bed contactors, 134–135, 138
Fixed bed reactors, 65, 66, 322–325
Fixed beds
adsorption rates for, 100–103, 106–120, 140, 418–420
vs. gravity beds, 134–135
mass transfer of carbon in, 120–121
models for, 107–108, 111, 138–141, 279
Fluoride, 299–300
Fractional capacity, 234–235
Free energy, 6, 41, 45, 53, 56, 57
Freundlich adsorption isotherm
and carbon dosage, 125–126, 128
description of, 18–20
and fixed bed adsorbers, 109, 112, 140
for multicomponent systems, 167–272

Freundlich adsorption isotherm (*cont.*)
  and phenol adsorption, 118, 128, 269, 494
  and THMs, 211, 213
Fulvic acids, 217–224, 379–381

GAC. *See* Granulated activated carbon
GAC columns, 128–130, 346–360
GAC systems
  costs of, 393–405, 461–469
  design of, 141–163, 399–405, 461
Gaseous films, 11–12, 13
Gas laws, 13–14
Gas-phase adsorption, 38
Gibbs, J. W., 7
Gibbs
  adsorption isotherm, 7–10, 55
  free energy, 6, 53
  thermodynamic analysis, 7
Gibbs-Duhem equation, 8
Gibbs-Helmholtz equation, 6
Granulated activated carbon (GAC). *See also* GAC columns; GAC systems
  adsorption of fulvic or humic acids on, 217–224
  adsorption of organic compounds on, 114–115, 208–211, 240, 247, 254–276
  adsorption of petrochemicals on, 238–240
  adsorption of phenols on, 224–237
  adsorption of THMs on, 73, 211–217, 354, 488–492
  comparative studies of, 338, 346–352, 355–360, 484–501
  detergent removal and, 68–70, 73, 254
  microbiology of, 405–412
  taste and odor removal and, 247–254
Graphite, 180
Gravity beds, 134–135
Groundwaters, 370–378

Haloforms, 342–346
Halogenated organic compounds. *See* Total organic halogens
Hearth furnaces, 414–417
Henry's Law, 21, 373
Herbicides, 242–243
Humic acids. *See also* Color removal
  coagulation to remove, 379–381
  column studies of, 219–224
  equilibrium studies of, 217–219
  resins to remove, 474–475
  salts and adsorption of, 22
  TOC associated with, 383–385
Hutchins' design model, 147–148, 154, 161–168
Hydrogen ion concentration, 74–75, 230
Hydrogen sulfide (H_2S), 325–326

Ideal adsorbed solution (IAS) model, 277–283
Ideal localized monolayer model. *See* Langmuir model
Infrared spectroscopy, 180
Inorganic compound removal, 287–327
Inorganic salts, 22, 42
Insoluble film, 10–13
Internal diffusion, 65–67, 85–88, 93–94, 100
Interruption tests, 89–93
Intraparticle diffusion, 66–67, 83–96, 104, 107, 111
Intraparticle kinetic parameters, 119–121
Isotherms. *See* Adsorption isotherms

Kinetics, 65–122
  batch studies of, 67–99, 118–119, 317
  of chloramine removal, 321–322
  of chlorine removal, 314–320
  in fixed beds, 100–122
  and the Langmuir equation, 16

Langmuir, I., 10
Langmuir
  adsorption isotherm, 16–18, 21, 88–89, 109, 296
  equation, 16–18, 266
  model, 16–18, 103, 267–272, 274, 276
Lead, 300–302
Ligin sulfonic acid (LSA), 272–273
Linear adsorption isotherm, 21
Liquid expanded films, 11
London, F., 3
London-force adsorption, 3-4, 42

Macropores, 116, 118–119, 433–434
Macroreticular resins. *See* Resins
Manes, M., 40–42, 44, 57

Mass balance, 101
Mass loading curves, 339–342
Mass transfer
  coefficients, 56, 103–104, 114
  parameters, 106–107, 110–112, 139–141
  resistances, 103–104, 138–139
  zone, 130–134, 137, 138, 364. *See also* Bed depth, critical; Breakthrough curves
Mass transfer models
  early, 100–106
  late, 107–121
  parameters of, 106–107
Maxwell-Boltzmann distribution law, 16
Mercury
  aqueous chemistry of, 302–304
  pore volume data on, 179–180
  removal of, 304–310
Methoxychlor removal, 245
MIB, 248–250, 498
Microbiology. *See* Bacterial activity
Microcrystallites, carbon, 178
Micropores, 116, 118–119, 434
Missouri River, 332–333
Mixed system adsorption
  column studies of, 276–283
  equilibrium studies of, 262
  kinetics of, 94–100
Monolayer model, ideal localized. *See* Langmuir model
Monolayers. *See* Surface films
Morris, J. C., 69
Moving bed contactors, 135–138
Mulheim, Germany, 449–450
Multiple-stage batch system, 124, 129

National Interim Primary Drinking Water Regulations, 393
National Secondary Drinking Water Regulations, 68, 325
Negative adsorption, 7
Net adsorption energy concept (NAEC)
  basis of, 26–33
  compared to other models, 58–62
  correlated to adsorption isotherms, 33–34
Nitrification, biological, 433
Nitro, West Virginia, 329–332
Nitrobenzene adsorption, 185
Nonaqueous solution regeneration, 471

Odor removal, 329–337
  chlorination and, 224–226
  in New Jersey, 337
  in Omaha, 332–336
  studies of, 247–251
  of sulfide, 336–337
  in West Virginia, 329–332
Organic carbons, nonspecific, *See* Total organic carbon removal; Total organic halogens
Organic compound removal, 338–370
  comparative study of, 370, 484–488
  cost of, 393–405, 435
  factors influencing, 256–258
  and ozonation, 386, 439–440
  pretreatment for, 379
Organic compounds
  analysis of, 193–206
  column studies of, 259–261, 276–283
  mixed. *See* Mixed system adsorption
  organoleptic, 247–251
Organic solvent removal, 352, 498–501
Oxidation
  as activation process, 168–170
  biological, 433, 449
  chemical, 193, 434–444
  as pretreatment, 434–435
Oxides, surface, 180–182, 185–187, 421–422
Ozonation
  by chemical coagulation, 207, 438–439
  costs of, 469
  and microbial activity, 408, 435–443
  by O_3, 385–386, 438, 443
  pilot studies of, 207–208, 385–387, 435–438
  and THM formation, 385–386, 436

Packed beds, 129
Paris, France, 452–453
Passaic River, 337
Pesticides, organic, 240–242, 246–247, 342–346
Petrochemicals, 238–240
Phenol adsorption, 224–237. *See also* Chlorophenols

Phenol adsorption (*cont.*)
and acidic surface oxides, 185
column studies of, 232-237
equilibrium studies of, 226-232
and Freundlich isotherm, 118, 128, 269, 494
kinetics of, 69, 76
rate of, 94, 96, 106-107
on resins, 479-482, 492-495
and solvents, 418-421
Physical adsorption, 1-2
Polyani adsorption potential theory, 34-45, 58-59
Polychlorinated biphenyls (PCBs), 210, 495
Polycyclic aromatic hydrocarbons (PAHs), 251-254
Polymeric adsorbents. *See* Resins
Pore diffusion mode, 106-107. *See also* Surface diffusion
Pore size, 116, 118, 167, 179
Porous carbon, 65-66, 178-179
Positive adsorption, 6
Postcontactor model, 338
Potential energy curves, 4-5
Powdered activated carbon (PAC)
and adsorption in aqueous systems, 65
and adsorption of inorganic compounds, 288-326
adsorption of organic compounds on, 73, 193-286, 370
and drinking water treatment, 329-336, 365-370, 381-383
use of, for batch systems, 123-124, 127
Preaeration. *See* Aeration
Pressurized systems, 134
Pretreatment. *See also* Ozonation
aeration as, 388-392
chemical coagulation as, 379-385
chlorination as, 392-393
ozonation as, 385-388
Pulsed beds, 135-138, 151-152
Pyrolysis, 168

Quadrapole-quadrapole interactions, 4-5
Quench tank, 416-417

Radionuclides, 310-312
Rate-limiting step
diffusion as, 80-86, 90, 93-94
in fixed-bed adsorption, 106-107
identification of, 65-67, 85
in phenol adsorption, 79, 91
transport as, 79, 85
Regeneration, 412-425
in BAC process, 433, 461
multiple hearth furnace, 414-417
by organic solvent, 417-424
of resins, 471-472
steam, 501
thermal, 424-425
Regeneration efficiency (RE), 421-424
Regeneration units, 152
Removal rates. *See* Adsorption rates
Resins
anion exchange, 358-359
characteristics of, 471-472
compared to GAC, 474-501
removal of various compounds with, 474-484
Rockaway, New Jersey, 371-374
Rotterdam, The Netherlands, 457-459

Sand replacement, 329, 338, 396-399
Secondary drinking water standards, 68, 325
Selenium, 312-313
Series beds, 151-152
Slurry. *See* Carbon slurry
Smith, J. M., 30
Solubility, 25-26, 33, 420
Solutes
affinity of, 30, 72, 73
concentration of, 76-80, 91-93
Solvency, maximum, 25
Solvents
regeneration, 417-425
removal of, 352, 498-501
Solvophobic theory of adsorption, 45-52, 61
Standard plate count (SPC), 374
Sulfide odors, 336-337. *See also* Odor removal
Sulfonates, 483-484. *See also* Alkylbenzenesulfonates
Supercritical fluids, 417
Surface
acidity, 182-185, 188-191

balance, 10
charges, 188–191
chemistry, 1–23
concentration, 9
diffusion, 66–67, 100, 108, 114
enthalpy, 6
energy, 5–6, 8, 14
films, 10–14. *See also* Film diffusion
functional groups, 180–191, 346
heterogeneity, 56
oxides, 180–182
phase, 7–8, 10
pressure, 10
tension, 5–7, 9, 12–13

Taste control, 224–226, 329–337
Temperature. *See* Kinetics
Thermodynamics, 7–10
3,4-benzpyrene (BP), 251
3-dodecylbenzenesulfonate (DBS), 68
Total chlorinated organic compounds (TOCl), 207
Total organic carbon (TOC) removal
analysis of, 193
and breakthrough, 208–210
by coagulation, 379–385
design of systems for, 338–342
and Freundlich isotherm, 211, 213
by GAC, 208–210, 339, 342–352, 386
mechanisms for, 342
by PAC, 369
Total organic halogens (TOX), 193, 222–226, 346–352, 359–360
Total plate counts (TPCs), 411
Total solubility parameter, 28–29
Total trihalomethanes (TTHMs), 214
Transport, rate of, 65
Traube's Rule, 66, 81
Treatment plant design, 393–395, 460–461
Trent, River, 342–343
Trichloroethanes, 210–211
Trihalogenated hydrocarbons, 210, 211–214
Trihalomethane (THM)
formation potential (THMFP), 222, 354–356, 360, 364–365, 383, 436, 438
precursors, 222, 353, 370, 392, 436, 474–477
Trihalomethane removal, 211–217, 353–370
by aeration, 392
by chlorine, 124, 222, 354, 392
by GACs, 73, 211–217, 354, 488–492
and ozonation, 436
by resins, 474–477, 488–492
2,4-D, 241–242
2,4-DCP, 91–94, 242
2,4,5-TP, 241
2-methylisoborneol (MIB), 248–250, 498
2-methylpyridine, 257–258

Unimolecular films, 10–14
Uranium, 312
Upflow contactors, 134–135

van der Waals interactions, 3–5
van't Hoff equation, 17
Volatile organic chemicals (VOCs), 371–372
Volatilization, 438

Wastewater, 123–165
Wave front rate, 130, 151, *See also* Mass transfer zone
Weber, W. J., 69, 86
Wohleher, D. A., 57
Waste systems, 374, 412

Zinc removal, 313
Zogorski, J. S., 70, 237